Ultrastructure of the Extraparietal Glands of the Digestive Tract

ELECTRON MICROSCOPY IN BIOLOGY AND MEDICINE

Current Topics in Ultrastructural Research

SERIES EDITOR: P.M. MOTTA

Already published in this series

Vol. 1 Motta, P.M. (ed.): Ultrastructure of Endocrine Cells and Tissues. ISBN: 0-89838-568-7.

Vol. 2 Van Blerkom, J. and Motta, P.M. (eds.): Ultrastructure of Reproduction: Gametogenesis, Fertilization, and Embryogenesis. ISBN: 0-89838-572-5.

Vol. 3 Ruggeri, A. and Motta, P.M. (eds.): Ultrastructure of the Connective Tissue Matrix. ISBN: 0-89838-600-4.

Vol. 4 Motta, P.M., Fujita, A. (eds.): Ultrastructure of the Digestive Tract. ISBN: 0-89838-893-7.

Vol. 5 Van Blerkom, J. and Motta, P.M. (eds.): Ultrastructure of Human Gametogenesis and Early Embryogenesis. ISBN: 0-89838-383-8.

Ultrastructure of the Extraparietal Glands of the Digestive Tract

edited by

Alessandro Riva, M.D.

Department of Cytomorphology, University of Cagliari, Cagliari, Italy

and

Pietro M. Motta, Ph.D., M.D.

Department of Anatomy, University "La Sapienza" of Rome, Rome, Italy

assistant editor

Francesca Testa Riva, Ph.D.

Department of Cytomorphology, University of Cagliari, Cagliari, Italy

Kluwer Academic Publishers
Boston / Dordrecht / London

Distributors

for North America:
Kluwer Academic Publishers
101 Philip Drive
Assinippi Park
Norwell, Massachusetts 02061 USA

for all other countries:
Kluwer Academic Publishers Group
Distribution Centre
Post Office Box 322
3300 AH Dordrecht, THE NETHERLANDS

Library of Congress Cataloging-in-Publication Data

Ultrastructure of the extraparietal glands of the digestive tract/
edited by Alessandro Riva and Pietro M. Motta
p. cm — (Electron microscopy in biology and medicine)
Includes bibliographies and index.

ISBN-13: 978-1-4612-8211-2 e-ISBN-13: 978-1-4613-0869-0
DOI: 10.1007/978-1-4613-0869-0

1. Salivary glands—Ultrastructure. 2. Liver—
Ultrastructure. 3. Pancreas—Ultrastructure. I. Riva,
Alessandro. II. Motta, Pietro M. III. Series.
[DNLM: 1. Gallbladder—ultrastructure. 2. Liver—
ultrastructure. 3. Pancreas—ultrastructure. 4. Salivary
Glands—ultrastructure. W1 EL33E / WI 101 U473]
QM325.U37 1989
611'.01893—dc20
DNLM/DLC
for Library of Congress

89-11144
CIP

Softcover reprint of the hardcover 1st edition 1990

Preface

This volume, the sixth of the series, represents the natural counterpart of the previous volume, *Ultrastructure of the Digestive Tract*. Unlike the latter, however, whose contents fell entirely within the domains of gastroenterology, *Ultrastructure of the Extraparietal Glands of the Digestive Tract* encompasses a few cognate sciences, such as hepatology, pancreatology, and even oral biology, which are usually dealt with separately. This allows, starting from cell biology, embryology, and comparative anatomy, a comprehensive survey of organs that have much in common both structurally and functionally.

The chapters of this book have been compiled by well-known experts in the field with the aim not only of reviewing and pointing out the state of the art of the subject covered, but also of giving directions for future work. Furthermore, through the integration of electron microscopy with immunocytochemistry, autoradiography, freeze fracture, maceration, enzymatic digestion, etc., and by providing superb illustrative material, the authors substantiate the pivotal role played by modern morphology in understanding human physiology and pathology. In fact, it must be stressed, that a consistent part of the tissues studied here are from human origin.

We believe that this volume should be read, not only by scientists and teachers active in the field, but also by a larger audience of students and professionals interested in knowing the scientific foundations of biomedicine.

Finally, we wish to express our gratitude to the authors for the quality of their contributions and their prompt cooperation. Warm thanks are due to Mr. J.K. Smith and his staff who, once again, have fully demonstrated their professional skill in the production of the book.

A. RIVA and P.M. MOTTA

Contents

Contributing authors

Adler, Guido, Department of Internal Medicine, Philipps University, Marburg/Lahn, Federal Republic of Germany

Bonvicini, Fiorenza, Institute of Clinical Electron Microscopy, Policlinico S. Orsola, University of Bologna, 40138 Bologna, Italy

Dardick, Irving, Department of Pathology, University of Toronto, and Head, Diagnostic Electron Microscopy, Department of Pathology, Toronto General Hospital, Toronto, Ontario, M5G1L5 Canada

Fumagalli, Guido, Department of Pharmacology, University of Milano, 20129 Milano, Italy

Hand, Arthur R., Clinical Investigations and Patient Care Branch, National Institute of Dental Research, National Institutes of Health, Bethesda, Maryland 20892, USA

Iino, Akihiro, Department of Anatomy, Tottori University, School of Medicine, Yonago, 683 Japan

Kern, Horst F., Institute für Zytobiologie und Zytopathologie, Robert-Koch-Strasse 5, D-3550 Marburg, Federal Republic of Germany

Lantini, Maria S., Department of Cytomorphology, University of Cagliari, Via Porcell 2, 09124 Cagliari, Italy

Laschi, Renzo, Institute of Clinical Electron Microscopy, Policlinico S. Orsola, University of Bologna, 40138 Bologna, Italy

Luciano, Liliana, Medizinische Hochschule Hannover, Abteilung Zellbiologie und Elektronenmikroskopie, Kostanty-Gutschow-Str. 8, 3000 Hannover 61, Federal Republic of Germany

Macchiarelli, Guido, Department of Anatomy, Faculty of Medicine, University "La Sapienza" of Rome, Via A. Borelli 50, 00161 Rome, Italy

Makabe, Sayoko, Department of Obstetrics and Gynecology, University of Toho, Tokyo, Japan

Mancini, Patrizia, Department of Experimental Medicine, University "La Sapienza" of Rome, Viale Regina Elena 324, 00161 Rome, Italy

Motta, Pietro M., Department of Anatomy, Faculty of Medicine, University "La Sapienza" of Rome, Via A. Borelli 50, 00161 Rome, Italy

Murakami, Takuro, Department of Anatomy, Okayama University Medical School, Okayama, 700 Japan

Nagato, Toshikazu, Department of Oral and Maxillofacial Surgery, Ehime University School of Medicine, Shizukawa, Shigenobu, Onsen-gun, Ehime 791–02, Japan

Naguro, Tomonori, Department of Anatomy, Tottori University, School of Medicine, Yonago, 683 Japan

Ohtani, Osamu, Department of Anatomy, Okayama University Medical School, Okayama, 700 Japan

Phillips, Carleton J., Department of Biology, Hofstra University, Hempstead, NY 11550, USA

Pinto da Silva, Pedro, Membrane Biology Section, Laboratory of Mathematical Biology, National Cancer Institute, Frederick Cancer Research Facility, Bldg 538, Rm 104, Frederick, MD 21701, USA

Reale, Enrico, Medizinische Hochschule Hannover, Abteilung Zellbiologie und Elektronenmikroskopie, Kostanty-Gutschow-Str. 8, 3000 Hannover 61, Federal Republic of Germany

Riva, Alessandro, Department of Cytomorphology, University of Cagliari, Via Porcell 2, 09124 Cagliari, Italy

Shimono, Masaki, Department of Pathology, Tokyo Dental College, 1–2–2 Masago, Chiba-city, Chiba, 260 Japan

Taguchi, Takehito, Department of Anatomy, Okayama University Medical School, Okayama, 700 Japan

Tandler, Bernard, Department of Oral Biology, School of Dentistry, Case Western Reserve University, Cleveland, OH 44106, USA

Testa Riva, Francesca, Department of Cytomorphology, University of Cagliari, Via Porcell 2, 09124 Cagliari, Italy

Torrisi, Maria Rosaria, Department of Experimental Medicine, University "La Sapienza" of Rome, Viale Regina Elena 324, 00161 Rome, Italy

Toyoshima, Kuniaki, Department of Oral Anatomy II, Kyushu Dental College, 2–6–1, Manazuru, Kokurakita-ku, Kitakyushu 803, Japan

Uchiyama, Yasuo, Department of Anatomy, Institute of Basical Medical Sciences, University of Tsukuba, Niihari-Gun, Ibaraki-Ken 305, Japan

Van Nostrand, A.W. Peter, Departments of Pathology and Otolaryngology, University of Toronto, Toronto, Ontario M561L5, Canada

Watanabe, Masahiko, Department of Anatomy, Institute of Basical Medical Sciences, University of Tsukuba, Niihari-Gun, Ibaraki-Ken 305, Japan

Ultrastructure of the Extraparietal Glands of the Digestive Tract

CHAPTER 1

The secretory process of salivary glands and pancreas

ARTHUR R. HAND

1. Introduction

The study of secretory cells and the definition of the secretory process have been at the forefront of modern cell biology since the inception of this field in the early 1950s. Secretory cells, particularly the acinar cells of the exocrine pancreas, presented a unique opportunity for cell biologists, who were in the process of marrying biochemistry and morphology: Here was a cell with virtually a single-minded mission, to produce and release large amounts of digestive enzymes, whose appearance in the microscope could be correlated with reactions in the test tube and that could be induced to perform its functions by physiological manipulations or the application of readily available drugs. Since almost all cells secrete proteins, these pioneering studies found wide application, and they spawned the development of methodological approaches and technological advances too numerous to mention.

Much has been learned in the 35 years that ensued after the first studies of secretory cells, and details of the progress made are available in several reviews [1–4]. The purpose of this chapter is not to reiterate the history of the development of this field, but to summarize the current status of the secretory process, with the intention of providing a framework for the subsequent chapters of this book. No attempt will be made to describe the mechanisms involved in fluid and electrolyte secretion by the salivary glands and pancreas; the interested reader may consult a number of recent reviews [5–7].

The secretory process consists of a series of interrelated steps, beginning with the synthesis of a secretory protein and culminating in the release of the mature product to the extracellular environment (Table 1-1). Although the process is continuous, and in one cell all of the different steps may occur simultaneously, it is customary to divide the secretory process into several discrete stages [1]. These stages are largely defined by the major biochemical processes occurring during secretory protein production and the cellular compartments in which they take place. Figure 1-1 shows the ultrastructure of a typical salivary-gland secretory cell, and Fig. 1-2 illustrates the basic pathway and general features of the secretory process: 1) synthesis of secretory proteins in the endoplasmic reticulum (ER) and their segregation in the lumen of the ER cisternae, 2) migration of the proteins to the Golgi apparatus, 3) concentration and packaging of the proteins in secretory granules, 4) intracellular storage, and 5) release of the granule content into the extracellular space.

2. Protein synthesis and segregation

The first step in the secretory process is the synthesis of secretory proteins. As with all other cellular proteins, secretory proteins are assembled on ribosomes from precursor amino acids under the direction of messenger ribonucleic acid (mRNA). Encoded within the mRNA, as part of the primary amino-acid sequence of the protein, is specific information that directs the protein to

Riva, A. and Motta, P.M., (eds.), Ultrastructure of the Extraparietal Glands of the Digestive Tract.
ISBN 978-1-4612-8211-2.

Table 1-1. Steps in the secretory process

Protein synthesis and segregation
Post-translational modifications
Intracellular transport and sorting
Secretory granule formation
Exocytosis

an appropriate cellular destination. In the case of secretory proteins, the most important information is contained in a region at the amino-terminal end of the protein, known as the *signal sequence* [9]. The signal sequence, consisting of about 15–30, mostly hydrophobic, amino acids, directs the protein to the ER and aids in its translocation across the ER membrane into the cisternal space of the ER. The signal sequences of different proteins differ in their amino acid sequences, but share common tertiary structural features, which most likely specify recognition and translocation of the nascent polypeptide chain by the ER. This essential process segregates those proteins destined for secretion from the cytoplasmic and many of the organellar proteins. The

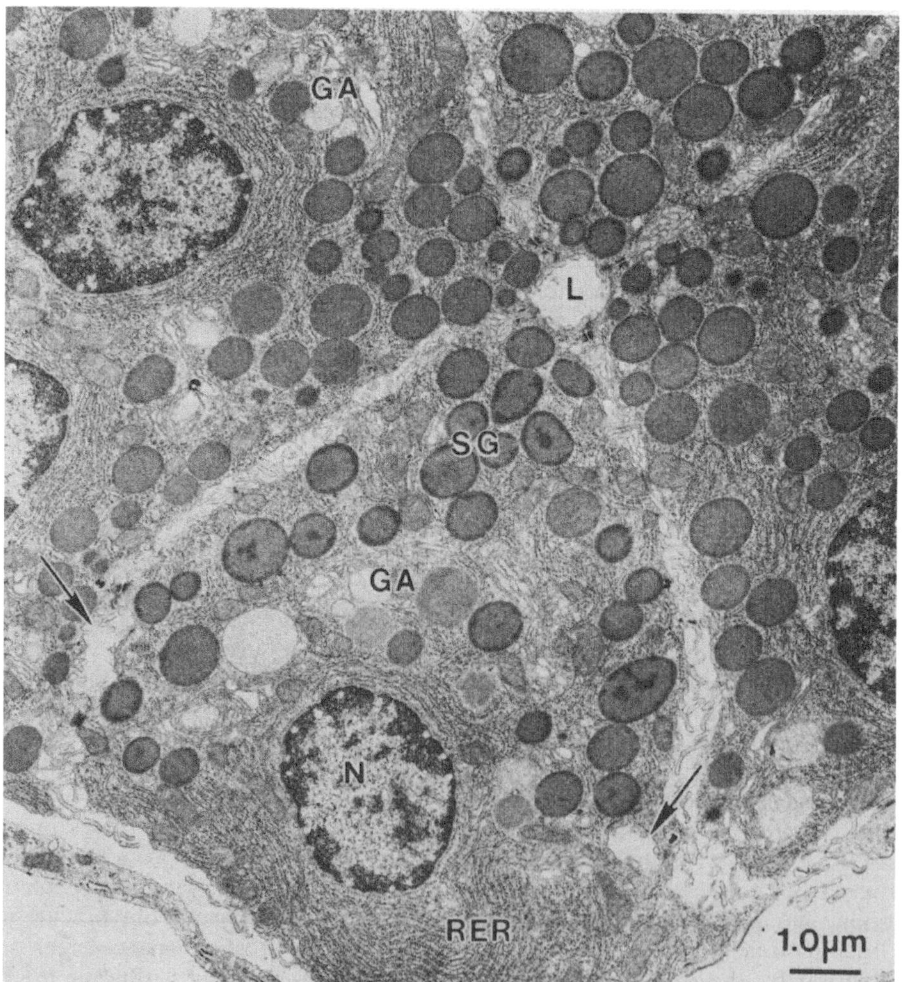

Fig. 1-1. Acinar cells of the rat parotid gland. A spherical nucleus (N) and abundant rough ER (RER) are present in the basal portion of the cell, a large Golgi apparatus (GA) is located apical or lateral to the nucleus, and numerous electron-dense secretory granules (SG) are stored in the apical cytoplasm. Lumen (L), intercellular canaliculi (arrows). (From [8]; reproduced with permission of Pergamon Press, Oxford.)

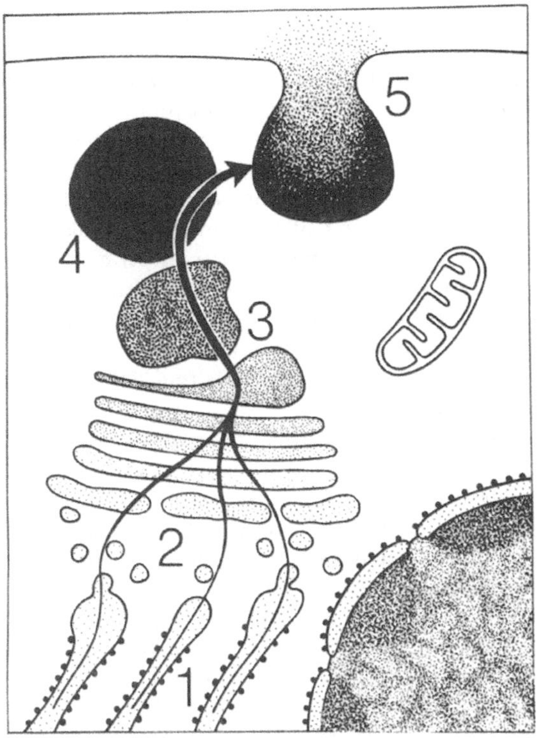

Fig. 1-2. Basic pathway (→) and general features of the secretory process in exocrine cells. 1) synthesis of proteins in the endoplasmic reticulum (ER) and segregation in the lumen of the ER; 2) migration to the Golgi apparatus; 3) concentration and packaging of the proteins in secretory granules; 4) intracellular storage; 5) discharge of the granule content by exocytosis.

mechanism by which this is achieved has been elucidated principally by Blobel and others [9–13] over the last dozen years. As the amino terminus of the protein emerges from the ribosome, when the nascent protein is about 70 amino acids long, it interacts with the signal recognition particle (SRP), an 11S cytoplasmic particle made up of proteins and RNA. Binding of the SRP to the signal sequence and the ribosome causes translation of the protein to stop. The SRP-ribosome complex then binds to an SRP receptor and (possibly) to other binding proteins or receptors on the ER membrane, the hydrophobic signal sequence inserts into the membrane, the SRP dissociates from the ribosome and SRP receptor, and translation resumes. Translocation of the protein across the ER membrane normally occurs cotranslationally (at least in most eukaryotic cells) and appears to require energy [14]. Soon after the amino-terminal end of the protein reaches the cisternal space, the signal sequence is removed by a specific signal peptidase.

This same basic mechanism is used to segregate membrane proteins and proteins destined for lysosomes from proteins resident in the cytoplasm or in other organelles. In the case of membrane proteins, other signals incorporated into the amino acid sequence of the protein prevent its complete transfer to the cisternal space and specify its positioning in the membrane [12]. Lysosomal proteins, which eventually must be removed from the secretory pathway, undergo a specific post-translational modification and are sorted from secretory proteins at a later stage [15, 16].

3. Post-translational modifications

After their synthesis, secretory, membrane, and lysosomal proteins undergo a number of covalent modifications. These modifications are essential for the protein to assume its proper configuration, to perform its biological functions, and to reach its final destination. The different modifications that various proteins may undergo are listed in Table 1-2. The first group of modifications are those that occur in the ER, many of which actually take place cotranslationally, i.e., as the protein is being synthesized and translocated across the ER membrane. These include signal-sequence cleavage, hydroxylation, disulfide bond formation, and N-linked glycosylation. Of these, the latter has been the most thoroughly investigated [17, 18] and begins with the assembly and subsequent transfer of an oligosaccharide precursor, $Glc_3Man_9GlcNAc_2$, from dolichol phosphate, a large ER-associated lipid, to asparagine residues in the acceptor glycoprotein. This precursor oligosaccharide is subsequently trimmed and rebuilt as the glycoprotein travels through the ER and Golgi apparatus by enzymatic removal of the glucose residues and some of the mannose residues, and addition of N-acetylglucosamine, galactose, fucose, and sialic acid. The final structure of the oligosaccharide groups varies among different proteins, as well as within the same protein, resulting in a microheterogeneity of glycoprotein structure.

Table 1-2. Post-translational modifications of secretory proteins

Modification	Intracellular Location
Signal sequence removal N-linked glycosylation Hydroxylation Disulfide bond formation	ER cisternae
Oligosaccharide trimming	ER and Golgi apparatus
Completion of N-linked oligosaccharides O-linked glycosylation Mannose-6-phosphate formation Fatty acylation Sulfation Phosphorylation	Golgi apparatus
Proteolytic processing	Immature and mature secretory granules Extracellular space

Subsequent modifications to the secretory proteins occur in the Golgi apparatus and may include O-glycosylation, fatty acylation, sulfation, and phosphorylation. The sequence of these modifications is specified by the topographic distribution of the modifying enzymes in the Golgi apparatus [18–20]. The structure of the Golgi apparatus (Fig. 1-3) reflects these biochemical requirements: a series of discrete, flattened, membranous saccules, providing maximum surface area for the interaction of membrane-bound enzymes and secretory proteins and providing separation of successive modification steps. Thus, as the proteins traverse the Golgi apparatus, they sequentially encounter the compartments containing the various enzymes, and their structure is modified accordingly. At present, it is not known how some proteins escape modification, even though their primary structure contains potential sites for the action of these enzymes.

The Golgi apparatus is also the location of modification and sorting steps involved in removal of lysosomal proteins from the secretory pathway and their routing to lysosomes. N-acetylglucosamine-1-phosphate is added to mannose residues on the N-linked oligosaccharides of the lysosomal proteins, then the N-acetylglucosamine residues are removed by a phosphodiesterase, leaving mannose-6-phosphate groups on the oligosaccharide [22–24]. Mannose-6-phosphate serves as a specific recognition marker of lysosomal enzymes [15, 16]; the mannose-6-phosphate-bearing enzymes bind to a receptor present in the Golgi apparatus and are subsequently routed to lysosomes [25–27]. Other, as yet unidentified, mechanisms for routing of proteins to lysosomes must also exist. For example, in I-cell disease, due to the absence of N-acetylglucosamine-1-phosphotransferase, lysosomal enzymes lack the mannose-6-phosphate recognition marker, yet a variable percentage of some enzymes reach the lysosomes [16].

Specific intracellular proteolytic cleavage of secretory protein precursors is an important post-translational modification in many cells, especially hormone-secreting cells [28, 29]. Many hormones, such as insulin, adrenocorticotrophic hormone, and vasopressin, are synthesized as larger, generally inactive, precursors (often called *prohormones*). Cleavage of the precursor by an endogenous protease is required to liberate the active hormone. Analysis of purified granules of endocrine cells [30] and comparison of the immunocytochemical localization of precursor and mature hormone [31], indicate that proteolytic processing probably begins at an early post-Golgi step and continues in the mature granule. In the exocrine pancreas, many of the digestive enzymes are synthesized as inactive proenzymes; under normal conditions their activation by proteolysis occurs only after they reach the lumen of the intestine. Specific intracellular proteolysis of exocrine secretory proteins has been less well studied, although a few examples may be cited. In the granular convoluted tubule cells of the male mouse submandibular gland, renin is synthesized as a larger precursor and appears to be cleaved to its active form intracellularly [32]. Nerve growth factor and epidermal growth factor are also synthesized as high-molecular-weight precursors by these cells and are found as complexes with proteases [33, 34]. The amino acid sequences of the human salivary acidic proline-rich proteins suggest that proteins III and IV are proteolytic products of proteins I and II, respectively [35]. A family of glutamic-acid/glutamine-rich proteins present in rat submandibular gland and saliva has recently been described [36]. Comparison of the

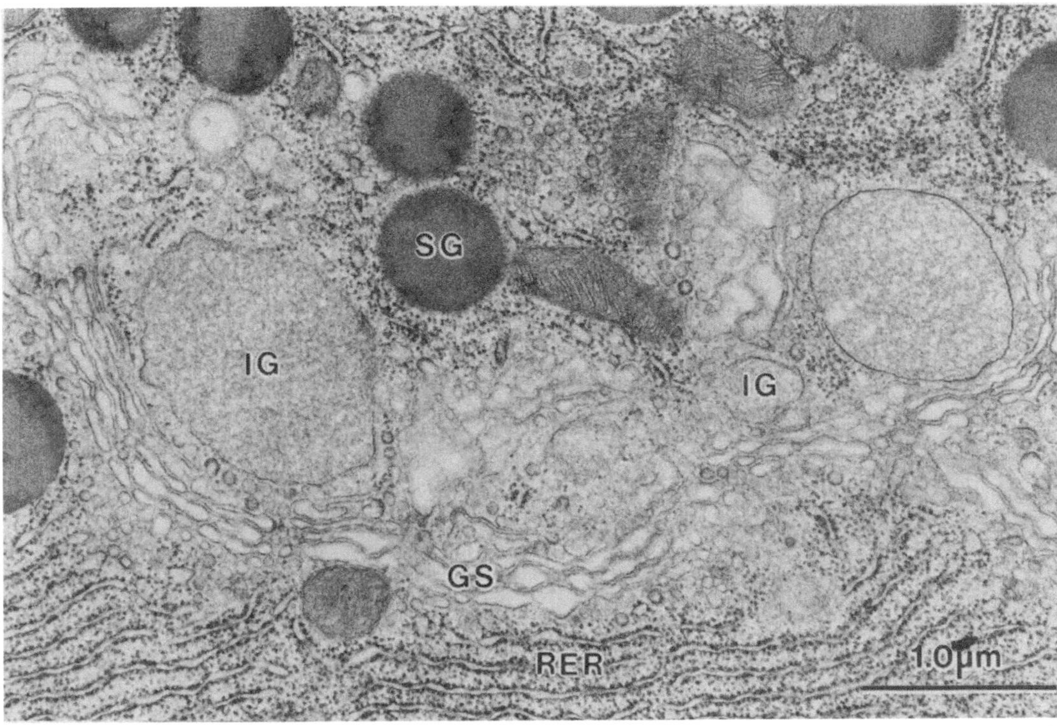

Fig. 1-3. Golgi apparatus of a rat parotid acinar cell. The Golgi apparatus is well developed and consists of 4–6 stacked saccules (GS), numerous vesicles, and immature granules (IG) of various sizes forming at the *trans* face. The rough ER (RER) closely approaches the *cis* face of the Golgi apparatus. SG = mature secretory granules. (From [21]; reproduced with permission of C.V. Mosby, St. Louis.)

protein species purified from glandular extracts with immunoprecipitated, in-vitro translation products suggests that these unusual secretory proteins may undergo post-translational proteolytic processing. Except for renin, however, the cellular location of the proteolytic processing reactions of these salivary secretory proteins has not been established. Secretory granules of rat parotid acinar cells contain a protease with similar specificity as those involved in prohormone processing in endocrine cells [37], but the protein substrates for this enzyme have not been identified.

4. Intracellular transport and sorting

Enzyme and immunocytochemical studies suggest that the entire complement of the cell's secretory proteins is synthesized and is present throughout the ER [38–40]. The movement of newly synthesized proteins within the ER probably occurs, for the most part, by simple diffusion through the cisternae. Regional morphological specialization of the ER is found principally in the transitional zone, where the ER approaches the *cis* face of the Golgi apparatus (Fig. 1-4). The transfer of proteins from the transitional ER to the Golgi apparatus occurs via small vesicles, which are abundant in this region and which apparently form by budding from the transitional ER [42–44]. The requirement of energy for movement of newly synthesized protein from the ER to the Golgi apparatus [45–47] and the disruption of transport by microtubule inhibitors [46, 48–50] are consistent with a vesicular transport mechanism.

Transport of secretory proteins through the Golgi apparatus occurs from the *cis* to the *trans* face, where the formation of secretory granules

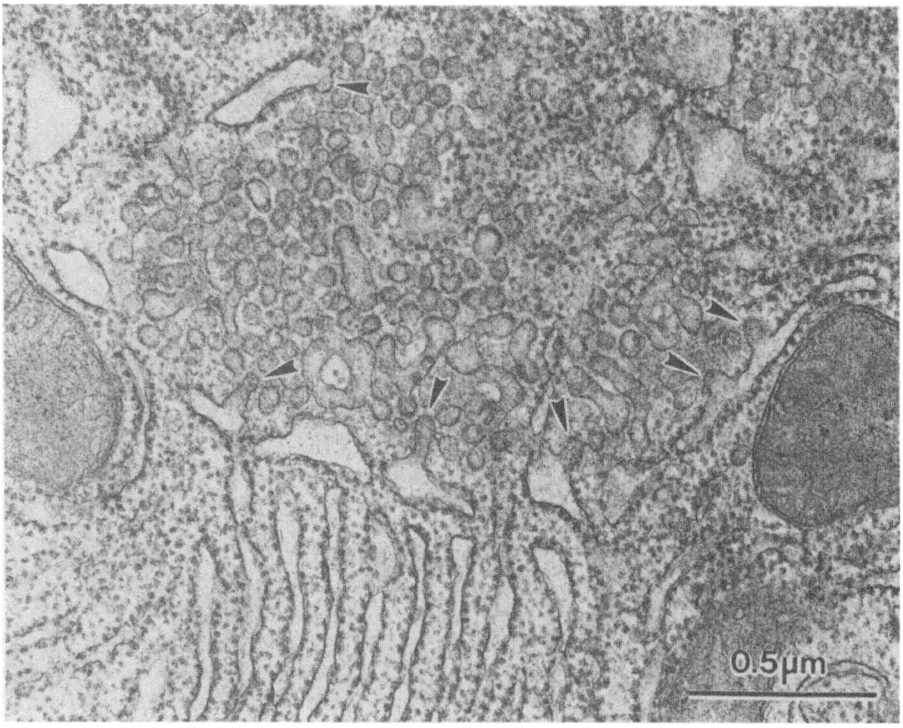

Fig. 1-4. Transitional ER in an acinar cell of the rat pancreas. Small vesicles appear to form (arrowheads) from dilated portions of the rough ER near the *cis* face of the Golgi apparatus (located out of the section plane). Numerous vesicles and tubules are present between the transitional ER and the Golgi apparatus. (From [41]; reproduced with permission of Lea & Febiger, Philadelphia.)

occurs. The mechanisms underlying this vectorial movement remain controversial. Morphological observations suggest that the Golgi saccules migrate intact toward the *trans* face, carrying their content of secretory proteins acquired at the *cis* face [39, 51–53]. In some cells that secrete large structures whose assembly begins in the *cis* Golgi saccules [54, 55], few alternatives seem possible. A more limited migration of saccules towards the *trans* face has been proposed for cells of Brunner's glands of the mouse [56]. The difficulty in reconciling the evidence for biochemical and functional heterogeneity of Golgi subcompartments, as well as the slow turnover of membrane proteins compared to that of the transported secretory proteins [57–60], with the model of saccule progression has led to the proposal that a system of vesicular shuttles operates between the Golgi saccules [1, 61]. Experiments demonstrating normal transport and processing of viral membrane proteins in recombinations of Golgi apparatus fractions from wild-type and glycosylation-deficient mutant cells [62, 63] provide additional evidence for a vesicular transport mechanism.

Vesicular transport of proteins between intracellular compartments, including the ER and Golgi apparatus, may involve receptor-mediated events rather than being a simple bulk-transfer process. Receptor recognition and binding would allow for efficient sorting and routing of proteins to different destinations. Several proteins have been shown to move between the ER and Golgi at different rates [64–66], suggesting that receptors for specific proteins either facilitate the transfer process or cause retention of the proteins in the ER. The transport of lysosomal proteins bearing the mannose-6-phosphate marker to lysosomes involves recognition and binding by a specific mannose-6-phosphate receptor [15, 16]. Biochemical and immunocytochemical evidence indicates that lysosomal proteins diverge from the secretory pathway at the *trans* face of the Golgi apparatus [20, 26, 27].

Clathrin-coated vesicles are often seen in the Golgi region, and it seems clear that they participate in the delivery of acid hydrolases to lysosomes [26, 27, 67]. Although in some cells clathrin-coated vesicles may contain secretory proteins, it has been postulated that this is related to the maturation of secretory granules [68]. In contrast, the "coat" of coated vesicles present in an in-vitro reconstituted Golgi transport system does not appear to consist of clathrin [69]. Microinjection of clathrin antibodies into cultured cells has little or no effect on the exocytotic pathway [70], and yeast mutants defective in clathrin synthesis exhibit near-normal growth rates, suggesting that this protein is not required for intracellular protein transport and secretion [71].

Recently, the concept of constitutive versus regulated secretion has been developed by Kelly and his colleagues [4, 72]. In cells that secrete by the regulated pathway, secretory proteins are typically concentrated and stored intracellularly in granules, and are released in response to a specific stimulus. Secretion via the constitutive pathway occurs continuously, little or no concentration of the proteins occurs, and the process is unaffected by cellular stimulation. Both constitutive and regulated secretion may occur in a single cell, such as an exocrine cell, with certain proteins following the constitutive pathway (e.g., basement membrane components and plasma membrane proteins) and others following the regulated pathway (e.g., secretory proteins). Disruption of the regulated pathway, for example, by certain ionophores or weak bases that dissipate pH gradients, results in diversion of regulated proteins into the constitutive pathway.

The divergence of the constitutive and regulated pathways occurs at the *trans* face of the Golgi apparatus (probably in the *trans* Golgi network [73, 74] and involves additional mechanisms for protein sorting and the targeting of vesicular carriers to different destinations. Expression of foreign proteins in cultured cell lines and their secretion by the correct pathway indicates that the sorting information is universal and is contained in the protein. It is believed that the sorting process is receptor mediated, but specific receptors or carrier proteins involved have not been identified.

5. Secretory granule formation

Exocrine glands, such as the salivary glands and pancreas, must secrete at very high rates for short periods of time. The protein synthetic capacity of these glands is insufficient to concurrently meet these demands. Thus, secretory proteins are synthesized more or less continuously and are stored in concentrated form within the cell until their release is required. The containers for the stored proteins, the secretory granules, are the characteristic feature of the acinar cells of exocrine glands. Exocrine secretory granules vary in size from about 0.5 μm to 2 μm, and their content may appear extremely electron dense or almost electron lucent, depending upon the cell type and the nature of the secretory product. In many cells, as described in other chapters, the granule content may appear extremely electron-dense or almost electron-lucent, depending upon the cell type and

Secretory granule formation occurs at the *trans* face of the Golgi apparatus (Figs. 1-3 and 1-5), where smaller, irregularly shaped, less-dense immature granules or condensing vacuoles are found in continuity with Golgi membranes. As the immature granules fill with secretory protein, they increase in size and density until they reach maturity, then they lose their connection with the Golgi apparatus. This simplified description, however, belies the fact that the mechanisms underlying granule formation and growth, and concentration of the content, remain poorly understood.

The source of the membrane surrounding the secretory granules has been a subject of intense interest. The continuity of Golgi membranes with the forming granules, and the similarity in enzyme cytochemical reactivity between the forming granule and the confluent saccule [39, 76–78], suggest that the granule membranes are derived from the Golgi apparatus. This scheme would require replacement of the membrane donated by components of the *trans* Golgi region; however, as noted above, the earlier models of saccule progression and Golgi apparatus turnover are inconsistent with data on the composition and

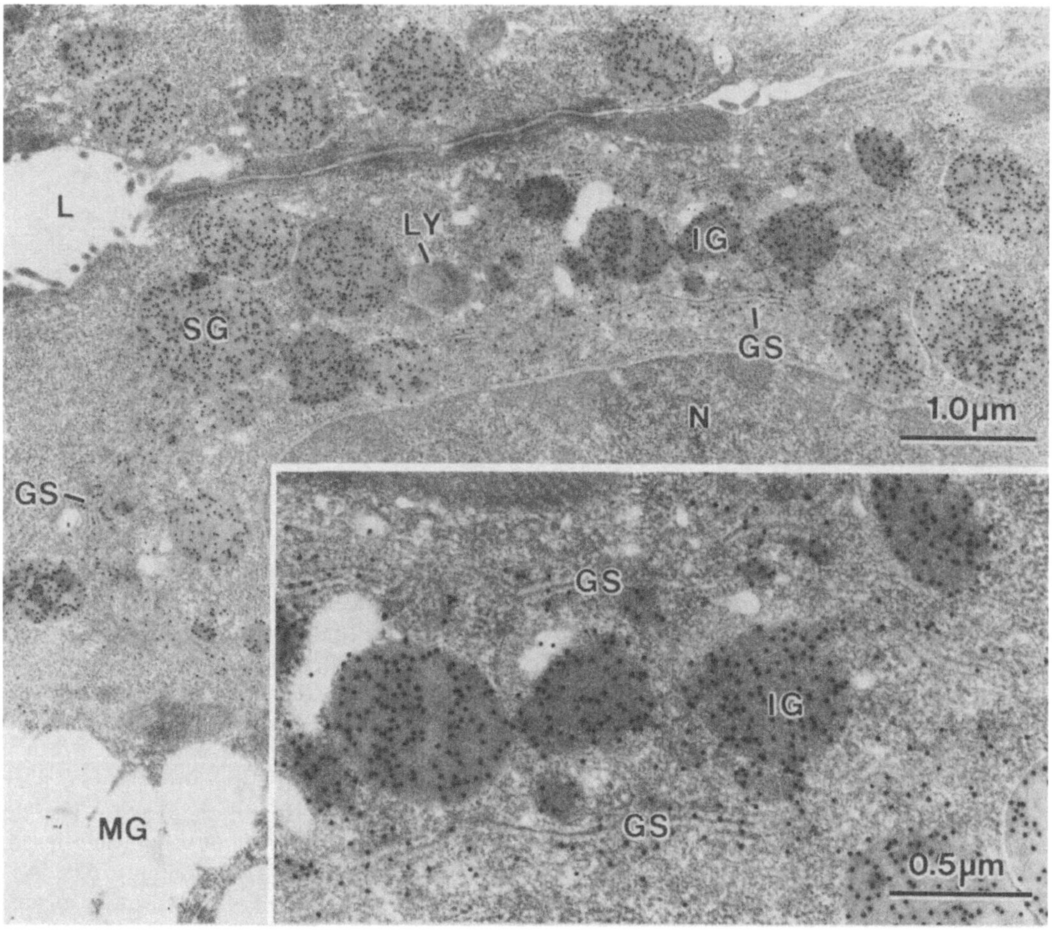

Fig. 1-5. Localization of secretory protein (B_1-immunoreactive protein) in serous demilune cells of the rat sublingual gland using the post embedding immunogold labeling technique. The thin section was incubated with an antibody to protein B_1, a secretory protein of the neonatal rat submandibular gland [75], and then with 15 nm protein A-gold. Gold particles indicating the presence of B_1-immunoreactive proteins are located over the Golgi saccules (GS) and over immature (IG) and mature (SG) secretory granules. The inset shows part of the Golgi apparatus at higher magnification. The labeling intensity (gold particles/unit area) of the granules is greater than that of the Golgi saccules. N = nucleus, LY = lysosome; L = lumen; MG = granules of adjacent mucous acinar cell.

turnover of cellular membranes. A number of studies demonstrating that membranes shuttle between the cell surface and various intracellular compartments [79–81] support the general notion that cells extensively recycle their membranes. Experiments employing electron-dense markers bound to secretory granule membranes transiently inserted into the plasma membrane during exocytosis have shown that, in some cells, the Golgi saccules and forming granules become labeled [82–85], implying that secretory granule membranes are reutilized during the formation and growth of new granules.

The pathway taken by the secretory proteins from the Golgi saccules to the forming granules has not been clearly defined. In some cells, granule formation appears to be initiated by the *trans* Golgi saccules, suggesting that the granules fill directly from these saccules. In acinar cells of the pancreas, salivary, and lacrimal glands, small forming granules are frequently seen in continuity with cisternal or tubular elements at the *trans*

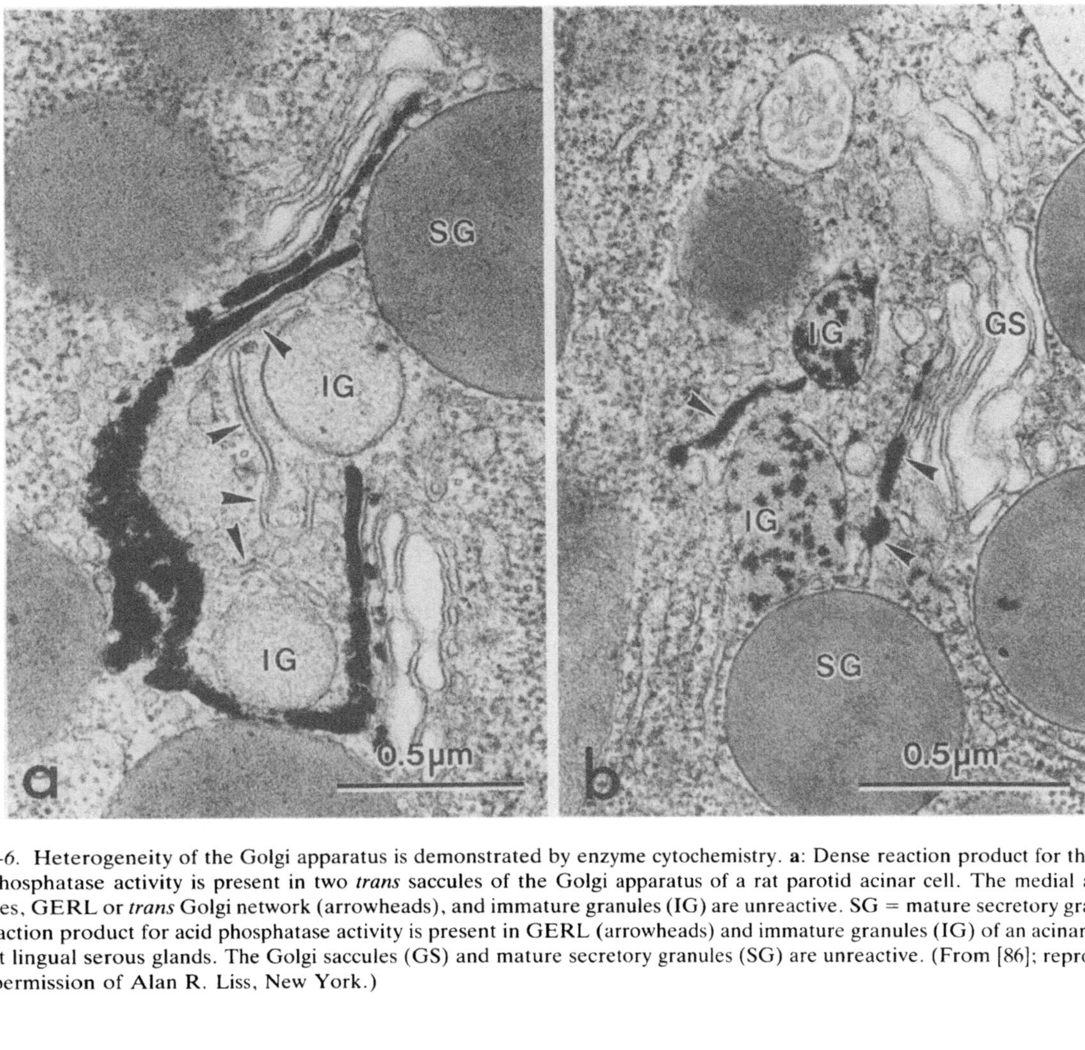

Fig. 1-6. Heterogeneity of the Golgi apparatus is demonstrated by enzyme cytochemistry. **a**: Dense reaction product for thiamine pyrophosphatase activity is present in two *trans* saccules of the Golgi apparatus of a rat parotid acinar cell. The medial and *cis* saccules, GERL or *trans* Golgi network (arrowheads), and immature granules (IG) are unreactive. SG = mature secretory granules. **b**: Reaction product for acid phosphatase activity is present in GERL (arrowheads) and immature granules (IG) of an acinar cell of the rat lingual serous glands. The Golgi saccules (GS) and mature secretory granules (SG) are unreactive. (From [86]; reproduced with permission of Alan R. Liss, New York.)

face (GERL [76] or *trans* Golgi network [73]; Fig. 1-6) that by enzyme [39] or immunocytochemistry [40, 87] lack detectable secretory proteins. In other cells, secretory proteins have been localized immunocytochemically in elements of the *trans* Golgi network that give rise to secretory granules [26, 27, 68]. Images of small vesicles either fusing with or budding from immature granules suggest that vesicular transport may also occur between the Golgi saccules and the forming granules.

In exocrine cells, the granules contain a mixture of all of the secretory proteins produced by the cells [88, 89]. The secretory proteins are concentrated during packaging by the Golgi apparatus, as indicated by increases in electron density associated with granule maturation, in immunocytochemical reactivity [87] (Fig. 1-5), and in the number of radioautographic grains seen in precursor incorporation studies [90,91]. Early studies demonstrating continued concentration in the absence of cellular respiration suggested that the process was passive rather than energy requiring [45], and directed attention toward the possible role of ionic interactions producing large aggregates with reduced osmotic activity [1]. Divalent cations, such as calcium, are present in relatively large amounts in some secretory granules [92–94] and could participate in the packaging process. Large anionic molecules, such as sulfated glycosaminoglycans [95, 96], which could form a structural basis for the intragranular aggregation of basic proteins, are also present in many secretory granules. In some granules, such as the insulin granule of the pancreatic B cells, crystallization of the secretory proteins also serves the function of concentration. The exact mechanisms involved in the concentration of secretory proteins in most cells have yet to be determined.

Studies of granules isolated from several cell types have revealed that the granule interior has an acidic pH and that a proton pump (H^+-ATPase) is present in the granule membrane [97]. Using a specific molecular probe to immunocytochemically identify acidic compartments, it was shown that a progressive fall in pH occurs from the *cis* to the *trans* face of the Golgi apparatus [98]. The low pH in the Golgi saccules and immature granules may be important for proper sorting of different secretory products [4]. In some cells, the acidic environment in secretory granules may be required for the accumulation and storage of certain components, e.g., biogenic amines [97]. Mature secretory granules of the rat parotid gland and exocrine pancreas, however, have little or no detectable H^+-ATPase, and their interior pH is near neutrality [99–101].

6. Exocytosis

Secretion of the content of stored secretory granules occurs by exocytosis (Fig. 1-7). During this process, granules approach the cell membrane, make close contact with it, and the lipid bilayers of the granule and cell membrane fuse; subsequent rupture of the fused portion of the membranes leaves the granule membrane in continuity with the cell membrane and results in the exteriorization of the content. The granule membrane is then retrieved by the cell by endocytosis. Although the regulation of exocytosis has been extensively studied (see below), the intracellular factors mediating the granule discharge process are less well defined.

Cytoskeletal elements, probably microtubules, seem to be involved in movement of the granules to the cell surface. Interactions between microtubules and granule membranes have been demonstrated in vitro [102], and microtubule poisons inhibit the release of newly synthesized proteins, although they do not block exocytosis of stored granules located near the plasma membrane [48, 50]. The role of actin and actin-associated proteins in exocytosis is poorly understood. Agents that inhibit actin filaments often cause an enhanced secretion [103], presumably due to disruption of the submembrane web of actin and related proteins usually present at the luminal surface [104] (Fig. 1-7). Actin also appears to be associated with the granule membranes [105].

Since granule discharge occurs only at the luminal surface of exocrine cells, there must be a specific recognition of the luminal membrane by components of, or associated with, the granule membrane. Studies of granule membranes isolated from several different cell types have revealed striking similarities in their protein composition [106]. Presumably, these proteins serve common functions related to exocytosis, such as recognition of luminal fusion sites or interaction with the cytoskeleton. Once a granule fuses with the cell membrane and the two membranes have become confluent, the granule membrane becomes an acceptable partner for subsequent fusion events, resulting in compound exocytosis [107, 108]. The mechanism by which the granule membrane develops fusion competency so rapidly is unknown.

Some freeze-fracture studies of exocytosis in stimulated cells suggest that there is a rapid clearing of intramembrane particles from the area of the apposed granule and cell membranes about to undergo fusion [109, 110]. Presumably, this facilitates the fusion of the lipid bilayers. Other studies of systems in which rapid freezing without chemical prefixation was employed indicate that the intramembrane particles are not cleared from the site of fusion and that the formation of a small pore in the apposed membranes is the first sign of fusion [111, 112]. Whether these contradictory observations are due to artifacts induced by the preparative methods or represent differences in the systems studied remains to be resolved. Observations of living cells (mast cells [113]) and in-vitro systems (isolated cortices of sea urchin eggs [114]) have shown that swelling of the granules occurs during exocytosis. Osmotic swelling was felt to be important in initiating membrane fusion, but evidence obtained from studies of chromaffin granule exocytosis [115] appears to be inconsis-

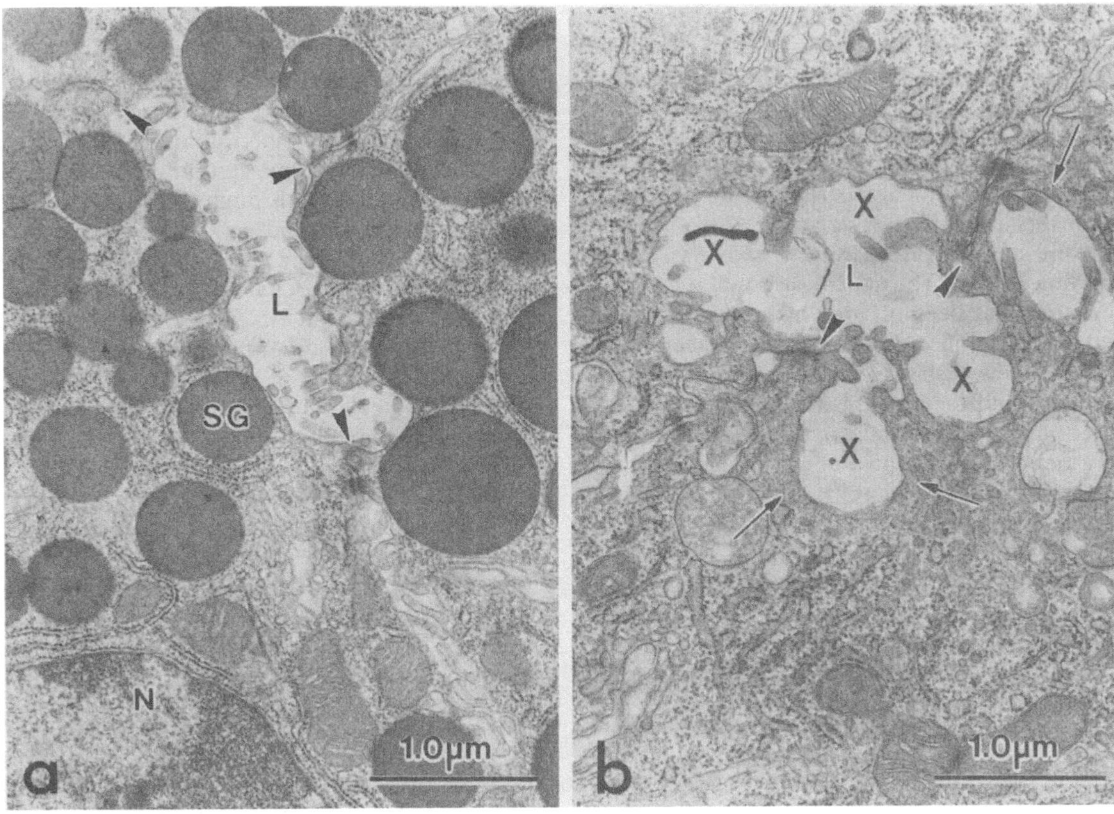

Fig. 1-7. Exocytosis in rat parotid acinar cells. **a**: The apical cytoplasm of unstimulated cells is filled with secretory granules (SG). N = nucleus; L = lumen. **b**: Stimulation with the β-adrenergic agonist isoproterenol causes exocytosis of all of the stored secretory granules. The lumen (L) is enlarged due to fusion with the granule membranes (X). A network of fine filaments (arrows) can be discerned beneath portions of the luminal membrane. Tight junctions (arrowheads). (From [41]; reproduced with permission of Lea & Febiger, Philadelphia.)

tent with this hypothesis. Recent observations of exocytosis of the giant granules of beige mouse mast cells indicate that fusion actually precedes the swelling of the granules [116]. Influx of extracellular molecules through the fusion pore may cause the subsequent swelling of the granule; enlargement and stabilization of the pore may depend upon granule swelling.

Following release of the granule content, the inserted membrane is retrieved from the cell surface by endocytosis. Small vesicles, frequently of the coated variety, but also some smooth-surfaced vesicles, pinch off of the granule membrane and fuse with either the Golgi saccules, immature granules, or lysosomes [82–85, 117, 118]. Fusion of the endocytic vesicles with Golgi saccules or immature granules suggests reutilization of the membrane in a subsequent round of granule formation. In contrast, fusion with lysosomes suggests that a portion of the membrane may be degraded and that new membrane must enter the cycle, probably at the level of the Golgi apparatus. Whether all secretory cells employ similar mechanisms for membrane recycling and the proportion of recycled versus new membrane contributing to granule formation remain to be established.

7. Regulation of the secretory process

Secretion by the pancreas and salivary glands is regulated physiologically by the autonomic nervous system and by circulating hormones. The latter have a much greater influence in regulating pancreatic secretion than salivary secretion. The neurotransmitters and hormones known to affect protein, water, and electrolyte secretion by the salivary glands and pancreas are listed in Table 1–3. It should be noted that wide variations in the pattern of innervation and in the response to those transmitters and hormones exist between species, and even among the different salivary glands within one species. The neurohormonal regulation of secretion has been an area of intense research activity, and significant advances have been made in the last few years. Several excellent reviews have been published on this topic [119–124].

The binding of a neurotransmitter released from a nerve terminal or a hormone to a specific receptor protein on the cell surface initiates the secretory response. The secretagogue receptors on pancreatic and salivary gland cells generally are of two classes, those whose effects are mediated by cyclic AMP and those that elevate intracellular calcium. The rapid increase in the concentration of these *second messengers* results in the activation of other intracellular enzymes and begins the cascade of reactions leading to exocytosis or secretion of fluid and electrolytes. Generation of elevated second-messenger levels is dependent upon the coupling of the receptor to a membrane enzyme, either adenylate cyclase or phospholipase C, through a guanine nucleotide binding protein, called a *G protein* [125]. In the case of receptors that elevate cyclic AMP, the G protein activates adenylate cyclase, which catalyzes the formation of cyclic AMP from ATP. Cyclic AMP subsequently activates the enzyme cyclic-AMP-dependent protein kinase, which phosphorylates other cellular proteins that presumably are involved in the secretory response. Several proteins undergo phosphorylative modification in response to receptor stimulation [126–129], but their identity and function are, for the most part, unknown. Recently, the phosphorylation of two rat parotid secretory granule membrane proteins after β-adrenergic stimulation was reported [130], but the role of these proteins in exocytosis has not been determined. Although

Table 1-3. Neurotransmitters and hormones regulating salivary and pancreatic secretion

Gland	Neurotransmitter or Hormone	Receptor	Intracellular Messenger	Process Regulated
Salivary	Norepinephrine	β-adrenergic	Cyclic AMP	Exocytosis
		α-adrenergic	Ca^{2+}	Fluid and electrolyte secretion
	Acetylcholine	Muscarinic cholinergic	Ca^{2+}	Fluid and electrolyte secretion, exocytosis
	Vasoactive intestinal peptide	VIP	Cyclic AMP	Exocytosis
	Substance P	Substance P	Ca^{2+}	Fluid and electrolyte secretion
Pancreas	Acetylcholine	Muscarinic cholinergic	Ca^{2+}	Exocytosis, fluid and electrolyte secretion
	Cholecystokinin Gastrin	CCK	Ca^{2+}	Exocytosis, fluid and electrolyte secretion
	Secretin	Secretin	Cyclic AMP	Fluid and electrolyte secretion
	Vasoactive intestinal peptide	VIP	Cyclic AMP	Exocytosis
	Bombesin	Bombesin	Ca^{2+}	Exocytosis
	Physalaemin Substance P	Physalaemin	Ca^{2+}	Exocytosis

cyclic-AMP-dependent protein kinase is widely distributed within these cells [131], the specificity of protein phosphorylation apparently results from its association with organelles and membranes, as well as stimulation-induced redistribution [132] and/or turnover of the enzyme subunits [133].

The G protein that is coupled to receptors elevating intracellular calcium activates phospholipase C, which hydrolyzes the membrane phospholipid, phosphatidylinositol-4,5-bisphosphate, to diacylglycerol and inositol-1,4,5-trisphosphate. Inositol-1,4,5-trisphosphate interacts with a receptor on an intracellular calcium storage compartment, believed to be the endoplasmic reticulum (possibly a specialized region of the ER) or a specific calcium-storage organelle [134], causing the release of calcium to the cytoplasm. The initial calcium release may be localized to sites near the cell membrane [135, 136]. A subsequent increase in cell-membrane permeability to calcium also occurs, resulting in the influx of extracellular calcium and a generalized elevation of cytoplasmic calcium concentrations. Diacylglycerol activates a calcium-dependent, membrane-associated protein kinase, protein kinase C [137]. Again, it is presumed that the phosphorylative modification of specific proteins by protein kinase C, as well as other calcium-activated protein kinases (e.g., calcium-calmodulin-dependent protein kinase) [138–140] is related to the secretory response observed after stimulation of this class of receptors.

8. Concluding remarks

This brief overview of protein secretion by the salivary glands and pancreas has focused on processes common to almost all secretory cells. These common features have enabled investigators from virtually all disciplines, using diverse technical approaches and a multitude of experimental systems, to contribute to our understanding of this universal property of eukaryotic cells. The increasing application of molecular biology techniques and the further development and refinement of in-vitro model systems promise to reveal, for many aspects of the secretory process, the molecular mechanisms underlying the problems that today remain unsolved.

References

1. Palade GE: Intracellular aspects of the process of protein secretion. *Science* 189: 347–358, 1975.
2. Hand AR, Oliver C (eds): Basic Mechanisms of Cellular Secretion. *Methods in Cell Biology*, Vol 23. New York: Academic Press, 1981.
3. Cantin M (ed): Cell Biology of the Secretory Process. Basel: Karger, 1984.
4. Burgess TL, Kelly RB: Constitutive and regulated secretion of proteins. *Ann Rev Cell Biol* 3: 243–293, 1987.
5. Shulz I: Electrolyte and fluid secretion in the exocrine pancreas. In: Physiology of the Gastrointestinal Tract, Vol 2, 2nd ed. LR Johnson (ed), New York: Raven Press, 1147–1171, 1987.
6. Martinez JR: Ion transport and water movement. *J Dent Res* 66: 638–647, 1987.
7. Young JA, Cook DI, van Lennep EW, Roberts M: Secretion by the major salivary glands. In: Physiology of the Gastrointestinal Tract, Vol 1, 2nd ed. LR Johnson (ed), New York: Raven Press, 773–815, 1987.
8. Hand AR, Ho B: Liquid-diet-induced alterations of rat parotid acinar cells studied by electron microscopy and enzyme cytochemistry. *Arch Oral Biol* 26: 369–380, 1981.
9. Blobel G, Dobberstein B: Transfer of proteins across membranes. I. Presence of proteolytically processed and unprocessed nascent immunoglobulin light chains on membrane bound ribosomes of murine myeloma. *J Cell Biol* 67: 835–851, 1975.
10. Walter P, Blobel G: Purification of a membrane associated protein complex required for protein translocation across the endoplasmic reticulum. *Proc Natl Acad Sci USA* 77: 7112–7116, 1980.
11. Walter P, Blobel G: Signal recognition particle contains a 7S RNA essential for protein translocation across the endoplasmic reticulum. *Nature* (London) 299: 691–698, 1982.
12. Sabatini DD, Kreibich G, Morimoto T, Adesnik M: Mechanisms for the incorporation of proteins in membranes and organelles. *J Cell Biol* 92: 1–22, 1982.
13. Walter P, Lingappa VR: Mechanism of protein translocation across the endoplasmic reticulum membrane. *Ann Rev Cell Biol* 2: 499–516, 1986.
14. Perara E, Rothman RE, Lingappa VR: Uncoupling translocation from translation: Implications for transport of proteins across membranes. *Science* 232: 348–352, 1986.
15. Sly WS, Fischer HD, Gonzalez-Noriega A, Grubb JH, Natowicz M: Role of the 6-phosphomannosyl-enzyme receptor in intracellular transport and adsorptive pinocytosis of lysosomal enzymes. In: Basic Mechanisms of Cellular Secretion. *Methods in Cell Biology*, Vol 23. AR Hand, C Oliver (eds), New York: Academic Press, 191–214, 1981.

16. von Figura K, Hasilik A: Lysosomal enzymes and their receptors. *Ann Rev Biochem* 55: 167–193, 1986.
17. Kornfeld R, Kornfeld S: Assembly of asparagine-linked oligosaccharides. *Ann Rev Biochem* 54: 631–664, 1985.
18. Hirschberg CB, Snider MD: Topography of glycosylation in the rough endoplasmic reticulum and Golgi apparatus. *Ann Rev Biochem* 56: 63–87, 1987.
19. Roth J: Subcellular organization of glycosylation in mammalian cells. *Biochim Biophys Acta* 906: 405–436, 1987.
20. Goldberg DE, Kornfeld S: Evidence for extensive subcellular organization of asparagine-linked oligosaccharide processing and lysosomal enzyme phosphorylation. *J Biol Chem* 258: 3159–3165, 1983.
21. Hand AR: Salivary glands. In: Orban's Oral Histology and Embryology, 10th ed. SN Bhaskar (ed), St. Louis: CV Mosby, 328–360, 1985.
22. Tabas I, Kornfeld S: Biosynthetic intermediates of β-glucuronidase contain high mannose oligosaccharides with blocked phosphate residues. *J Biol Chem* 255: 6633–6639, 1980.
23. Varki A, Kornfeld S: Identification of a rat liver α-N-acetylglucosaminyl phosphodiesterase capable of removing "blocking" α-N-acetylglucosamine residues from phosphorylated high mannose oligosaccharides of lysosomal enzymes. *J Biol Chem* 255: 8398–8401, 1980.
24. Pohlmann R, Waheed A, Hasilik, A, von Figura K: Synthesis of phosphorylated recognition marker in lysosomal enzymes is located in the *cis* part of Golgi apparatus. *J Biol Chem* 257: 5323–5325, 1982.
25. Brown WJ, Farquhar MG: The mannose-6-phosphate receptor for lysosomal enzymes is concentrated in *cis* Golgi cisternae. *Cell* 36: 295–307, 1984.
26. Geuze HJ, Slot JW, Strous GJAM, Hasilik A, von Figura K: Ultrastructural localization of the mannose-6-phosphate receptor in rat liver. *J Cell Biol* 98: 2047–2054, 1984.
27. Geuze HJ, Slot JW, Strous GJAM, Hasilik A, von Figura K: Possible pathways for lysosomal enzyme delivery. *J Cell Biol* 101: 2253–2262, 1985.
28. Steiner DF, Quinn PS, Patzelt C, Chan SJ, Marsh J, Tager HS: Proteolytic cleavage in the posttranslational processing of proteins. In: Cell Biology: A Comprehensive Treatise, Vol 4. DM Prescott, L Goldstein (eds), New York: Academic Press, 175–202, 1980.
29. Gluschankof P, Cohen P: Proteolytic enzymes in the post-translational processing of polypeptide hormone precursors. *Neurochem Res* 12: 951–958, 1987.
30. Gainer H, Sarne Y, Brownstein MJ: Biosynthesis and axonal transport of rat neurohypophysial proteins and peptides. *J Cell Biol* 73: 366–381, 1977.
31. Orci L, Ravazzola M, Amherdt M, Madsen O, Vassalli J-D, Perrelet A: Direct identification of prohormone conversion site in insulin-secreting cells. *Cell* 42: 671–681, 1985.
32. Bing T, Poulsen K, Hackenthal E, Rix E, Taugner R: Renin in the submaxillary gland. A review. *J Histochem Cytochem* 28: 874–880, 1980.
33. Isackson PJ, Dunbar JC, Bradshaw RA, Ullrich A: The structure of murine 7S nerve growth factor: Implications for biosynthesis. *Int J Neurosci* 26: 95–108, 1985.
34. Stoscheck CM, King LE Jr: Functional and structural characteristics of EGF and its receptor and their relationship to transforming proteins. *J Cell Biochem* 31: 135–152, 1986.
35. Bennick A: Salivary proline-rich proteins. *Mol Cell Biochem* 45: 83–99, 1982.
36. Mirels L, Bedi GS, Dickinson DP, Gross KW, Tabak LA: Molecular characterization of glutamic acid/glutamine-rich secretory proteins from rat submandibular glands. *J Biol Chem* 262: 7289–7297, 1987.
37. von Zastrow M, Tritton TR, Castle JD: Exocrine secretion granules contain peptide amidation activity. *Proc Natl Acad Sci USA* 83: 3297–3301, 1986.
38. Herzog V, Miller F: Die Lokalisation endogener Peroxydase in der Glandula parotis der Ratte. *Z Zellforsch Mikrosk Anat* 107: 403–420, 1970.
39. Hand AR, Oliver C: Relationship between the Golgi apparatus, GERL, and secretory granules in acinar cells of the rat exorbital lacrimal gland. *J Cell Biol* 74: 399–413, 1977.
40. Geuze JJ, Slot JW, Tokuyasu KT: Immunocytochemical localization of amylase and chymotrypsinogen in the exocrine pancreatic cells with special attention to the Golgi complex. *J Cell Biol* 82: 697–707, 1979.
41. Hand AR: Salivary glands. In: Oral Histology: Inheritance and Development, 2nd ed. DV Provenza, W Seibel (eds), Philadelphia: Lea and Febiger, 388–417, 1986.
42. Zeigel RF, Dalton AJ: Speculations based on the morphology of the Golgi systems in several types of protein secreting cells. *J Cell Biol* 15: 45–54, 1962.
43. Caro LG, Palade GE: Protein synthesis, storage, and discharge in the pancreatic exocrine cell. A autoradiographic study. *J Cell Biol* 30: 473–495, 1964.
44. Jamieson JD, Palade GE: Intracellular transport of secretory proteins in the pancreatic exocrine cell. I. Role of the peripheral elements of the Golgi complex. *J Cell Biol* 34: 577–596, 1967.
45. Jamieson JD, Palade GE: Intracellular transport of secretory proteins in the pancreatic exocrine cell. IV. Metabolic requirements. *J Cell Biol* 39: 589–603, 1968.
46. Chu LLH, MacGregor RR, Cohn DV: Energy-dependent intracellular translocation of proparathormone. *J Cell Biol* 72: 1–10, 1977.
47. Tartakoff A: Temperature and energy dependence of secretory protein transport in the exocrine pancreas. *EMBO J* 5: 1477–1482, 1986.
48. Patzelt C, Brown D, Jeanrenaud B: Inhibitory effect of colchicine on amylase secretion by rat parotid glands. Possible localization in the Golgi area. *J Cell Biol* 73: 578–593, 1977.
49. Malaisse-Legae F, Armherdt M, Ravazzola M, Sener A, Hutton JC, Orci L, Malaisse WJ: Role of microtubules in the synthesis, conversion, and release of (pro)insulin. A biochemical and radioautographic study in rat islets. *J Clin Invest* 63: 1284–1296, 1979.
50. Williams JA: Effects of antimitotic agents on ultrastructure and intracellular transport of protein in pancreatic acini. In: Basic Mechanisms of Cellular Secretion. Methods in Cell Biology, Vol 23. AR Hand, C Oliver

(eds), New York: Academic Press, 247–258, 1981.

51. Neutra M, Leblond CP: Synthesis of the carbohydrate of mucus in the Golgi complex as shown by electron microscope radioautography of goblet cells from rats injected with glucose-H^3. *J Cell Biol* 30: 119–136, 1966.
52. Slot JW, Geuze JJ, Poort C: Synthesis and intracellular transport of proteins in the exocrine pancreas of the frog (*Rana esculenta*). I. An ultrastructural and autoradiographic study. *Cell Tissue Res* 155: 135–154, 1974.
53. Bennett G: Role of the Golgi complex in the secretory process. In: Cell Biology of the Secretory Process. M Cantin (ed), Basel: Karger, 102–147, 1984.
54. Brown RM Jr: Observations on the relationship of the Golgi apparatus to wall formation in the marine chrysophycean alga, *Pleurochrysis scherffelii* Pringsheim. *J Cell Biol* 41: 109–123, 1969.
55. Marchi F, Leblond CP: Radioautographic characterization of successive compartments along the rough endoplasmic reticulum-Golgi pathway of collagen precursors in foot pad fibroblasts of [^{3}H]proline-injected rats. *J Cell Biol* 98: 1705–1709, 1984.
56. Rambourg A, Clermont Y, Hermo L, Segretain D: Tridimensional architecture of the Golgi apparatus in mucous cells of Brunner's glands of the mouse. *Am J Anat* 179: 95–107, 1987.
57. Meldolesi J: Dynamics of cytoplasmic membranes in guinea pig pancreatic acinar cells. I. Synthesis and turnover of membrane proteins. *J Cell Biol* 61: 1–13, 1974.
58. Meldolesi J: Membranes and membrane surfaces. Dynamics of cytoplasmic membranes in pancreatic acinar cells. *Phil Trans Roy Soc Lond* Ser B 268: 39–53, 1974.
59. Wallach D, Kirshner N, Schramm M: Non-parallel transport of membrane proteins and content proteins during assembly of the secretory granule in rat parotid gland. *Biochim Biophys Acta* 375: 87–105, 1975.
60. Winkler H: The biogenesis of adrenal chromaffin granules. *Neuroscience* 2: 657–683, 1977.
61. Farquhar MG: Progress in unraveling pathways of Golgi traffic. *Ann Rev Cell Biol* 1: 447–488, 1985.
62. Balch WE, Dunphy WG, Braell WA, Rothman JE: Reconstitution of the transport of protein between successive compartments of the Golgi measured by the coupled incorporation of N-acetylglucosamine. *Cell* 39: 405–416, 1984.
63. Rothman JE, Miller R, Urbani LJ: Intercompartmental transport in the Golgi complex is a dissociative process: Facile transfer of membrane protein between two Golgi populations. *J Cell Biol* 99: 260–271, 1984.
64. Fitting T, Kabat D: Evidence for a glycoprotein "signal" involved in transport between subcellular organelles. Two membrane glycoproteins encoded by murine leukemia virus reach the cell surface at different rates. *J Biol Chem* 257: 14011–14017, 1982.
65. Lodish, HF, Kong N, Snider M, Strous GJAM: Hepatoma secretory proteins migrate from the RER to Golgi at characteristic rates. *Nature* (London) 304: 80–83, 1983.
66. Scheele G, Tartakoff A: Exit of non-glycosylated secretory proteins from the RER is asynchronous in the exocrine pancreas. *J Biol Chem* 260: 926–931, 1985.
67. Friend DS, Farquhar MG: Functions of coated vesicles during protein absorption in the rat vas deferens. *J Cell Biol* 35: 357–376, 1967.
68. Tooze J, Tooze SA: Clathrin-coated vesicular transport of secretory proteins during the formation of ACTH-containing secretory granules in AtT20 cells. *J Cell Biol* 103: 839–850, 1986.
69. Orci L, Glick BS, Rothman JE: A new type of coated vesicular carrier that appears not to contain clathrin: Its possible role in protein transport within the Golgi stack. *Cell* 46: 171–184, 1986.
70. Doxsey S, Helenius A, Blank G, Brodsky F: Inhibition of endocytosis by anti-clathrin antibodies. *J Cell Biol* 103: 53a, 1986.
71. Payne GS, Schekman R: A test of clathrin function in protein secretion and cell growth. *Science* 230: 1009–1014, 1985.
72. Kelly RB: Pathways of protein secretion in eukaryotes. *Science* 230: 25–32, 1985.
73. Griffiths G, Simons K: The *trans* Golgi network: Sorting at the exit site of the Golgi complex. *Science* 234: 438–443, 1986.
74. Tooze J, Tooze SA, Fuller SD: Sorting of progeny coronavirus from condensed secretory proteins at the exit from the *trans*-Golgi network of AtT20 cells. *J Cell Biol* 105: 1215–1226, 1987.
75. Ball WD, Hand AR, Johnson AO: Secretory proteins as markers for cellular phenotypes in rat salivary glands. *Dev Biol* 125: 265–279, 1988.
76. Novikoff AB: The endoplasmic reticulum: A cytochemist's view (a review). *Proc Natl Acad Sci USA* 73: 2781–2787, 1976.
77. Broadwell RD, Oliver C: Golgi apparatus, GERL, and secretory granule formation within neurons of the hypothalamoneurohypophysial system of control and hyperosmotically stressed mice. *J Cell Biol* 90: 474–484, 1982.
78. Hand AR, Oliver C: Effects of secretory stimulation on the Golgi apparatus and GERL of rat parotid acinar cells. *J Histochem Cytochem* 32: 403–412, 1984.
79. Schneider Y-J, Tulkens P, de Duve C, Trouet A: Fate of plasma membrane during endocytosis. II. Evidence of recycling (shuttle) of plasma membrane constituents. *J Cell Biol* 82: 466–474, 1979.
80. Muller WA, Steinman RM, Cohn ZA: The membrane proteins of the vacuolar system. II. Bidirectional flow between secondary lysosomes and plasma membrane. *J Cell Biol* 86: 304–314, 1980.
81. Steinman RM, Mellman IS, Muller WA, Cohn ZA: Endocytosis and the recycling of plasma membrane. *J Cell Biol* 96: 1–27, 1983.
82. Herzog V, Farquhar MG: Luminal membrane retrieved after exocytosis reaches most Golgi cisternae in secretory cells. *Proc Natl Acad Sci USA* 74: 5073–5077, 1977.
83. Farquhar MG: Recovery of surface membrane in anterior pituitary cells. Variations in traffic detected with anionic and cationic ferritin. *J Cell Biol* 77: R35–R42, 1978.
84. Ottosen PD, Courtoy PJ, Farquhar MG: Pathways followed by membrane recovered from the surface of plasma cells and myeloma cells. *J Exp Med* 152: 1–19,

1980.
85. Herzog V, Reggio H: Pathways of endocytosis from luminal plasma membrane in rat exocrine pancreas. *Eur J Cell Biol* 21: 141–150, 1980.
86. Hand AR: Morphology and cytochemistry of the Golgi apparatus of rat salivary gland acinar cells. *Am J Anat* 130: 141–158, 1971.
87. Bendayan M: Concentration of amylase along its secretory pathway in the pancreatic acinar cell as revealed by high resolution immunocytochemistry. *Histochem J* 16: 85–108, 1984.
88. Kraehenbuhl JP, Racine L, Jamieson JD: Immunocytochemical localization of secretory proteins in bovine pancreatic exocrine cells. *J Cell Biol* 72: 406–423, 1977.
89. Bendayan M, Roth J, Perrelet A, Orci L: Quantitative immunocytochemical localization of pancreatic secretory proteins in subcellular compartments of the rat acinar cell. *J Histochem Cytochem* 28: 149–160, 1980.
90. Jamieson JD, Palade GE: Condensing vacuole conversion and zymogen granule discharge in pancreatic exocrine cells: Metabolic studies. *J Cell Biol* 48: 503–522, 1971.
91. Castle JD, Jamieson JD, Palade GE: Radioautographic analysis of the secretory process in the parotid acinar cell of the rabbit. *J Cell Biol* 53: 290–311, 1972.
92. Wallach D, Schramm M: Calcium and the exportable protein in rat parotid gland. Parallel subcellular distribution and concomitant secretion. *Eur J Biochem* 21: 433–437, 1971.
93. Clementi F, Meldolesi J: Calcium and pancreatic secretion. I. Subcellular distribution of calcium and magnesium in the exocrine pancreas of the guinea pig. *J Cell Biol* 65: 88–102, 1975.
94. Verdugo P, Deyrup-Olsen I, Aitken M, Villalon M, Johnson D: Molecular mechanism of mucin secretion: I. The role of intragranular charge shielding. *J Dent Res* 66: 506–508, 1987.
95. Reggio HA, Palade GE: Sulfated compounds in the zymogen granules of the guinea pig pancreas. *J Cell Biol* 77: 288–314, 1978.
96. Giannattasio G, Zanini A, Rosa P, Meldolesi J, Margolis RK, Margolis RU: Molecular organization of prolactin granules. III. Intracellular transport of sulfated glycosaminoglycans and glycoproteins of the bovine prolactin granule. *J Cell Biol* 86: 273–279, 1980.
97. Mellman I, Fuchs R, Helenius A: Acidification of the endocytic and exocytic pathways. *Ann Rev Biochem* 55: 663–700, 1986.
98. Anderson RGW, Pathak RK: Vesicles and cisternae in the *trans* Golgi apparatus of human fibroblasts are acidic compartments. *Cell* 40: 634–643, 1985.
99. Arvan P, Castle JD: Osmotic properties and internal pH of isolated rat parotid secretory granules. *J Biol Chem* 259: 13567–13572, 1984.
100. Arvan P, Rudnick G, Castle JD: Relative lack of H^+ translocase activity in isolated parotid secretory granules. *J Biol Chem* 260: 14945–14952, 1985.
101. Orci L, Ravazzola M, Anderson RGW: The condensing vacuole of exocrine cells is more acidic than the mature secretory vesicles. *Nature* (London) 326: 77–79, 1987.
102. Sherline P, Lee Y-C, Jacobs LS: Binding of microtubules to pituitary secretory granules and secretory granule membranes. *J Cell Biol* 72: 380–389, 1977.
103. Orci L, Gabbay KH, Malaisse WJ: Pancreatic beta-cell web: Its possible role in insulin secretion. *Science* 175: 1128–1130, 1972.
104. Drenckhahn D, Mannherz HG: Distribution of actin and the actin-associated proteins myosin, tropomyosin, alpha-actinin, vinculin, and villin in rat and bovine exocrine glands. *Eur J Cell Biol* 30: 167–176, 1983.
105. Bendayan M: Ultrastructural localization of actin in muscle, epithelial and secretory cells by applying the protein A-gold immunocytochemical technique. *Histochem J* 15: 39–58, 1983.
106. Cameron RS, Cameron PL, Castle JD: A common spectrum of polypeptides occurs in secretion granule membranes of different exocrine glands. *J Cell Biol* 103: 1299–1313, 1986.
107. Amsterdam A, Ohad I, Schramm M: Dynamic changes in the ultrastructure of the acinar cell of the rat parotid gland during the secretory cycle. *J Cell Biol* 41: 753–773, 1969.
108. Hand AR: The fine structure of von Ebner's gland of the rat. *J Cell Biol* 44: 340–353, 1970.
109. Orci L, Perrelet A, Friend DS: Freeze-fracture of membrane fusion during exocytosis in pancreatic B-cells. *J Cell Biol* 75: 23–30, 1977.
110. Tanaka Y, De Camilli P, Meldolesi J: Membrane interactions between secretion granules and plasmalemma in three exocrine glands. *J Cell Biol* 84: 438–453, 1980.
111. Chandler DE, Heuser JE: Arrest of membrane fusion events in mast cells by quick-freezing. *J Cell Biol* 86: 666–674, 1980.
112. Ornberg RL, Reese TS: Beginning of exocytosis captured by rapid-freezing of *Limulus* amebocytes. *J Cell Biol* 90:40–54, 1981.
113. Curran MJ, Brodwick MS: Direct visualization of exocytosis in mast cells. *Biophys J* 45: 170a, 1984.
114. Zimmerberg J, Sardet C, Epel D: Exocytosis of sea urchin egg cortical vesicles *in vitro* is retarded by hyperosmotic sucrose: Kinetics of fusion monitored by quantitative light-scattering microscopy. *J Cell Biol* 101: 2398–2410, 1985.
115. Holz RW: The role of osmotic forces in exocytosis from adrenal chromaffin cells. *Ann Rev Physiol* 48: 175–189, 1986.
116. Zimmerberg J, Curran M, Cohen FS, Brodwick M: Simultaneous electrical and optical measurements show that membrane fusion precedes secretory granule swelling during exocytosis of beige mouse mast cells. *Proc Natl Acad Sci USA* 84: 1585–1589, 1987.
117. Oliver C, Hand AR: Uptake and fate of luminally administered horseradish peroxidase in resting and isoproterenol stimulated rat parotid acinar cells. *J Cell Biol* 76: 207–220, 1978.
118. Patzak A, Winkler H: Exocytotic exposure and recycling of membrane antigens of chromaffin granules: Ultrastructural evaluation after immuno-labeling. *J Cell Biol* 102: 510–515, 1986.
119. Butcher FR, Putney JW Jr: Regulation of parotid gland

function by cyclic nucleotides and calcium. *Adv Cyclic Nucleotide Res* 13: 215–249, 1980.
120. Williams JA: Regulatory mechanisms in pancreas and salivary acini. *Ann Rev Physiol* 46: 361–375, 1984.
121. Putney JW Jr: Identification of cellular activation mechanisms associated with salivary secretion. *Ann Rev Physiol* 48: 75–88, 1986.
122. Baum BJ: Regulation of salivary secretion. In: The Salivary System. LM Sreebny (ed), Boca Raton, FL: CRC Press, 123–134, 1987.
123. Solomon TE: Control of exocrine pancreatic secretion. In: Physiology of the Gastrointestinal Tract, 2nd ed. LR Johnson (ed), New York: Raven Press, 1173–1207, 1987.
124. Gardner JD, Jensen RT: Secretagogue receptors on pancreatic acinar cells. In: Physiology of the Gastrointestinal Tract, 2nd ed. LR Johnson (ed), New York: Raven Press, 1109–1127, 1987.
125. Stryer L, Bourne HR: G proteins: A family of signal transducers. *Ann Rev Cell Biol* 2: 391–419, 1986.
126. Dowd FJ, Watson EL, Horio B, Lau Y-S, Park K: Phosphorylation of rabbit parotid microsomal protein occurs only with β-adrenergic stimulation. *Biochem Biophys Res Commun* 101: 281–288, 1981.
127. Baum BJ, Freiberg JM, Ito, H, Roth GS, Filburn CR: β-adrenergic regulation of protein phosphorylation and its relationship to exocrine secretion in dispersed rat parotid gland acinar cells. *J Biol Chem* 256: 9731–9736, 1981.
128. Freedman SD, Jamieson JD: Hormone-induced protein phosphorylation. II. Localization to the ribosomal fraction from rat exocrine pancreas and parotid of a 29,000-dalton protein phosphorylated *in situ* in response to secretagogues. *J Cell Biol* 95: 909–917, 1982.
129. Quissell DO, Deisher LM, Barzen KA: The rate-determining step in cAMP-mediated exocytosis in the rat parotid and submandibular glands appears to involve analogous 26-kDa integral membrane phosphoproteins. *Proc Natl Acad Sci USA* 82: 3237–3241, 1985.
130. Marino CR, Gorelick FS, Castle JD: Isoproterenol induced phosphorylation of granule membrane proteins in the rat parotid gland. *J Cell Biol* 105: 57a, 1987.
131. Mednieks MI, Jungmann RA, Hand AR: Ultrastructural immunocytochemical localization of cyclic AMP-dependent protein kinase regulatory subunits in rat parotid acinar cells. *Eur J Cell Biol* 44: 308–317, 1987.
132. Mednieks MI, Hand AR: Cyclic AMP-dependent protein kinase in stimulated rat parotid gland cells: Compartmental shifts after in vitro treatment with isoproterenol. *Eur J Cell Biol* 28: 264–271, 1982.
133. Schwoch G: Selective regulation of the amount of the catalytic subunit of cyclic AMP-dependent protein kinases during isoprenaline-induced growth of the rat parotid gland. *Biochem J* 248: 243–250, 1987.
134. Volpe P, Krause K-H, Hashimoto S, Zorzato F, Pozzan T, Meldolesi J, Lew PD: "Calciosome," a cytoplasmic organelle: The inositol 1,4,5-trisphosphate-sensitive Ca^{2+} store of non-muscle cells? *Proc Natl Acad Sci USA* 85: 1091–1095, 1988.
135. Putney JW Jr: A model for receptor-regulated calcium entry. *Cell Calcium* 7: 1–12, 1986.
136. Foskett JK, Gunter-Smith PJ, Melvin JE, Turner RJ: Physiological localization of an agonist-sensitive pool of Ca^{2+} in parotid acinar cells. *Proc Natl Acad Sci USA* 86:167–171, 1989.
137. Kikkawa U, Nishizuka Y: The role of protein kinase C in transmembrane signalling. *Ann Rev Cell Biol* 2: 149–178, 1986.
138. Burnham DB, Williams JA: Effects of carbachol, cholecystokinin, and insulin on protein phosphorylation in isolated pancreatic acini. *J Biol Chem* 257: 10523–10528, 1982.
139. Wrenn RW: Phosphorylation of a pancreatic zymogen granule membrane protein by endogenous calcium/phospholipid-dependent protein kinase. *Biochim Biophys Acta* 775: 1–6, 1984.
140. Dowd F, Watson EL, Lau Y-S, Justin J, Pasieniuk J, Jacobson KL: Calcium-dependent protein kinase reactions associated with parotid gland secretory granule membranes. *J Dent Res* 66: 557–563, 1987.

Author's address:
Dr. A.R. Hand
NIDR, NIH
Bldg. 10, Rm. 1A13
Bethesda, MD 20892
USA

CHAPTER 2

Intercellular junctions in salivary glands

GUIDO FUMAGALLI & MASAKI SHIMONO

1. Introduction

Human and mammalian salivary glands are made up of combinations of acinar, ductal, and myoepithelial cells, with each cell type playing a defined role in salivary secretion. Thus, acinar cells are devoted to production and release of proteins and electrolytes, duct cells constitute the walls of the pipes that convey secreted material from acini to the oral cavity, and myoepithelial cells facilitate the movement of the secreted material. In these cells three types of intercellular junctions are present: tight junctions, gap junctions, and desmosomes.

Although abundant information on the structure and function of these intercellular junctions has been obtained from various tissues, few studies have been published on their morphology, distribution, and development in salivary glands. This chapter will focus on the distribution and the morphological pattern of the junctions in both adult and developing salivary glands under physiological and pathological conditions. A schematic drawing of the freeze-fracture and thin-section appearance of the three junctions is provided in Fig. 2-1.

2. Tight junctions

Tight junctions have been described in rat parotid [1,2], submandibular [3], and sublingual glands [4], as well as in the developing submandibular gland [5]. The morphology of tight junctions in salivary gland is analogous to that seen in other tissues. In rat sublingual gland, tight junction is the most luminal element of the junctional complexes (see schematic drawing in Fig. 2-1) and can be detected at the apical plasma membrane of acinar and ductal, but not myoepithelial, cells [4]. They are identified as multiple focal fusions of adjacent plasma membranes in thin sections of tissue fragments stained en bloc with uranyl acetate (Fig. 2-2A). The number of these focal fusions per junction is variable and tends to be lower in acinar cells. In lanthanum-treated specimens, the tracer does not penetrate beyond the most abluminal focal fusion; occasionally it appears to pass through one or more abluminal focal fusions and to be stopped at a more luminal level (Fig. 2-2B). In ruthenium-red-treated specimens, the intercellular space is occupied by electron-dense material and appears outlined in negative contrast at the points of focal fusion (Fig. 2-2C). In freeze-fracture replicas of the rat submandibular gland, tight junctions can be observed at the boundary between the luminal and lateral plasma membranes (Fig. 2-2D) as a continous beltlike network of branching and anastomosing ridges of the PF face and as a series of interconnecting grooves on the complementary EF face. In acinar cells, the junctions consist of a shallow, loosely linked, and occasionally interrupted network composed of only one to four strands (mean ± S.E. = 2.4 ± .4) running perpendicularly to the major axis of the cells. The depth of the junction ranges from

Riva, A. and Motta, P.M., (eds.), Ultrastructure of the Extraparietal Glands of the Digestive Tract.
ISBN 978-1-4612-8211-2.

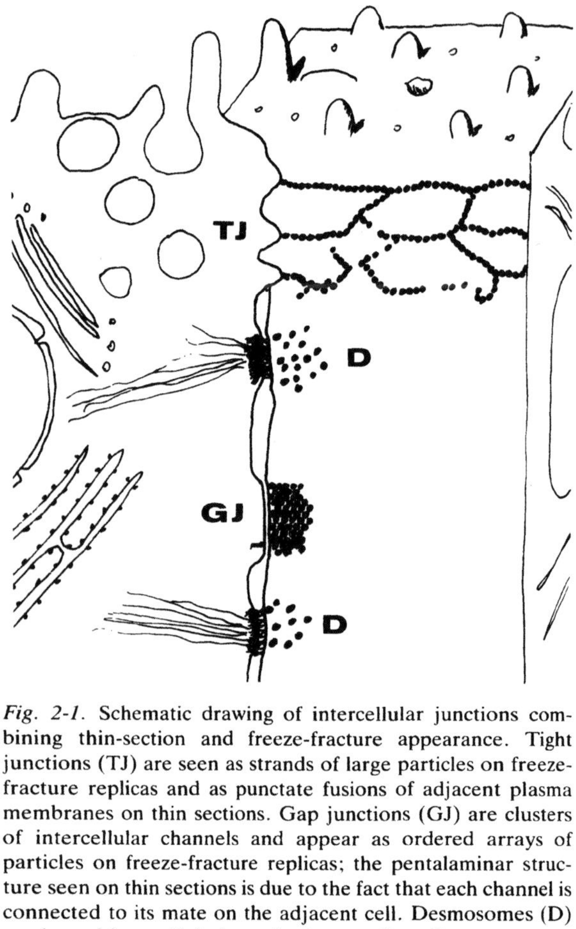

Fig. 2-1. Schematic drawing of intercellular junctions combining thin-section and freeze-fracture appearance. Tight junctions (TJ) are seen as strands of large particles on freeze-fracture replicas and as punctate fusions of adjacent plasma membranes on thin sections. Gap junctions (GJ) are clusters of intercellular channels and appear as ordered arrays of particles on freeze-fracture replicas; the pentalaminar structure seen on thin sections is due to the fact that each channel is connected to its mate on the adjacent cell. Desmosomes (D) are bound intracellularly to the intermediate filament system and are anchored extracellularly to the desmosomal plaque (irregular cluster of large particles on freeze-fracture replicas) of the adjacent cell via a thin fibrillar net.

.03 μm to .58 μm (.28 ± .08 μm). There are no obvious differences in junction appearance between serous and mucous cells.

At the level of the (granular) striated ducts, tight junctions consist of five or more (6.7 ± .7) closely anastomosing strands of large particles, and the depth ranges from .25 μm to 1.0 μm (.51 ± .11 μm). In intercalated ducts, tight junctions appear intermediate between those of acini and striated ducts. Similar results have been obtained in rat parotid [6, 7].

A correlation between tight-junction morphology and permeability function has been proposed by Claude and Goodenough [8]. The general principle that has emerged from their studies and other studies is that the tightness of the junction (in terms of tracer diffusion or electrical conductance) is regulated by the number of interconnected strands. In rat parotid acinar cells, where only two or three strands are present, tight junctions are "leaky" [6]; the loose network of strands present in these cells could allow apical expansion of the cytoplasm during accumulation of secretory granules [9]. Some authors, however, have suggested that the number of strands is not the only factor controlling the permeability function of tight junctions [10].

Both physiological and morphological studies have suggested that secretory stimulation produces changes in junctional permeability in the salivary gland and pancreas [11, 12]. Sympathetic nerve stimulation results in the appearance of endovenously injected horseradish peroxidase in the saliva [13]. Tracers of various molecular weights retrogradely injected into the main excretory duct of the rat parotid gland are usually restricted to the luminal spaces of acini and ducts; upon administration of the β-adrenergic agonist isoproterenol, there is a transient increase of junctional permeability, allowing passage of tracers as large as 34,500 daltons [7]. On the other hand, Yamane et al. [14] have shown a transient increase in the number of strands and in the depth of tight junctions of acinar cells after isoproterenol treatment (Fig. 2-2E); the presence of free endings and endings with loops, similar to those seen in developing submandibular gland [5], suggests that dynamic changes of junction arrangement occur during secretion. A similar morphological appearance has been demonstrated after chronic treatment with isoproterenol (Fig. 2-2F) [3]. Tracer studies suggest that acini become less permeable in these animals. From these observations, the authors conclude that free endings and endings with loops are indicative of tight-junction formation and remodeling.

3. Gap junctions

Gap junctions have been demonstrated between secretory cells in rat parotid [15], submandibular [14, 16], and sublingual [4, 17] glands. They generally occur between homologous cells,

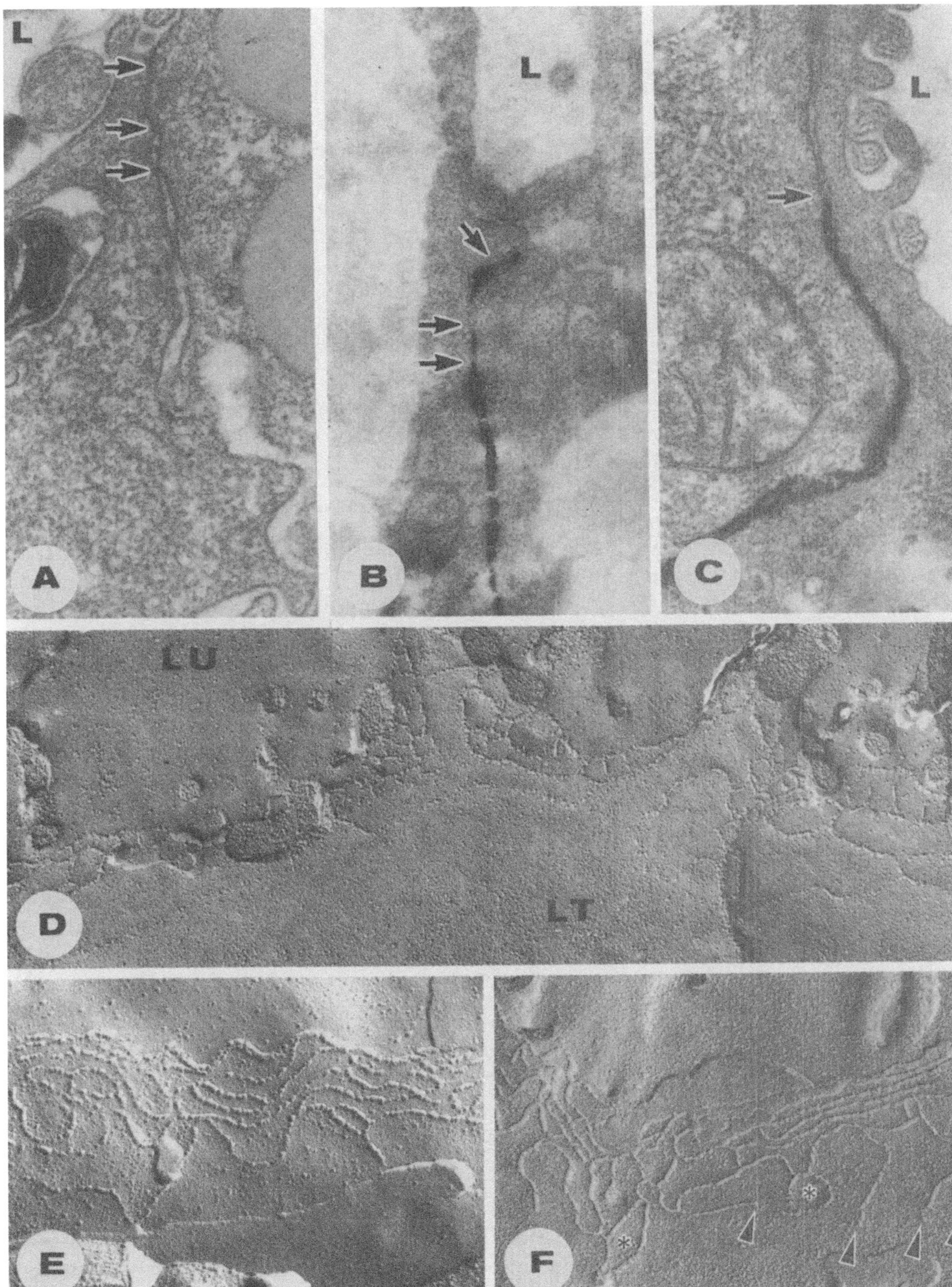

Fig. 2-2. Tight junctions between acinar cells of rat sublingual (A-D) and submandibular (E-F) glands. Specimens stained en bloc with: **A**: uranyl acetate; **B**: colloidal lanthanum: **C**: ruthenium red. Tight junctions are detected in proximity to the acinar lumen (L) as series of focal fusions of adjacent membranes (arrows) that are not infiltrated by the extracellular space tracers. **D**: freeze-fracture replica. Tight junction is identified as a beltlike meshwork composed of two to four anastomosed strands of large particles. **E**: freeze-fracture replica of rat submandibular acinar cells after 10 minutes of isoproterenol treatment. There is an increase in the number of strands, which are frequently interrupted. **F**: freeze fracture replica of submandibular acinar cells after long-term treatment with isoproterenol. Tight junctions undergo remodeling, as indicated by the presence of terminal loops (asterisks) and free endings (arrowheads). A: x 43,600; B: x 42,500; C: x 45,000; D: x 40,000; E: x 52,000; F: x 35,000. A–D: from Shimono et al., 1980 [4] with permission.

i.e., serous-to-serous, mucous-to-mucous, and myoepithelial-to-myoepithelial [17], and are missing between ductal cells [4].

The ultrastructure of gap junctions in salivary glands is essentially similar to that in other tissues (Fig. 2-3) [16, 18–21]. In thin sections fixed with modified Karnowsky's fixative, gap junctions consist of two apposed unit membranes separated by a 3 nm gap (Fig. 2-3A). Colloidal lanthanum penetrates the central gap, and the electron-dense material appears to be interrupted by regularly spaced, thin, electron-lucent bridges crossing the intercellular space (Fig. 2-3B). Tangential sections of tracer-infiltrated junctions reveal hexagonal arrays of electron-lucent subunits measuring 9 nm in diameter; the distance between adjacent globules is approximately 11 nm (Fig. 2-3B). The globules frequently contain an electron-opaque central dot about 2 nm in diameter. A similar morphological pattern is seen after ruthenium-red infiltration (Fig. 2-3C). In freeze-fracture replicas, the junctions appear on the PF face as clusters of particles measuring 8–10 nm in diameter, often arranged in a distorted hexagonal pattern. With a center-to-center spacing of 9–11 nm (Fig. 2-3D). The EF face of the gap junction shows small complementary pits (Fig. 2-3E). The diameter of individual gap junctions between acinar cells varies between .2 μm and 1.2 μm (.5 ± .12 μm). Gap junctions between myoepithelial cells are larger (.88 ± .27 μm).

It is well established that the biological role of the gap junction is to mediate the transfer of electrochemical signals and metabolites. Gap junctions are made up by closely packed transmembrane channels, called *connexons*, which allow cell-to-cell diffusion of small molecules with a molecular weight lower than 1.2 kD [22]. The major polypeptide obtained from liver gap junctions has a relative molecular mass of 27 kD [23] and is related to the cardiac gap-junction protein CJP47 [24]; its sequence has been deduced from rat and human cDNAs [25, 26]. When reconstituted into lipid bilayers, the proteins confer channel properties similar to those of the intact junction [27]. A less abundant component of the gap junction is a protein of 21 kD that is related to but distinct from the 27 kD protein [28]; both proteins are present in the same gap junctions.

The functional activity of the gap junction in salivary glands has been studied in voltage-clamped cell pairs [29]; conductance is regulated by voltage, pH, and calcium ions. Functional uncoupling also occurs with alcohol; it has been suggested that these modulators induce morphological changes in the hexameric subunit [30]. The number and size of gap junctions increases when secretion is stimulated in rat pancreatic β cells [31]; however, this finding has not been confirmed in rat submandibular glands after either short- [14] or long- [3] term stimulation with isoproterenol.

4. Desmosomes

Desmosomes are junctions that confer mechanical stability to adjacent cells. They are found between homologous cells as well as myoeithelial and adjacent acinar or intercalated duct cells. They are more frequent in ductal cells and are often observed on the lateral plasma membrane directly below tight junction [4].

In thin sections, desmosomes appear as parallel portions of cell membranes separated by an interspace approximately 30–40 nm wide (Figs. 2-1 and 2-4A). Attachment plaques and typical intercellular substructures are usually well resolved in ductal cells only. In ruthenium-treated specimens, desmosomes are clearly outlined (Fig. 2-4B), the positive material reveals the existence of connections between adjacent membranes that appear as electron-lucent bridges crossing the intercellular space. On the cytoplasmic side, tonofilament (intermediate filament) bundles terminate near or pass through the electron-dense plaque.

In freeze-fracture replicas, desmosomes of salivary glands appear as round or elongated clusters of particles measuring 8–11 nm in diameter on both the PF and EF faces (Fig. 2-4C). The clusters are more clearly identified on the EF face. The size of desmosomes of salivary glands is smaller than in covering epithelia [21, 32].

The biochemical characterization of desmosomes has been carried out mainly on desmosomes from bovine muzzle epidermis [33]; nonglycosylated desmosomal proteins (DP), also called *desmoplaquins*, are present and have molecular weights ranging from 25 kD to 83 kD. Three

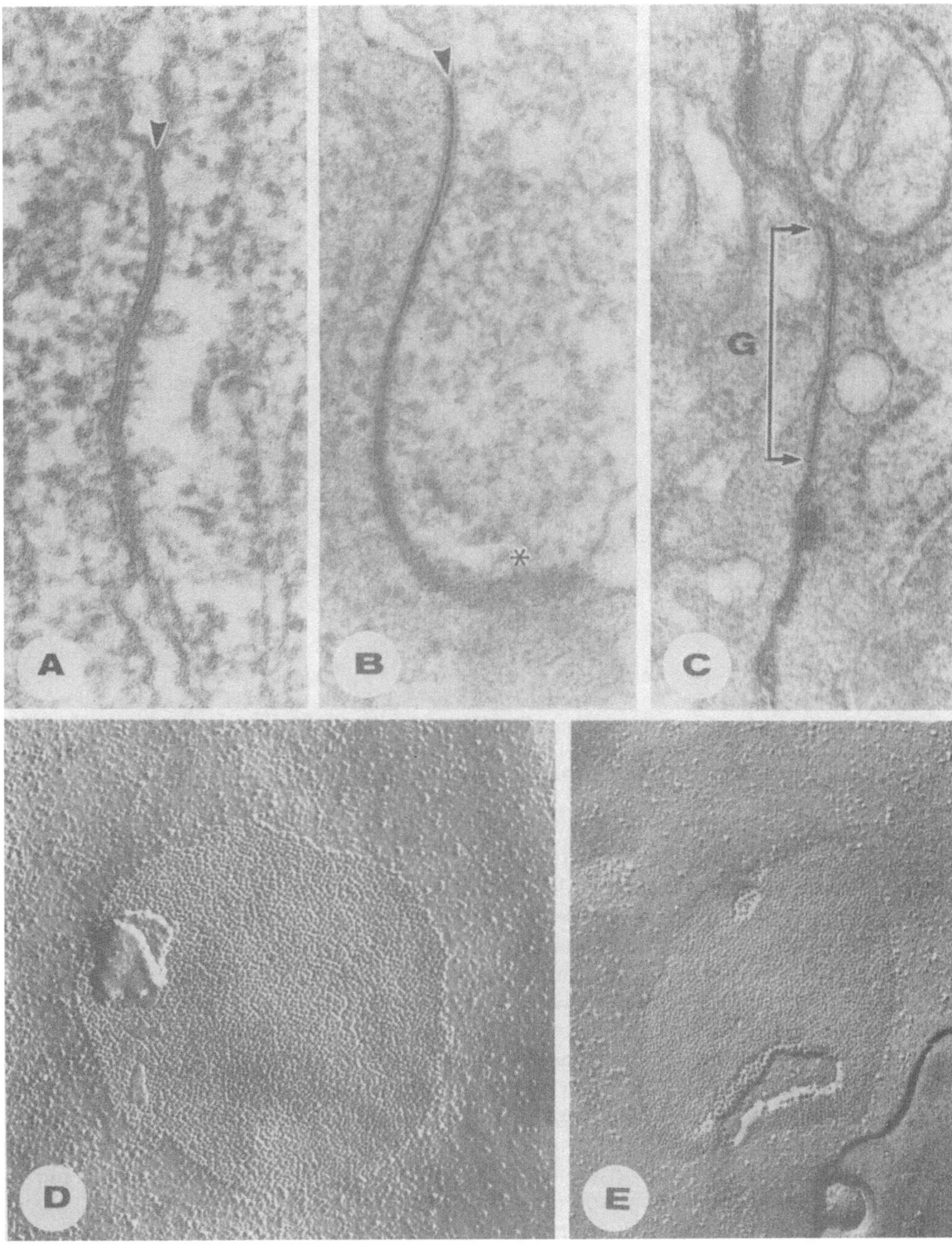

Fig. 2-3. Gap junctions in acinar cells of the rat sublingual gland. Specimens stained en bloc with: **A**: uranyl acetate; **B**: colloidal lanthanum; **C**: ruthenium red. The gap junction is composed of two opposed unit membranes separated by a 3 nm central gap (G; arrowheads), which can be infiltrated by extracellular space tracers. In B, a portion of the junction is tangentially cut (asterisk) and lanthanum outlines an ordered array of globules. **D** and **E**: freeze-fracture replicas displaying PF and EF face, respectively. A: x 126,000; B: x 100,000; C: x 76,100; D: x 79,000; E: x 56,800. From Shimono et al., 1980 [4] with permission.

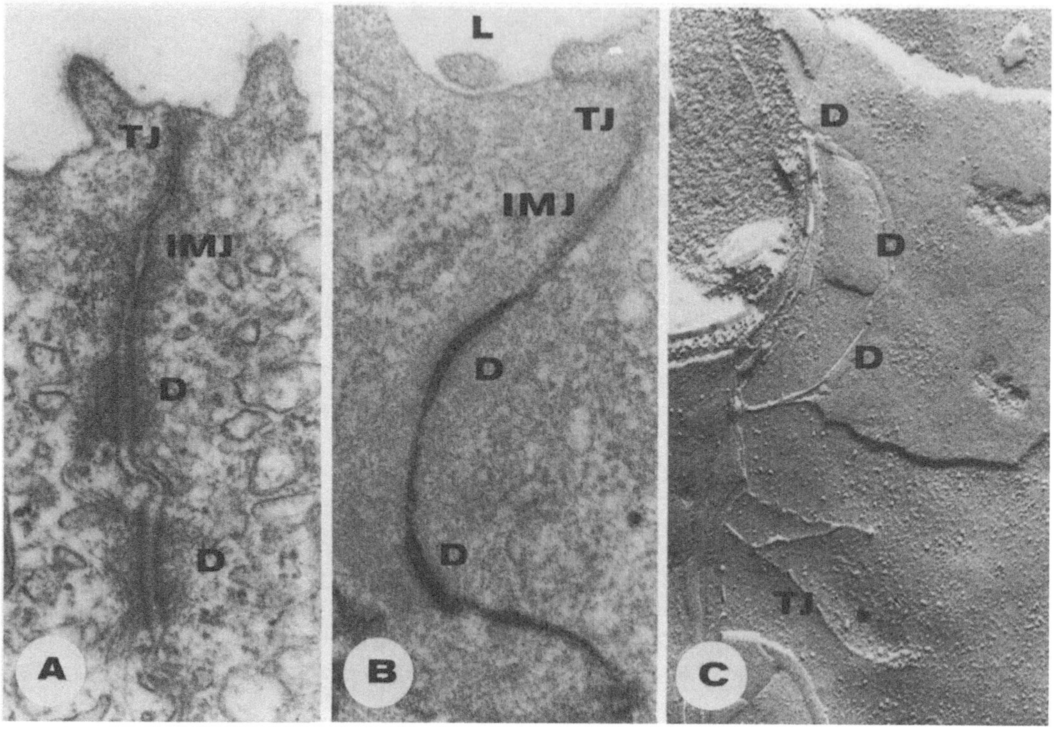

Fig. 2-4. Desmosomes in rat sublingual gland. **A–B**: ductal cells stained en bloc with: **A**: uranyl acetate; **B**: ruthenium red. **C**: freeze-fracture replica of acinar cells. Desmosomes (D) are more easily recognized on the EF face. TJ-tight junction; IMJ-intermediate junction. A: x 79,000; B: x 56,800; C: x 52,600. From Shimono et al., 1980 [4] with permission.

major DPs, named DPI, DPII, and DPIII, in order of molecular mass, and a minor calmodulin-binding protein (descalmin), have been described. DPI and DPII have similar amino-acidic composition, but DPII seems to be missing in non-stratified epithelia. DPs are probably involved in cytokeratin (the epithelial intermediate filaments) bundles binding to desmosomal plaques [34]. Glycosylated desmosomal proteins (DG) are also classified according to their molecular weight, which ranges from 15 kD to 22 kD. Epitopes conservation among species is more prominent for DPs and DGI, as indicated by immuno-cross reactivity and mutual desmosome formation between different cell types and species [35]. The desmosomal protein electrophoretic pattern obtained from tissues other than stratified epithelia shows variations in the spectrum of DPs expressed and in the molecular mass of DGs [36].

Assembly of desmosome junctions in culture is regulated by extracellular Ca^{2+} levels: in low Ca^{2+} medium, desmosome are not assembled and DGs are degraded at a faster speed [37]; in this situation, DGs are uniformly distributed on the cell surface and DPs are located throughout the cytoplasm. Raising Ca^{2+} to physiological levels rapidly activates desmosome assembly: concentration of desmosomal proteins at focal sites of the cell periphery, metabolic stabilization of DGs, and assembly of cytokeratin bundles all contribute to desmosome formation and stabilization.

Desmosome assembly in vitro can also be inhibited by culturing cells in the presence of Fab' fragments directed against the extracellular domains of some DGs (desmocollins) [38]. Anti-desmosome antibodies have been suggested to play a pathogenic role in human pemphigus [39].

5. Intercellular junctions during salivary gland development

Developmental studies of salivary glands have been extensively carried out in various animals [40–43], but a correlative study between tight-junction formation and gland maturation has been carried out on the rat submandibular gland only [5].

At the 16th day of gestation, when rudiments of the gland consist of cuboidal or pleomorphic cells arranged in cords and anchored by desmosomes, lanthanum freely penetrates the intercellular space and early signs of tight-junction formation are visible on freeze-fracture replicas in the form of numerous *maculae occludentes*. Three types of maculae occludentes can be recognized at this stage [5]: the first type is represented by elevations on the PF face (Fig. 2-5A) and by shallow furrows on the EF face, deprived of obvious particle strands. The second type is characterized by discontinous rows of particles on the ridges of the linear elevations on the PF face and by sharp furrows on the EF face (Fig. 2-5B). The third type consists of strands forming ridges on the PF face (Fig. 2-5C). *Zonulae occludentes*, in the form of beltlike structures circumscribing the juxtaluminal region, are common at birth but were already recognizable at the 16th day of embryonic life. At the 18th and 19th gestational day, acini and ducts are present, microvilli appear in their lumen, lanthanum freely penetrates the intercellular space, and the majority of the tight junctions show simple, interrupted junctional elements consisting of only one to three strands (Fig. 2-5D). As in the mature gland, the number of strands and the depth of the junction are greater between ductal cells; *fasciae occludentes*, with free endings and terminal loops, are already visible (Fig. 2-5D). At birth, lanthanum diffusion is restricted and freeze-fracture replicas show tight junctions, almost similar to those present in the mature gland.

To conclude, we can draw guidelines for tight-junction development. First, small furrows on the EF face and elevations on the PF face appear [see also 44 and 45]; next, discontinuous rows of particles are assembled on the elevations of the PF face; these then fuse and form smooth ridges. Maculae and zonulae occludentes appear and fuse together to form fasciae occludentes, possessing strands with free endings and terminal loops. In acini these fasciae may be remodeled and moved more luminally to be organized into zonulae occludentes, similar to the junctions of the mature gland.

Few studies have been carried out on gap junctions during salivary gland development. Based on our unpublished data, the first definite gap junction can be seen on the lateral plasma membrane of glandular cells at the 17th gestational day. They are very small and consist of about 20 particles. Associations of small gap junctions with maculae occludentes are frequently seen; this is a characteristic feature of gap junctions in developing salivary glands and is absent in adult animals. On the 19th day of embryonic life, small gap junctions containing small, smooth, particle-free areas are seen, as well as larger ones occupying vast areas of the lateral plasma membrane. The appearance of large gap junctions is temporally associated with the first signs of evoked secretion of saliva [40]. The junction increases in size and frequency during the course of postnatal development.

Small clusters of particles of FF replicas revealing desmosomes are already present at the 16th day of embryonic life and are usually close to newly formed zonulae occludentes; this suggests that formation of desmosomes is concurrent with that of tight junctions. Intermixing of desmosomes between tight-junction strands forming macula occludentes is frequent at the 19th gestational day (Fig. 2-5D); this organization is missing at birth and in the mature gland. The size of individual desmosomes tends to become larger during development.

6. Intercellular junctions in neoplastic disorders

The morphological changes of intercellular junctions have been described in two cases of acinic cell carcinoma arising in the glossopalatine glands [46]. Acinic cell carcinoma was originally regarded as an adenoma, but it has been shown that it can recur and metastasize, despite its slow rate of

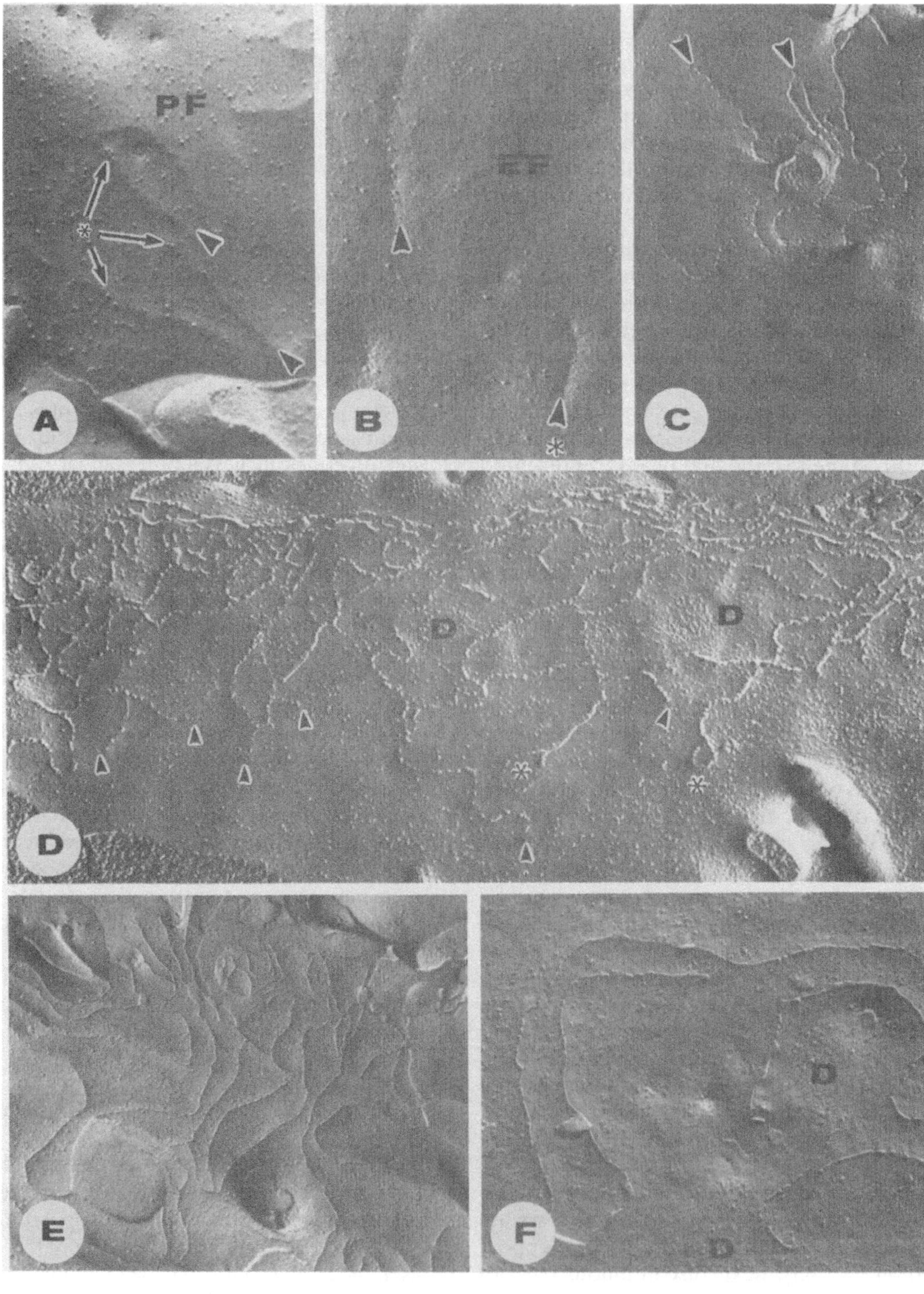

PF
A
EF
B
C
D
D
D
E
D
F
D

growth [47]. Due to its degree of cellular differentiation, this neoplasia occupies an intermediate position between benign and malignant tumors. Inoue et al. [46] have found tight junctions composed of ten or more anastomosed strands at the boundary between the lateral and luminal plasma membranes of tumor cells; elongation of basal frontier strands with free endings and terminal loops were frequently recognizable in these tight junctions. The authors have concluded that tight-junction morphology of acinic cell carcinoma resembles that of developing salivary glands and corresponds to the degree of cellular differentiation.

The morphology of tight junctions has been studied in an undifferentiated carcinoma of the parotid gland by Shimono et al. [48]. The undifferentiated carcinoma is regarded as a tumor with a poor level of cytodifferentiation [47]. In freeze-fracture replicas, the carcinoma cells show proliferation and fragmentation of tight junctions (Fig. 2-5E–2-5F). Scattered and irregularly interrupted junctional strands can be observed on the plasma membranes, indicating that the organization of the junctional strands has not yet been established. Tight junctions appear as spotlike or sheetlike, rather than beltlike, networks and are analogous to those seen in developing salivary glands. From these results, the authors have concluded that the plasma membrane of carcinoma cells can not be fully differentiated into luminal, lateral, and basal compartments and that the organization of junctional strands does not occur completely. Tight junctions in an undifferentiated carcinoma seem to be morphologically different from those seen in an acinic cell carcinoma: the junctional strands of the former are more irregular, scattered, fragmented, and poorly organized, and are similar to those seen in salivary glands at early stages of development. In contrast, tight junctions of acinic cell carcinoma are relatively well organized and are analogous to those present in the late stages of salivary gland development. These observations further support the idea of a correlation between the morphology of tight junctions and their degree of cellular differentiation.

The morphology and physiology of intercellular junctions in acinic cell carcinoma of the canine pancreas are analogous to that described in salivary gland tumors [49]. The frequency of tight junctions is related to the degree of cellular differentiation and to neoplastic behavior; cell proliferation and fragmentation of tight junctions are prominent, and a decrease in the number and size of gap junctions has been demonstrated in this tumor [50]. A similar pattern of modification of tight-junction morphology has been demonstrated in human thyroid tumors [51]: in this tissue, tight junctions of adenomas are morphologically identical to normal controls, whereas carcinomas with low malignancy display only focal attenuations of the junctional belt, and anaplastic carcinomas of high malignancy are completely devoid of tight junctional ridges. These results suggest that tight-junction alterations are related to tumor differentiation.

7. Concluding remarks

The data reviewed in this chapter indicate that intercellular junctions are dynamic structures that regulate and define the morphological architecture and the functional specialization of salivary glands.

Desmosomes, which are connected to the intermediate filament cytoskeletal network, provide the mechanical stability necessary to maintain the spatial organization of acini and ducts.

←

Fig. 2-5. Freeze-fracture replicas of tight junctions in salivary gland development and carcinoma. **A–B**: 16th gestational day; **C–D**: 19th gestational day; **E–F**: undifferentiated carcinoma of human parotid. Early signs of macula occludens formation are linear elevations (arrowheads in A) with a few aggregated particles (arrows in A) on the PF face and sharp grooves with discontinuous rows of particles on the EF face (arrowheads in B). Strands with ridges on the PF face are indicated by arrowheads in C. In D, a zonula occludens presents elongation and branching of the basal frontier strands, free endings (arrowheads), and terminal loops (asterisks). Desmosomes (D) intermix with the junction at this stage of development. In E, a tight junction reveals that the fascia occludens is composed of ten or more anastomosed strands. In F, a macula occludens with intermixed desmosomes occupies a relatively large area. A: x 55,000; B: x 50,000; C: x 41,000; D: x 59,000; E: x 28,000; F: x 49,000. A–D: from Shimono et al., 1981 [5]; E–F: from Shimono et al., 1987 [48], with permission.

Gap junctions are clusters of voltage-regulated intercellular channels connecting homologous cells; they allow simultaneous activation of a number of cells in response to parasympathetic or orthosympathetic stimulation. Gap junctions are developmentally regulated, and their size and frequency correlate with the functional activity of salivary glands and with the degree of cellular differentiation in neoplastic disorders.

Tight junctions limit solute diffusion between adjacent cells. Their permeability function is regulated by the number and the morphological pattern of the strands of large intramembranous particles that allow identification of the junction in freeze-fracture replicas. This type of junction appears late during gestational life; its development and morphological pattern correlates with the establishment of electrochemical tightness between luminal and intercellular spaces, and the location, function, and degree of differentiation of the epithelial cells in salivary glands.

References

1. De Camilli P, Peluchetti D, Meldolesi J: Dynamic changes of the luminal plasmalemma in stimulated parotid acinar cells. *J Cell Biol* 70: 59–74, 1976.
2. Mazariengos MR, Tice LW, Hand AR: Alteration of tight junctional permeability in the rat parotid gland after isoproterenol stimulation. *J Cell Biol* 98: 1865–1877, 1984.
3. Inoue T, Yamane H, Yamamura T, Shimono M: Morphological changes of intercellular junctions in the rat submandibular gland treated by long-term repeated administration of isoproterenol. *J Dent Res* 66: 1303–1309, 1987.
4. Shimono M, Yamamura T, Fumagalli G: Intercellular junctions in salivary glands: Freeze-fracture and tracer studies of normal rat sublingual gland. *J Ultrastruct Res* 72: 286–299, 1980.
5. Shimono M, Nishihara K, Yamamura T: Intercellular junctions in developing rat submandibular glands. (I) Tight junctions. *J Electron Microsc* 30: 29–45, 1981.
6. Simson JAV, Bank HL: Freeze-fracture and lead ion tracer evidence for a paracellular fluid secretory pathway in rat parotid glands. *Anat Rec* 208: 69–80, 1984.
7. Mazariengos MR, Hand AR: Regulation of tight junctional permeability in the parotid gland by autonomic agonists. *J Dent Res* 63: 1102–1107, 1984.
8. Claude P, Goodenough DA: Fracture faces of zonulae occludentes from "tight" and "leaky" epithelia. *J Cell Biol* 58: 390–400, 1973.
9. Hull BE, Staehelin LA: Functional significance of the variations in the geometrical organization of the tight junction networks. *J Cell Biol* 68: 688–704, 1976.
10. Mollgard K, Malinowska DH, Saunders NR: Lack of correlation between tight junction morphology and permeability properties in developing choroid plexus. *Nature* 264: 293–294, 1976.
11. Martin K, Burgen ASV: Changes in the permeability of the salivary gland caused by sympathetic stimulation and by catecholamines. *J Gen Physiol* 46: 225–243, 1962.
12. Jansen JW, DePont JJ, Bonting SL: Transepithelial permeability in the rabbit pancreas. *Biochim Biophys Acta* 551: 95–108, 1979.
13. Garrett JR, Klinger AH, Parsons PA: Permeability of canine submandibular glands to blood-borne horseradish peroxidase during parasympathetic secretion. *J Exp Physiol* 67: 31–39, 1982.
14. Yamane H, Yamamura T, Satoh T, Shimono M: Tight junctions in submandibular acinar cells during secretion (abstr). *J Dent Res* 65: 833, 1986.
15. Dunn J, Revel JP: Association of gap junctions with endoplasmic reticulum in rat parotid glands. *Cell Tiss Res* 238: 589–594, 1984.
16. Friend DS, Gilula NB: Variations in tight and gap junctions in mammalian tissues. *J Cell Biol* 53: 758–776, 1972.
17. Nagato T, Tandler B: Gap junctions in rat sublingual gland. *Anat Rec* 214: 71–75, 1986.
18. Goodenough DA, Revel JP: A fine structural analysis of intercellular junctions in the mouse liver. *J Cell Biol* 45, 272–290, 1970.
19. Peracchia C: Gap junctions. Structural changes after uncoupling procedures. *J Cell Biol* 72: 628–641, 1977.
20. Staehelin LA: Structure and function of intercellular junctions. Int Rev Cytol 39: 191–283, 1974.
21. Shimono M, Clementi F: Intercellular junctions of oral epithelium. I. Studies with freeze-fracture and tracing methods of normal rat keratinized oral epithelium. *J Ultrastruct Res* 56: 121–136, 1976.
22. Simpson I, Rose B, Loewenstein WR: Size limit of molecules permeating the functional membrane channels. *Science* 195: 294–296, 1977.
23. Hertzberg EL, Gilula NB: Isolation and characterization of gap junctions from rat liver. *J Biol Chem* 254: 2138–2147, 1979.
24. Nicholson BJ, Gros DB, Kent SBH, Hoos LE, Revel JP: The Mr 28,000 gap junction protein from rat heart and liver are different but related. *J Biol Chem* 260: 6514–6517, 1985.
25. Paul D: Molecular cloning of cDNA from rat liver gap junction protein. *J Cell Biol* 103: 123–134, 1986.
26. Kumar M, Gilula NB: Cloning and characterization of human and rat liver cDNAs coding for a gap junction protein. *J Cell Biol* 103: 767–776, 1986.
27. Young JD, Cohn ZA, Gilula NB: Functional homology of gap junction conductance in lipid bilayer: Demonstration that the major 27 Kd protein forms the junctional channel. *Cell* 48: 733–743, 1987.
28. Nicholson B, Dermietzel R, Teplow D, Traub O, Willecke K, Revel JP: Two homologous protein components of hepatic gap junctions. *Nature* 329: 732–733, 1987.
29. Obaid AL, Socolar SJ, Rose B: Cell to cell channels with two independently regulated gates in series: Analysis of

functional conductance modulation by membrane potential, calcium and pH. *J Membr Biol* 73: 69–89, 1983.
30. Unwin PMT, Ennis PD: Calcium-mediated changes in gap junction structure: Evidence from the low angle X-ray pattern. *J Cell Biol* 97: 1459–1466, 1983.
31. Meda P, Perrelet A, Orci L: Increase of gap junctions between pancreatic B-cell during stimulation of insulin secretion. *J Cell Biol* 82: 441–448, 1979.
32. Kelly DE, Shienvold FL: The desmosome: Fine structure with freeze-fracture replication and tannic acid staining of sectioned epidermis. *Cell Tiss Res* 172: 309–323, 1976,
33. Penn EJ, Hobson C, Rees DA, Magee AI: Structure and assembly of desmosome junctions: Biosynthesis, processing and transport of the major protein and glycoprotein components in cultured epithelial cells. *J Cell Biol* 105: 57–68, 1987.
34. Jones JCR, Goldman RD: Intermediate filaments and the initiation of desmosome assembly. *J Cell Biol* 101: 506–517, 1985.
35. Mattey DL, Garrod DR: Mutual desmosome formation between all binary combination of human, bovine, canine, avian and amphibian cells: Desmosome formation is not tissue or species specific. *J Cell Sci* 75: 377–399, 1985.
36. Giudice GJ, Cohen SM, Patel NH, Steinberg MS: Immunological comparison of desmosomal components from several bovine tissues. *J Biol Chem* 26: 35–45, 1984.
37. Penn EJ, Burdett IDJ, Hobson C, Magee AI, Rees DA: Structure and assembly of desmosome junctions: Biosynthesis and turnover of the major desmosome components of Madin–Darby canine kidney cells in low calcium medium. *J Cell Biol* 105: 2327–2334, 1987.
38. Cowin P, Mattery D, Garrod DR: Identification of desmosomal surface components (desmocollins) and inhibition of desmosome formation by specific fab'. *J Cell Sci* 70: 41–60.
39. Jones JCR, Arnn J, Staehelin LA, Goldman RD: Human autoantibodies against desmosomes: Possible causative factors in pemphigus. *Proc Natl Acad Sci USA* 81: 2781–2785, 1984.
40. Cutler LS, Chaudhry AP: Intercellular contacts at the epithelial-mesenchymal interface during the prenatal development of the rat submandibular gland. *Develop Biol* 33: 229–240, 1973.
41. Cutler LS, Chaudhry AP: Cytodifferentiation of the acinar cells of the rat submandibular gland. *Develop Biol* 41: 31–41, 1974.
42. Cutler LS, Chaudhry AP: Cytodifferentiation of striated duct cells and secretory cells of the convoluted granular tubules of the rat submandibular gland. *Am J Anat* 143: 201–217, 1975.
43. Redman RS, Sreebny LM: Proliferative behavior of differentiating cells in the developing rat patotid gland. *J Cell Biol* 46: 81–87, 1970.
44. Montesano R, Friend DS, Perrelet A, Orci L: In vivo assembly of tight junctions in fetal rat liver. *J Cell Biol* 67: 310–319, 1975.
45. Luciano L, Thiele J, Reale E: Development of follicles and of occluding junctions between the follicular cells of the thyroid gland. *J Ultrastruct Res* 66: 164–181, 1979.
46. Inoue T, Shimono M, Yamamura T, Saito I, Watanabe O, Kawahara H: Acinic cell carcinoma arising in the glossopalatine glands: A report of two cases with electron microscopic observations. *Oral Surg Oral Med Oral Pathol* 57: 398–407, 1984.
47. Lucas RB: Pathology of Tumours of the Oral Tissues, 4th ed. Edinburgh: Churchill Livingstone, 1984.
48. Shimono M, Sato T, Yamane H, Katayanagi T, Osada K, Watanabe I, Yamane G, Noma H: Tight junctions in an undifferentiated carcinoma originating in the parotid gland: A study with freeze-fracture. *Bull Tokyo Dent Coll* 28: 63–73, 1987.
49. Alroy J, Banner BF, Pauli BU, Leav I: Alterations of intercellular junctions in acinic cell carcinoma of the canine pancreas. *Virchows Arch* (Cell Pathol) 28: 21–30, 1978.
50. Banner BF, Alroy J, Pauli BU, Carpenter JL: An ultrastructural study of acinic cell carcinomas of the canine pancreas. *Am J Pathol* 93: 165–182, 1978.
51. Kerjaschki D, Krisch K, Sleyter UB, Umrath W, Jakesz R, Depisch R, Kokoschka R, Hörandner H: The structure of tight junctions in human thyroid tumors. *Am J Pathol* 96: 207–226, 1979.

Author's address:
Dr. M. Shimono
Department of Pathology
Tokyo Dental College
1-2-2 Masago, Chiba-City
Chiba, 260, Japan

CHAPTER 3

Ultrastructural diversity in chiropteran salivary glands

BERNARD TANDLER, CARLETON J. PHILLIPS,
TOSHIKAZU NAGATO & KUNIAKI TOYOSHIMA

1. Introduction

Mammalian salivary glands, which produce a huge variety of glycoproteins and mucosubstances, as well as engaging in intense transport of electrolytes, have an obvious appeal for morphologists, physiologists, and biochemists. Despite the versatility of these organs, studies on salivary glands have been restricted largely to a few commonly available laboratory animals. Comparative studies, though still relatively sparse, have shown that salivary gland histology is generally similar throughout the class Mammalia, but that both the acinar and duct cells exhibit considerable ultrastructural variation between species [1–3]. To illustrate this morphological variability, the present chapter surveys ultrastructural aspects of the parotid, submandibular, and accessory submandibular salivary glands in bats, order Chiroptera.

It is worthwhile to briefly consider the rationale for the comparative study of salivary gland cell structure. Examination of cells at an ultrastructural level provides a means of visualizing macromolecular events in situ. Ultrastructural comparisons can best be made between homologous, regulated, polarized secretory cells, because the surface and interior of such cells are highly structured and predictable within cell type, functional state, and/or species [3]. Salivary glands therefore hold much promise for comparative studies.

The secretory process is now fairly well known, and many aspects of the process have been visualized by means of transmission electron microscopy [4–6]. Little is known, however, about the range of variation or evolutionary plasticity in the process. Are all secretory cells really the same? Is the process highly conserved, or, if it varies, then what biological factors are involved? Finally, which components of the process have beeen selectively altered during the course of secretory cell evolution? Systematic, comparative ultrastructural analysis of regulated, polarized secretory cells can serve as a foundation for molecular comparisons [3].

Comparisons of commonly studied mammals, such as the various domesticated rodents, carnivores, lagomorphs, and ungulates, are unreliable because the species under comparison are often the products of separate and distant evolutionary origins. By contrast, the nearly 850 species of the order Chiroptera (nearly one quarter of the living species of mammals) are genetically more closely related; their extraordinary phenotypic differences are related to diversification in feeding habits and diet rather than to extreme genetic distance [3, 7].

Bats in the suborder Microchiroptera are thought to have originated from a common insectivorous ancestor [7]. As a result of adaptive radiation, many existing species specialize in feeding on plant materials or flesh or blood, while others have retained the ancestral insectivorous diet. Bats in the other suborder, the Megachiroptera, are strictly frugivorous or nectarivorous. The latter may have evolved from an insectivorous ancestor in common with the Microchiroptera,

Riva, A. and Motta, P.M., (eds.), Ultrastructure of the Extraparietal Glands of the Digestive Tract.
ISBN 978-1-4612-8211-2.

although some recent evidence suggests that the Megachiroptera might share ancestry with the primates [8,9]. The anatomy, physiology, and behavior that typify each chiropteran genus are tightly linked to the successful exploitation of particular nutrient resources available in a particular habitat. Features such as dentition, stomach structure, vision, brain size, and the shape and aspect ratio of the wings in bats are all correlated with their general dietary categories [10–13].

The diets of bats encompass virtually all of the nutrient resources used by all other species of mammals. Not only are there species of bats that feed on insects in general, but some also specialize on Lepidoptera, Coleoptera, or ground-dwelling Arthropoda [7]. There also are species that possibly restrict their feeding to certain fruits or the nectar and pollen of certain flowers [14, 15]. Some species mostly eat fish, while others feed on terrestrial vertebrates; a few species feed exclusively on the blood of either mammals or birds [7]. Because saliva is such a significant component of the digestive process, the nature of the salivary glands and their secretory products must relate to these various diets. The structure of salivary glands in bats, therefore, may vary in a way that is correlated with diet.

Previous interspecific investigations have demonstrated that the salivary glands of bats are exceedingly diverse at the cellular level [3]. The acinar secretory cells are especially variable, particularly with respect to the ultrastructure of those mature secretory granules temporarily stored in the cell apex. Indeed, virtually every species exhibits morphologically unique secretory granules. The striking variation in secretory granule ultrastructure implies that there is an extraordinary evolutionary plasticity in acinar cells. The interspecific variations in ultrastructure between homologous cells are not random; correlations can be made with genetic history [16, 17], diet [18], and species isolation [19].

Fixation and processing are critical to projects involving transmission electron microscopy (TEM). This is especially true in comparative studies where wild animals are used and where the ultrastructural data are employed as tests of hypotheses regarding the secretory process or the relationship between certain cellular features and dietary habits. Repeatability, an essential part of the scientific process, requires a well-defined method for TEM fixation. Selection of suitably fixed representative cells from within a tissue block is also crucial, because subtle but significant artifacts can easily be introduced by hypoxia and changes in osmolality [3].

We obtained our salivary gland specimens from bats collected in India, Thailand, Mexico, Suriname, and the Antilles, and fixed nearly all of them by immersion in a buffered triple-aldehyde mixture under field conditions. This fixative originally was developed by Kalt and Tandler [20] and then was modified by Phillips for use in the field [21]; it consists of 1% paraformaldehyde, 3% glutaraldehyde, 0.5% acrolein, 2.5% dimethysulfoxide (DMSO), and 1 mM $CaCl_2$ in 0.05 M cacodylate buffer with sucrose (pH 7.2). The tissue samples were fixed overnight and then transferred to fresh 0.05 M cacodylate buffer and stored until refrigeration was available. Afterwards, the tissues were transferred to 3% glutaraldehyde in 0.05 M cacodylate buffer and stored at 4°C. Additional details about the preparation techniques used can be found in Tandler and Walter [22], Tandler et al. [23], and Forman and Phillips [24]. For the following survey of salivary gland ultrastructure, we selected 33 species from among the more than 120 that we have studied. The species illustrated were chosen for convenience in documenting ultrastructural diversity. We have examined at least several specimens, and in some cases as many as 20 individuals, from each species.

2. Acinar cell shape

Typical salivary gland acini consist of a single layer of wedge-shaped secretory cells arranged around a central lumen. In mixed glands (those that contain two distinct types of secretory cells), the acini often are capped by *demilunes*. Coursing between adjacent cells are delicate prolongations of the lumen, the *intercellular canaliculi* (formerly called secretory capillaries), which extend nearly to the cell base. Where a demilune is present, the canaliculi extend between adjacent demilunar cells and provide an avenue whereby water and the secretory products of the demilune cells can gain access to the acinar lumen [25].

The basal surfaces of those acinar cells that are not shrouded by demilune cells are covered by basal lamina; the latter extends over the outer surface of the demilune cells; there is no basal lamina separating demilune cells from the overlying acinar cells. In both acinar and demilune cells, the cell surface that contacts the basal lamina may show some structural modification in the form of basal folds. These structures, first described in acinar cells of human submandibular glands, are slender, usually organelle-free plications that are several micrometers in height [26]. They extend beyond the lateral margins of each cell and interfoliate in a regular manner with folds under their neighbors.

In bats, the number and prominence of basal folds varies from species to species and from gland to gland. Basal folds are rather infrequent in the parotid acinar cells in the common vampire bat, *Desmodus rotundus* (Fig. 3-1), while they are extremely abundant in the homologous cells in species of *Myotis* [27] and *Eptesicus* (Figs. 3-2 and 3-3), which are two geographically widespread insectivorous genera. In these insectivorous species, the crest of each fold (where it contacts the basal lamina) contains a density that extends longitudinally within the fold; these densities are distinctly different from hemidesmosomes. In *Myotis lucifugus* (and several other bat species), gap junctions occasionally are located between adjacent folds. In the parotid gland acini of *Pteronotus parnellii*, another insectivorous bat, a moderately dense material fills the intercellular space between folds (Fig. 3-4); similar material has been found in sheep parotid gland [28].

The precise function of basal folds has not been determined, but their presence obviously increases the surface area of the basal plasma membrane. It has been calculated that basal folds serve to increase basal surface area ten times in the human submandibular gland, and four to five times in the human parotid gland [29]. In both canine and human glands the basal folds harbor a Na^+-K^+-ATPase, as shown by cytochemistry [29, 30]; those in feline submandibular salivary glands have an associated ATPase, as shown by autoradiography using labeled ouabain as a probe [31]. It seems likely that the folds, together with their complement of enzymes, are involved in a variety of transport processes that ultimately are of importance in the elaboration of the initial saliva [29].

In most species of bats, the luminal surfaces of the acinar cells have at least a few stubby microvilli. In contrast, the acinar luminal surfaces in submandibular glands of vampire bats (*Desmodus rotundus*) bear so many microvilli that the acinar lumen appears to be almost occluded. A similar abundance of acinar microvilli occurs in some other mammals, particularly some of the herbivores that produce copious amounts of saliva [32]. The relative number of microvilli might be related to the concentration of electrolytes in the initial saliva.

The lateral borders of acinar cells sometimes appear to be smooth and straight but can sometimes exhibit variable degrees of plication and interleafing. Gap junctions often are present and in some types of bats may be quite large. Many bats have prominent mitochondrion-desmosome complexes along the lateral borders of their acinar cells. These consist of single desmosomes of typical morphology flanked on one or both sides by a mitochondrion that is closely appressed to the desmosomal cytokeratin filaments. Similar complexes have been described in a variety of tissues and organisms, and rather fanciful schemes have been concocted, without a shred of supporting evidence, relating the complexes to energy transfer and intercellular communication. It seems more probable that these complexes are the result of nonspecific adhesivity of the desmosomal filaments and attachment plaques and that they have no functional significance [33].

3. Organelles in serous and seromucous acinar cells

Serous and seromucous cells are engaged in intensive production of secretory glycoproteins (both enzymatic and nonenzymatic) for export and, therefore, possess the structural hallmarks of all protein-secreting cells, i.e., an extensive and regular rough endoplasmic reticulum (RER) and a prominent Golgi complex. In some species of bats, the acinar cells show variation in the structure of the RER. For example, intracisternal granules are present in acinar cells in both the little brown bat (*Myotis lucifugus*) [27] and the

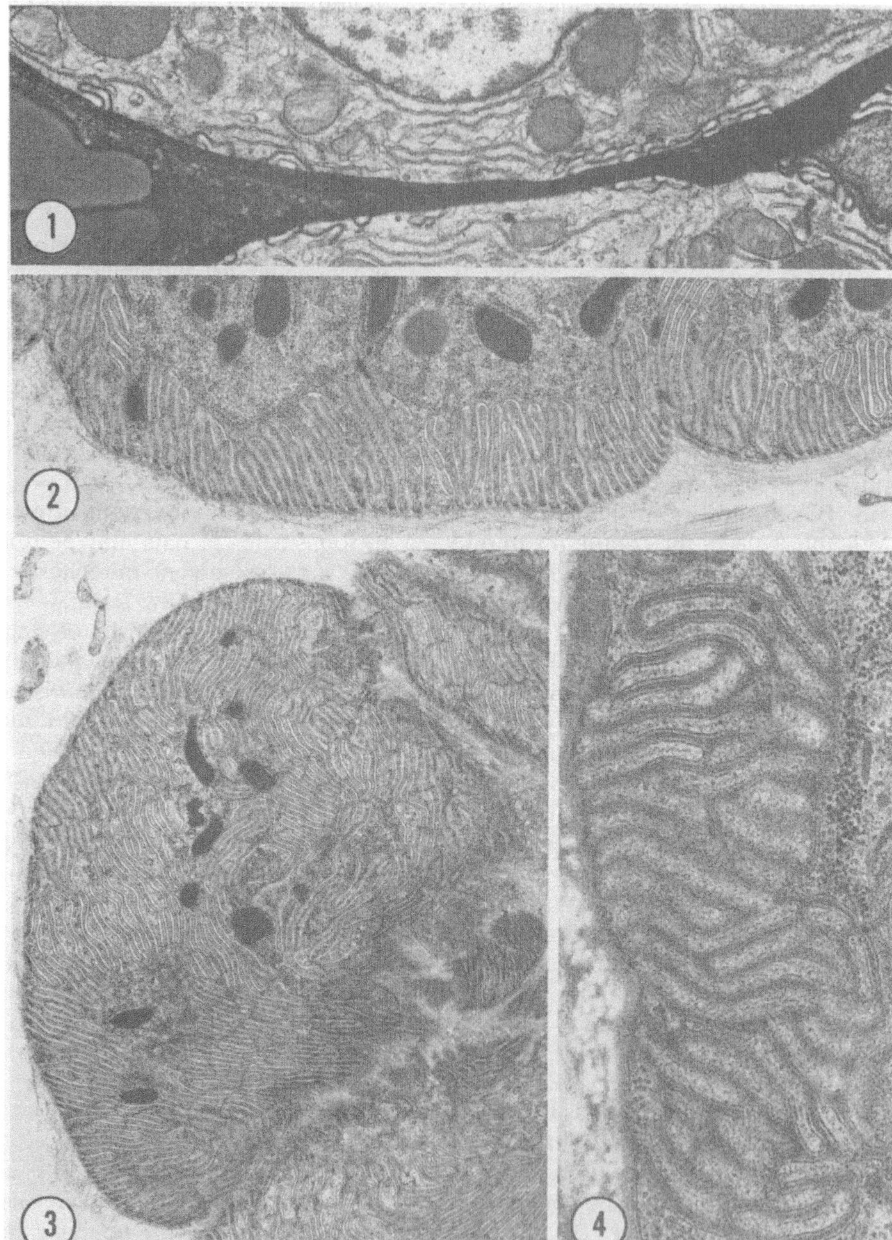

Fig. 3-1. Desmodus rotundus. The base of a serous cell in the principal submandibular gland, showing the paucity of basal membrane specializations in this species. x 10,600.
Fig. 3-2. Eptesicus brasiliensis. A vertical section through the base of serous cells in the parotid gland. The basal surface consists of a series of slender folds. x 11,000.
Fig. 3-3. Eptesicus brasiliensis. A section through the cell base of parotid serous cells, taken in a plane parallel to that of the basement membrane. The extent and close packing of the basal folds are evident. x 7,200.
Fig. 3-4. Pteronotus parnellii. The well-developed system of basal folds on the serous cells of the parotid gland is characterized by the presence of dense material in the intercellular spaces between adjacent folds. x 26,500.

vampire bat (Fig. 3-5). The presence of such granules may be predicated on the fact that in these animals protein is being synthesized at a rate faster than it can be packaged by the Golgi complex and that when the intracisternal protein reaches a critical concentration and local environmental conditions are favorable, it precipitates in visible form. Another example of unusual accumulation of secretory material in RER can be seen in the seromucous lingual glands of vampire bats; in these acinar cells, RER cisternae in the basal cytoplasm can be greatly distended by a homogeneously dense secretory product (Fig. 3-6). One of the most striking modifications to the RER is found in one of the insectivorous bats that we have examined. In *Tylonycteris* sp., the RER cisternae appear to undergo segmental densification. In this bat, individual cisternae can consist of normal-appearing areas interspersed with regions where their membranes have an augmented number of ribosomes on their cytoplasmic aspect; in these latter regions there is a sheet of dense material within the cisterna (Fig. 3-7). This patchy construction of the RER in *Tylonycteris* suggests the possibility of regional differentiation and might prove to be a reflection of the synthesis of different proteins in different sections of the RER.

Acinar cells in bat salivary glands generally contain only a few scattered elements of smooth endoplasmic reticulum (SER). An exception to this rule is provided by the parotid glands in male round-eared bats, *Tonatia sylvicola* [19]. In these animals, the acinar cell SER forms large aggregates of parallel, hexagonally arrayed tubules that are continuous with conventional SER at their ends [3, 19]. We have hypothesized that these aggregates produce a steroid that influences some aspect of behavior such as mate selection and sexual behavior in this species [19].

Mitochondria in acinar cells of bats usually are typical in appearance, that is, the mitochondrial profiles closely resemble one of the conventional morphologies. However, in *Tonatia sylvicola*, cup-shaped mitochondria often are nested one inside the other. Cut crosswise, such mitochondrial aggregates appear as a series of concentric ring-shaped organelles (Fig. 3-8). In *Carollia perspicillata*, which is an omnivorous neotropical species, a significant percentage of the acinar-cell mitochondria have numerous, parallel, prismatic cristae, each periodically attached to its immediate neighbors by intercristal bridges [23] (Fig. 3-9). Mitochondria with paracrystalline plates within their cristae are present in the very unusual accessory submandibular glands of the neotropical frog-eating species, *Trachops cirrhosis* (Fig. 3-10). Yet another variation is found in the flower bat, *Erophylla sezekorni*. The inner compartment of mitochondria in the peculiar green salivary gland (*vide infra*) of male *Erophylla sezekorni* has a plaited or paracrystalline appearance (Fig. 3-11). This unique gland also contains accumulations of peroxisomes in a spiral or circular arrangement (Fig. 3-12); almost identical peroxisomal clusters have been found in the uropygial glands of ducks, where they may play a special role in lipid metabolism [34].

4. Secretory granules

Serous and seromucous cells of bat salivary glands belie the widely held notion that secretory granules are homogeneously dense bodies without discernible substructure. In bats, depending on the species, matrices of acinar cell secretory granules range from the structureless to the dazzlingly complex. The stored secretory product, which accumulates in the apical cytoplasm before release from the acinar cell, is of special interest in comparative studies. We now know that the ultrastructural appearance of the secretory granules varies interspecifically in bats [3, 17]. Indeed, as our examination of the salivary glands in more than 120 chiropteran species indicates, secretory granule ultrastructure is virtually an identifying signature for each species. In some instances, when species are grouped according to granule ultrastructure, the groupings match those that result from independently derived data on the genetics of the animals [16]. Consequently, it appears that secretory-granule ultrastructure is to some extent meaningfully correlated with systematics. This finding is noteworthy because it illustrates the evolutionary versatility of the acinar cells and raises the question of what causes differences in secretory granule ultrastructure. The striking species specificity in secretory granule ultrastructure that we have found in bats might

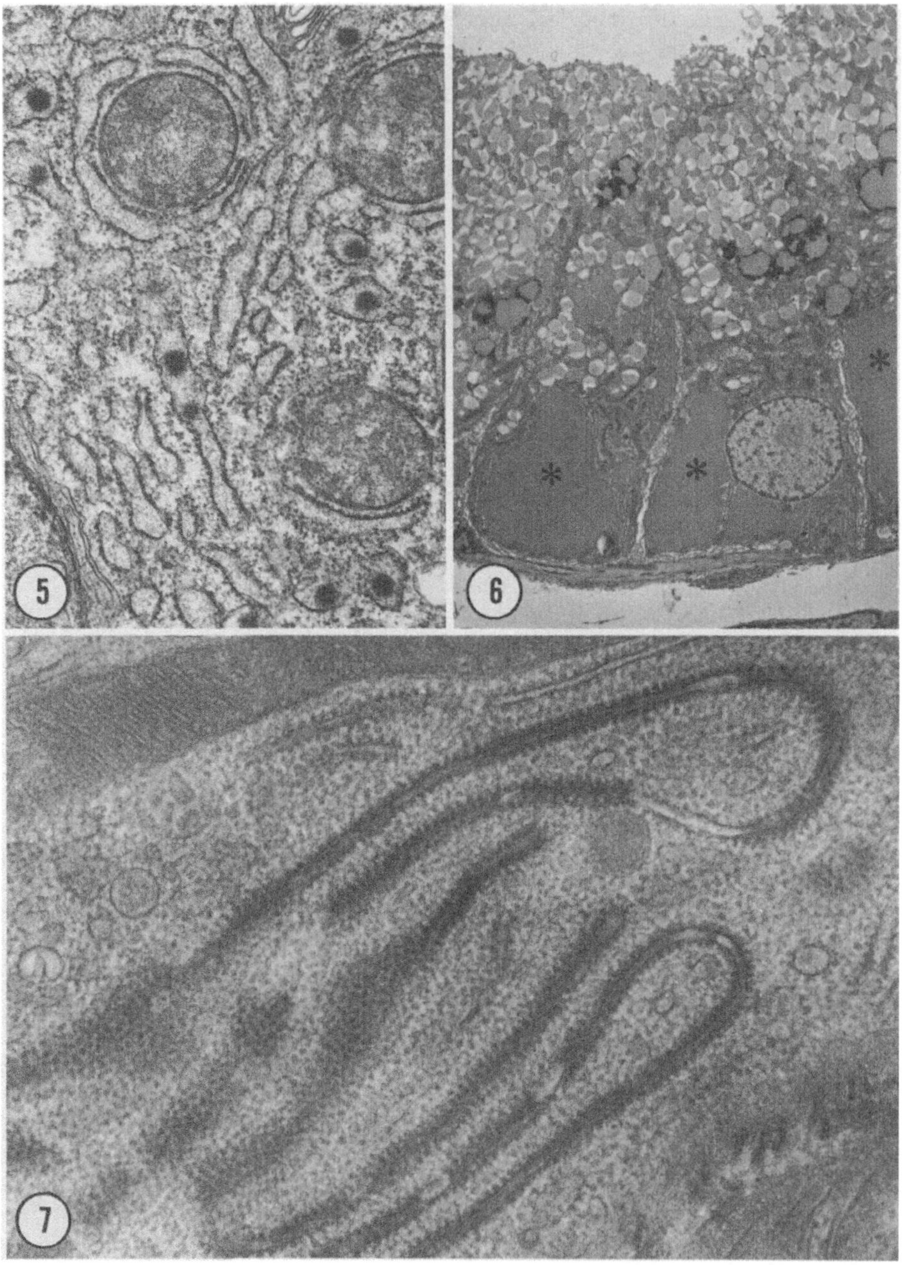

Fig. 3-5. Desmodus rotundus. The RER in serous cells of the principal submandibular gland contains dense intracisternal granules. x 23,700.

Fig. 3-6. Desmodus rotundus. Secretory cells in the lingual minor salivary glands have large RER cisternae (*) containing a homogeneous dense material. x 3,900.

Fig. 3-7. Tylonycteris sp. The RER in serous cells of the parotid gland has a peculiar "patchy" appearance. A single cisterna consists of dense segments with closely approximated ribosomes, alternating with normal-appearing segments. x 33,600.

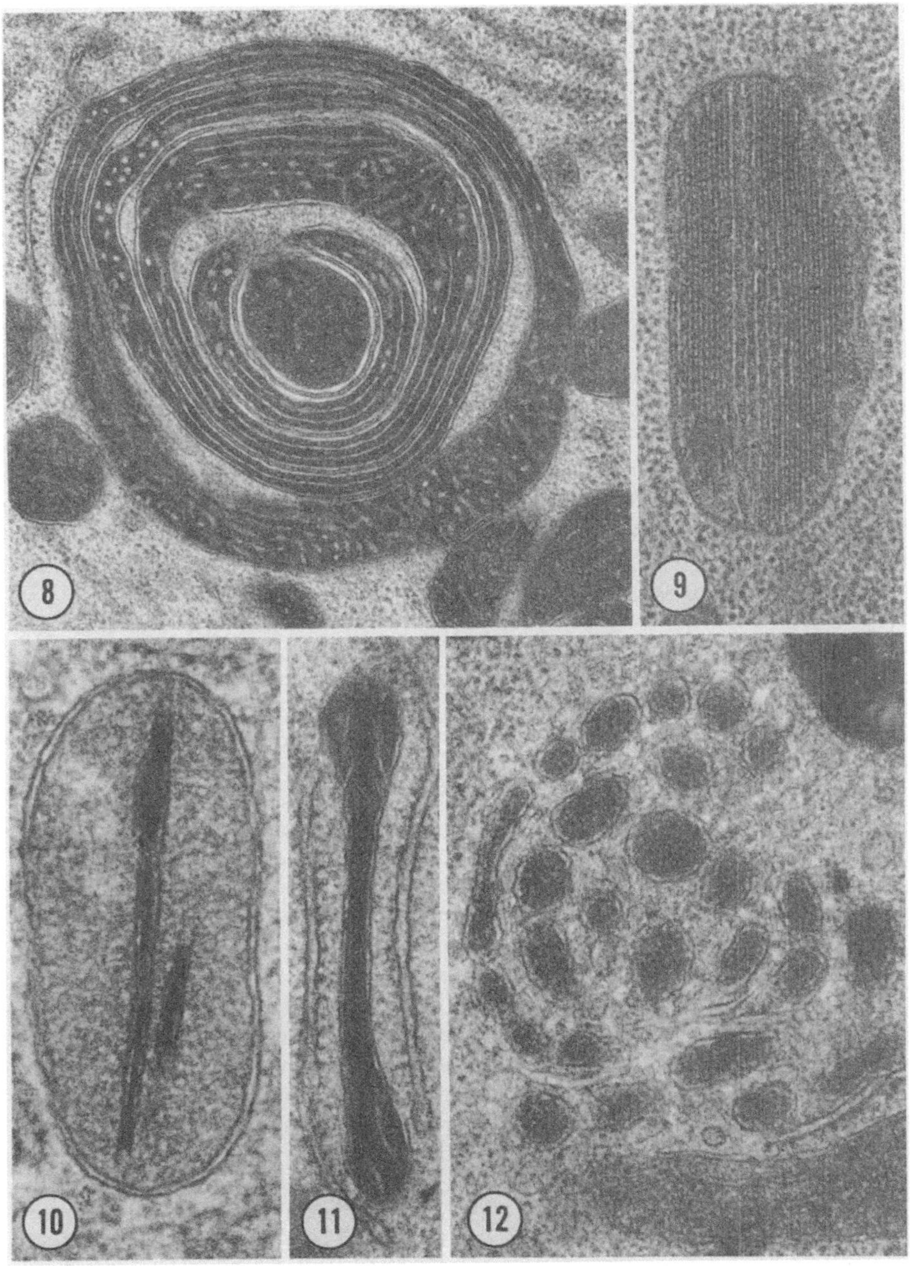

Fig. 3-8. Tonatia sylvicola. Mitochondria in the seromucous cells of the submandibular gland occasionally form complex aggregates. That shown in this micrograph is of a series of cup-shaped mitochondria sitting one inside the other; a section passing parallel to the rims of the cups gives the spurious impression of a series of concentric ring-shaped organelles. x 32,400.
Fig. 3-9. Carollia perspicillata. A mitochondrion with parallel, rigid-appearing cristae in a seromucous cell of the parotid gland. x 31,200.
Fig. 3-10. Trachops cirrhosis. A mitochondrion within a secretory cell of the accessory submandibular gland. Dense crystalloids are present within the sparse cristae. x 66,000.
Fig. 3-11. Erophylla sezekorni. A mitochondrion with a "honeycomb" interior in a secretory cell of the "green gland." x 56,100.
Fig. 3-12. Erophylla sezekorni. A spiral aggregate of small peroxisomes in a secretory cell of the "green gland." x 50,700.

partly be the consequence of structural differences in some of the protein molecules contained within the granules. Elsewhere we have presented our argument that secretory granule substructure is the result of molecular differences in the chemical products and their distribution within the granules [3]. Species-specific differences in secretory granules produced by homologous cells in congeneric species of bats suggest that a rapid evolutionary divergence has taken place in chiropteran salivary gland acinar cells. The great diversity in bats might have been possible because the secretory process offers numerous options for evolutionary modification [3]. The latter explanation is supported by the fact that considerable intraspecific variation in proline-rich proteins (PRP) and glycoproteins is obtained through post-translational modification of these acinar-cell products [35, 36]. Indeed, mRNA splicing and postribosomal proteolytic cleavage account for the production of a greater number of varieties of proline-rich proteins than are encoded by a small number of nuclear genes [35]. Some of the post-translational modifications, such as glycosylation of N-linked glycoproteins, can also be influenced by extracellular signals [37]. The extensive variation in secretory granule ultrastructure in bats might be an indication of species differences in post-translational modification of a relatively small number of basic products. Finally, species-specific differences in granule ultrastructure might be indicative of functional differences in saliva. For example, differences in the structure of glycoproteins could correlate with the oral ecology characteristic of each species [38].

Aside from vague characterizations such as *enzyme-rich* or *mucous*, the chemical nature of secretory products is largely unknown. While it is not surprising that we do not know much about wild mammals, it also is true for many domestic species that otherwise have been investigated in considerable detail. This is partly the result of technical complexities in isolating and then characterizing the products. For the present, we can only categorize acinar cell products by their ultrastructural patterns, with the idea that grouping according to species or diets, or both, might serve as a foundation for future biochemical analyses.

4.1. Serous and seromucous granules

4.1.1. Parotid gland.

4.1.1.1. Insectivorous species. Many species of bats in several families feed on species of insects. The parotid gland secretory product of six such species is compared in this section. Typically their granules are *serous* and possibly enzyme–rich [3, 17].

Serous granules in *Pteronotus parnellii* consist of short, randomly disposed, platelike densities, each with a thin, lucent covering (Fig. 3-13). These inclusions rarely abut the limiting membrane, usually being surrounded by a peripheral zone of moderate density. Granules in *Phyllostomus latifolius* have an extremely dense matrix in which is embedded an electron-lucent inclusion that can exhibit a variety of designs, but generally has a twisted pretzel shape (Fig. 3-14). In the round-eared bat, *Tonatia sylvicola*, the parotid granules contain a large, moderately dense inclusion that has one or several less-dense lacunae (Fig. 3-15).

In the related species, *Tonatia bidens*, the granules have a dense matrix in which are embedded vermicular lucencies, which sometimes become hexagonally packed — fortuitous sections passing transversely through a sheaf of aligned lucencies show an array of lucent spots in a precise honeycomb pattern (Fig. 3-16). The structure of granules in the Japanese horseshoe bat, *Rhinolophus ferrumequinum*, varies with the state of activation of the animal. During the bat's functional period, acinar granules are homogeneously dense, while during hibernation the granules may acquire a bulls-eye or honeycombed internum [39]. In species of *Eptesicus*, the granules are bisected by a very thick, homogeneously dense central zone, leaving the poles occupied by moderately dense material. These polar regions frequently contain a series of dense, parallel, riblike structures (Fig. 3-17). The seromucous granules in the related species, *Myotis lucifugus*, are quite variable, but generally fall into one of two classes [27]. In one type, there appears to be a dense ribbon with axial periodicity underlying the limiting membrane. In reality, these "ribbons" are shell-like affairs consisting of parallel curved slats whose short axes are radially

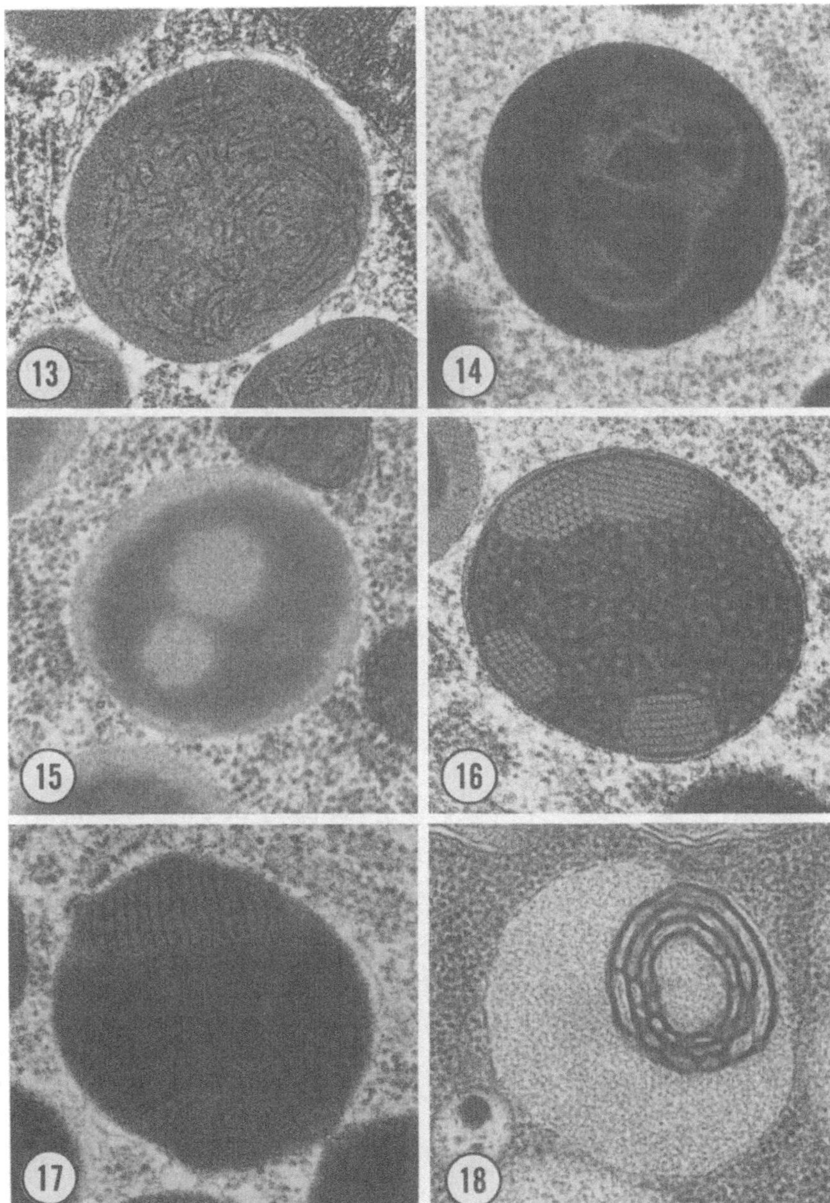

Fig. 3-13 to 3-17. Serous granules in the parotid gland. Fig. 3-13: *Pteronotus parnelli*, x 39,600. Fig. 3-14: *Phyllostomus latifolius*, x 38,900. Fig. 3-15: *Tonatia sylvicola*, x 40,300. Fig. 3-16: *Tonatia bidens*, x 40,700. Fig. 3-17: *Eptesicus brasiliensis*, x 45,600. *Fig. 3-18. Carollia perspicillata.* Parotid seromucous granule. x 45,600.

oriented and whose long axes follow the contour of the membrane. The second type of granule also possesses the peripheral slatted shell, but instead of a structureless matrix, this matrix consists of randomly arranged packets of short, parallel tubules. Individual little brown bats usually have only one of the two principal categories of granules; the matrix configuration is independent of the sex of the animal or of the quality of cytoplasmic fixation.

4.1.1.2. Omnivorous and frugivorous species. In bats that feed mainly on fruit, the parotid secretory granules generally are seromucous instead of serous. Seromucous granules in the parotid gland of the neotropical omnivorous species, *Carollia perspicillata*, have a lucent matrix and an inclusion that has the form of a haphazardly crumpled sheet [23]. Because these inclusions are intersected in random fashion in thin sections, a huge variety of patterns is apparent. In certain granules, the dense inclusions assume a cagelike, geodesic conformation, which in some planes of section takes on the appearance of a poorly constructed target (Fig. 3-18). In *Ariteus flavescens*, a small fruit-eating bat found only in the Antillean Islands, the granules have a very light matrix with small, flakelike cystalloids of moderate density. Instead of the small crystalloids, a few granules have a single, large, extremely dense rhomboidal or quadrilateral crystalloid that lacks discernible periodicity and that is surrounded by vermicular structures of moderate density (Fig. 3-19). The granules in *Sturnira lilium* have a large central dense body surrounded by a light halo that is peppered with punctate densities (Fig. 3-20).

4.1.2. Submandibular gland.

4.1.2.1. Insectivorous species. Serous granules in the submandibular glands of the insectivorous bat, *Mormoops blainvillii*, have a very dense, structureless matrix, with a series of light and dark laminae underlying the limiting membrane. These laminae appear to have considerable rigidity, giving the granules a polygonal silhouette rather than a smoothly rounded one (Fig. 3-21). In stark contrast to their counterparts in the parotid gland, the submandibular granules in *Tonatia bidens* are uniformly dense, completely devoid of any hint of substructure (Fig. 3-22). Granules in *Pipistrellus coromandra* have a finely speckled internum and a peripheral band of filaments (Fig. 3-23).

4.1.2.2. Frugivorous species. The granules in the Old World megachiropteran fruit bat, *Rousettus leschenaulti*, display an assortment of internal patterns. In some, there is a crumpled dense sheet reminiscent of those in *Carollia*, while in others there may be targetlike inclusions (Fig. 3-24). There is a large hexagonal crystalloid plate in each granule in another megachiropteran, *Megaerops caudatus* (Figs. 3-25 and 3-26). Secretory granules in *Macroglossus sabrinus*, which is a nectarivorous species, are extremely variable in size, but, regardless of diameter, they contain a mottled matrix (Fig. 3-27).

In order to test for species-specific patterns in granule substructure, we compared secretory granules in the submandibular gland in five species of neotropical fruit bats of the genus *Artibeus* [11, 16, 40]. The serous cells in all five have identical granules; these are uniformly dense and have no apparent substructure. In contrast, the seromucous demilune cells have highly structured granules whose characteristic morphologies divide the studied species into three distinct groups. In *A. cinereus* and *A. phaeotis*, the granules consist of a series of thin, dark rings (shells in three dimensions) separated by thicker pale rings. Up to 16 rings of each type may be present in the largest granules (Fig. 3-28). In *A. jamaicensis* and *A. lituratus*, the light and dense components usually form a mazelike array; in a few granules an imperfect "target" may be generated. In *A. concolor*, the granules are simpler in structure than are those in the other studied *Artibeus* species, consisting of a pale matrix with small foci of fibrillogranular material and one or two moderately dense, acentrically placed inclusions [16].

Finally, another pattern is found in granules in *Lonchophylla thomasi*, which is another neotropical species. In this species the granules have a rigid, lucid corona that imparts an angulated outline to each granule (Fig. 3-29). The lucent material in places invades the dense central matrix. A few granules have a jellyroll appearance.

4.2. Mucous granules

Mucous granules in chiropteran salivary glands in general are less complex in structure than are their serous or seromucous counterparts. They are present only in mixed salivary glands; this section deals with mucous granules in bat principal or accessory submandibular glands and is arranged according to the diet of the animals.

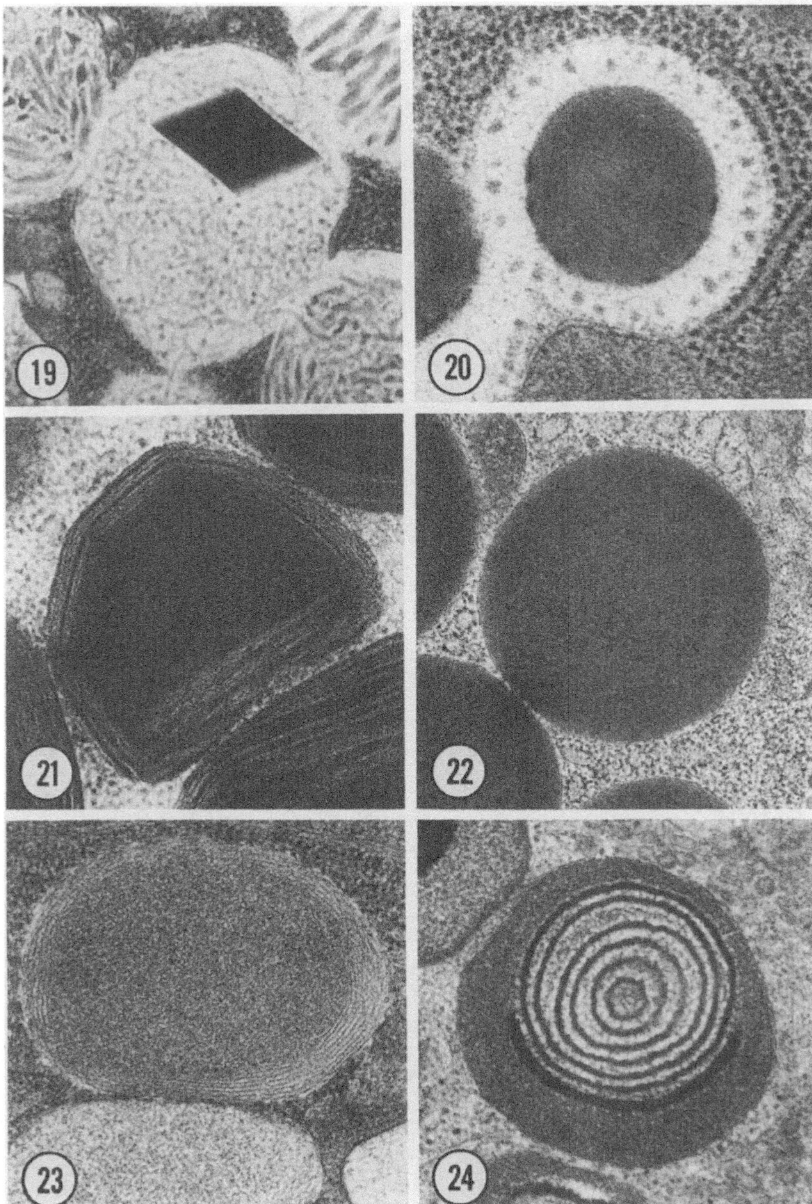

Fig. 3-19. *Ariteus flavescens*. A seromucous granule in the parotid gland containing a dense crystalloid and vermicular structures. x 28,800.
Fig. 3-20. *Sturnira lilium*. Parotid serous granule. x 61,400.
Fig. 3-21. *Mormoops blainvillii*. Serous granule in the principal submandibular gland. x 38,900.
Fig. 3-22. *Tonatia bidens*. Serous granule in the submandibular gland. x 25,200.
Fig. 3-23. *Pipistrellus coromandra*. Serous granule in the submandibular gland with a submembranous band of filaments. x 33,600.
Fig. 3-24. *Rousettus leschenaulti*. Submandibular serous granule. x 35,400.

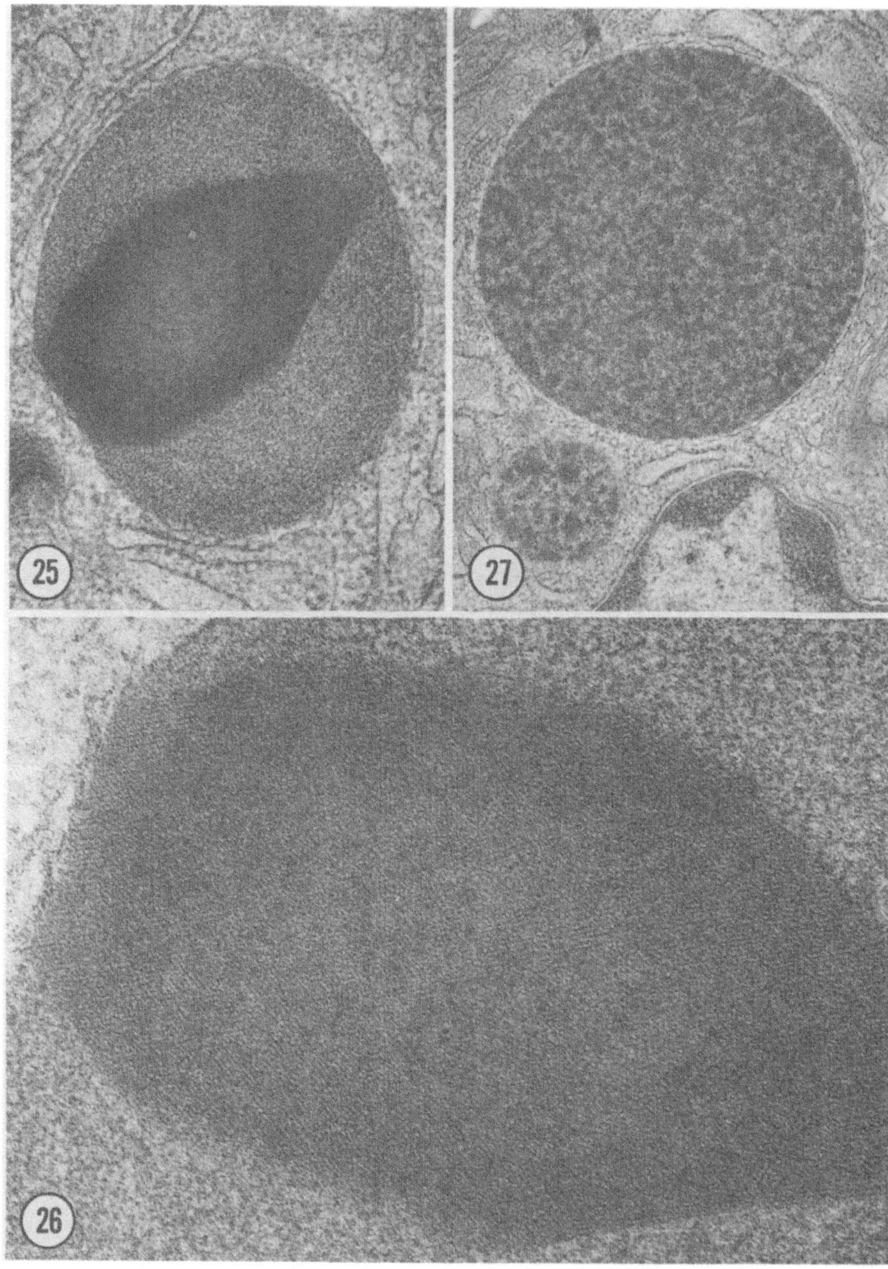

Fig. 3-25. Megaerops caudatus. Submandibular serous granule containing a hexagonal crystalline platelike inclusion. x 48,000.
Fig. 3-26. Higher magnification of Fig. 3-25 showing the crystalline nature of the inclusion. x 103,200.
Fig. 3-27. Macroglossus sabrinus. A giant serous granule in the submandibular gland. The degree of enlargement of such granule can be gauged by comparison with the usual-sized granule at the lower left. x 21,600.

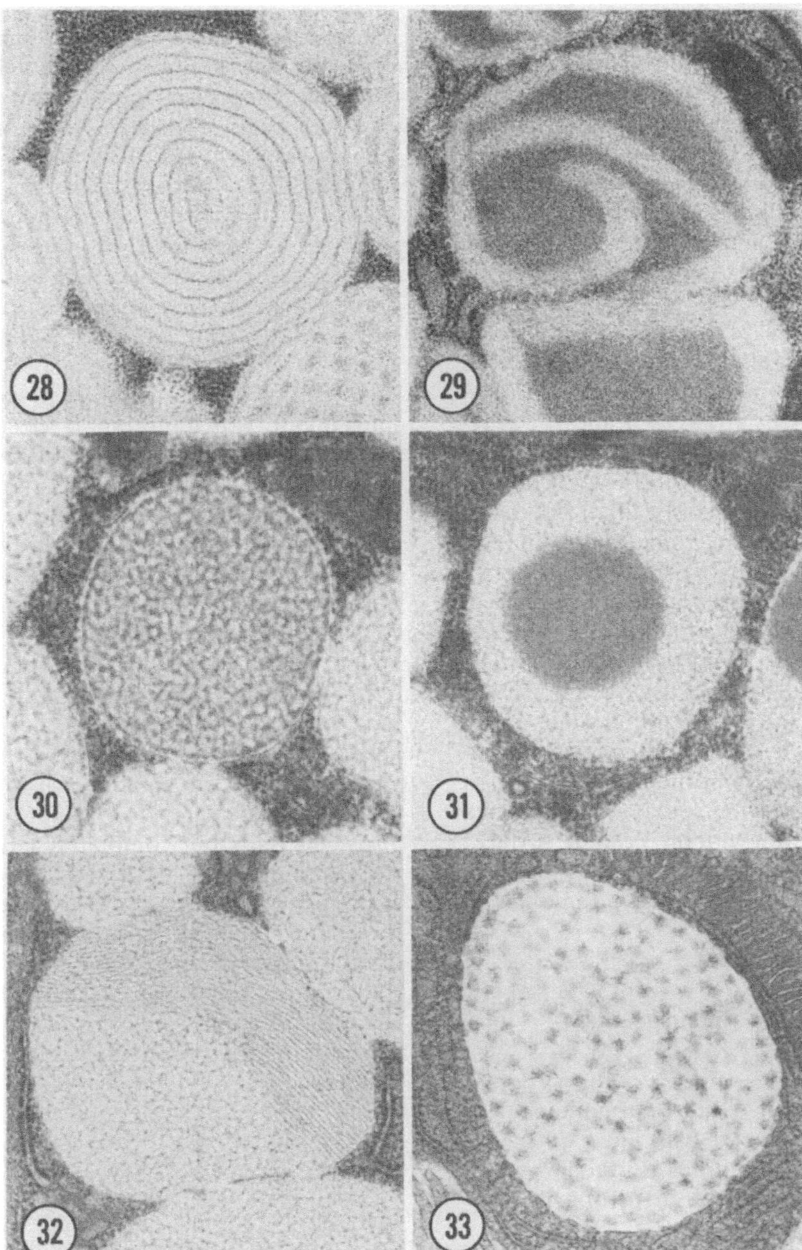

Fig. 3-28. Artibeus cinereus. A seromucous granule with a targetlike internal configuration in the submandibular gland. x 25,000.
Fig. 3-29. Lonchophylla thomasi. Serous granule in the submandibular gland. x 25,200.
Fig. 3-30 to 3-33. Mucous granules in the submandibular gland. Fig. 3-30: *Noctilio leporinus*, x 27,500. Fig. 3-31: *Phyllostomus hastatus*, x 26,600. Fig. 3-32: *Monophyllus redmani*, x 23,400. Fig. 3-33: *Hipposideros armiger*, x 28,800.

In the fish-eating bat, *Noctilio leporinus*, mucous granules have a finely dotted content (Fig. 3-30). The granules in the carnivorous bat, *Phyllostomus hastatus*, have a pale matrix with a large, dense inclusion (Fig. 3-31). In the insectivorous/frugivorous bat, *Monophyllus redmani*, the mucous granules have a finely speckled matrix and a large bundle of filaments (Fig. 3-32). Stellate densities characterize the mucous granules of *Hipposideros armiger* (Fig. 3-33) and *Rhinolophus ferrumequinum* [41], both insectivorous; the granules closely resemble those in acinar cells of the submandibular gland of the cat [42]. The granules in *Miniopterus schreibersi* and *M. magnator*, both insectivorous, are the most unusual in the gallery of granules, whether mucous or serous. Regardless of species-specific patterns of granules in most bats, they all have single limiting membranes. However, those in the accessory submandibular glands of *Miniopterus* are characterized by the presence of multiple limiting membranes (up to a dozen) at their surface [3, 43]. These surface membranes often extend as a packet into the interior of the granule in the manner of cristae in mitochondria. In certain planes of section, packets of membranes lying in the granule matrix give the spurious appearance of having no connection with the granule surface. The *Miniopterus* granules with supernumerary membranes appear to be unique among *all* exocrine cells studied so far.

5. Atypical aggregates of secretory cells

In the vast majority of bats, the salivary gland secretory cells are arranged in the familiar histological fashion, but several bats have unusually constructed secretory endpieces. In the accessory submandibular gland of the neotropical frog-eating species, *Trachops cirrhosis*, the secretory cells form large, thyroidlike follicles with capacious lumina rather than typical acini [18]. The cells that make up the walls of the follicles are also cytologically atypical. Although they possess an extensive RER, they lack conventional secretory granules. Instead, they contain near their luminal surface a small number of lucent, amorphous to ribbonlike bodies that appear to discharge their contents into the follicular lumen by exocytosis. Many of the cells are charged with lipid droplets of unusual appearance [3]. An interesting comparison can be made to two species of Asian bats, *Megaderma lyra* and *M. spasma*, which also eat frogs. This feeding habit evolved independently in these bats, so comparison of their accessory submandibular to that of *Trachops* provides us with data on convergent evolution. The accessory submandibular glands in *Megaderma lyra* and *M. spasma* also have follicles rather than acini, as in typical glands [18]. A common feature of all three species is that some of the tropical frogs that they eat have a toxic integument. It is tempting to speculate that these unique salivary glands, which constitute a prime histological example of evolutionary convergence, produce a substance or substances that neutralize the toxins in the skin of their prey [3, 18].

We have recently discovered a heretofore unknown gland in sexually mature males of *Erophylla sezekorni*. These glands, which in living animals have a pronounced greenish hue, have a main excretory duct that drains into the same kind of duct originating in the submandibular gland, thus marking the green gland as a salivary gland. This green gland consists of long tubes rather than of acini. The secretory cells contain abundant RER but completely lack conventional secretory granules. There are, however, many small, lucent vesicles in the apical cytoplasm, and these may function as repositories and carriers of secretory material. The basis for the greenish coloration is unknown. It seems likely that this gland is in some way associated with sexual behavior in this bat, which occurs only in the Antilles.

6. Ducts

Once liberated into the lumina of the secretory endpieces, saliva makes its way into the mouth by way of a series of ducts of progressively increasing caliber. These ducts are, in order, the intercalated ducts, the striated ducts, the excretory ducts, and the main excretory ducts. With the exception of the last-named ducts, each of the other ducts may be highly branched. Striated ducts, which are formed by anastomosis of intercalated ducts, in turn coalesce to form excretory ducts. In this section, we will confine our attention to the intercalated and striated ducts.

6.1. *Intercalated ducts*

In most bat species, as in most other mammals that have been studied, intercalated ducts consist of pyramidal cells that show almost no cytological specialization. In the parotid gland of *Rousettus leschenaulti*, for example, the duct cells have relatively smooth and uncomplicated surfaces and contain a few sparse organelles (Fig. 3-34). In the little brown bat, *Myotis lucifugus*, the parotid duct cells contain extensive RER with an abundance of intracisternal granules [27] (Fig. 3-35). A few structureless secretory granules are present in parotid duct cells of the frugivorous/nectarivorous *Glossophaga soricina*, but intracisternal granules are absent. In contrast, parotid intercalated duct cells in *Phyllostomus latifolius* have an abundance of small granules with surface protrusions and a variegated matrix consisting of a moderately dense substance in which fragmented, amorphous, denser material is scattered (Fig. 3-36). Based on these and similar observations in other mammals, it may be concluded that intercalated ducts not only act as conduits of saliva, but in some animals, including bats, they may make a significant contribution to the organic content of the saliva.

6.2. *Striated ducts*

In terms of general morphology, striated ducts in bats resemble those in other groups of mammals. By light microscopy, they show prominent basal striations and a fairly large lumen [44]. By electron microscopy, the striations are seen to consist principally of numerous vertically oriented mitochondria (Fig. 3-37). Striated duct cells have extremely complex basal contours; the basal portions of adjacent cells are intricately interlocked; and the basal mitochondria are lodged within cell processes resulting from the irregular configuration of the cell base [25, 29]. In most bats, for example, *Rousettus leschenaulti*, the basal mitochondria appear to be extremely long, rod-shaped organelles with transverse cristae and prominent matrix granules (Fig. 3-38). In fortuitous sections, however, where the mitochondria are grazed rather than cut in profile, it becomes apparent that these organelles have considerable breadth and that they are not virgulate. The large diameter of the basal mitochondria is readily observed in sections that pass through the base of a striated duct in a plane parallel to that of the basal lamina. Based on such three-dimensional reconstructions, it may be concluded that each basal mitochondrion is shaped like an irregular disk. Striated ducts have been implicated in modification of the electrolyte composition of saliva [1], and it is more than likely that the basal mitochondria provide the ATP necessary to power the transport processes.

The luminal surface of striated duct cells in most species of bats is fitted with a few microvilli. In the vampire bat, the submandibular striated duct cells, like the acinar cells in the same animal, posses a huge complement of microvilli, forming the salivary equivalent of a renal brush border (Fig. 3-39). The microvilli are long and closely packed, but lack bundles of actin filaments and the fuzzy coat (glycocalyx) found in the striated border microvilli of intestinal absorptive cells. In many fruit-eating bats, the apical surface of the striated duct cells is thrown into a series of folded processes (Fig. 3-40). The plasma membrane covering these frondose processes has on its cytoplasmic side a periodic series of small projections (Fig. 3-41). Similar membrane projections, termed *portasomes* [45], have been described in several invertebrate organs, where they have been implicated in the mass movement of sodium [46]. Of the two types of bats that show deviations from the usual structure of the luminal surface of striated duct cells, both types are confronted by unusual problems in electrolyte homeostasis. Vampire bats ingest huge quantities of sodium chloride in the blood they consume, while fruit bats ingest food that is extremely rich in potassium but extremely poor in sodium content. The respective apical specializations of the ducts in these bats may be necessary to regulate loss or retention of the critical electrolytes.

In addition to their role in ion and water translocation, striated ducts may have a secretory function. Most bats have some dense secretory granules in the apical cytoplasm of the duct cells.

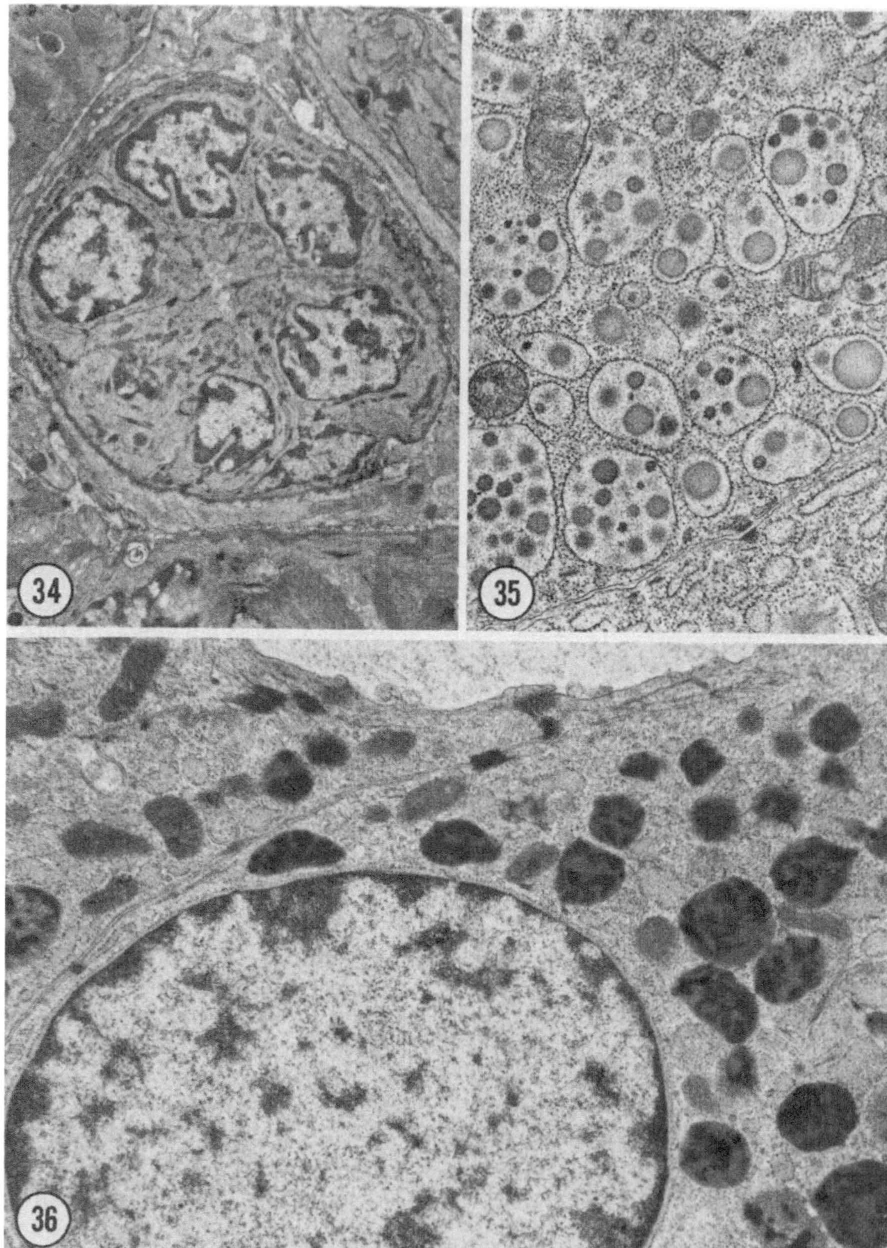

Fig. 3-34. Rousettus leschenaulti. A transversely sectioned intercalated duct in the parotid gland. The duct cells lack both secretory granules and plasma membrane specializations. x 5,300.

Fig. 3-35. Myotis lucifugus. Dilated RER cisternae with abundant intracisternal granules in an intercalated duct cell of the parotid gland. x 17,300.

Fig. 3-36. Phyllostomus latifolius. Parotid intercalated duct cell with abundant serous-type secretory granules, which have a variegated substructure x 12,600.

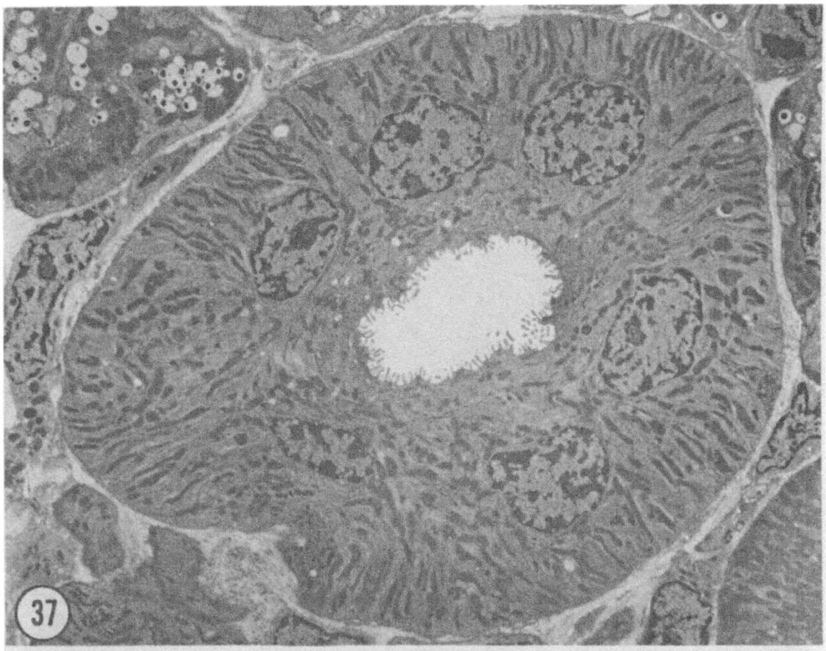

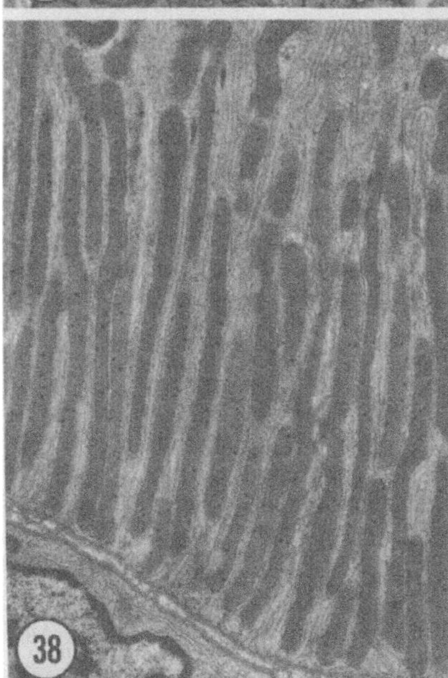

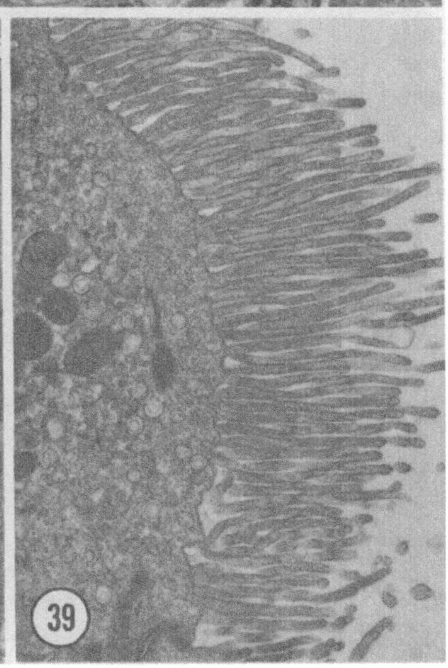

Fig. 3-37. Rousettus leschenaulti. Transversely sectioned submandibular striated duct showing the basal striations and a modest number of microvilli. x 3,000.

Fig. 3-38. Rousettus leschenaulti. Basal, apparently rod-shaped mitochondria in submandibular striated duct. x 8,600.

Fig. 3-39. Desmodus rotundus. Microvilli lining the luminal surface of striated duct cells in the principal submandibular gland. x 16,600.

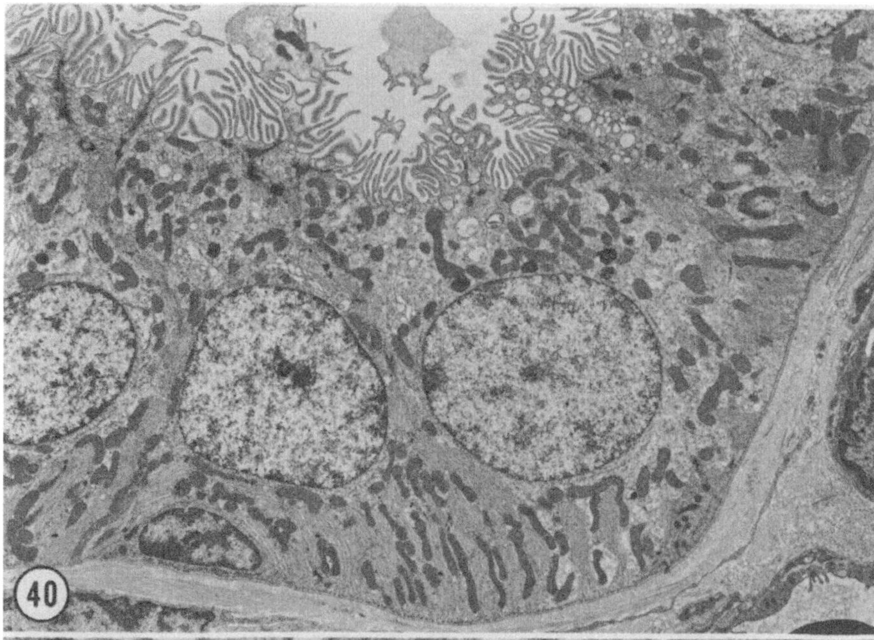

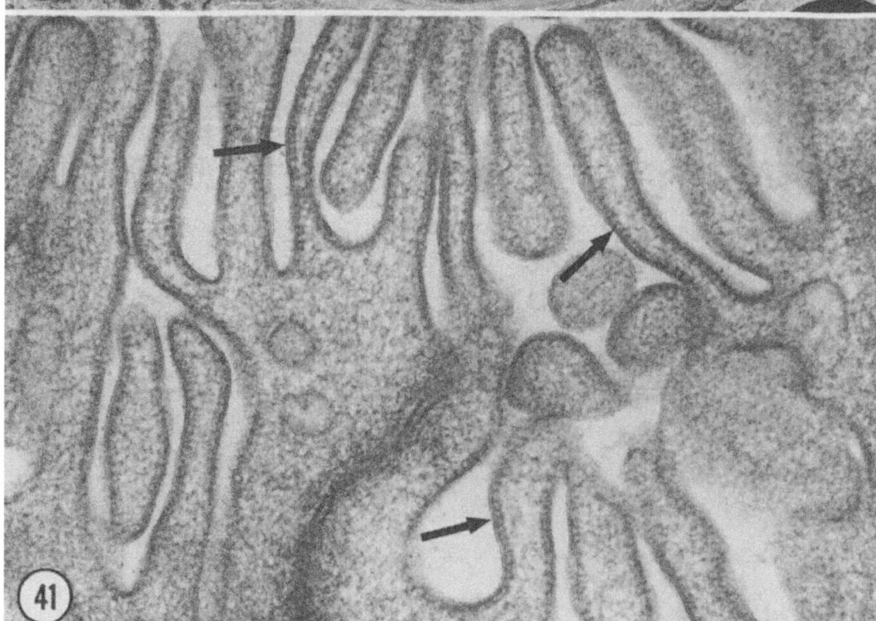

Fig. 3-40. Artibeus jamaicensis. Striated duct in the submandibular gland. The structure of the cell base is identical to that of striated ducts in most other mammalian species, but the luminal surface is characterized by irregular frondose processes, rather than by microvilli. x 4,100.

Fig. 3-41. Megaerops caudatus. High magnification of frondose processes in a parotid striated duct. Repeating subunits (= portasomes) are present on the cytoplasmic aspect of the plasmalemma and are particularly evident at the arrows. x 66,200.

In the laboratory rat, *Rattus norvegicus*, they contain glycoproteins that are used to coat the cell surface [47]. In cats, the granules contain the vasoactive substance, kallikrein [48]. In the parotid gland of the insectivorous bats, *Pipistrellus coromandra*, there is a thin band of small secretory granules immediately below the apical surface of the striated duct cells (Fig. 3-42). Similar cells in *Carollia perspicillata* contain a somewhat greater number of such granules. In *Myotis lucifugus*, the duct granules are more abundant. These granules have a paracrystalline substructure [27]; the only other mammal in which paracrystalline granules are present in striated duct cells is the mongoose, *Herpestes edwardsi* [49]. The most extreme example of duct cell granulation is found in the parotid gland of the vampire bat, *Desmodus rotundus*. The duct cells, which are quite tall, have a basally displaced nucleus, and the supranuclear cytoplasm is almost completely filled with granules (Fig. 3-43). These cells appear to be specialized for secretion rather than for electrolyte transport, since they have virtually no basal striations.

In one species of bat, *Miniopterus magnator*, polygonal crystalloids are abundant in the striated duct-cell cytoplasm; these structures lack limiting membranes, so they are not lysosomal in nature. Crystalloids are not present in the striated duct cells of the closely related bat, *Miniopterus schreibersi*.

7. Innervation

Salivary gland innervation is complex; the relative influence of parasympathetic and sympathetic nerves and their various neuropeptides can regulate the composition of the saliva [50]. The degree of innervation of chiropteran salivary glands varies according to the gland and species. Innervation is accomplished by means of hypolemmal nerve terminals, which are naked axons that have penetrated the basal lamina to end blindly between adjacent parenchymal cells. The terminus of these structures is expanded or bulbous and contains the organelles usually associated with synaptic endings, i.e., synaptic vesicles and mitochondria, as well as some elements of SER. There are no membrane modifications such as postsynaptic thickenings on the contiguous parenchymal cells to identify which cells are actually being directly innervated. Most terminals contain small, clear synaptic vesicles, but some larger dense-cored vesicles may be present. In some terminals, most, if not all, vesicles are of the dense-cored variety.

Nerve terminals in typical bat salivary glands are confined largely to acini and intercalated ducts; there is a paucity of terminals in the striated and excretory ducts. A major exception to this rule is found in the principal submandibular gland of *Myotis lucifugus* (Fig. 3-44). In this gland, the number of terminals increases steadily from acini through the intralobular duct system. In other words, there are significantly more terminals in the intercalated ducts than in the acini, and the striated ducts are the most heavily innervated of all.

8. Conclusions

This brief survey of chiropteran salivary glands serves to dramatically document variation in cell structure. It is clear that no single, simple statement would adequately describe the structure of most types of salivary gland cells. The correlations among structural features of homologous cells, diet, and systematics show that this variation is not random. The ultrastructural differences seen in salivary gland acinar and duct cells provide a starting point for tracing the evolutionary history of the secretory process in mammals. These differences also serve as phenotypic markers of functional diversity.

Our present textbook view of cell structure in mammals is strongly biased, having been largely based on data from fewer than one quarter of 1% of the living species. Some of the most intensely studied species probably are not even "typical" for mammals. The inadequacy of this small sample is readily apparent from our data. This electron microscope survey of salivary glands provides an overview of only 33 of the more than 190 species that we have examined in the order Chiroptera. To our knowledge, no other organ has been studied in as many species within a single mammalian order. Nevertheless, we have barely scratched the surface, because more than 700

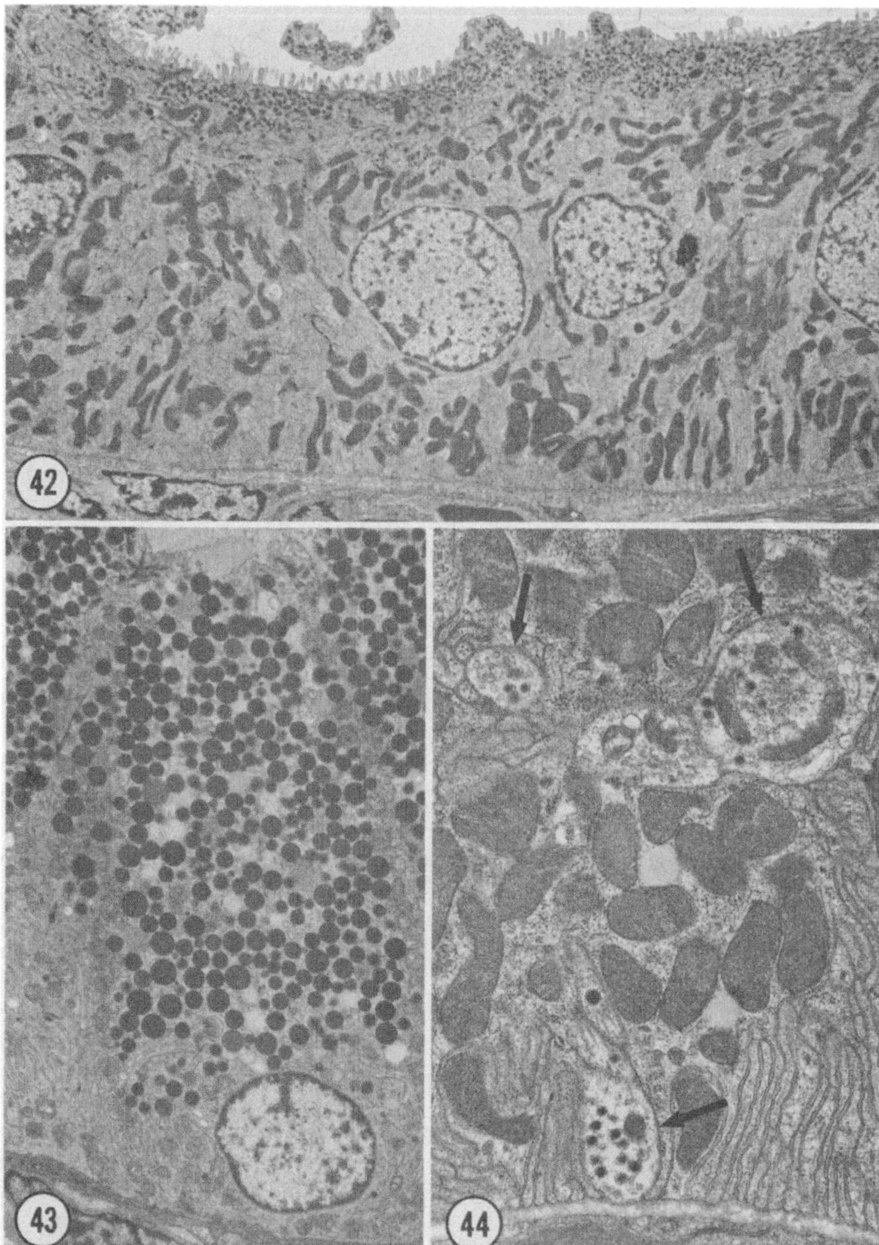

Fig. 3-42. Pipistrellus coromandra. Parotid striated duct showing the thin band of small, dense secretory granules underlying the luminal surface. x 4,300.

Fig. 3-43. Desmodus rotundus. Striated duct cells in the parotid gland with an abundance of secretory granules. x 3,700.

Fig. 3-44. Myotis lucifugus. Striated duct in the principal submandibular gland. Some of the numerous hypolemmal nerve terminals are indicated by arrows. x 14,900.

species of bats remain to be investigated. As these and other species of mammals become available for study, the lacunae in the story of exocrine gland evolution will become progressively smaller, and the structural patterns visible in the electron microscope will, we trust, take on new meaning. Meanwhile, it already is clear that some conventional assumptions about secretory-cell structure will require modification.

Acknowledgments

Financial support for the research reported here came from NIH grant R01 DE 07648-01A1 (Tandler and Phillips), NIH grant DE 03455-02 (Phillips), NSF grant CDP-801-8653 (Phillips), Research Corporation grant, c-1251 (Phillips), Hofstra University HCLAS grants (Phillips), and NIH grant AM-08305 (Tandler). We are very grateful to the many people who have participated in our field and laboratory work. Among these we especially wish to thank Hugh H. Genoways, University of Nebraska, and Duane A. Schlitter, Carnegie Museum of Natural History, without whom much of the field work could not have been undertaken. We also thank Janice Tandler for her editorial assistance and Anne Tassone for typing our manuscript.

References

1. Young JA, van Lennep EW: The Morphology of Salivary Glands. New York: Academic Press, 1978.
2. Pinkstaff CA: The cytology of salivary glands. *Int Rev Cytol* 63: 141–261, 1980.
3. Phillips CJ, Tandler B: Mammalian evolution at the cellular level. In: Current Mammalogy. HH Genoways (ed), New York: Plenum Press, 1–66, 1987.
4. Palade GE: Intracellular aspects of the process of protein secretion. *Science* 189; 347–358, 1975.
5. Case RM: Synthesis, intracellular transport and discharge of exportable proteins in the pancreatic acinar and other cells. *Biol Rev* 53: 211–354, 1978.
6. Tartakoff AM: The Secretory and Endocytic Paths. Mechanism and Specificity of Vesicular Traffic in the Cell Cytoplasm. New York: John Wiley & Sons, 46–170, 1987.
7. Hill JE, Smith JD: Bats — A Natural History. Austin, TX: Univ Texas Press, 1984.
8. Pettigrew JD: Flying primates? Megabats have the advanced pathway from eye to midbrain. *Science* 231: 1304–1306, 1986.
9. Rouse GW, Robson SK: An ultrastructural study of megachiropteran (Mammalia: Chiroptera) spermatozoa: implications for chiropteran phylogeny. *J. Submicrosc Cytol* 18: 137–152, 1986.
10. Findley JS, Wilson DE: Ecological significance of chiropteran morphology. In: Ecology of Bats. TH Kunz (ed), New York: Plenum Press, 243–260, 1982.
11. Phillips CJ, Grimes GW, Forman GL: Oral Biology. In: Biology of Bats of the New World Family Phyllostomatidae. Part II. RJ Baker, JK Jr Jones, DC Carter (eds), Lubbock, TX: Texas Tech Univ Press 121–246, 1977.
12. Phillips CJ, Studholme KM, Forman GL: Comparative ultrastructure of gastric mucosae in four genera of bats (Mammalia: Chiroptera), with comments on gastric evolution. *Ann Carnegie Mus* (Pittsburgh) 53: 71–117, 1984.
13. Studholme KM, Yazulla S, Phillips CJ: Interspecific comparisons of immunohistochemical localization of retinal neurotransmitters in four species of bats. *Brain Behav Evol* 30: 160–173, 1987.
14. Fleming TH: Foraging strategies of plant-visiting bats. In: Ecology of Bats. TH Kunz (ed), New York: Plenum Press, 327–367, 1982.
15. Heithuas ER: Coevolution between bats and plants. In: Ecology of Bats. TH Kunz (ed), New York: Plenum Press, 327–367, 1982.
16. Tandler B, Nagato T, Phillips CJ: Systematic implications of comparative ultrastructure of secretory acini in the submandibular salivary gland of *Artibeus* (Chiroptera: Phyllostomidae). *J Mammal* 67: 81–90, 1986.
17. Phillips CJ, Nagato T, Tandler B: Comparative ultrastructure and evolutionary patterns of acinar secretory product of parotid salivary glands in neotropical bats. In: Studies in Neotropical Mammalogy: essays in Honor of Philip Hershkovitz, BD Patterson, RM Timm (eds), Chicago: Field Mus Nat Hist, 213–229, 1987.
18. Phillips CJ, Tandler B, Pinkstaff CA: Unique salivary glands in two genera of tropical microchiropteran bats: An example of evolutionary convergence in histology and histochemistry. *J Mammal* 68: 235–242, 1987.
19. Nagato T, Tandler B, Phillips CJ: Unusual smooth endoplasmic reticulum in submandibular acinar cells of the male round-eared bat, *Tonatia sylvicola*. *J Ultrastruct Res* 87: 275–284, 1984.
20. Kalt, MR, Tandler B: A study of fixation of early amphibian embryos for electron microscopy. *J Ultrastruct Res* 36: 633–645, 1971.
21. Phillips CJ: Field fixation and storage of museum tissue collections suitable for electron microscopy. *Acta Zool Fenn* 170: 87–90, 1985.
22. Tandler B, Walter, RJ: Epon-Maraglas embedment for electron microscopy. *Stain Technol* 52: 238–239, 1977.
23. Tandler B, Phillips CJ, Nagato T: Evolutionary aspects of the ultrastructure of the parotid gland in an omnivorous neotropical bat, *Carollia perspicillata*. *Zool Scripta* 17: 419–427, 1988.
24. Forman GL, Phillips CJ: Preparation and fixation of tissues for histological, immunohistochemical, and electron microscopic studies. In: Ecological and Behavioral Me-

thods for the Study of Bats. TH Kunz (ed), Washington, DC: Smithsonian Inst Press, 405–424, 1988.
25. Tandler B: Salivary glands and the secretory process. In: A Textbook of Oral Biology. JH Shaw, EA Sweeney, CC Cappucino, SM Meller (eds), Philadelphia: WB Saunders Company, 547–592, 1978.
26. Tandler B: Ultrastructure of the human submaxillary gland. I. Architecture and histological relationships of the secretory cells. *Am J Anat* III: 287–307, 1962.
27. Tandler B, Cohan RP: Ultrastructure of the parotid gland in the little brown bat. *Anat Rec* 210: 491–502, 1984.
28. Van Lennep EW, Kennerson AR, Compton JS: The ultrastructure of the sheep parotid gland. *Cell Tissue Res* 179: 377–392, 1977.
29. Tandler B, Riva A: Salivary glands. In: Human Oral Embryology and Histology. IA Mjör, O Fejerskov (eds), Copenhagen: Munksgaard International Publishers, 243–284, 1986.
30. Nakagaki I, Goto T, Sasaki S, Imai Y: Histochemical and cytochemical localization of (Na^+-K^+)-activated adenosine triphosphatase in the acini of dog submandibular glands. *J Histochem Cytochem* 26: 835–845, 1978.
31. Bundgaard M, Moller M, Poulsen JH: Localization of sodium pump sites in cat salivary glands. *J Physiol* 273: 339–353, 1977.
32. Shackleford JM, Wilborn WH: Ultrastructure of bovine parotid glands. *J Morphol* 127: 453–474. 1969.
33. Tandler B, Hoppel CL: Peroxisome-desmosome complexes in mouse hepatic cells. *Z Zellforsch* 110: 166–172, 1970.
34. Zaar K, Gorgas K: Peroxisome-endoplasmic reticulum aggregates in the duck uropygial gland. *Eur J Cell Biol* 38: 322–327, 1985.
35. Bennick A: Structural and genetic aspects of proline-rich proteins. *J Dent Res* 66: 457–461, 1987.
36. Schulte BA: Genetic and sex-related differences in the structure of submandibular glycoconjugates. *J Dent Res* 66: 442–450, 1987.
37. Kousvelari EE, Fox PC, Baum BJ: Regulatory aspects of N-linked glycoproteins. *J Dent Res* 66: 552–556, 1987.
38. Levine MJ, Reddy MS, Tabak LA, Loomis RE, Bergey EJ, Jones PC, Cohen RE, Stinson MW, Al-Hashimi I: Structural aspects of salivary glycoproteins. *J Dental Res* 66: 436–441, 1987.
39. Mineda T: An electron microscopic study on the parotid gland of the Japanese horseshoe-bat (*Rhinolophus ferrumequinum nippon*) in the functional and hibernating period. *Aichi-Gakuin J Dent Sci* 16: 139–146, 1978.
40. Phillips CJ: Salivary glands: comparative ultrastructure of secretory cells. *Proc Nassau County Med Cent* 3: 162–167, 1976.
41. Mineda T: An electron microscopic investigation on the submandibular gland of the Japanese horseshoe-bat (*Rhinolophus ferrumequinum nippon*) — Histological and histochemical studies. *Aichi-Gakuin J Dent Sci* 15: 35–47, 1977.
42. König B Jr, Kühnel W: Licht- und elektronenmikroskopische Untersuchunger an der Glandula parotis und der Glandula submandibularis der Hauskatze. *Z Mikrosk-Anat Forsch* 100: 469–483, 1986.
43. Tandler B, Phillips CJ: Unusual secretory granules in the accessory submandibular gland of long-winged bats. *Anat Rec* 211: 192A, 1985.
44. Pinkstaff CA, Tandler B, Cohan RP: Histology and histochemistry of the parotid and principal and accessory submandibular glands of the little brown bat. *J Morphol* 172: 271–285, 1982.
45. Harvey WR, Cioffi M, Wolfersberger MG: Portasomes as coupling factors in active ion transport and oxidative phosphorylation. *Am Zool* 21: 775–791, 1981.
46. Kikuchi S: A unique cell membrane with a lining of repeating subunits on the cytoplasmic side of presumably ion-transporting cells in the gill epithelium of *Daphnia magna* (Crustacea: Cladocera). *J Submicrosc Cytol* 14: 711–715, 1982.
47. Hand AR: Synthesis of secretory and plasma membrane glycoproteins by striated duct cells of rat salivary glands as visualized by radioautography after ^{3}H-fucose injection. *Anat Rec* 195: 317–340, 1979.
48. Schachter M, Wheeler GD, Matthews RW, Peret MW, Moriwaki C: Ultrastructural immunolocalization of kallikrein in apical granules of striated duct cells of cat submandibular gland. *J Histochem Cytochem* 31: 345–347, 1983.
49. Schramm U, Flegler C, Ginsbach G: Zur Ultrastruktur der Glandula parotis und der Glandula submandibularis der Schleichkatze *Herpestes edwardsi* (Viverridae). *Anat Anz* 146: 113–132, 1979.
50. Garrett JR: The proper role of nerves in salivary secretion: a review. *J Dent Res* 66: 387–397, 1987.

Author's address:
Dr. Bernard Tandler
School of Dentistry
Case Western Reserve University
Cleveland, OH 44106
USA

CHAPTER 4

Normal human salivary glands

ALESSANDRO RIVA, MARIA S. LANTINI & FRANCESCA TESTA RIVA

1. Introduction

The human salivary glands comprise the major and minor salivary glands. The major salivary glands are encased by a fibrous capsule and are connected to the oral cavity by long excretory ducts. The minor salivary glands are abundant in the tongue and in the walls of the oral cavity; their excretory ducts are short. There are three paired sets of major glands: parotid, submandibular, and sublingual and several hundreds of minor salivary glands, which are named according to their location: buccal, labial, lingual, palatine, and glossopalatine.

During embryogenesis [1] salivary glands originate from a proliferation of the epithelium into the underlying mesenchyme. The resultant epithelial buds are surrounded by connective tissue of mesenchymal origin, which, in the developed glands, forms a series of septa that compartmentalize the parenchyma into lobes and lobules. The septa carry the vessels and nerves, and contain, in addition to their complement of connective tissue cells and fibers, groups of lymphoid cells. These lymphoid cells are particularly numerous in association with the ducts of minor salivary glands, where they play an important role [2] in the local immunoresponse.

The secretory endpieces of human salivary glands consist of an aggregate of cells, called an *acinus* (or *alveolus*) if spherical, or a *tubule* if elongate, that opens, within the lobule, into a system of branching and coalescing ducts of progressively larger caliber. These intralobular ducts are particularly developed in the major glands, where they differentiate into the intercalated (or intercalary) and striated (or salivary) ducts. Outside the lobules the salivary ducts coalesce to give rise to the excretory (extralobular) ducts, which, at least in the major glands, unite to form the main excretory duct.

The finding that salivary glands consist of lobes and lobules separated by septa has been known since the seventeenth century. In fact Steno [3], adopting the terminology proposed by his erstwhile teacher Sylvius [4], described the parotid gland in 1661 as *glandula conglomerata* (made up of parts), in contradistinction to the adjacent lymph nodes, which he called *glandulae conglobatae* (fused masses). In mature glands the intraglandular connective tissue can be totally removed by maceration and digestion, enabling direct visualization under scanning electron microscopy (SEM) of the epithelial component of the glandular parenchyma. With these techniques the lobules of the parotid and of the other serous glands (Figs. 4-1 and 4-2) resemble a bunch of grapes, while the mucous glands (Figs. 4-3 and 4-4) resemble a loose skein of tubules. In general, our preparations are strikingly similar to the models constructed by Maziarski [5] at the very beginning of this century utilizing the wax reconstruction method.

SEM studies carried out with the corrosion cast method in rat salivary glands [6] have shown that capillaries of excretory (extralobular) ducts are derived from extralobular arteries and constitute a circulatory system separated from that of lobu-

Riva, A. and Motta, P.M., (eds.), Ultrastructure of the Extraparietal Glands of the Digestive Tract.
ISBN 978-1-4612-8211-2.

les. At the acinar level, in fact, there is a capillary plexus, originating from intralobular arterioles, that drains into portal veins, which in turn empty into the capillaries which invest the striated ducts.

Nerve fibers reach the parenchymal elements by following the arterial ramifications. They form epilemmal and hypolemmal nerve terminals [1, 7–9] containing a mixed population of small, clear or dense vesicles. In human major salivary glands, hypolemmal terminals, though present in the acini of all glands, are more numerous in the intercalated ducts, where they are particularly observed in the interfaces between the ductal and myoepithelial cells. Hypolemmal fibers are more frequently present between the acinar cells of human minor salivary glands [1, 10] than between acinar cells of human major glands.

For almost two centuries [11], it has been known that two kinds of saliva, albuminous (or serous) and mucous, are produced by human salivary glands. On the basis of these findings, Heidenhain [12] classified the salivary glands and their secretory cells as serous or mucous. Those glands that contained a mixture of the two kinds of secretory cells were considered to be mixed. With the advent [13–16] of the new staining techniques of histochemistry and transmission electron microscopy (TEM), attempts to better characterize the morphological and cytological features of secretory cells led to what Pinkstaff [9] has termed the *serous-seromucous controversy*, an issue that is still unsettled [17]. In this chapter we will use the terms *serous* and *mucous* in the general sense [10] and mainly with reference to the appearance that secretory cells have in sections stained with hematoxylin and eosin [17]. The serous acini usually have a very small lumen and consist of pyramidal cells endowed with a rather large euchromatic nucleus and basophilic cytoplasm. In contrast, mucous cells have a basally placed heterochromatic nucleus and clear cytoplasm; they are arranged in tubuli that often have a large lumen. In mixed glands, the serous cells may cap the blind ends of mucous tubules, forming demilunes [18]. TEM studies have shown that although mucous cells in all human glands exhibit the same ultrastructural appearance, serous cells have characteristics peculiar to each gland [1, 10, 19, 20].

Besides secretory cells in acini and tubuli, there are also the basally located myoepithelial cells (me). These cells, originally called *basket cells* [21], have a polygonal body and numerous tapering processes. The me cells seem to be particularly numerous in minor salivary glands and are also present in intercalated ducts and, with less frequency, in striated ducts. In the excretory ducts, there are small cuboidal cells, rather than true myoepithelial cells [22], resting upon the basal lamina. Another cell type sometimes encountered both in secretory endpieces and ducts of salivary glands of older people is represented by large eosinophilic cells known as *oncocytes* [23], which have also been studied with TEM [24].

Contrary to what has been observed in the salivary glands of certain rodents [9, 25], there is no sexual dimorphism in humans. It has been shown, however, that a sex-related difference exists [26] in the histochemical reactivity of some steroid dehydrogenases.

2. Parotid gland

Secretory cells of the parotid gland (Fig. 4-5) are pyramidal and have a height of 12–20 μm and a width at the cell base that, in cells distended by secretion, may surpass the cell's vertical axis. The cells are arranged around a small lumen that is continuous with long, intercellular canaliculi. Cell surfaces bordering the acinar lumen and the intercellular canaliculi are lined with microvilli. The canaliculi, which seem to have an important role in fluid and electrolyte secretion [27], are particularly evident in SEM fractured preparations, where they can be seen in both longitudinal and transverse sections (Fig. 4-6) or after the removal of adjacent cells. The luminal junctional complexes extend the length of the canaliculi, sealing these structures from the intercellular spaces. Below the complexes, lateral cell surfaces are folded and cytoplasmic extensions project into the intercellular spaces. As can be seen in SEM, these folds, both in fractured and macerated specimens, appear mostly as frondlike processes and are very numerous at the point where acinar cells come in contact with a myoepithelial cell. Toward the cell base, these processes become more numerous and constitute a series of short, parallel lamellae that correspond to the basolateral organelle-free folds, first noticed in TEM by Tandler [28] in serous cells of human submandibular glands. These processes are intimately interlocked with those of neighboring cells and greatly increase the basal surface area [1,28]. In addition, cytochemical studies [29] have demonstrated a strong reactivity for the transporting enzyme Ca^{2+}-ATPase at the level of the basolateral folds.

Nuclei are placed in the basal half of the cell, and their morphology varies according to their chromatin pattern. As a rule, the nuclei with condensed chromatin belong to cells completely filled with secretion [30]. On the contrary, cells rich in cytoplasmic organelles and containing only a number of secretory granules [30, 31] have a large nucleus. The latter shows a distinct fibrous lamina with patent nuclear pores and also has large nucleoli. Dense intranuclear inclusions formed by several laminae arranged concentrically [32] and contained within a vacuole may also be present. Inclusions of this kind have been detected in scattered nuclei of acinar cells of nearly all human salivary glands [1]. Cells of salivary glands also contain nuclear bodies similar to those observed in a variety of human and animal tissues [32].

In its most typical form, the rough endoplasmic reticulum (RER) consists of long, compact layers of parallel cisternae, preferentially located in the perinuclear region (Fig. 4-5). In the other portions of the cytoplasm, the RER is often segmented into smaller units. The elements of the RER contain a finely granular material of intermediate density. In cells showing signs of high secretory activity, the Golgi apparatus may form a crescentic area in the supranuclear zone. It consists of microvesicles, condensing vacuoles, and elongated saccules, often located near transitional elements of endoplasmic reticulum. Some saccules have a terminal distension, which contains a dense material similar to that forming the spherule of the secretory granules. Mitochondria are rather small and their matrix seems to be less dense than the ground cytoplasm.

Secretory granules (Fig. 4-5) have a diameter between 1 μm and 3μm, and in many cases they displace and sometimes compress most of the cytoplasmic organelles and even the nucleus. The granules show a bipartite structure [30], being composed (Figs. 4-5 and 4-12) of a matrix and of one or more, seldom two or three, dense bodies named *spherules* [1]. The density of the granule matrix varies greatly and seems to depend on the penetration rate of the fixative. The spherules, which have a more or less eccentric position (Fig. 4-12), have a diameter between 0.1 μm and 0.6 μm, and are sometimes noticeable even in SEM. Immunocytochemical investigations performed on the secretory granules of the parotid and submandibular glands of the monkey, *Macaca fascicularis*, [33] have shown that the spherules consist of or contain proline-rich proteins (p.r.p.), the most abundant salivary proteins. According to cytochemical studies, secretory granules of the human parotid gland are reactive for the enzymes peroxidase [34] and amylase [35, 36]. In addition, immunocytochemical investigations carried out at the light microscopic (LM) level [37] have shown that serous cells of human parotid and submandibular glands contain lactoferrin, lysozyme, and the secretory component of IgA; lysozyme is better demonstrated in cells of intercalated ducts.

Images of exocytosis are seldom seen in our specimens, since they have been obtained at surgery from patients subjected to general anesthesia involving preoperative medication with small doses of atropine. In a few cases, however, images related to exocytosis have even been seen under SEM. Moreover, undissolved dark spherules have been observed in the lumen among amorphous material of secretory origin.

3. Submandibular gland

The submandibular gland is mixed, consisting of serous and mucous secretory cells. Serous cells constitute 92% of the parenchyma [38] and are contained in either pure serous or mixed endpieces. The latter may consist of an admixture of the two kinds of secretory cells or of a mucous unit capped by serous cells, which constitute the demilunes. A few pure mucous units may also be encountered. In SEM the external configuration of macerated serous acini of the submandibular gland (Fig. 4-1) differs little from that observed in the parotid gland (Fig. 4-2). In contrast (Figs. 4-3 and 4-4), the mixed and mucous units are easily distinguishable, since they are made up of tubules. Even in TEM, the serous secretory cells of submandibular acini are fairly similar in morphology to those of the parotid gland. However, a quantitative ultrastructural study based on the pattern of organelle association in seven different types of human exocrine glands [39] has shown that the cytological organization of the submandibular gland is more similar to that of the pancreas than to that of parotid gland. A further differences between the serous secretory cells of the two glands resides in the higher degree of infolding [28] of the basolateral plasmalemma in the submandibular gland (Figs. 4-7 to 4-10). The intercellular canaliculi (Fig. 4-9) also seem to be longer and more numerous. The larger surface area of the basolateral aspects of acinar cells and the higher development of the intercellular canaliculi may be related to the finding [40] that it is the submandibular gland that produces the higher volume of saliva in the absence of stimulation.

The nucleus is generally ovoid, with dispersed chromatin and prominent nucleoli. Nuclear inclusions are less frequently observed than in secretory cells of other human major salivary glands. Mitochondria, the RER, and the Golgi apparatus are usually well represented, since the secretory granules do not accumulate to the extent noted in other exocrine cells. The secretory granules of the submandibular gland, which never reach the dimensions (up to 3 μm) sometimes observed in those of the parotid, exhibit a distinctive substructure (Fig. 4-14). They consist [41, 42] of a matrix containing a dense material that may be segregated into several foci as spherical and/or lenticular areas. Riva et al. [19] noted that, while the bulk of the dense components forms a crescent, the rest of it may extend as a thin layer that underlies the inner aspect of the granule membrane. This form of granule occurs only in the submandibular gland. Although the serous cells forming a single acinus are usually endowed with similar characteristics, there are acini composed of secretory cells whose secretory granules vary greatly [43] with respect to their substructures and to the density of their matrices. Sometimes, in fact, the dense material is so scarce and the matrix so uniformly pale that they, apart from their tendency to remain discrete, resemble mucous droplets. In other instances, their matrices are distinctly granular, or they have a bipartite structure and are indistinguishable from those of the parotid gland. In general, the secretory granules of human submandibular gland exhibit the same histochemical characteristic as do those of the parotid gland, and therefore they may possess the same constituents. There are, however, two notable exceptions. Both the reactivity of the granules for peroxidase and their stainability for mucins seem to be at variance in the two glands. Contrary to what is observed in the parotid gland, a substantial population of submandibular serous secretory granules remains unstained by techniques for the ultrastructural demonstration of mucins [17] and also seems to be devoid of peroxidase activity [34].

Immature granules have a flocculent matrix and may contain a single spherule, similar to those discernible in the saccules of the Golgi apparatus. The vacuoles are usually bigger than mature granules and have irregular outlines. Subsequently there is a gradual increase in the density of the matrix, a third component may appear, and the granule reaches its final size and shape (Fig. 4-14). As in parotid glands, images of direct extrusion of secretory material into the lumen are seldom encountered.

The distribution of cell organelles in serous cells can be seen to advantage in specimens treated with the metal-impregnation technique of Thiery and Rambourg [44], a staining method that has also proven its usefulness in the demonstration of myoepithelial cells and of the cytoarchitecture of acinar (Fig. 4-7) and ductal (Fig. 4-17) cells.

Unlike serous cells, which vary in structure among the different salivary glands, mucous cells, even by TEM, are identical, regardless of the gland in which they occur. Therefore, mucous cells will be described in the part of this chapter devoted to the sublingual gland, where we will also deal with myoepithelial cells.

4. Sublingual gland

In contrast to the other two major glands, the human sublingual gland is a compound organ consisting, in the majority of cases [1, 45–47] of a large segment, the major gland, and of a cluster of 8–30 small minor glands. The major and each of the minor glands have their own excretory duct and, even from the point of view of their embryological origin [48], they are independent organs. According to Imai et al. [46], each of the several constituents of the sublingual gland may differ histologically in that the minor glands usually contain more mucous cells than does the major sublingual gland. When observed in TEM [47], secretory cells from specimens obtained at surgery show extensive variations, which are probably dependent on the portion of the gland from which the samples were excised. Some specimens consist entirely of mucous cells, others are mixed, and still others are entirely serous. However, a number of specimens taken from different individuals are consistent in ultrastructure, being composed of mucous cells, which slightly outnumber serous cells. In this chapter, therefore, we base our description on such specimens, which correspond to the "classic" sublingual gland of histology textbooks and probably represent [47] the major sublingual gland.

Following maceration (Fig. 4-3), the mixed nature of the gland is evident in SEM because there are numerous tubules that, as stated before, are characteristic of mixed or mucous glands. Contrary to the report by Espinal et al. [49], we found it difficult to recognize serous and mucous cells in fractured specimens on the basis of the morphology of their secretory products. In our hands, both serous granules and mucous droplets look spherical, whereas the mucous droplets, as a rule, do not coalesce.

In standard TEM preparations (Fig. 4-11), serous cells look smaller than mucous cells and are usually distinguishable by the greater density of their secretory granules and by the fact that the latter never coalesce, whereas mucous droplets do. When grouped together in serous acini or mixed tubuli, serous cells are pyramidal, with truncated apices occasionally bulging into the lumen. Demilune cells (Fig. 4-11), which exhibit the same cytological features as the other serous cells, are flattened and cap the mucous cells on the blind end of the tubule. Demilune cells may reach the lumen by means of elongated processes, which often contain secretory granules. Long and tortuous intercellular canaliculi, which communicate with the lumen, lie between the various serous cells. The basolateral portions of serous cells possess a system of basal folds that is more developed than that observed in the parotid gland and is nearly as extensive as that of serous cells of the submandibular gland.

Nuclei are distended and oval in shape; their finely dispersed chromatin tends to be marginated along the inner aspect of the nuclear envelope, where nuclear pores are patent. Nucleoli are prominent, whereas nuclear bodies and nuclear inclusions seem to be less developed than in mucous cells. Secretory granules are discrete and consist of a pale and a dense component. The latter may assume the form of a crescent or may be fragmented into several foci of different shapes and densities. There are also secretory granules, where dark rods are surrounded by short ribbons of lesser density, all in a clear matrix [47]. In all cases, the boundaries between the granule components are ill defined. Secretory discharge occurs by a typical merocrine mechanism and is more frequently observed in specimens from patients operated on under local anesthesia.

The fine structure of mucous cells varies according to the stage of their secretory cycle. Cells with little mucus are smaller than "mature," cells i.e., cells that are filled with secretion. The "immature cells" are cuboidal, and, as usual, adjacent cells membranes are straight apically and possess junctional complexes and desmosomes; cells interdigitate extensively toward the cell base. Nuclei, which are large and ovoid, have finely dispersed chromatin. They are located in the basal half of the cell but are usually separated from the basal plasmalemma by a strip of cytoplasm containing well-developed RER. Besides nucleoli and nuclear bodies, the nucleoplasm may show large nuclear inclusions of the kind observed in other human salivary glands. Cellular organelles are well represented, their prominence being inversely proportional to the quantity of mucus present in the cytoplasm. Newly formed granules are small and often irregular in shape; they may contain myelin figures and groups of microvesicles.

Mature cells (Fig. 4-11) are conical and have a wide base, a convex apex, and rather straight outlines. Nuclei, which are small and indented, are often placed close to the basal plasmalemma. They are very heterochromatic, their high density masking the nucleoli. There are no nuclear inclusions. Ground cytoplasm is scarce and is located mainly around the nucleus and along the cellular borders. It also forms thin trabeculae that surround the individual droplets of the fused masses of mucus. There are only a few small mitochondria, and the scarce elements of RER and of the Golgi apparatus are interspersed among the mucous droplets. The latter, which are, as a rule, uniformly electron lucent, may contain a fibrillar material or even a small, dense spherule. In specimens treated with Thiery's silver proteinate method for the demonstration of glycoproteins [50], the main portion of the droplet reacts heavily with silver, while the spherule remains unstained. The mucous droplets are also reactive for acid mucosubstances [42]. Frequently the individual mucous droplets seem to fuse into large masses. The mucous droplets located in the apical portion of the cell often bulge into the lumen and presumably undergo exocytosis. This process can involve almost all the mucous droplets in a given cell, leaving the cell in a truncated condition.

In addition to the mucous droplets, a second secretory component (Figs. 4-11 and 4-13) originates in the Golgi apparatus. It consists of membrane-bound filamentous inclusions that react with silver proteinate but that show a different texture from mucous droplets [47]. They do not react when tested with methods for detecting acid mucins [17], nor do they contain blood-group substances (Fig. 4-13). These inclusions [1, 47, 51], which are present in most human mucous cells (with the exception of goblet cells), have been named *filamentous bodies* by Vidić and Tandler [52]. Most of these filamentous bodies fuse with the ordinary mucous droplets before exocytosis. Some filamentous bodies, however, seem to be discharged individually by a merocrine mechanism.

5. Myoepithelial cells

In conventionally stained sections, when viewed by TEM, myoepithelial cells are sometimes detected with difficulty [53]. In humans, contrary to some animals [9, 54, 55], myoepithelial cells are present in all major and minor salivary glands. The area around the nucleus houses organelles, whereas the processes contain cytofilaments. With various immunohistochemical and cytochemical techniques, it has been shown that these filaments consist either of actin [56], myosin [57], protein S100 [58], vimentin [59], or different kinds of cytokeratins [60].

A peculiar feature of myoepithelial cells, first noted by Tandler [61] in human salivary glands and later observed in various mammals [54], is the presence of a single cilium that extends across the intercellular space to contact the plasmalemma of an adjacent perenchymal cell. Although the significance of this finding is still unknown, Garrett and Emmelin [54] speculate that the cilium may have a role as a chemoreceptor or a mechanoreceptor.

In laboratory animals [62], it has been shown that adjacent myoepithelial cells are joined by gap junctions. Furthermore, Tandler [61], Nagashima and Ono [63], and Geiger et al. [60] have reported the occurrence of desmosomes in the human submandibular gland, both between myoepithelial cells and between myoepithelial cells and parenchymal cells. Such junctions [22] ensure that the pressure exerted by myoepithelial cell contraction will force salivary secretions into the ducts. Unlike the same cells in other animals, human salivary myoepithelial cells do not react with alkaline phosphatase but are positive for ATPase [54]. The latter finding has been confirmed by ultrastructural cytochemistry [29], with all surfaces of myoepithelial cells showing a strong Ca^{2+}-ATPase activity.

As shown in the rat by Brocco and Tamarin [53] and by Nagato et al. [64], myoepithelial cells of salivary glands become evident by SEM only after the maceration of the connective tissue and the enzymatic digestion of the basal lamina with collagenase. To remove these obscuring envelopes, we have employed the methods of both Evan et al. [65] and of Nagato et al. [64]. In addition to the hydrolysis with 8 N HCl used in both methods, we employed [66], alternatively or in sequence, a treatment with 6 N NaOH or with 30% KOH followed by a brief sonication of the specimens at 50 Hz [67]. In the acini, myoepithelial cells seem to be located mostly in the sulci between secretory cells. In fact, in specimens subjected to prolonged sonication, myoepithelial cells are removed from the acini, revealing grooves bordered by short folds that were formerly occupied by their processes. As seen in SEM (Figs. 4-15 and 4-16), myoepithelial cells have a roughly triangular body and long tapering processes. In rare instances, these processes are seen penetrating deeply into the acinus, while in other cases they pass from an acinus to a nearby one. Processes from one myoepithelial cell occasionally overlap other myoepithelial cells. In general, the processes do not ramify to the extent noticed in rat salivary glands [53, 64]. Moreover, in contrast to the findings of Nagato et al. [64], who observed that the shape of myoepithelial cells varies according to the salivary gland considered, the morphology of me cells located in the acini does not seem to vary from gland to gland.

In addition to myoepithelial cells (Fig. 4-15) with a smooth surface [68], there are cells that show parallel ridges, usually oriented at right angles to the long axis of the cellular body. These plicated cells may represent myoepithelial cells in the contracted state [66]. Although TEM studies [61, 69] have shown that myoepithelial cells are most numerous in intercalated ducts, the authors [53, 64, 68] who have studied myoepithelial cells in SEM failed to demonstrate them in this location [70]. We have been successful in visualizing myoepithelial cells on intercalated ducts only in specimens where, before digestion with collagenase, a brief exposure to 30% KOH followed the HCl (40') treatment. In keeping with the data obtained with TEM [54, 69, 70], myoepithelial cells are oriented along the length of the ducts, with long, ribbon-shaped processes extending to the acini.

6. Ducts

Ducts of human salivary glands consist of both intralobular and extralobular (excretory) ducts. Each segment of the duct system has its own unique character, which is more important in duct identification than is their histological location. For example, whereas excretory ducts may have an intralobular position [1, 9], ducts having the typical features of intralobular striated ducts may be found outside the lobules as well [10, 71]. In major salivary glands, excretory ducts join to form the main excretory duct.

Unlike serous cells, the ductal cells of the human major salivary glands do not possess features peculiar to a particular gland. For this reason, a single description will be given here. In general, however, intralobular ducts are less differentiated in the sublingual glands, where intralobular striated ducts are scanty [47]. In addition, the transition between secretory endpieces and intercalated ducts, and between the latter and striated ducts, seems to be more gradual in the parotid gland than in the submandibular gland.

6.1. *Intralobular ducts*

Each intralobular duct has two parts: the first is an intercalated duct and the second is a striated duct. In the proximal portion of the intercalated duct, the lumen is quite narrow and extends among the cells as prominent intercellular canaliculi. Cells of intercalated ducts, which are low cuboidal, have rather straight contours and possess small secretory granules [42, 72]. These granules react at the TEM level [17] in the parotid and the submandibular glands to methods for demonstrating acid and neutral mucins. In addition, cytochemical studies carried out with the LM have shown [73] that they are strongly reactive for lactoferrin, while staining for lysozyme and amylase is less intense, especially for the latter. More distally, the lumen becomes wider, and the membrane bordering the lumen is covered by a thick glycocalyx and exhibits a few microvilli. The cell contains a large nucleus, some lipid inclusions, dense bodies, and scanty cytoplasmic organelles. In SEM, intercalated ducts are difficult to identify. In the few instances where it has been possible to expose their lateral surfaces, the cells appear to be cuboidal and rather smooth. As already noted, the ducts are ensheathed by numerous myoepithelial cells.

Striated ducts are of larger caliber than are intercalated ducts. Their epithelium (Figs. 4-17 to 4-19) consists of principal cells, which are columnar, and of sparse, cuboidal basal cells. A few myoepithelial cells may be present. The luminal surfaces of principal cells show some cytoplasmic protrusions, similar to those present in excretory ducts, and many microvilli, which are especially evident in SEM. A thick glycocalyx is present [74] in specimens treated by methods for histochemical demonstration of glycoproteins. The nuclei are ovoid, with dispersed chromatin. In SEM preparations of fractured specimens fixed primarily with osmium tetroxide, these nuclei look like a tightly wound skein. Mitochondria

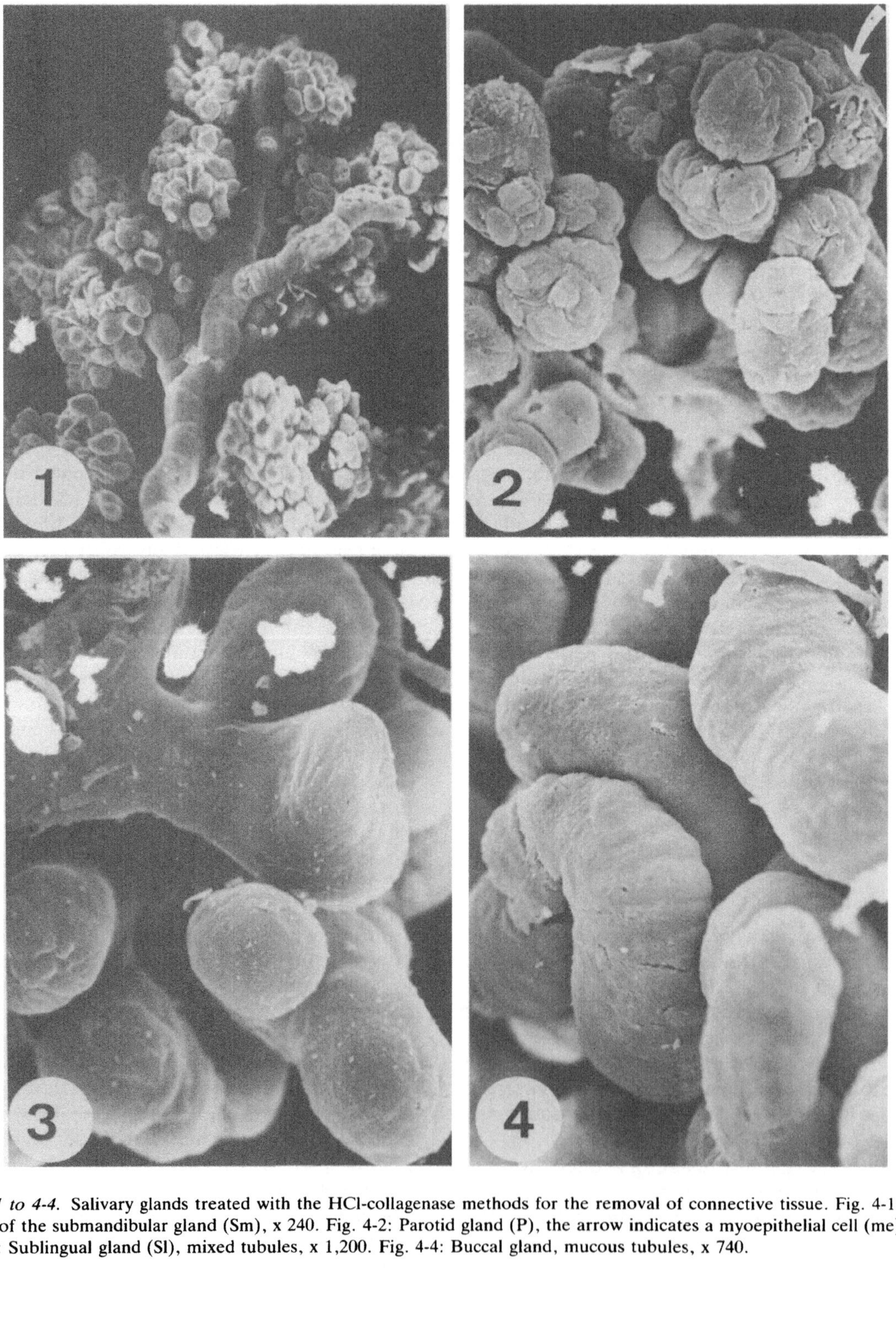

Figs. 4-1 to 4-4. Salivary glands treated with the HCl-collagenase methods for the removal of connective tissue. Fig. 4-1: Serous portion of the submandibular gland (Sm), x 240. Fig. 4-2: Parotid gland (P), the arrow indicates a myoepithelial cell (me), x 980. Fig. 4-3: Sublingual gland (Sl), mixed tubules, x 1,200. Fig. 4-4: Buccal gland, mucous tubules, x 740.

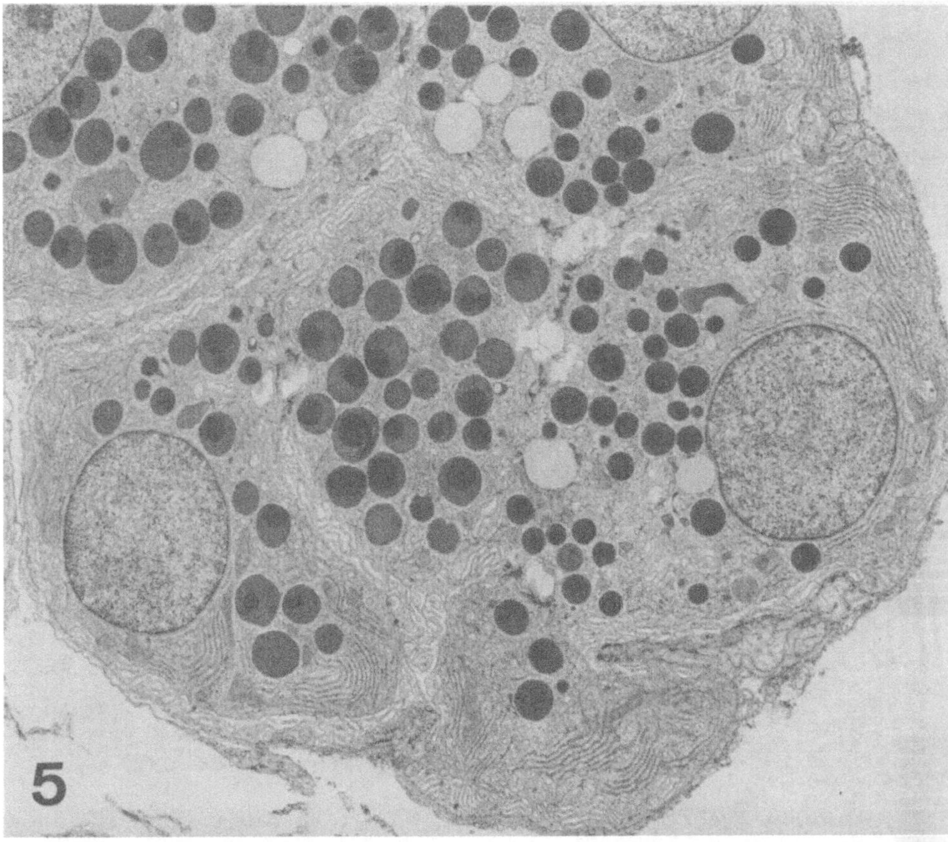

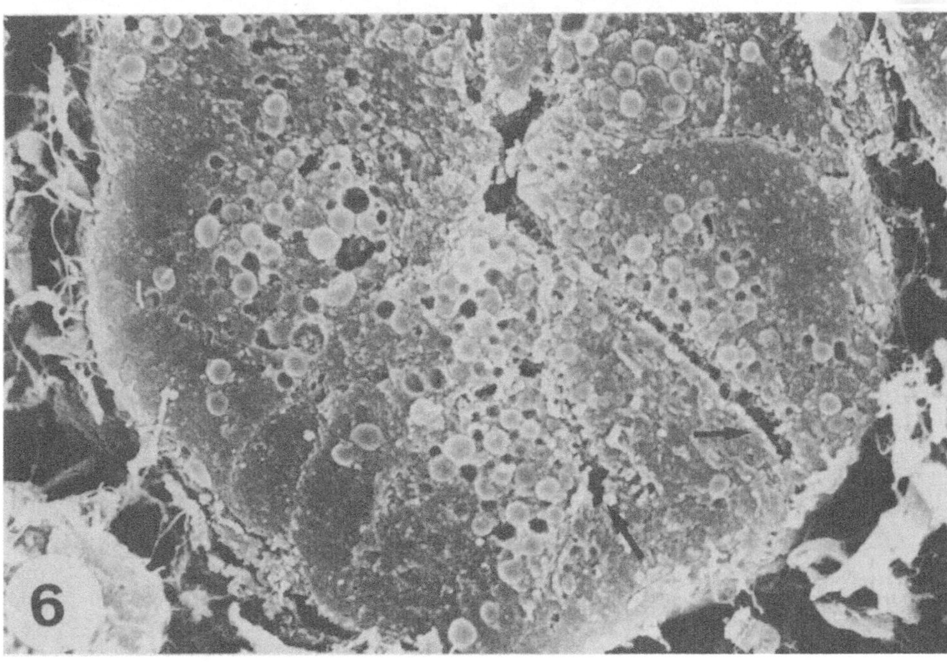

Fig. 4-5. Survey micrograph of an acinus. P, x 4,500.
Fig. 4-6. Fractured portion of an acinus. Note the spherical secretory granules and the intercellular canaliculi (arrows). P, x 6,000.

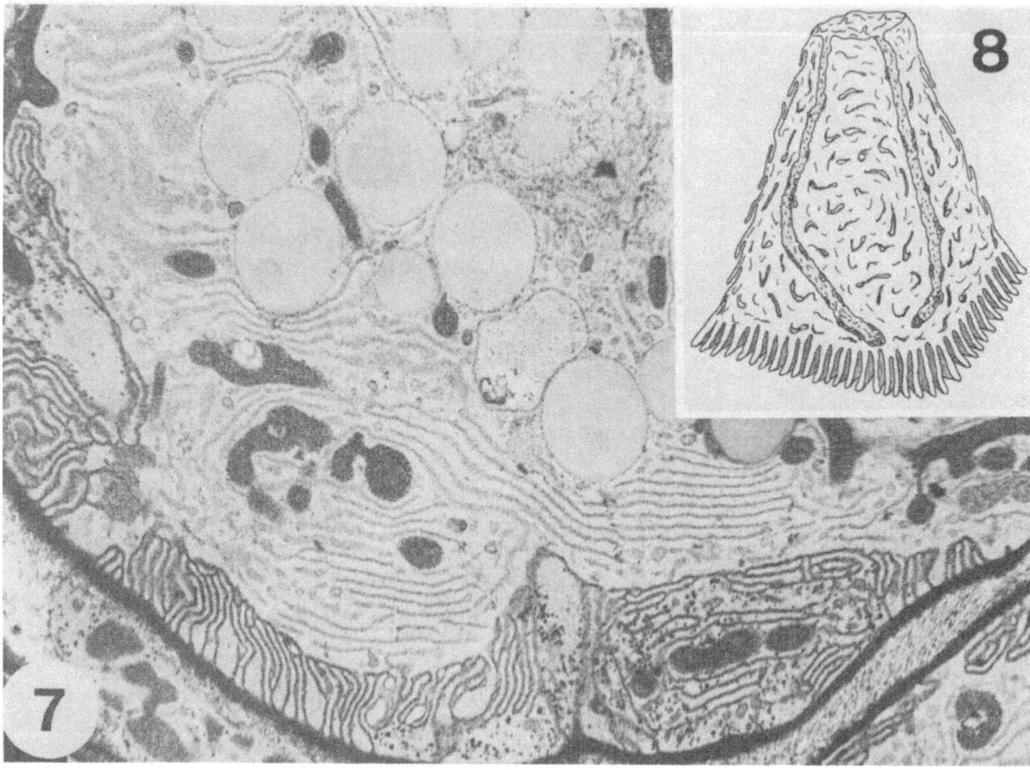

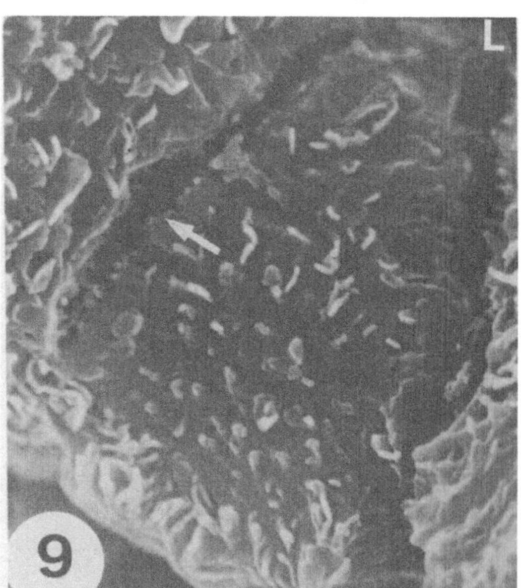

Fig. 4-7. Basal portion of a serous cell treated with the metal impregnation method of Thiery and Rambourg. Note the arrangement of the folds of the basal plasmalemma. Sm, x 8,800.

Fig. 4-8. The drawing summarizes the three-dimensional shape of a serous secretory cell of the human major salivary glands.

Fig. 4-9. External surfaces of serous cells exposed from the basal lamina to the lumen (L). The arrow indicates an intercellular canaliculus. Sm, x 15,000. From [66].

Fig. 4-10. Higher magnification of the basal folds of the plasmalemma. Sm, 37,000.

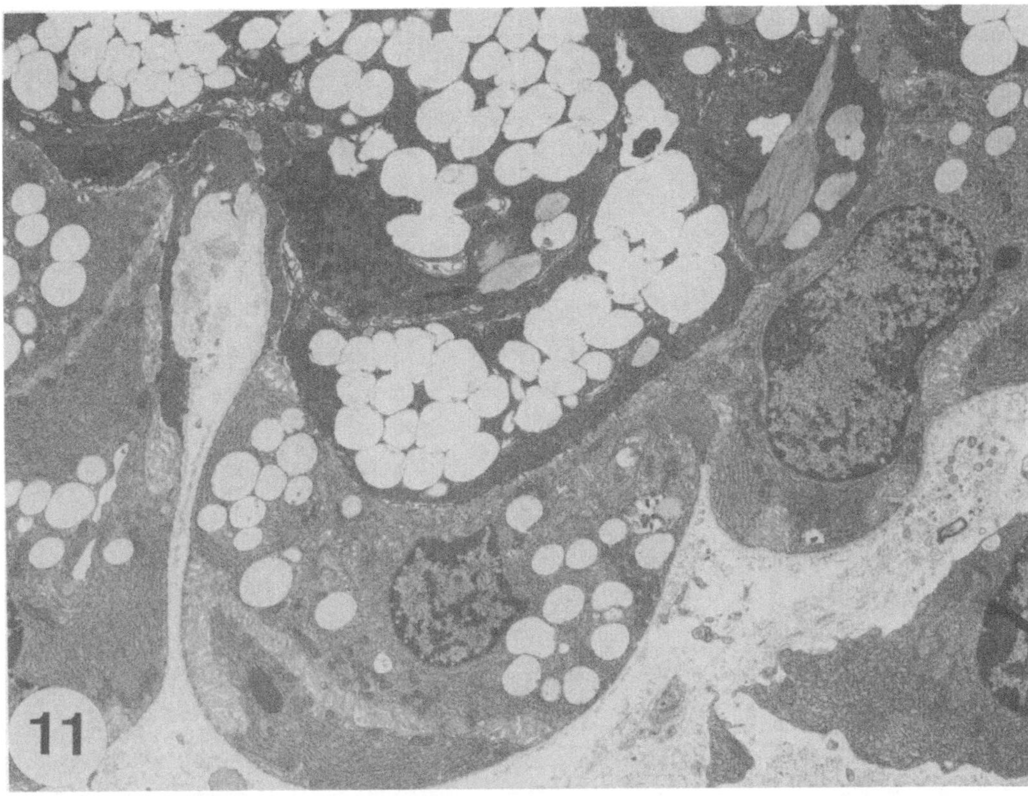

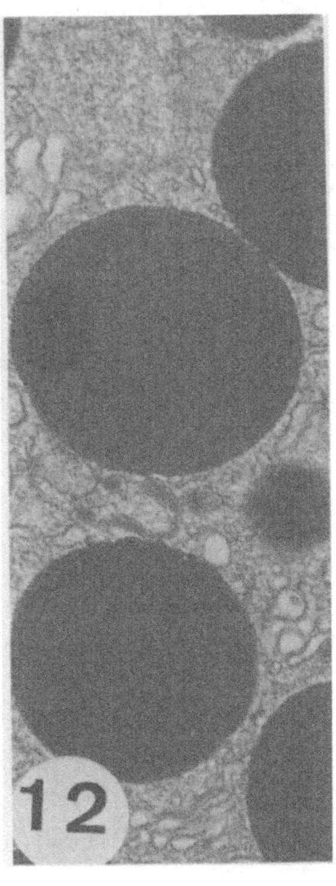

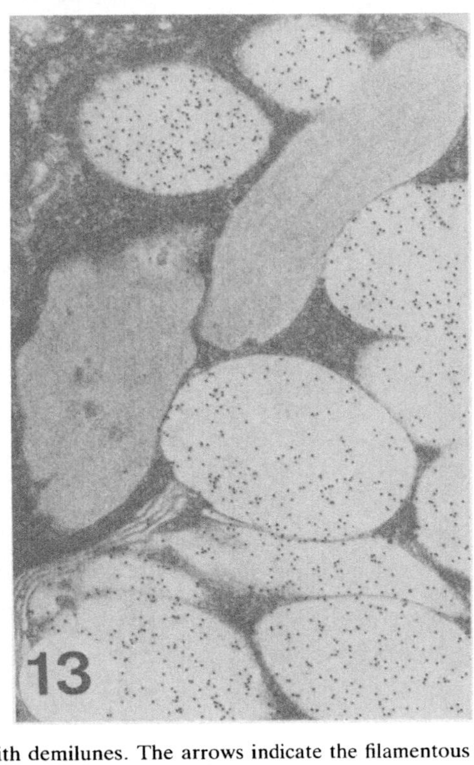

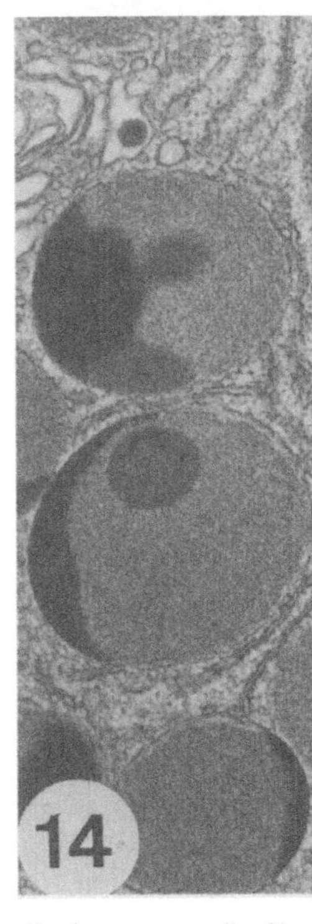

Fig. 4-11. Portion of a mixed acinus with demilunes. The arrows indicate the filamentous bodies in mucous cells. Sl, x 3,000.
Fig. 4-12. Secretory granules. P, x 21,000.
Fig. 4-13. Portion of a mucous cell treated with the immunogold method for the demonstration of blood group substances. Note that the filamentous bodies are unreactive. Sm, x 30,000.
Fig. 4-14. Secretory granules. Sm, x 21,000.

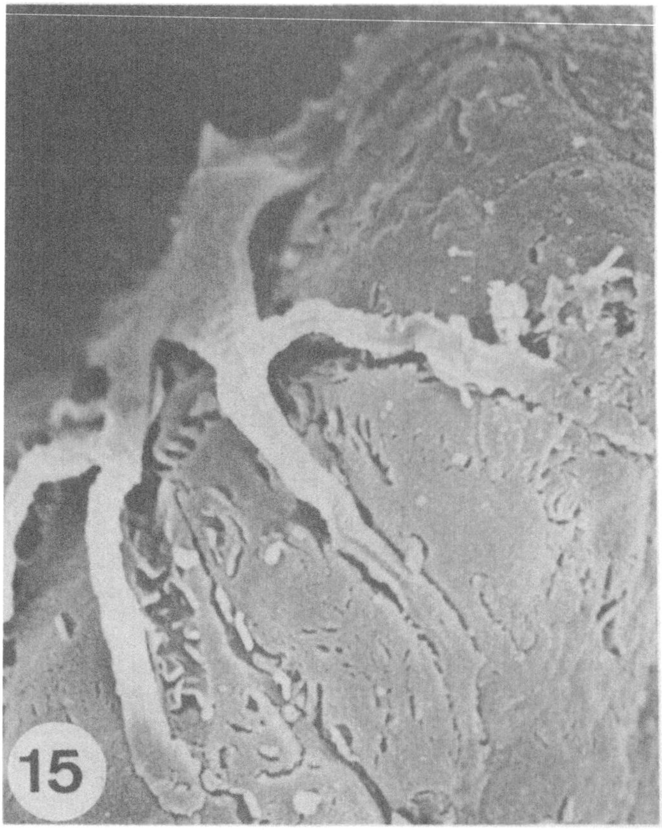

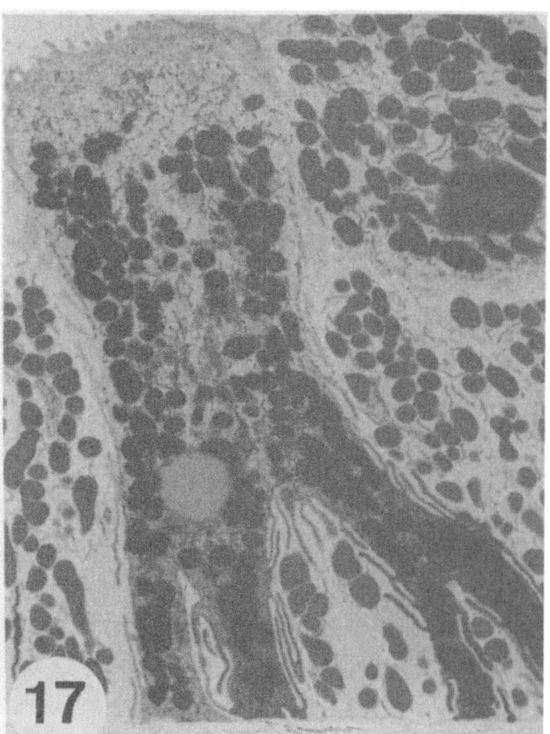
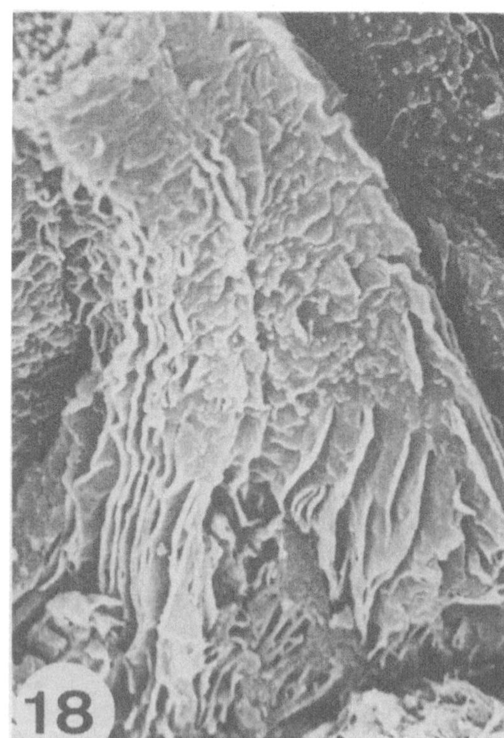

Fig. 4-15. Me cell on the surface of an acinus. HCl-collagenase method. P, x 9,500. From [66].
Fig. 4-16. Acini capped by me cells (arrows). HCl-collagenase method. Sm, x 2,000. From [66].
Fig. 4-17. Principal cells of a striated duct impregnated with the method of Thiery and Rambourg. Sm, x 4,000.
Fig. 4-18. Striated duct. Principal cells exposed from the basal lamina to the lumen. Sm, x 7,000.

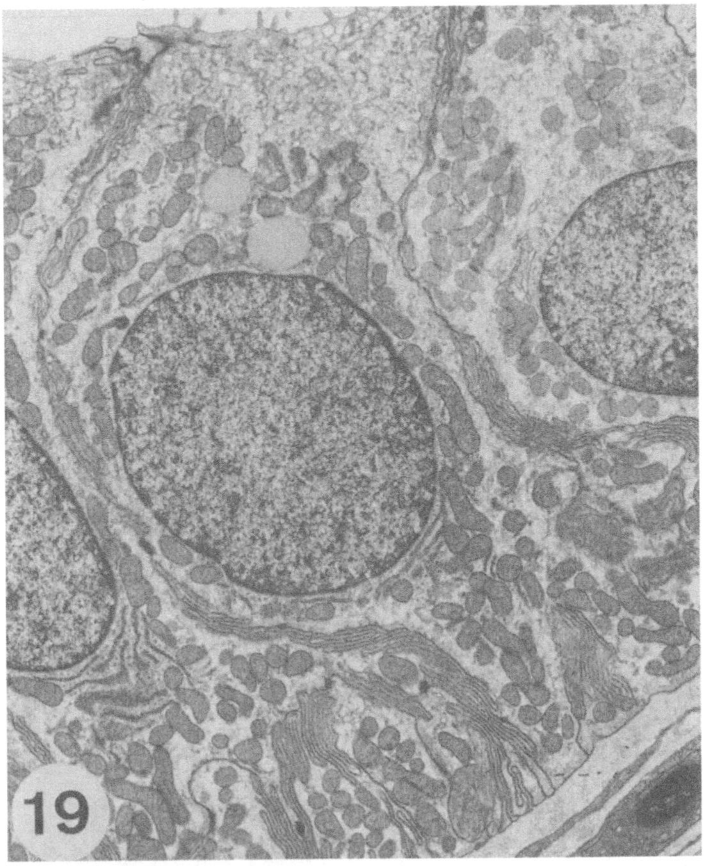
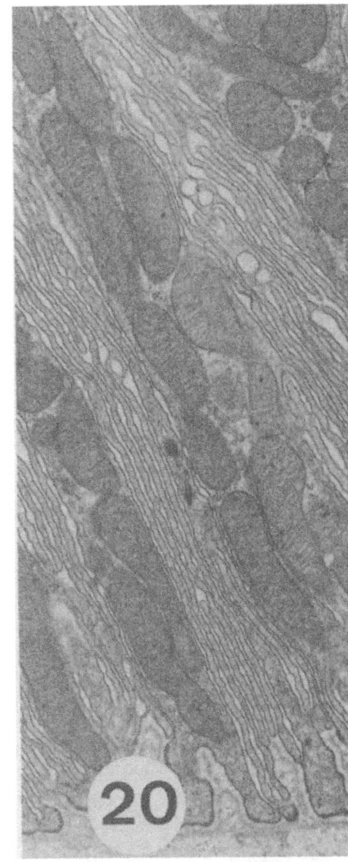
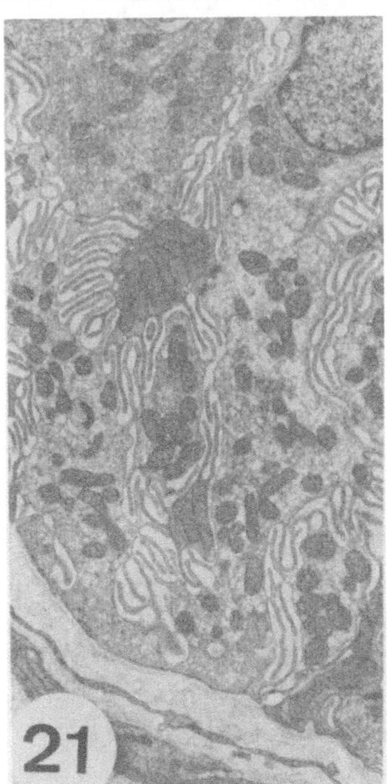
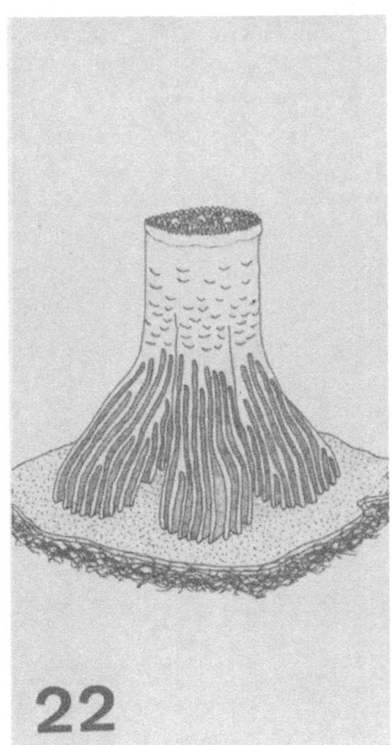
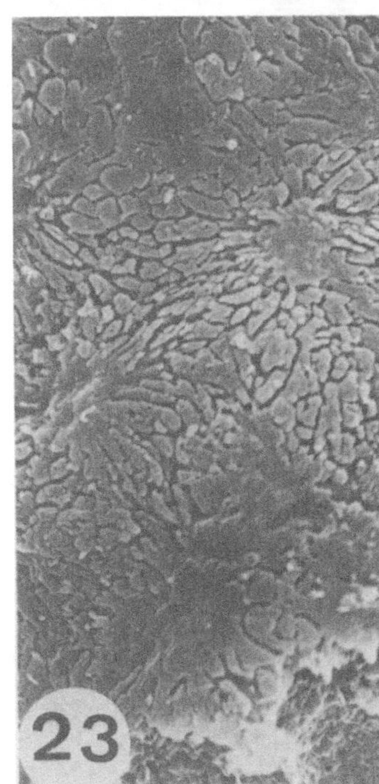

Fig. 4-19. Survey micrograph of the wall of a striated duct. Sm, x 3,100.
Fig. 4-20. Detail of the basal region of a principal cell. Striated duct. Sm, x 15,000.
Fig. 4-21. Principal cells cross sectioned near the cell base. Compare with Fig. 4-23. Striated duct. Sm, x 4,000. From [66].
Fig. 4-22. Diagrammatical representation of a principal cell. Striated duct.
Fig. 4-23. Base of a striated duct seen from below after treatment with the HCl-collagenase method. Sm, x 3,000.

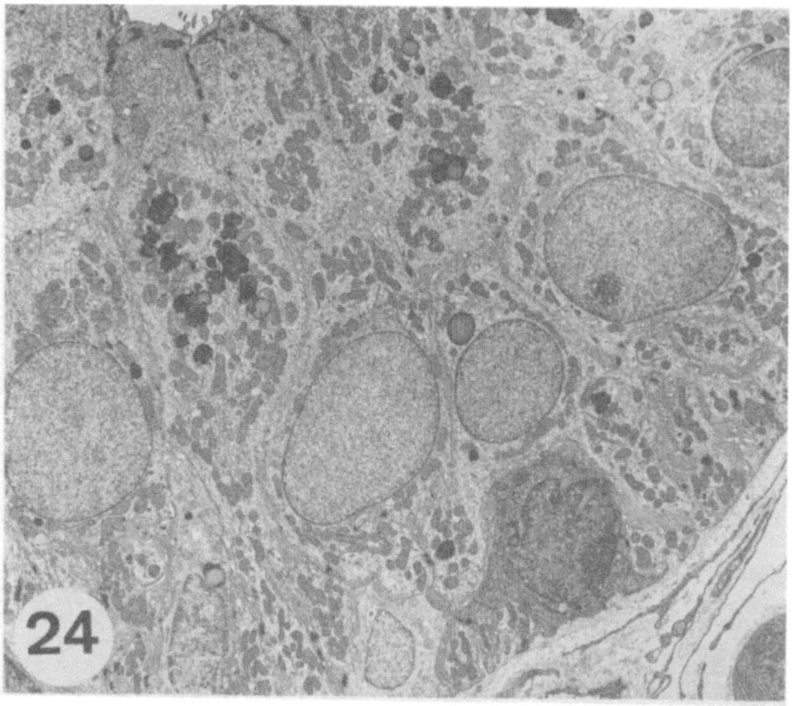

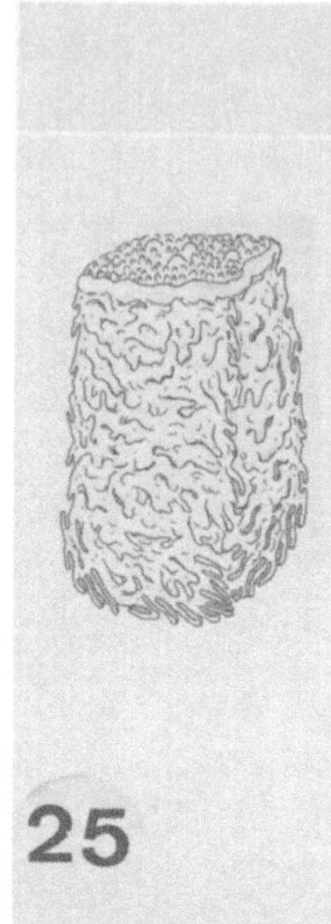

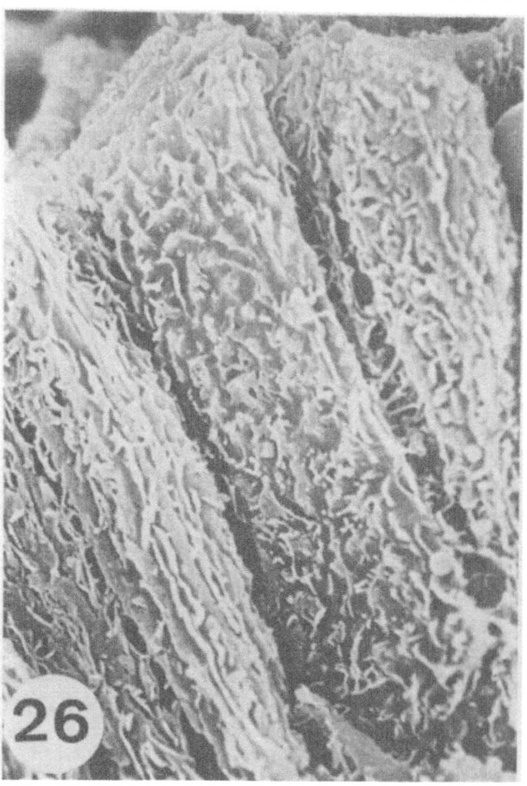

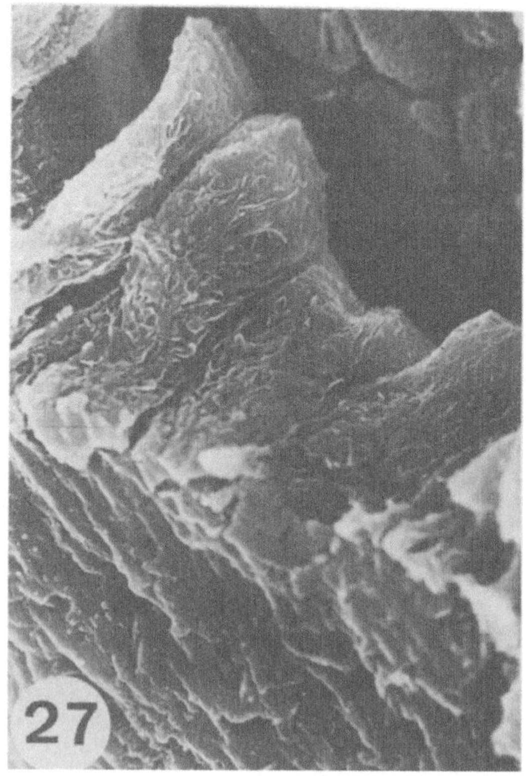

Fig. 4-24. Survey micrograph of the wall of an excretory duct. Sm x 3,300.

Fig. 4-25. The drawing illustrates the three-dimensional shape of a cell of an excretory duct of the human major salivary glands.

Fig. 4-26. Lateral surfaces of the cells of an excretory duct. Sm, x 7,100. From [66].

Fig. 4-27. Excretory duct treated with the HCl-collagenase method. P, x 4,700.

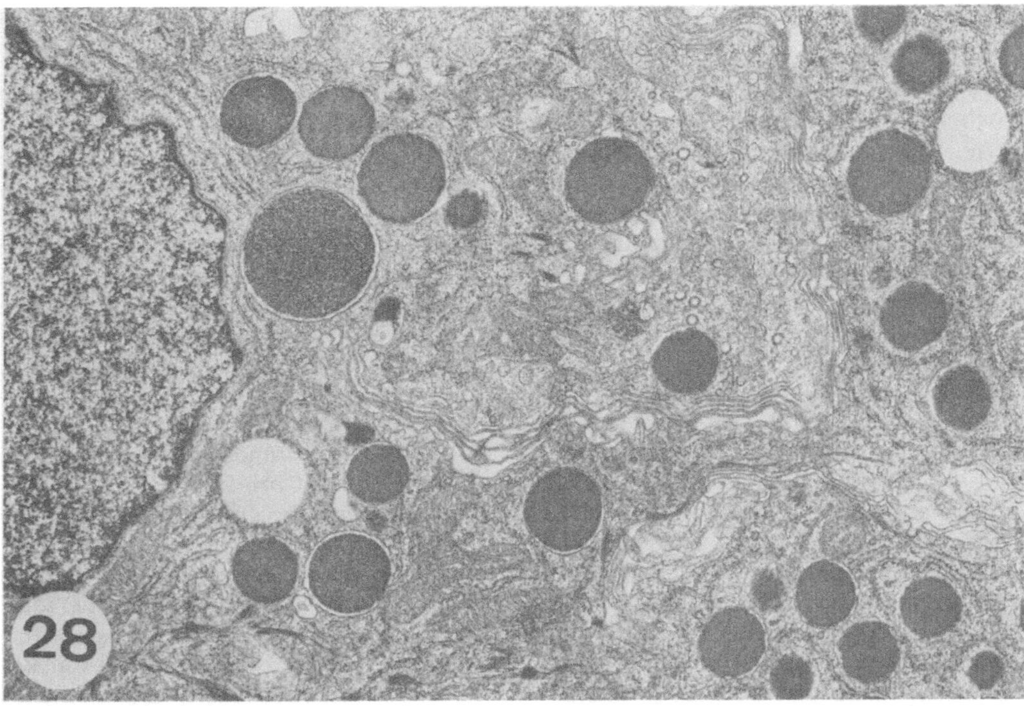

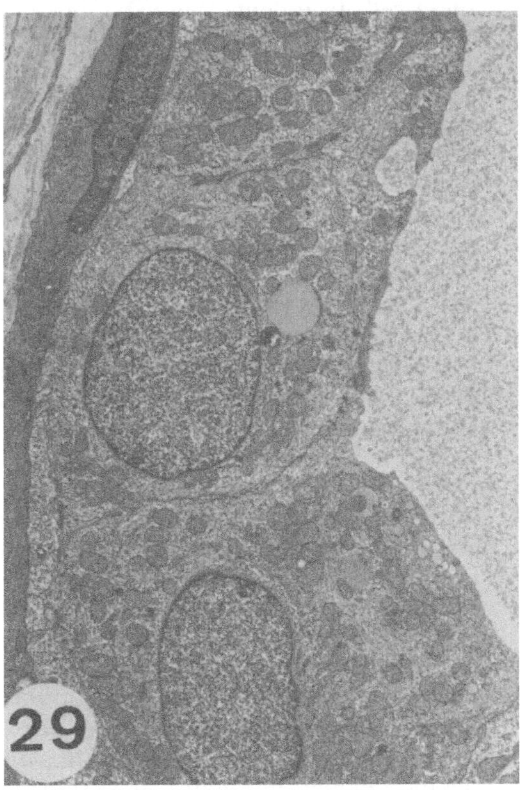

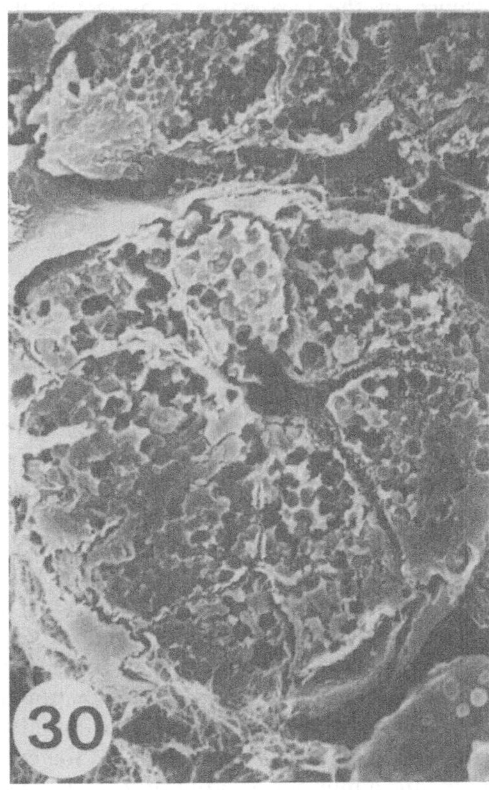

Fig. 4-28. Supranuclear cytoplasm of secretory cells of the deep posterior lingual (Ebner's) glands. x 14,000.
Fig. 4-29. Survey micrograph of an intercalated duct of the deep posterior lingual glands. x 4,500.
Fig. 4-30. Fractured tubule of a buccal gland. x 3,300.

are the most prominent organelles. They are rod shaped, with transversae cristae and a finely granular matrix. The mitochondria (Figs. 4-20 and 4-21) are vertically aligned between the vertical portions of the basal plasmalemma and are responsible for the striae observed with the LM; above the nucleus, mitochondria are disposed at random. RER is scarce and consists of a few cisternae dispersed among the mitochondria; free ribosomes are numerous. A few vesicles and dilated saccules of the Golgi apparatus are present near the nucleus. In this site, there are also large lipofuscins deposits, scattered glycogen particles, and a few peroxisomes [74]. The smooth reticulum is quite well represented and is in the form of irregular tubules, located just beneath the areas of the apical cytoplasm, that are loaded with small granules and vesicles. The latter are membrane bound and contain a material of variable electron density that is reactive to silver proteinate [17]. Immunohistochemical studies have shown the presence of the vasoactive substance kallikrein [75, 76]. Experimental studies carried out in laboratory animals [77] have demonstrated that they are heterogeneous in nature, because, together with vesicles that are excreted into the lumen, there are other vesicles that are involved in the reabsorption of proteins. Besides the cells described above, columnar cells, which are characterized by a very dark cytoplasm, are occasionally seen. While some authors [71, 78] state that they represent a second kind of ductal cell, we believe [1] that they are due to inadequate fixation. As noted by Tandler [79], these dark cells can be useful in revealing the shape of ductal cells. This can be seen to advantage in specimens treated with the metal-impregnation technique of Thiery and Rambourg [44] (Fig. 4-17). The complex cytoarchitecture of striated ducts becomes fully appreciable when the fractured epithelium is observed in SEM (Fig. 4-18). Just beneath the luminal plasmalemma, there is a smooth band that surrounds the cellular apex and probably corresponds to the portion of the upper lateral plasmalemma occupied by the junctional complexes. Below it there is a zone showing only some irregular folds, while, more basally, the cell body splits into a number of large processes. The surfaces of the latter (Fig. 4-18) bear a series of vertical grooves alternating with laminae that become wider as they approach the cell base and interfoliate with those belonging to processes of neighboring cells. The extensive interlocking of minor and major processes of ductal cells is also seen when the ducts are observed following removal of the extracellular material and digestion of the basal lamina. In these preparations (Fig. 4-23), the major processes exhibit a stellate outline when viewed from below, with laminae from one cell fitting into grooves on processes of adjacent cells.

As can be seen from Fig. 4-22, which summarizes the results reported above, the three-dimensional shape of principal cells of human striated ducts seems to be quite different from that represented in the model of cells of rat striated ducts made by Tamarin and Sreebny [80] and based on data obtained in TEM by the serial-section reconstruction method. Although a direct comparison is not possible, since, to our knowledge, there are no SEM data on the cytoarchitecture of these cells in rodents, the variations may reflect species-related differences. It seems, from TEM studies [79], that basal plications are far more complex in human salivary ducts than in other species that have been studied. The great number of desmosomes [74, 79] present between the plications is [9] a prominent feature of human glands. Our SEM findings also emphasize the differences existing between the cytoarchitecture of cells of salivary ducts and that of kidney convoluted tubules, to which they bear a general resemblance [77], a fact that was pointed out as early as 1909 [81]. Cells of striated ducts, as shown here (Fig. 4-22), look quite different from those of the convoluted tubules of the rat. Even the arrangement of the minor and major processes of the cells seems to be at variance in the two organs. While in the convoluted tubule the cell basal processes are oriented [65] perpendicular to the long axis of the tubule, in human salivary ducts they have a radial arrangement.

6.2. *Excretory ducts*

The excretory ducts comprise a hierarchy of ducts of a progressively larger caliber that run within the interlobular, intralobar, and interlobar septa. In the parotid, submandibular, and major sublingual glands, excretory ducts coalesce to give rise to a single main excretory duct. Interlobular ducts may maintain the morphological features of striated ducts, from which they originate, for some distance. When this happens, they are recognizable as interlobular only because they are surrounded by a pronounced layer of connective tissue containing relatively large interlobular blood vessels. More distally (Fig. 4-24), the ducts become larger in diameter and, as the number of the basal cells increases, the plications of the basal plasmalemma of the principal cells become progressively less prominent. Instead, there is an extensive infolding of the lateral plasma membranes. The cytoplasm of the principal cells (Fig. 4-24) contains many randomly disposed mitochondria. Above the nucleus there are large lipofuscin granules, while in the most apical region a number of small granules and vesicles are present that are similar to those in cells of striated ducts. Both in TEM and SEM, the principal cells of these ducts show large protrusions which bulge into the lumen and may represent a kind of apocrine secretion [10].

SEM (Figs. 4-26 and 4-27) corroborates the differences existing in the cytoarchitecture between principal cells of excretory ducts and those of striated ducts [10, 82]. Below the smooth band corresponding to the area occupied by the junctional complex, the lateral cell surfaces are completely covered by a great number of short, irregular processes oriented at random. In addition, in contrast to the cells of striated ducts, the cell body does not give rise to major basal processes. When it rests upon a basal cell, its convex base is inserted in the concave upper surface of the latter. The extralobular ducts differ from striated ducts, even when they are observed from their base following the removal of extracellular material (Fig. 4-27). They do not show the radially oriented processes seen in cells of striated ducts but have only a few short ridges, which may correspond to the boundaries of basal cells. The morphological variations reported here (Fig. 4-25) may have a functional significance. Owing to the small caliber of striated ducts, micropuncture studies that test the ionic composition of saliva are performed solely in excretory ducts [83, 84]. In light of our findings, it seems hazardous to assume [83, 84] that such data are valid for striated ducts as well.

6.3. Main excretory ducts

The main excretory ducts (MED) have received little attention from electron microscopists [9, 25]. To our knowledge, the normal ducts of major salivary glands have been extensively studied only in the submandibular glands of rats [85–87], cats [88], rabbits [89], and humans [90], and in the parotid glands of rats [87]. We have also examined the MED (Stensen's duct) of the human parotid gland by electron microscopy. The human ducts are lined by a stratified columnar epithelium that exhibits the same morphological characters in both human glands. It consists of principal cells resting on a continuous row of cuboidal basal cells. A few goblet and ciliated cells [90], which greatly increase in number in inflammatory states [91], are also seen. The principal cells are columnar and possess, on their lateral aspects, a series of short folds that are nearly as numerous as those described in cells of excretory ducts. The nucleus, which is often indented, has a central position and its chromatin is marginated. The small Golgi apparatus shares the supranuclear cytoplasm with large lipofuscin granules. RER is scarce, whereas free ribosomes, either isolated or clustered in rosettes, are frequent. Numerous mitochondria are dispersed throughout the cytoplasm. Near the cell apex, there is a network of tubules belonging to the smooth endoplasmic reticulum. Just beneath the plasmalemma bordering the lumen, there is also a number of small granules and vesicles that are similar to those observed in striated and excretory ducts. These vesicles occasionally are fused with both the apical plasmalemma and the membrane bordering the intercellular spaces. While no significant spatial relationship is noted between the elements of the Golgi apparatus and these vesicles, the latter are often intermingled with the tubules of the smooth reticulum. The luminal surfaces is provided with short microvilli and is covered by a thick glycocalyx. There also are apical protrusions of the kind noted in excretory ducts. The basal cells, which form a continuous layer, are linked by desmosomes to the principal cells and interdigitate with the latter by means of thin cytoplasmic expansions. Their basal surfaces are dotted with hemidesmosomes and show frequent indentations that are closely followed by the underlying basal lamina. Although, from a morphological point of view, these cells are indistinguishable from basal cells of striated and excretory ducts, their functional role might be different, since some immunocytochemical differences have been reported [92].

At their opening, the lining of the main excretory ducts becomes stratified and assumes the characteristics of the oral epithelium.

7. Labial salivary glands

The glands of the lips are predominantly mucous; there are, however, some tubuli capped by serous demilunes [1]. In general, their mucous and serous cells look similar to those of the sublingual gland. Mucous cells are seen in all stages of their secretory cycle [93]. Cells that have released their secretion exhibit well developed cytoplasmic organelles, including an extensive RER. In about half of these cells [94], which are at the beginning of their secretory cycle, nuclei contain membrane-bound inclusions of the kind described in major salivary glands.

Intercalated ducts [95] vary greatly in caliber and length, and their epithelial lining may be flattened or low columnar. In many cases ductal cells contain a large number of mitochondria, but, as a rule, their basolateral plasmalemma does not possess the vertical plications typical of cells of striated ducts. Excretory ducts have basal and principal cells, similar to those present in the analogous ducts of human major salivary glands.

In keeping with the presence of lymphoid tissue associated with the ducts [96], the saliva produced by minor salivary and labial glands, in particular [97], has a greater concentration of IgA than that found in the saliva of human major salivary glands. In addition, the secretory enzyme, lysozme, has been detected with the immunocytochemical method [98, 99], both in the demilunar and the intercalated duct cells of the gland.

8. Buccal glands

The buccal glands, which also comprise the larger malar glands located near the stoma of Stensen's duct, are mixed. Though our TEM findings confirm that they closely resemble labial glands [1, 100], it seems that they possess a larger population of serous cells. In addition to the demilunes, these serous cells may form a more substantial part of the mixed unit or may be intermingled between mucous cells. In fractured specimens, as clearly seen in SEM (Fig. 4-30), long intercellular canaliculi extend from the central lumen. The tubular nature of these glands is readily observed by SEM after the connective tissue is removed (Fig. 4-4).

9. Palatine and glossopalatine glands

The palatine and glossopalatine glands are mucous glands formed by secretory cells, similar to those found in the other salivary glands. From studies carried out with the LM [101], it seems, however, that their ductal system includes many highly convoluted and dilated intercalated ducts.

10. Anterior lingual glands

Anterior lingual glands (of Blandin and Nuhn) are mixed glands located in the ventral part of the tongue on either side of frenulum [1]. Their excretory ducts (five to eight in number) open under the tongue medially to the anterior part of the plica fimbriata. Their serous cells differ from those of the sublingual glands in that they possess only a limited number [1] of basolateral folds. Their secretory granules, however, resemble those observed in serous cells of the sublingual and of the other minor salivary glands. As usual, mucous cells do not exhibit distinctive ultrastructural characteristics.

11. Posterior deep lingual glands

The posterior deep lingual salivary glands (of von Ebner) are tubuloacinar serous glands situated deep beneath the vallate papillae on the dorsum of the tongue and, to a lesser extent, beneath the foliate papillae on the lateral borders of the organ. Their ducts open into the sulci of the vallate papillae or into the furrows between the foliate papillae. The shape of the secretory cells [20] varies from pyramidal to cylindrical according to their location in an acinus or in a tubule. Their lateral borders exhibit thin interdigitating folds and, unlike the situation in the major salivary glands, their basolateral plasma membranes are usually devoid of folds [20]. The lumen of the secretory units is bordered by microvilli and is continuous with intercellular canaliculi. Secretory granules (Fig. 4-28), although sometimes very numerous, never compress the nucleus and the other cellular organelles to the extent noticed in serous cells of the parotid gland. The granules range in size from 0.4 μm to 1.7 μm and, in the form that may be regarded as typical [20], they consist of an indented, dense core and of a lucent marginal zone filled with strands of dense material. RER is well developed, and at the cell base it forms regular arrays of parallel cisternae. The Golgi is usually extensive and is located above the nucleus. Of particular interest is the finding obtained with both biochemical and cytochemical methods [102] that their secretory granules contain a potent lipase. Following exocytosis of the granules, the enzyme is swallowed with the saliva and presumably acts [102] in the acid environment of the stomach. In addition, studies carried out with the LM have shown that the secretory cells also contain lysozyme [99] and peroxidase [103]. The epithelial lining of the ducts which follow the secretory units varies from cuboidal to low columnar. The ductal cells, which are often ensheathed by myoepithelial cells, may contain a number of small secretory granules in their apical cytoplasm [20]. Though striated ducts are absent, some ductal cells may be rich in mitochondria (Fig. 4-29) and may show a degree of lateral folding. The excretory ducts [104] are lined by a bistratified epithelium consisting of small, cuboidal basal cells and of tall principal cells endowed with small secretory granules. In the MED [104], there are several rows of basal cells, while, among the principal cells, groups of goblet and ciliated cells are encountered.

12. Posterior superficial lingual glands

The posterior superficial lingual glands (of Weber) are the most numerous [1] among the lingual glands. They are located on the lateral margins of the tongue in front of the foliate papillae and in the pharyngeal portion of the organ, posterior to the vallate papillae. Anteriorly, they may extend far enough to intermingle with the deep posterior serous glands [1]. The glands are associated with the lingual tonsil, and their ducts open into the crypts or between the tonsillar follicles. Although they are mostly composed of mucous tubuli, there are also a few mixed units. Ultrastructurally their secretory cells do not differ from those of the other mixed minor salivary glands.

13. Concluding remarks

The use of the new techniques of fracture and maceration in SEM allows the direct visualization of the parenchymal components of the human salivary glands. There are, however, some interpretation problems, depending mainly on the relative crudeness of the preparative methods, which need further refinements. Nevertheless, SEM findings, especially when combined with data obtained with TEM, reveal the complexity of the cytoarchitecture of the glandular cells, as well as their interrelationships.

The most interesting results are a better definition of the three-dimensional shape of acinar and ductal cells, with the possible consequence of a better understanding of their relevant function and role in the mechanisms of saliva production.

From our studies carried out on human material, it also emerges that it is not safe to extrapolate to humans observations made on other mammals. Salivary glands exhibit a great species variability, which is surely related to their histophysiology.

Acknowledgments

Financial support for this work came from the C.N.R. and the M.P.I. We are very grateful to Dr. Bernard Tandler of Case Western Reserve University of Cleveland, Ohio for reading the text and for helpful criticism. Technical assistance was provided by Mrs. S. Bernardini Foddis and Messrs. A. Cadau and A. Locci. Mr. Alessandro Cadau drew figures 4-8, 4-22, and 4-25.

References

1. Tandler B, Riva A: Salivary glands. In: Human Oral Embryology and Histology. IA Mjör, O Fejerskov (eds).

Copenhagen: Munksgaard, p 243–284, 1986.
2. Nair PNR, Schroeder HE: Duct associated lymphoid tissue (DALT) of minor salivary glands and mucosal immunity. *Immunology* 57: 171–180, 1986.
3. Steno N (Stensen N): Observationes anatomicae de glandulis oris & novis earundem vasis. Leyden, 1661.
4. Sylvius (de la Boe, F): De lienis et glandularum usu. Leyden, 1660.
5. Maziarski S: Über den Bau der Speicheldrüsen. *Bull L Acad Sci Cracovie* 41: 22S, 1900.
6. Ohtani O, Ohtsuda A, Lipsett Y, Gannon B: The microvasculature of rat salivary glands. A scanning electron microscopic study. *Acta Anat* 115: 345–356, 1983.
7. Emmelin N: Nervous control of salivary glands. In: Handbook of Physiology, Sect 6, Alimentary Canal II. CF Code, W Heidel (eds), Baltimore: Williams and Wilkins, 595–632, 1967.
8. Garrett JR: Innervation of salivary glands: neurohistological and functional aspects. In: The Salivary system. LM Sreebny (ed), Boca Raton, FL: CRC Press, 69–93, 1987.
9. Pinkstaff CA: The cytology of salivary glands. *Int Rev Cytol* 63: 141–261, 1980.
10. Riva A, Tandler B: Salivari ghiandole (anatomia). In: Enciclopedia Medica Italiana, 2nd ed, Vol 13. Firenze, Italy: USES, 1729–1744, 1986.
11. Garrett JR: Changing attitudes on salivary secretion. A short history on spit. *Proc Roy Soc Med* 68: 553–560, 1975.
12. Heidenhain A: Über die Acinösen Drüsen der Schleimhäute, insbesondere der Nasenschleimhäut. Doctoral thesis, Breslav, 1870.
13. Stormont DL: The salivary glands. In: Special Cytology, 2nd ed, Vol 1. New York: Hoeber, 152–196, 1932.
14. Aureli G, Rizzotti M, Ferri G, Castellani A: Studio dei sialopolisaccaridi nelle ghiandole retrolinguale e sottomandibolare di *Erinaceus europaeus*. Analisi biochimica e caratteri istochimici differenziali. *Acta Histochem* (Jena) 16: 109–115, 1963.
15. Munger BL: Histochemical studies on seromucous-and mucous-secreting cells of human salivary glands. *Am J Anat* 115: 411–429, 1964.
16. Shackleford JM, Wilborn WH: Structural and histochemical diversity in mammalian salivary glands. *Ala J Med Sci* 5: 180–203, 1968.
17. Harrison JD, Auger W, Paterson KL, Rawley PSA: Mucin histochemistry of submandibular and parotid salivary glands of man: light and electron microscopy. *Histochem J* 19: 555–564, 1987.
18. Giannuzzi G: Von den Folgen des beschleunigten Blutstroms fur die Absonderung des Speichels. *Ber Kgl Sachs. Ges Wiss Math-Phys KL* 17: 68–84, 1865.
19. Riva A, Motta G, Riva Testa F: Ultrastructural diversity in secretory granules of human major salivary glands. *Am J Anat* 139: 293–298, 1974.
20. Testa Riva F, Cossu M, Lantini MS, Riva A: Fine structure of human deep posterior lingual glands. J Anat 142: 103–115, 1985.
21. Krause W: Über die Drüsennerven. II. Die Nervenendi gung in den Drüsen. *Z Rat Med* 23: 46–62, 1865.
22. Tandler B: Structure of the human parotid and submandibular gland. In: The Salivary System. LM Sreebny (ed), Boca Raton, FL: CRC Press, 21–41, 1987.
23. Shaffer J: Beiträge zur Histologie menschlicher Organe. IV. Zunge. V. Mundhöhle-Schlundkopf. VI. Oesophagus. VII. Kardia. *S Akad Wiss Wien, Math Nat Kl* 106: 353–455, 1897.
24. Tandler B; Fine structure of oncocytes in human salivary glands. *Virchows Arch* (Pathol Anat) 341: 317–326, 1966.
25. Young JA, van Lennep EW: The morphology of salivary glands. London: Academic Press, 1987.
26. Sirigu P, Cossu M, Perra MT, Puxeddu P: Histochemistry of the 3β-hydroxysteroid, 17β-hydroxysteroid and 3α-hydroxysteroid dehydrogenases in human salivary glands. *Archs Oral Biol* 27: 547–551, 1982.
27. Sakai T: Intercellular canaliculi of salivary glands serve as a device for secretion of electrolytes and fluids. *Biomed Res* 5: 433–440, 1984.
28. Tandler B: Ultrastructure of the human submaxillary gland. I. Architecture and histological relationships of the secretory cells. *Am J Anat* 111: 287–307, 1962.
29. Cossu M, Lantini MS, Puxeddu P, Riva A: Cytochemical localization of ouabain-sensitive K^+-dependent p-nitrophenylphosphatase and Ca^{++}-stimulated adenosine triphosphatase activities in human parotid and submandibular glands. *Histochemistry* 81: 221–225, 1984.
30. Riva A, Riva Testa F: Fine structure of acinar cells of human parotid gland. *Anat Rec* 176: 149–165, 1973.
31. Seifert G, Donath K: Die Morphologie der Speicheldrüsenerkrankungen. *Arch Oto-Rhino-Laryng* 213: 111–208, 1976.
32. Motta G, Riva A: Contributo alla conoscenza della ultrastruttura delle ghiandole salivari. Le inclusioni nucleari nelle cellule della ghiandola parotide. *Oto Rino Laring It* 38: 247–265, 1969.
33. Kousvelari EE, Oppenheim FJ, Cutler LS: Ultrastructural localization of salivary acidic proline-rich proteins from *Macaca fascicularis*. *J Histochem Cytochem* 30: 274–278, 1982.
34. Riva A, Puxeddu P, Del Fiacco M, Testa Riva F: Ultrastructural localization of endogenous peroxidase in human parotid and submandibular glands. *J Anat* 127: 181–191, 1978.
35. Zimmer KP, Caselitz S, Seifert G, Grenner J: Immunoelectron-microscopy of amylase in the human parotid gland. Ultrastructural localization by use of both the protein A-gold and the biotin-avidin technique. *Virchows Arch* (Pathol Anat) 404: 187–196, 1984.
36. Machino M, Morioka H, Tachibana M: Amylase and lysozyme differentiate their localization within the serous secretory granule of the human salivary glands. *Acta histochem cytochem* 19: 329–332, 1986.
37. Korsrud FR, Brandtzaeg P: Characterization of epithelial elements in human major salivary glands by functional markers: localization of amylase, lactoferrin, lysozyme, secretory component and secretory immunoglobulins by paired immunofluorescence staining. *J Histochem Cytochem* 30: 657–666, 1982.
38. Scott J: The proportional volume of mucous acinar cells in normal human submandibular glands. *Archs Oral Biol*

24: 479–481, 1979.
39. Diaz G, Cossu M, Riva A: Quantitative ultrastructural approach to the study of the spatial relationship among cell organelles. I. Cytological organization of human exocrine epithelia. *J Electr Micr Techn* 7: 167–175, 1987.
40. Schneyer LH, Levin LK: Rate of secretion by individual salivary gland pairs of man under conditions of reduced exogenous stimulation. *J Appl Physiol* 7: 508–512, 1955.
41. Tandler B, Erlandson RA: Ultrastructure of the human submaxillary gland. IV. Serous granules. *Am J Anat* 135: 419–439, 1972.
42. Harrison JD, Auger DW, Badir, MS, Paterson KL: Ultrastructural morphology of secretory granules of submandibular and parotid salivary glands of man. *Archs Oral Biol* 32: 229–234, 1987.
43. Testa Riva F: Ultrastructure of the human submandibular gland. *Submicr Cytol* 9: 251–266, 1977.
44. Thiery S, Rambourg A: A new staining technique for studying thick sections in the electron microscope. *J Microscopie Biol Cell* 26: 103–106, 1976.
45. Leppi TJ: Gross anatomical relationship between primate submandibular and sublingual glands. *J Dent Res* 46: 359–365, 1967.
46. Imai M, Shibata T, Shimano J: An opinion on the minor sublingual gland. *Okajimas Folia Anat Jpn* 58: 635–647, 1982.
47. Riva A, Tandler B, Testa Riva F: Ultrastructural observations on human sublingual gland. *Am J Anat* 181: 385–392, 1988.
48. Schulte H von W: The development of the salivary glands in man. In: Studies in Cancer and Allied Subjects, Vol 4. GS Huntington, HW von Schulte (eds), New York: Columbia Univ Press, 25–72, 1913.
49. Espinal ES, Ubios AM, Pradier M, Cabrini RL: Scanning electron microscopy of human submandibular gland. *Acta Anat* 122: 22–28, 1985.
50. Riva A, Puxeddu P, Testa Riva F: Sulla localizzazione al microscopio elettronico delle mucine nelle ghiandole salivari maggiori dell'uomo. *Riv Istoc Norm Pathol* 21: 203–205, 1977.
51. Shimono M, Hasegawa E, Yamane H, Kawahara H, Edà S, Yamamura T, Yamane J, Kato T, Okuyama T: Ultrastructure of human labial salivary glands. First report: the acinar cells. *Shikwa Gakuho* 71: 369–380, 1971.
52. Vidić B, Tandler B: Ultrastructure of the secretory cells of the submucosal glands in the human maxillary sinus. *J Morphol* 150: 167–182, 1976.
53. Brocco SL, Tamarin A: The topography of rat submandibular gland as observed with SEM. *Anat Rec* 194: 445–460, 1979.
54. Garrett JR, Emmelin N: Activities of salivary myoepithelial cells: a review. *Med Biol* 57: 1–28, 1979.
55. Van Niekerk JP, Raubenheimer EJ: Ultrastructure of myoepithelium in salivary glands of African elephant (*Loxodonta africana*). *Proc XI Int Congr Electron Microsc*, Kyoto, 2869–2870, 1986.
56. Nilsen R, Donath K: Actin containing cells in normal human salivary glands. *Virchows Arch* (Pathol Anat) *391*: 315–322, 1981.
57. Longtine JA, Pinkus JS, Fujiwara K, Corson JM: Immunohistochemical localization of smooth muscle myosin in normal human tissues. *J Histochem Cytochem* 33: 179–184, 1985.
58. Nakazato Y, Ishida Y, Takahashi K, Suzuki K: Immunohistochemical distribution of S-100 protein and glial fibrillar protein in normal and neoplastic salivary glands. *Virchows Arch* (Pathol Anat) 405: 299–310, 1985.
59. Seifert G, Miehlke A, Haubrich J, Chilla R: Diseases of the salivary glands. New York: Thieme, 1986.
60. Geiger S, Geiger B, Leitner O, Marshak G: Cytokeratin polypeptides expression in different epithelial elements of human salivary glands. *Virchows Arch* (Pathol Anat) 410: 403–414, 1987.
61. Tandler B: Ultrastructure of the human submaxillary gland. III. Myoepithelium. *Z Zellforsch* 68: 852–863, 1965.
62. Nagato T, Tandler B: Gap junctions in rat sublingual gland. *Anat Rec* 214: 71–75, 1986.
63. Nagashima Y, Ono K: Myoepithelial cell ultrastructure in the submandibular gland of man. *Anat Embryol* 171: 259–265, 1985.
64. Nagato T, Yoshida H, Yoshida A, Uehara Y: A scanning electron microscope study of myoepithelial cells in exocrine glands. *Cell Tissue Res* 209: 1–10, 1980.
65. Evan AP, Dail WG, Damrose D, Palmer C: Scanning electron microscopy of cell surfaces following removal of extracellular material. *Anat Rec* 185: 433–447, 1976.
66. Riva A, Valentino L, Lantini MS, Cotti E, Testa Riva F: Surface microanatomy of human major salivary glands. In: Cells and Tissues: A Three Dimensional Approach by Modern Techniques in Microscopy. PM Motta (ed), New York: Alan Liss, 235–241, 1989.
67. Highison GJ, Low FN: Microdissection by ultrasonication after prolongued OsO_4 fixation: a technique for scanning electron microscopy. *J Submicr Cytol* 14: 161–170, 1982.
68. Nagai T, Nagai M: Scanning electron microscopy of the human salivary glands. *Arch Otholaryngol* 241: 265–266, 1985.
69. Tamarin A: Myoepithelium of the rat submaxillary gland. *J Ultrastruct Res* 16: 320–338. 2: (Suppl) 317–323, 1966.
70. Tandler B: Unusual myoepithelium in the retrolingual salivary gland of the European hedgehog. *J Submicrosc Cytol* 18: 261–270, 1986.
71. Chaudry AP, Cutler LS, Yamane GM, Labay GR, Sunderraj M, Manak JR Jr: Ultrastructure of normal human parotid gland with special emphasis on myoepithelial distribution. *J Anat* 152: 1–11, 1987.
72. Lantini MS, Valentino L, Riva A: A granular cell in the proximal intercalated duct of human parotid and submandibular glands. *J Submicrosc Cytol Pathol* 20: 147–152, 1988.
73. Takata T, Ogawa I, Mikai H, Ogura M, Ijuhin N, Ito H: Immunohistochemical studies on salivary gland tumors. I. Localization of various secretory epithelial markers in the normal human salivary gland tissues. *Jpn J Oral Biol* 26: 1190–1199, 1984.
74. Riva A, Testa Riva F, Del Fiacco M, Lantini MS: Fine structure and cytochemistry of the intralobular ducts of

the human parotid gland. *J Anat* 122: 627–640, 1976.
75. Garrett JR, Harrison JD, Kidd A, Kiriacou K, Smith RE: Kallikrein-like activity in human salivary glands and colon, including mast cells. *J Physiol* 334, 78, 1983.
76. Kimura K, Moriya H: Enzyme and immunohistochemical localization of kallikrein. I. The human parotid gland. *Histochemistry* 80: 367–372, 1984.
77. Hand AR, Coleman R, Mazariegos MR, Lustman J, Lotti LV: Endocytosis of proteins by salivary gland duct cells. *J Dent Res* 66: 412–419, 1987.
78. Wilborn WH, Shackleford JM: Microanatomy of human salivary glands. In: The Biologic Basis of Dental Caries. L Menaker (ed), Hagerstown, Harper and Row: 3–62, 1980.
79. Tandler B: Salivary glands and the secretory process. In: A Textbook of Oral Biology. JH Shaw, EA Sweeney, CC Cappuccino, SM Meller (eds), Philadelphia: WB Saunders, 547–592, 1978.
80. Tamarin A, Sreebny LM: The rat submaxillary salivary gland. A correlative study by light and electron microscopy. *J Morphol* 117: 295–352, 1965.
81. Regaud C, Mawas J: Sur les mitochondries des glandes salivaires chez les mammiferes. *CR Soc Biol* 66: 97–100, 1909.
82. Riva A, Zaccheo D, Testa Riva F: A SEM study of the human parotid and submandibular glands. *Proc XI Int Congr Electron Miscrosc*, Kyoto, 2863–2864, 1986.
83. Schneyer LH, Young, JA, Schneyer CA: Salivary secretion of electrolytes. *Physiol Rev* 82: 720–777, 1972.
84. Martinez JR: Ion transport and water movement. *J Dental Res* 66: 638–647, 1987.
85. Shackleford JM, Schneyer LH: Ultrastructural aspects of the main excretory duct of rat submandibular gland. *Anat Rec* 169: 679–696, 1971.
86. Sato A: Fine structure of the main excretory duct of rat submandibular gland. *Biol Cell* 39: 237–240, 1980.
87. Sato A, Miyoshi S: Ultrastructure of the main excretory duct epithelia of the rat parotid and submandibular glands with a review of the literature. *Anat Rec* 220: 239–251, 1988.
88. Tandler B, Poulsen JH: Ultrastructure of the main excretory duct of the cat submandibular gland. *J Morphol* 149: 183–198, 1976.
89. Garrett JR, Parsons PA: Changes in the main submandibular ducts of rabbits, resulting from ductal ligation. *Z Micr Anat Forsch* 935: 593–608, 1979.
90. Testa Riva F, Puxeddu P, Riva A, Diaz G: The epithelium of the excretory duct of the human submandibular gland: a transmission and scanning electron microscopic study. *Am J Anat* 160: 381–393, 1981.
91. Testa Riva F, Riva A, Puxeddu P: Ciliated cells in the main excretory duct of the submandibular gland in obstructive sialadenitis: A SEM and TEM study. *Ultrastruct Pathol* 11: 1–10, 1987.
92. Dardick I, Parks WR, Little J, Brown DL: Characterization of cytoskeletal proteins in basal cells of human parotid salivary gland ducts. *Virchows Arch* (Pathol Anat) 412: 525–532, 1988.
93. Tandler B, Denning CR, Mandel ID, Kutscher AH: Ultrastructure of human labial salivary glands. I. Acinar secretory cells. *J Morphol* 127: 384–407, 1969.
94. Tandler B, Denning CR, Mandel ID, Kutscher AH: Ultrastructure of human labial salivary glands. II. Intranuclear inclusions in the acinar secretory cells. *Z Zellforsch* 94: 555–564, 1969.
95. Tandler B, Denning CR, Mandel ID, Kutscher AH: Ultrastructure of human labial salivary glands. III. Myoépithelium and ducts. *J Morphol* 130: 227–245, 1970.
96. Scott J: Qualitative and quantitative observations on the histology of human labial salivary glands obtained post mortem. *J Biol Buccale* 8: 107–200, 1980.
97. Smith SD, Taurman MA, King WF: Immunological features of minor salivary gland saliva. *J Cli Immunol* 7: 449–455, 1987.
98. Moro I, Umemura S, Crago SS, Mestecky J: Immunohistological distribution of immunoglobulins, lactoferrin and lysozyme in human minor salivary glands. *J Oral Pathol* 13: 97–104, 1984.
99. Reitamo S, Klockars M, Raeste AM: Immunohistochemical identification of lysozyme in the minor salivary glands of man. *Arch Oral Biol* 22: 515–519, 1977.
100. Hand AR: Salivary glands. In: Orban's Oral Histology and Embryology, 9th ed. SB Bhaskar (ed), 336–370, 1980.
101. Black JB: The structure of salivary glands of the human soft palate. *J Morphol* 153: 107–117, 1977.
102. Hamosh M, Burns WA: Lipolytic activity of human lingual glands (Ebner). *Lab Invest* 37: 603–608, 1977.
103. Taylor T, Erlandson SL: Peroxidase localization in von Ebner's glands of man. *J Dent Res* 52: 635, 1973.
104. Azzali G, Bucci G, Gatti R, Ferrari G: Fine structure of the excretory system of the deep posterior (Ebner's) salivary glands of the human tongue. *Acta Anat*, in press, 1989.

Author's address:
Prof. Alessandro Riva
Dipartimento di Citomorfologia
Universita' di Cagliari
Via Porcell 2
09124 Cagliari
Italy

CHAPTER 5

Recent contributions of electron microscopy to salivary gland pathology

IRVING DARDICK & A.W. PETER VAN NOSTRAND

1. Introduction

Judging from the numerous well-written articles about various aspects of salivary gland pathology that have appeared in the literature in the past 5 to 10 years, there has been a resurgence of interest in these glands. Ultrastructural investigations and observations have been the focal point of some of these studies. However, as pointed out by van Nostrand [1], pathology has entered an era where a multiplicity of techniques can be applied to diagnostic problems. Although immunoperoxidase, immunofluorescent, flow cytometric, and cytogenetic investigations, along with cell culture and tumor implantation in nude mice, have all been successfully used in studies of salivary gland neoplasms and other lesions, we will concentrate on the contribution that electron microscopy has made to the pathology of these glands in the past decade.

Electron microscopy has only a minor role in the routine diagnosis of salivary gland tumors. Yet, as we will detail in this chapter, ultrastructural studies have been a major component in the study of the histogenetic aspects, the myoepithelial-cell structural modifications, the contribution of neoplastic myoepithelium, and the developmental and cellular differentiation processes in salivary gland tumors. Experimental studies of this gland have also benefited from this technique.

2. Histogenetic concepts

Many aspects of the classification of salivary gland tumors have been based on observations of the organization of ductal cells in the developing human salivary gland [2]. The concept that specific reserve or stem cells in normal ducts were central to the induction and histogenesis of salivary gland tumors was developed further by Batsakis and associates [3, 4]. These were conceived as basal cells in the case of the excretory duct and luminal cells in the intercalated ducts; acinar cells were excluded from a role in parenchymal regeneration and tumor induction [3, 4]. However, the nature and exact location of such reserve cells has not been documented. Even prior to the establishment of this concept, there was experimental evidence, in the form of autoradiographic studies in the rat, that acinar cells were not only capable of dividing but were the principal dividing cells in salivary glands stimulated by adrenergic drugs [5, 6]. In such studies, dividing cells associated with ducts were initially limited, and minimal isotopic labeling of the nuclei of salivary duct cells occurred only after acinar cells were the major cell type labeled [5, 6]. Both autoradiographic experiments (Fig. 5-1b) and assessment of mitotic figures (Fig. 5-2) in the ducts and terminal tubules of the neonatal rat salivary gland reveal no evidence for the specific localization of cycling cells in a basal position.

Riva, A. and Motta, P.M., (eds.), Ultrastructure of the Extraparietal Glands of the Digestive Tract.
ISBN 978-1-4612-8211-2.

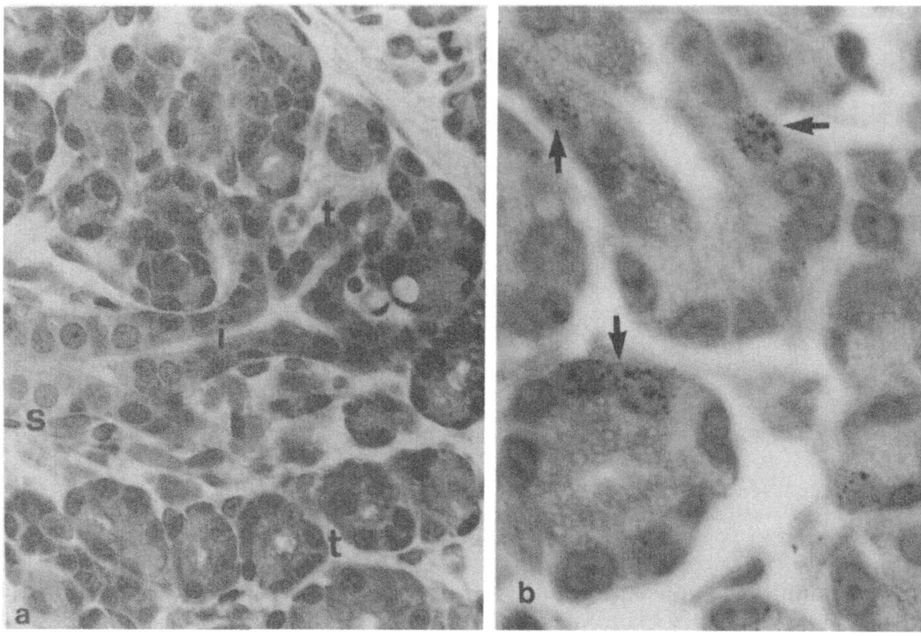

Fig. 5-1. Rat submandibular gland, 3 days postpartum. **a**. Glycol methacrylate section showing that both the developing striated (s) and intercalated (i) ducts, as well as the terminal tubules (t), are all layers of single cells at this stage of development. **b**: An autoradiogram of tritiated-thymidine-labeled salivary gland tissue (black silver grains (arrows) over the nuclei) reveals that terminal tubule lining cells (full of secretory granules) are the principal ones undergoing cell division. Hematoxylin and eosin. **a**: x 550; **b**: x 1,300.

The major DNA synthetic and mitotic activities in the developing salivary gland occur in duct luminal epithelial cells. The potential for duct luminal cells to divide persists in the adult animal [5, 6]. In one ultrastructural study, dividing cells on the outer aspect of intercalated ducts of a 3-week-old rat were interpreted as immature myoepithelial cells [7]. A similar feature was shown in an electron micrograph of an acinus of the parotid gland from a 5-day-old rat [8]. In repeating the classical experiments of Barka [5] and Schneyer [6], we can only confirm their observations that, following isoproterenol stimulation, acinar cells of the adult sublingual and submandibular glands are the main dividing population, and that, in the case of the ducts, the minimal labeling detected is confined to luminal cells, even in the excretory duct. It should be appreciated that in normal rat and human salivary glands, there are extremely few acinar or ductal cells that are actually in cell cycle. Perhaps this is a major factor in the infrequency of neoplasia in these organs.

Ultrastructural observations provide additional support for luminal cells acting as stem cells. A recent study of normal human breast tissue revealed that all mitotically active cells observed in ducts and ductules were joined to adjacent cells by zonulae occludentes and bore microvilli on their apical surface, i.e., were distinctly luminal epithelial cells [9]. A parallel situation exists in the neonatal rat salivary gland (Fig. 5-2). In the case of normal human breast tissue, there is ultrastructural evidence that what have been termed *basal clear cells* may be an independent precursor cell for myoepithelial cells [10]. To date, therefore, there is no direct evidence for a specific ductal stem or "reserve" cell giving rise to both luminal and myoepithelial cells.

At least in the early stages of development of the rat submandibular gland, terminal tubules are composed of a single layer of cells (Figs. 5-1d and 5-2). As illustrated above, at least a proportion of such cells are mitotically active. Despite this simple structure, this region gives rise to multiple acini and eventually forms the intercalated and

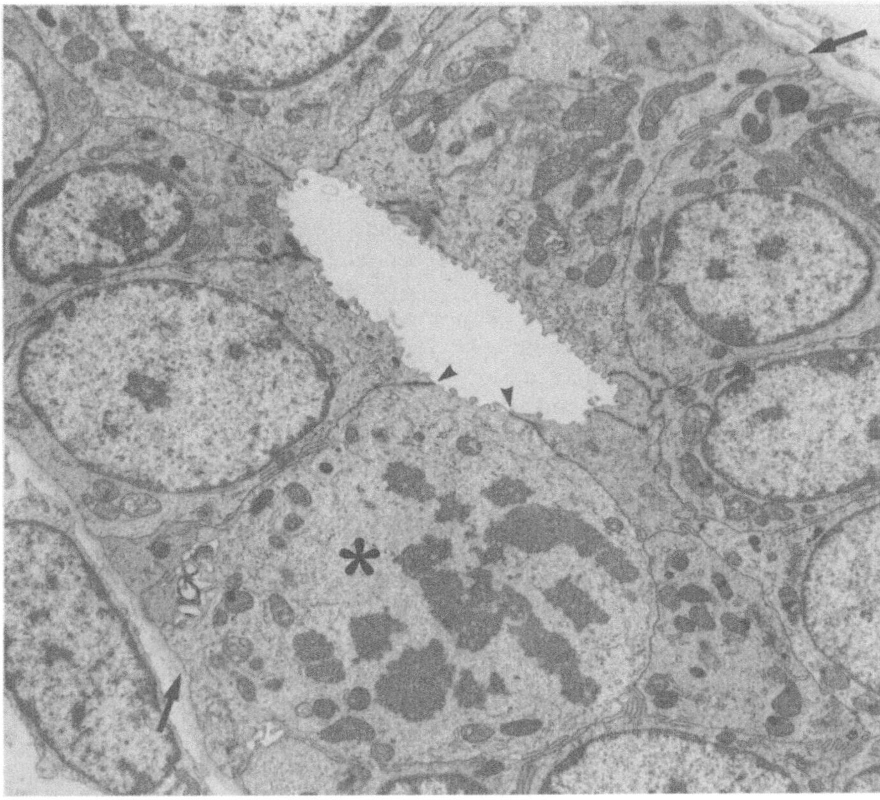

Fig. 5-2. Rat submandibular gland, 3 days postpartum. Electron microscopy of an intercalated duct shows that a cell in metaphase (asterisk) bears microvilli and is joined to adjacent ductal cells by apical tight junctions (arrowheads). Note that, at this stage of development, there are no cells interposed between the base of the duct cells and the enveloping basal lamina (arrows). Uranyl acetate and lead citrate. x 6,800.

striated ducts [11–13]. However, the development of human salivary glands may be more complex or may differ from that in the rat. At least the principal ductal system in fetal palatal salivary glands are bilayered, a feature that was central to the hypothesis that the outer layer of cells acted as "reserve" cells for the formation of the inner or luminal layer of ductal cells [2, 3]. However, whether regions of human fetal salivary gland comparable to the terminal tubules of the rat salivary gland are unilayered or bilayered remains to be determined. The statement that the regenerative capacity of human salivary gland parenchyma is confined to the intercalated duct [4] is unsupported by experimental evidence. As noted above, there is ultrastructural evidence that in the rat submandibular gland and the human breast, luminal epithelial cells can divide. If this feature applies to human salivary gland, there is no requirement for a specific basally located precursor cell, even in the excretory duct. At least a certain proportion of salivary gland cells, whether acinar, duct luminal, or myoepithelial, may be capable of acting as stem cells and, therefore, are potential elements in the histogenesis of salivary gland tumors.

Perhaps, not all salivary gland tumors recapitulate exact stages occurring in the developing mammalian salivary gland, as has been proposed [4]. However, based on recent ultrastructural and immunohistochemical observations, many subtypes of tumors in this gland have common morphogenetic mechanisms that involve one or more of the cell types found in the normal salivary gland acini and ducts [14]. Light [15] and electron [16] microscopic assessments of salivary gland tumors indicate that many subtypes involve a coordinated proliferation of luminal- or acinar-like cells and classical or modified myoepithelial cells. Recent ultrastructural studies of salivary gland tumors particularly emphasize that the organization of these two basic types of tumor cells resembles that of the normal salivary gland acinus and duct [14, 17, 18]. This ducto-acinar concept of the morphogenesis of salivary gland tumors has practical advantages over the conventional histogenetic approach to classifying these tumors and obviates the need to hypothesize specific reserve cells. The basis for the ducto-acinar model and its application to other salivary gland neoplasms [14] is best developed in relation to pleomorphic adenoma in a subsequent section.

3. Morphological modifications of myoepithelium

Electron microscopy has an important role in delineating the morphology of the myoepithelial cells in neoplastic and non-neoplastic lesions of the salivary glands. It is important for detecting the fine structural characteristics of myoepithelium, since modifications from the typical features of normal myoepithelial cells are the rule in tumors. Without an awareness of this fact and the way modified myoepithelial cells are organized in salivary gland tumors, it is difficult to properly identify individual cells. The nature of many salivary gland tumor cells illustrated in the literature is incorrectly portrayed on this basis.

3.1. Modifications in tumors

Batsakis and colleagues [15] have indicated the range of tumors in which neoplastic myoepithelium plays a role in producing the characteristic histology. Even by light microscopy, the range of cytological appearance of myoepithelium is quite variable [15]. Ultrastructural review further illustrates the myoepithelial cell component in these tumors and extends the classes of salivary gland lesions within these cells [16]. The differing morphology of the tumor cells comprising salivary gland tumors and the organization of these cells is best appreciated in low-magnification, survey-type electron micrographs [14, 16, 19]. Many previous ultrastructural studies have been primarily descriptive of individual or small groups of cells and have often not reliably reflected the interrelationship of cells comprising salivary gland tumors. In attempting to discern the role of myoepithelial cells, the latter approach is essential.

In pleomorphic adenoma, the angular, slightly separated tumor cells, surrounding well-defined ducts composed of luminal cells, are conventionally designated as the *myoepithelial cell component* (Fig. 5-3). Although somewhat variable at the ultrastructural level, Figure 5-4 illustrates the usual features of such neoplastic myoepithelium. Even such specializations as intercellular junctions differ from case to case; some display desmosomes, while other cases have few if any poorly formed junctions. Intermediate filaments may be a major cytoplasmic complement, but microfilaments are only seen in the occasional cell in most cases of pleomorphic adenoma [17, 18, 20]. However, the rare example may contain a substantial number of well-differentiated myoepithelial cells [21]. Apparently, the characteristic ultrastructural features of normal myoepithelium associated with acini and intercalated ducts of salivary glands are infrequent in pleomorphic adenoma.

Such a finding is an important contribution of electron microscopy to understanding the many facets of salivary gland pathology and has considerable significance to other types of salivary gland tumors, including myoepitheliomas. In the group of monomorphic adenomas, including basal cell, tubular, and trabecular adenomas, the counterpart of normal myoepithelium is an outer layer of cuboidal to angular tumor cells [19]. These cells may or may not contain tonofilament bundles, and desmosomes may join adjacent neoplastic myoepithelial cells to each other, as well as to the inner zone of luminal cells [19, 21]. Angular, somewhat separated, multilayered cells, which have proliferated more extensively than the gland-forming luminal cells, are the counterpart of myoepithelium in the cribriform and solid regions of adenoid cystic carcinoma [22–25]. In lesions such as epithelial-myoepithelial (clear-cell) carcinoma of the intercalated duct, the modified myoepithelium consists of an outer row of tumor cells in which the "clear" cytoplasm seen in histological sections results from excessive accumulation of cytoplasmic glycogen [26–29].

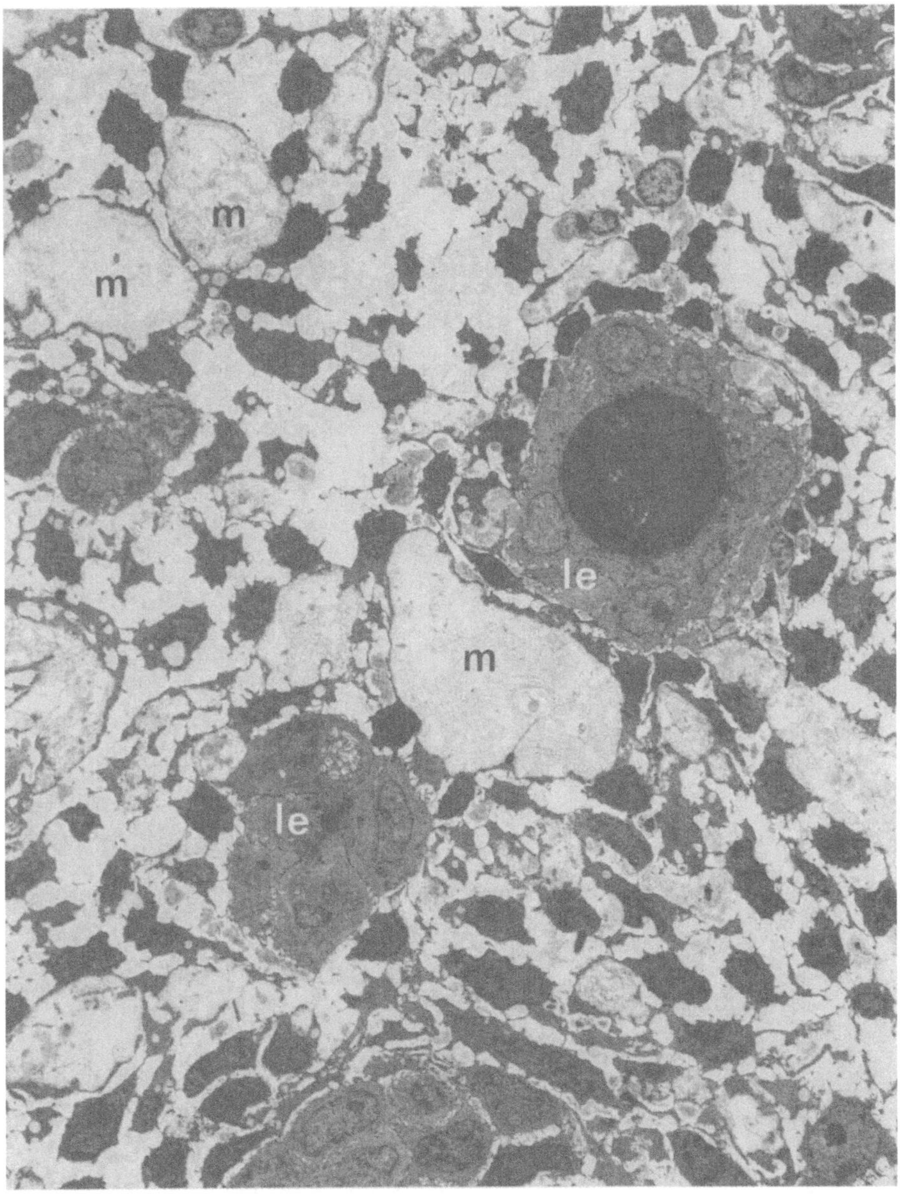

Fig. 5-3. Pleomorphic adenoma, cellular region. In a survey-type electron micrograph, it is possible to appreciate the differentiation of luminal epithelial cells (le) surrounded by the more numerous angular and irregularly shaped myoepithelial cells separated by the glycosaminoglycan-rich intercellular matrix. Note that myoepithelial cells and their attenuated processes enclose specific pools of matrix materials (m), a process that is instrumental in the formation of myxoid areas in this tumor. Uranyl acetate and lead citrate. x 1,600.

Based on electron microscopic studies, modified myoepithelial calls with a range of cytological features may also play a central role in mucoepidermoid carcinoma [14, 30]. Even the rare sialadenoma papilliferum reveals ultrastructural evidence for a mixture of classical and modified myoepithelial cells, as well as duct epithelial cells [31].

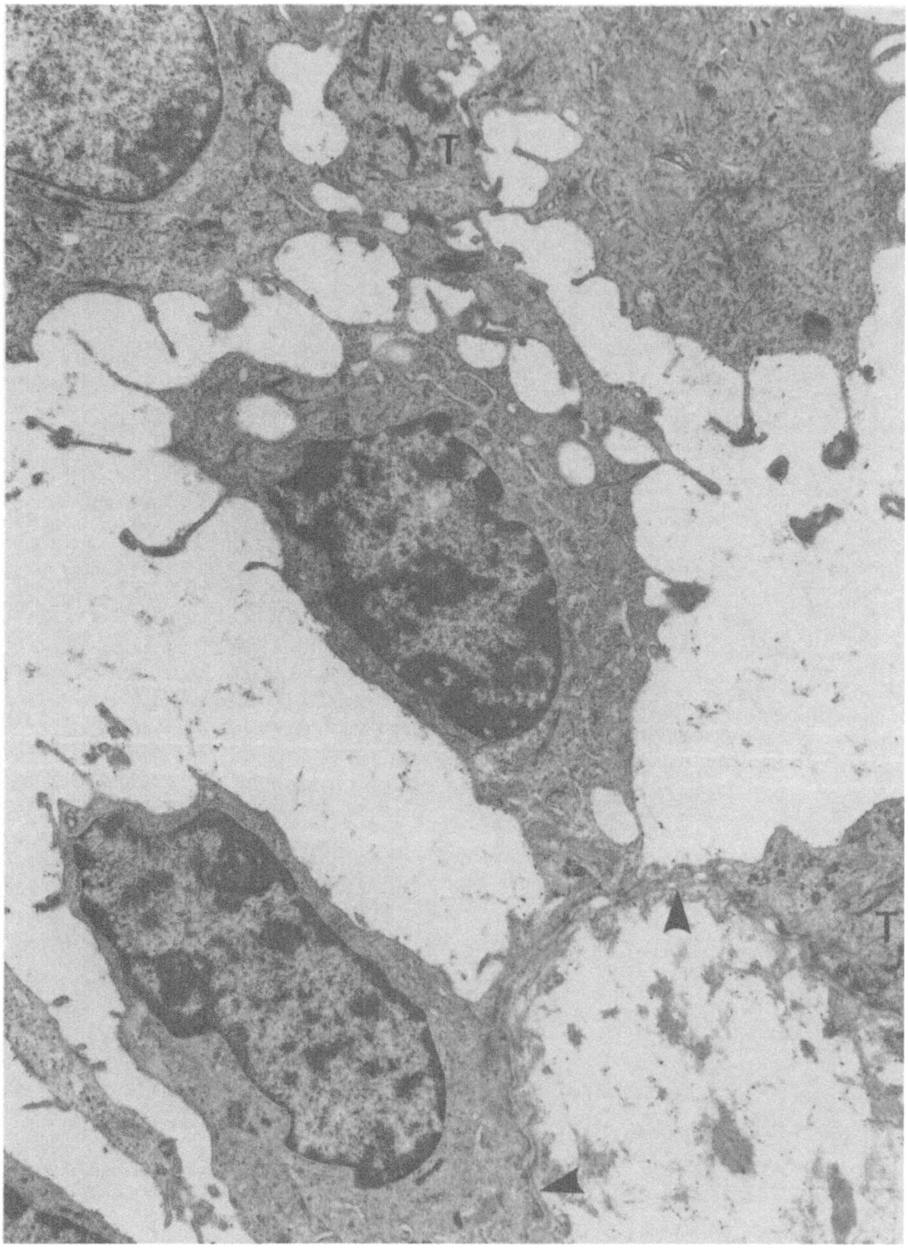

Fig. 5-4. Pleomorphic adenoma, developing myxoid region. A group of considerably separated, modified myoepithelial cells have portions of their cytoplasm aligned around an extracellular space that is lined by basal lamina (arrowheads). This type of specific extracellular space corresponds to the cell-enclosed and matrix-containing areas designated in Fig. 5-3. The tumor cells contain tonofilament bundles (T). Uranyl acetate and lead citrate. x 10,300.

3.2. *Modifications in non-neoplastic disorders*

Ultrastructural experimental studies of rat salivary gland show that both luminal and myoepithelial cells can rapidly undergo squamous metaplasia following an ischemic episode [32]. Perhaps an ischemic insult and resulting squamous metaplasia of myoepithelium of minor salivary glands is the mechanism for necrotizing sialometaplasia. Squamous metaplasia occurring in myoepithelial cells is a primary mechanism for this process in pleomorphic adenoma [18, 33].

Myoepithelial cells also undergo alterations in chronic sialadenitis. Along with the involution of acinar cells and increasing ductulization of salivary gland parenchyma, myoepithelial cells appear to undergo hyperplasia, or at least increased prominence. This is best appreciated using a monoclonal antibody specific for cytokeratin 14 [34, 35] (Fig. 5-5a). Ultrastructurally, this is seen as myoepithelial cells, ringing duct-like structures, with an increased complement of tonofilaments and decreased myofilaments (Fig. 5-5b).

Despite persisting use of the term *epimyoepithelial islands* for the residual salivary gland tissue in a benign lymphoepithelial lesion (Sjögren's syndrome), there is continuing argument over the role of the myoepithelium in the development of these structures. Other immunohistochemical and ultrastructural studies have indicated little or no role for myoepithelium in epimyoepithelial islands [36–40], while other studies suggest a major contribution from this tissue [41–43]. Other ultrastructural studies have noted the considerable complement of tonofilaments and prominent desmosomes of the cells comprising epimyoepithelial islands and have consequently attributed these modifications to squamous metaplasia of luminal cells of salivary ducts [38, 40]. Such studies have failed to appreciate that squamous metaplasia is a common result when salivary gland physiology and structure is altered. Ultrastructurally, the cell component of epimyoepithelial islands with an increased complement of tonofilaments has an organizational arrangement differing from the residual luminal cells that can be identified in electron micrographs of some epimyoepithelial islands [43]. Extensive proliferations of myoepithelial cells are the principal populations of the epimyoepithelial islands of benigh lymphoepithelial lesions [42, 43]. The major role of modified myoepithelial cells in benign lymphoepithelial lesions is supported by immunohistochemistry using monoclonal antibodies [42, 43].

Electron microscopy clearly shows that features such as microfilaments, dense bodies, and micropinocytotic vesicles, while characteristic of normal myoepithelium, are infrequently expressed in many of the non-neoplastic lesions of the salivary gland. Therefore, the absence of such features in tumor cells, otherwise localized like the myoepithelium of normal salivary gland, does not exclude their designation as neoplastic, but modified, myoepithelial cells. Previous ultrastructural criteria for identifying tumor cells as myoepithelial have been too restrictive. In no other type of tumor is there a requirement that the neoplastic cells be the exact counterpart of the normal cell of the involved organ or tissue.

4. Myoepithelioma

The principles outlined above for the type and degree of structural modifications evident in myoepithelial cells in various diseases of the salivary glands also apply to the tumor class, myoepithelioma. The classical ultrastructural characteristics of normal myoepithelium are not found in a proportion of these tumors.

In histological sections, myoepitheliomas are usually considered to consist of two main cell types, spindle (myoid) and hyaline (plasmacytoid) cells [44, 45]. The latter type has been recognized on the basis of epithelial cells with hyaline cytoplasm and eccentrically positioned nuclei that are found in the myoepithelial cell regions of some pleomorphic adenomas occurring in salivary glands, particularly the minor glands [46]. Ultrastructurally, hyaline cells have a major portion of their cytoplasm occupied only by intermediate filaments and lack evidence of myofilaments [44, 46]. Sciubba and Brannon [44] found that the pure hyaline cell form of myoepithelioma accounted for 17% of their series of 23 cases, while a mixed hyaline-spindle cell type was present in another 13%. It is the more common spindle-cell variant of myoepithelioma

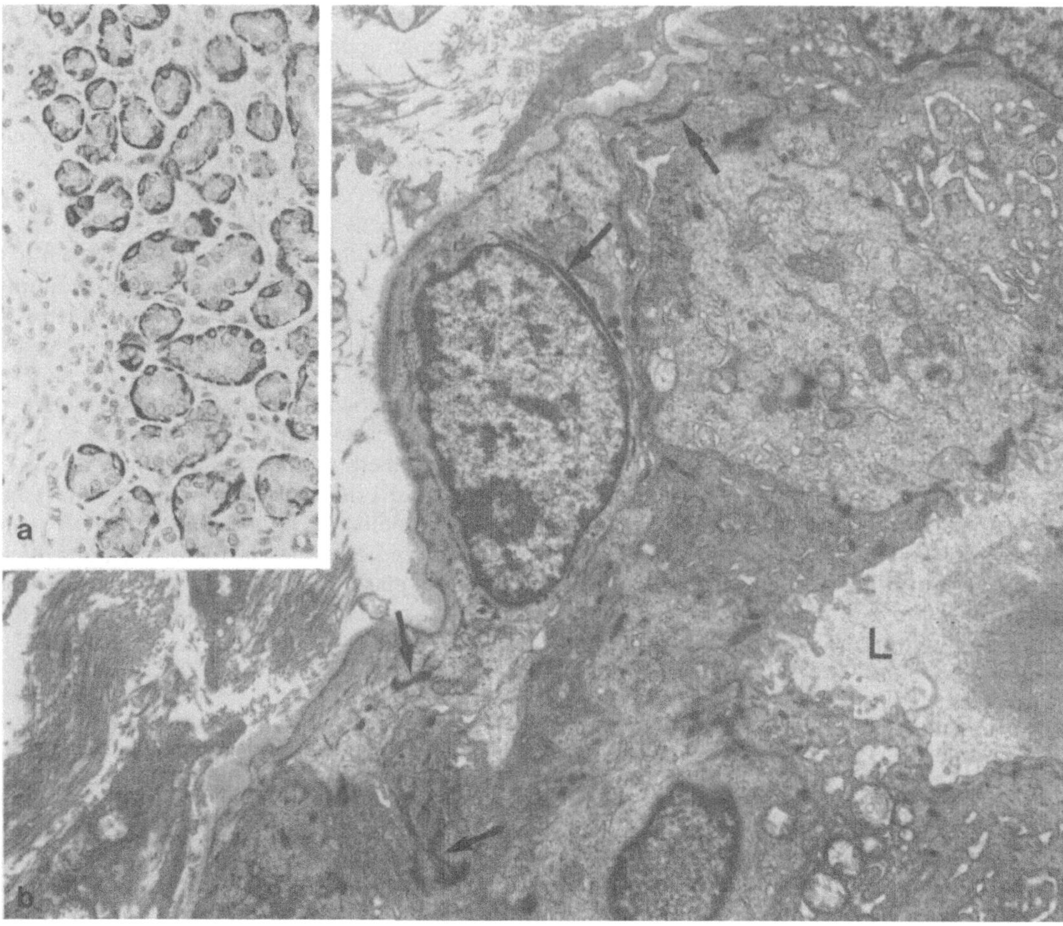

Fig. 5-5. Chronic sialadenitis, submandibular gland. **a**: Atrophic acini and ducts are almost completely encompassed by myoepithelial cells in this methacarn-fixed tissue that has been immunostained with a monoclonal antibody specific for cytokeratin 14; only myoepithelial cells are stained. **b**: An electron micrograph of an intercalated duct from this case shows that the peripherally situated myoepithelium has many tonofilament bundles (arrows) but little or no myofilaments. L = duct lumen. **a**: Immunoperoxidase anti-cytokeratin with hematoxylin counterstain. x 250. **b**: Uranyl acetate and lead citrate, x 7,300.

that may contain a complement of tumor cells, often small in number, containing myofilaments associated with dense bodies, micropinocytotic vesicles, desmosomes, and basal lamina [16, 44, 47, 48]. Two such spindle-cell myoepitheliomas, confirmed by electron microscopy, were shown to contain S100 protein immunohistochemically [48]. This spectrum of differentiation indicates that, even in myoepitheliomas, the tumors cells are considerably modified compared to normal myoepithelium.

Using the myoepithelial cell population of pleomorphic adenoma as a model, potentially there is a third form of myoepithelioma, the epithelial type [15]. Some pleomorphic adenomas have regions in which closely associated polygonal cells, with central nuclei and moderate amounts of cytoplasm, surround and separate ductal structures. Those neoplasms, composed exclusively of such tumor cells, i.e., without ductal or glandular structures and with or without myxoid regions, are an epithelial variant of myoepithelioma (Fig. 5-6); such lesions are usually classified as cellular mixed tumors or monomorphic adenomas. Electron microscopy of such lesions (Fig. 5-7) reveals that this variant

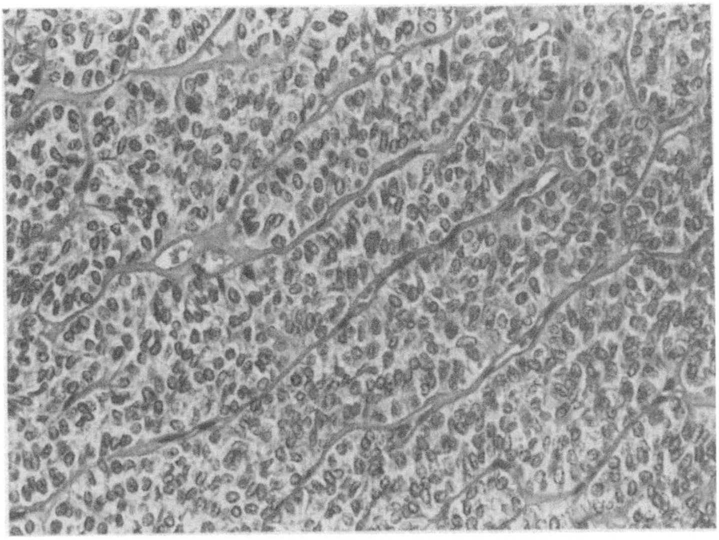

Fig. 5-6. Myoepithelioma (epithelial-type), parotid gland. This lesion lacked the ductal component of a cellular pleomorphic adenoma and did not have the architectural or cytological features associated with monomorphic lesions such as basal cell adenoma. The elongated nests of tumor cells are compactly organized, polygonal shaped, and have moderate amounts of cytoplasm, which in certain regions appears "clear." Hematoxylin and eosin. x 250.

shares some features with the hyaline cell form. Some examples may have prominent accumulations of cytoplasmic intermediate filaments, but these are more uniformly distributed and do not displace the nucleus within the cell to a peripheral location (Fig. 5-7). Myofilaments are generally absent.

Although many myoepitheliomas present histologically as solid lesions, a few have been illustrated that have tumor cells separated by myxoid stroma or microcystic spaces [44, 49]. In those cases associated with a myxoid stroma, a well-defined basement membrane separates the cords of tumor cells from the extracellular proteoglycans and collagen. Electron microscopically, the microcystic spaces of a myoepithelioma consist of basal-lamina-lined intercellular spaces containing proteoglycans [44]. Similar intercellular microcystic spaces are an integral part of the histologic patterns of pleomorphic adenoma [14, 17].

5. Pleomorphic adenoma

Electron microscopy has played a major role in the understanding of developmental processes involved in the production of the characteristic histology of pleomorphic adenoma [17, 18, 20, 50, 51]. This involves coordinated proliferation of luminal epithelial and myoepithelial cells and/or extensive regions consisting almost exclusively of neoplastic myoepithelial cells displaying varying forms of cellular modification, including squamous metaplasia [17, 18, 20]. Superimposed on the cellular aspects of pleomorphic adenoma is the controlled synthesis of basal lamina, elastic tissue, and glycosaminoglycans by the modified myoepithelial cell component [50, 52]. Figure 5-8 illustrates the main components of these processes in a model form, a method useful in the appreciation of the mechanisms underlying the range of histology, the complex morphological patterns, and the formation of the myxoid and chondroid in pleomorphic adenoma.

In this model (Fig. 5-8), the basic premise is that tumor cell differentiation and organization in pleomorphic adenoma recapitulates similar aspects of the acini and ducts in normal salivary glands. The bicellular arrangement of acini and ducts becomes reflected in cellular regions of pleomorphic adenoma where tubular elements composed of epithelial and modified myoepithelial cells may dominate or be admixed with

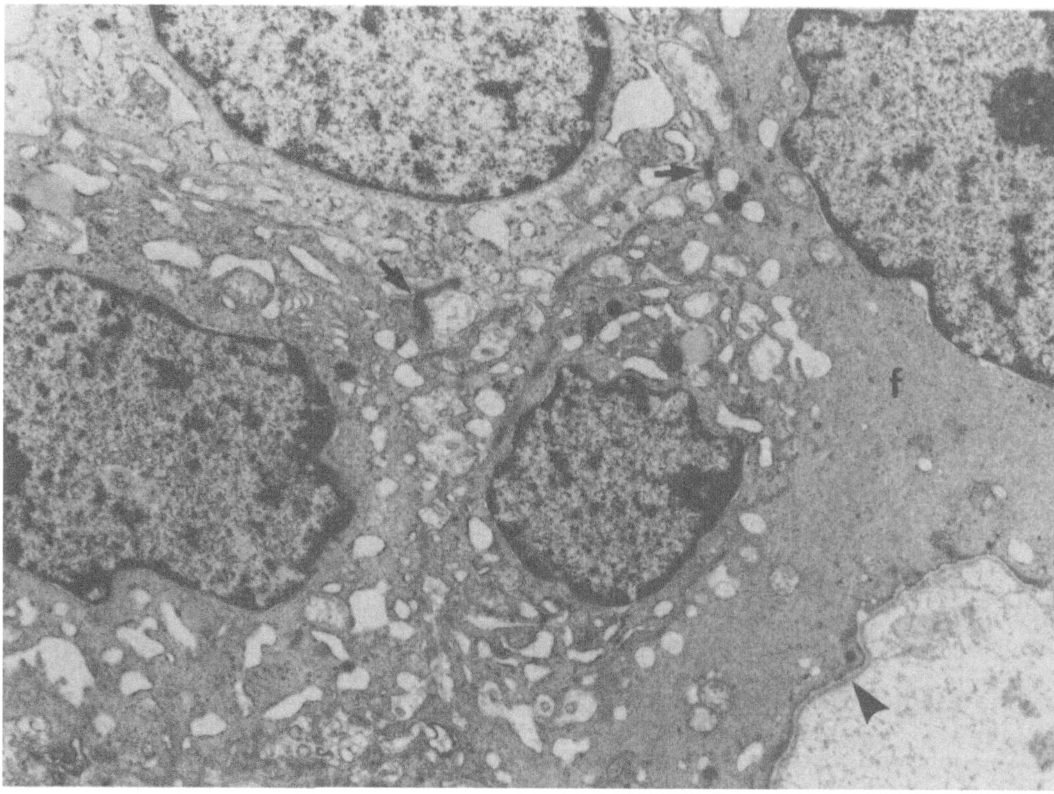

Fig. 5-7. Myoepithelioma (epithelial-type), parotid gland. Small desmosomes (arrows) join tumor cells that contain regions filled with intermediate filaments (f). A distinct basal lamina (arrowhead) separates clusters of tumor cells from adjacent glycosaminoglycan-containing stroma. Uranyl acetate and lead citrate. x 9,000.

variable proliferations of myoepithelial cells. In other area, neoplastic myoepithelial cells may predominate, or the coordinated proliferation of luminal epithelial and myoepithelial cells can result in solid sheets of cellular tumor. Gradual development of intercellular accumulations of glycosaminoglycans, basal lamina, other collagens, and elastic tissue between the myoepithelial cell component is instrumental in the development of the myxoid and chondroid components of pleomorphic adenoma (Fig. 5-8). Major aspects of this concept were derived from ultrastructural studies of this salivary gland tumor [17, 18, 20, 50].

These observations are reflected in the cellular regions of pleomorphic adenomas (Fig. 5-3), where well-organized duct luminal cells form ductlike structures enclosed within the more numerous, angular, irregularly shaped, slightly separated myoepithelial cell component. This pattern of differentiation is reflected in the right-hand portion of the model sequence of Fig. 5-8. These ultrastructural features are the basis for considering that differentiation in pleomorphic adenoma involves a coordinated growth of multiple units of luminal epithelial and myoepithelial cells, basically organized like the acini and intercalated ducts of the normal salivary gland, i.e., mimicking the ducto-acinar unit [14, 17]. Although difficult to appreciate in histological sections, electron microscopy has clearly shown that acinar cell development does occur in pleomorphic adenoma [20]. In such a model, there is no requirement to hypothesize a specific reserve or stem cell. Neoplastic induction of any cell within the acinar or ductal system of the salivary gland would be capable of differentiating acinar, ductal, or myoepithelial cells, either alone or in any combination. Such a model removes the emphasis

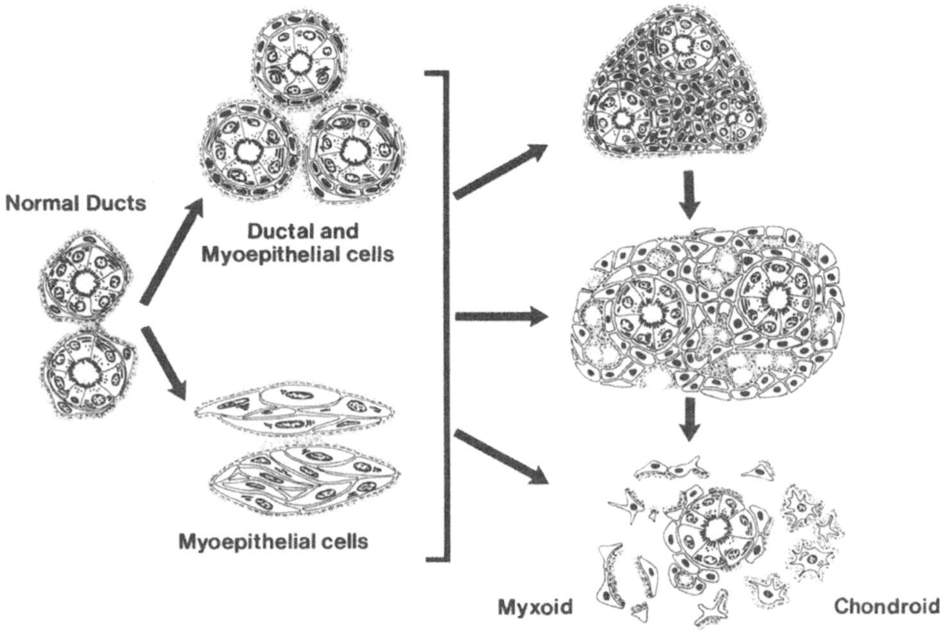

Fig. 5-8. Model diagram of morphogenetic processes in pleomorphic adenoma. Proliferation in the tumor mimics the elements in the normal ducto-acinar unit so that some pleomorphic adenomas will have regions with tubular or glandlike structures composed of both luminal and myoepithelial cells, while other regions will be formed almost exclusively of neoplastic myoepithelial cells (left side of diagram). At the right of the model, cellular regions (either with luminal and myoepithelial or myoepithelial cells alone) may or may not develop varying amounts of intercellular proteogylcans (stippled regions) and basal lamina (dashed lines). Myxochondroid regions result when proteoglycan production is excessive, forcing separation of the modified myoepithelial cells.

from determining histogenesis of salivary gland tumors and places it on understanding the morphogenesis of these lesions, a concept that has more practical application to improving the classification of this histologically diverse group. The patterns of tumor cell differentiation and the formation of the basal-lamina-lined intercellular spaces in relation to the modified myoepithelial cells are evident in survey-type and higher magnification electron micrographs of pleomorphic adenoma (Figs. 5-3 and 5-4). The gradual enlargement of specific intercellular regions and the role they play in producing the myxoid and chondroid patterns (Fig. 5-8) have been documented by electron microscopy [14, 17, 18, 50]. Taken together, these differentiation and synthetic factors allow an understanding of the marked variation in histology both within and between cases of pleomorphic adenoma.

Although, at the ultrastructural level, luminal cells are usually quite well organized and differentiated (they may even show acinar cell differentiation in the form of serous-type granules [14, 20]), the myoepithelial cell component infrequently expresses myofilaments [16–18, 20, 33]. However, typical myoepithelial cells with the same relationship to luminal cells, as evident in Figs. 5-3 and 5-8, were illustrated in electron micrographs by Chaudhry et al. [21].

Two ultrastructural studies of pleomorphic adenoma [53, 54] have attempted to quantitate the various types of tumor cells seen in this tumor. A principal conclusion of both studies is that myoepithelial cells have a minimal role in pleomorphic adenoma — the major cell type being a modified intercalated duct epithelial cell — since classical myoepithelial cells are rarely observed. When individual tumor cell assessment is based solely on its cytological characteristics, as was the case in these two studies [53, 54], this is the expected result. For example, by simply categorizing a tonofilament-rich cell as a ductal cell

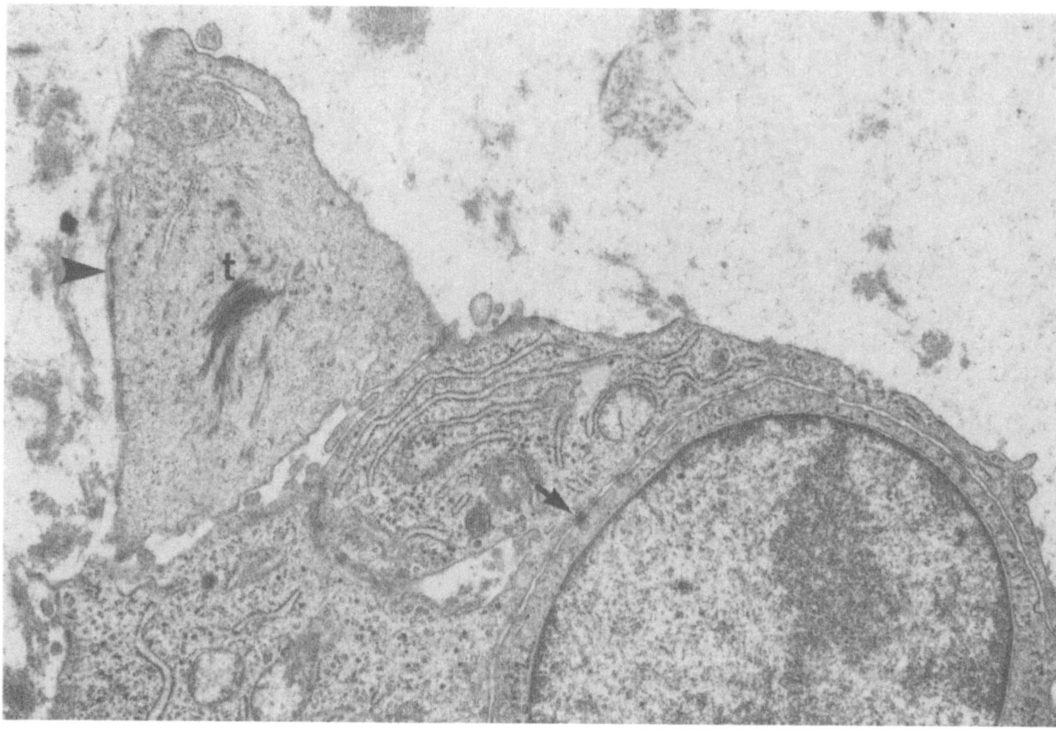

Fig. 5-9. Pleomorphic adenoma, myxochondroid region. What appears to be an isolated mesenchymal-like cell by light microscopy is in reality a small group of adjacent tumor cells joined by tonofilament-associated demosomes (arrow), partially encased with basal lamina (arrowhead) and containing tonofilaments (t). Such features, along with the processes illustrated in Figs. 5-3, 5-4, and 5-8, are the type of evidence indicating a myoepithelial cell origin for the majority of tumor cells in myxoid zones. Uranyl acetate and lead citrate. x 15,900.

without noting the organizational aspects of tumor cells in pleomorphic adenoma, as well as the modifications common to neoplastic myoepithelial cells [17, 18, 20], the role of this type of cell can be overemphasized. Such results do not correlate with light and electron microscopic observations that considerable regions of modified myoepithelium (often exhibiting squamous metaplasia as well as chondrocytic modification) surround a lesser component of duct-cell-lined lumens [14, 15, 17, 18, 20, 33, 55]. We concur with Erlandson and associates [20] that the mesenchymal-like regions of pleomorphic adenoma are formed of metaplastic epithelial, primarily myoepithelial, cells.

Evidence from both immunohistochemical [56, 57] and ultrastructural [17, 18, 20, 51] studies further indicates that the "mesenchymal" component of pleomorphic adenoma derive from metaplastic myoepithelial cells. Electron microscopy of myxoid and chondroid areas in some pleomorphic adenomas reveals tumor cells with tonofilaments [51], desmosomes, and focal basal lamina [Fig. 5-9]. Such features are absent in normal chondrocytes. Both the epithelial and mesenchymal-like cells in pleomorphic adenoma are capable of cell division [58].

Appreciation of the developmental processes underlying the varied histomorphology of pleomorphic adenoma has more than theoretical value. This is best illustrated by the biopsy of a parapharyngeal mass incidentally discovered in an 86-year-old woman being treated for a pulmonary embolus. On the basis of histology showing both a glandular and stromal component that was keratin positive and vimentin positive, respectively, the tumor was diagnosed as a biphasic synovial sarcoma (Fig. 5-10). When elec-

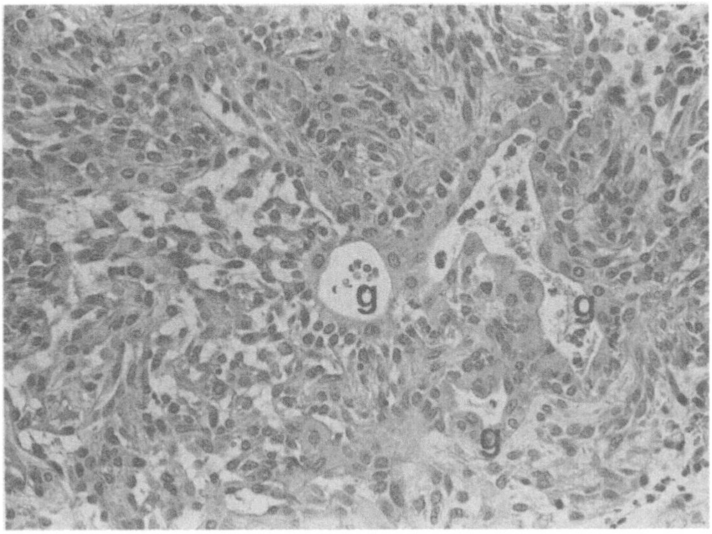

Fig. 5-10. Pleomorphic adenoma, parapharyngeal area. The tumor is composed of cuboidal cells forming glandular structures (g) surrounded by irregularly shaped and somewhat separated, mesenchymal-appearing cells, features initially suggesting a diagnosis of synovial sarcoma. Electron microscopy (Fig. 5-11) supported a diagnosis of pleomorphic adenoma. Hematoxylin and eosin. x 250.

tron micrographs revealed a two-cell population with one cell type related to basal-lamina-lined and glycosaminoglycan-containing intercellular spaces (Fig. 5-11), and a monoclonal antibody (anti-cytokeratin 14) selective for myoepithelium [34, 35] showed that the vimentin positive "stromal" cells were also keratin positive, the diagnosis was altered to a pleomorphic adenoma arising in the deep lobe of the parotid salivary gland.

6. Monomorphic adenoma

As described by Batsakis and associates [59], this is a histologically diverse group of lesions encompassing such tumors as basal cell, tubular, trabecular, canalicular, and other adenomas. By light microscopy, these tumors are usually described as being isocellular, hence their designation as monomorphic adenomas [60]. However, by electron microscopy, such tumors fall into two main groups, one of which is composed only of luminal epithelial cells, e.g., canalicular adenoma [61–63], while the other form has a complement of both luminal and modified myoepithelial cells [19]. Luminal cells in both the monocellular and bicellular tumors are cuboidal to columnar in form; closely apposed on their lateral aspects, which are joined by well-developed tight junctions and desmosomes; and bear apical microvilli. In contrast, electron microscopy of bicellular monomorphic adenomas reveals that the myoepithelial cell population is varied in form and arrangement, is frequently separated by widened intercellular spaces, may be associated with basal-lamina-lined and proteoglycan-containing intercellular regions, and only infrequently displays typical features of myoepithelium in the form of myofilaments and micropinocytotic vesicles [16, 19, 64, 65]. However, the organization of myoepithelial-like cells in monomorphic adenoma and their relationship to specific basal-lamina-lined spaces mirrors that described for pleomorphic adenoma [14]. These ultrastructural features are evident in a tubulo-trabecular adenoma that by light microscopy appeared isocellular (Fig. 5-12).

Ultrastructural studies of monomorphic adenomas have resulted in an appreciation of the range of differentiation possible in this group [14] and the rationale behind the varied morphology so often responsible for difficulties in the differential diagnosis and classification of this category of salivary gland tumors. The complexity of dif-

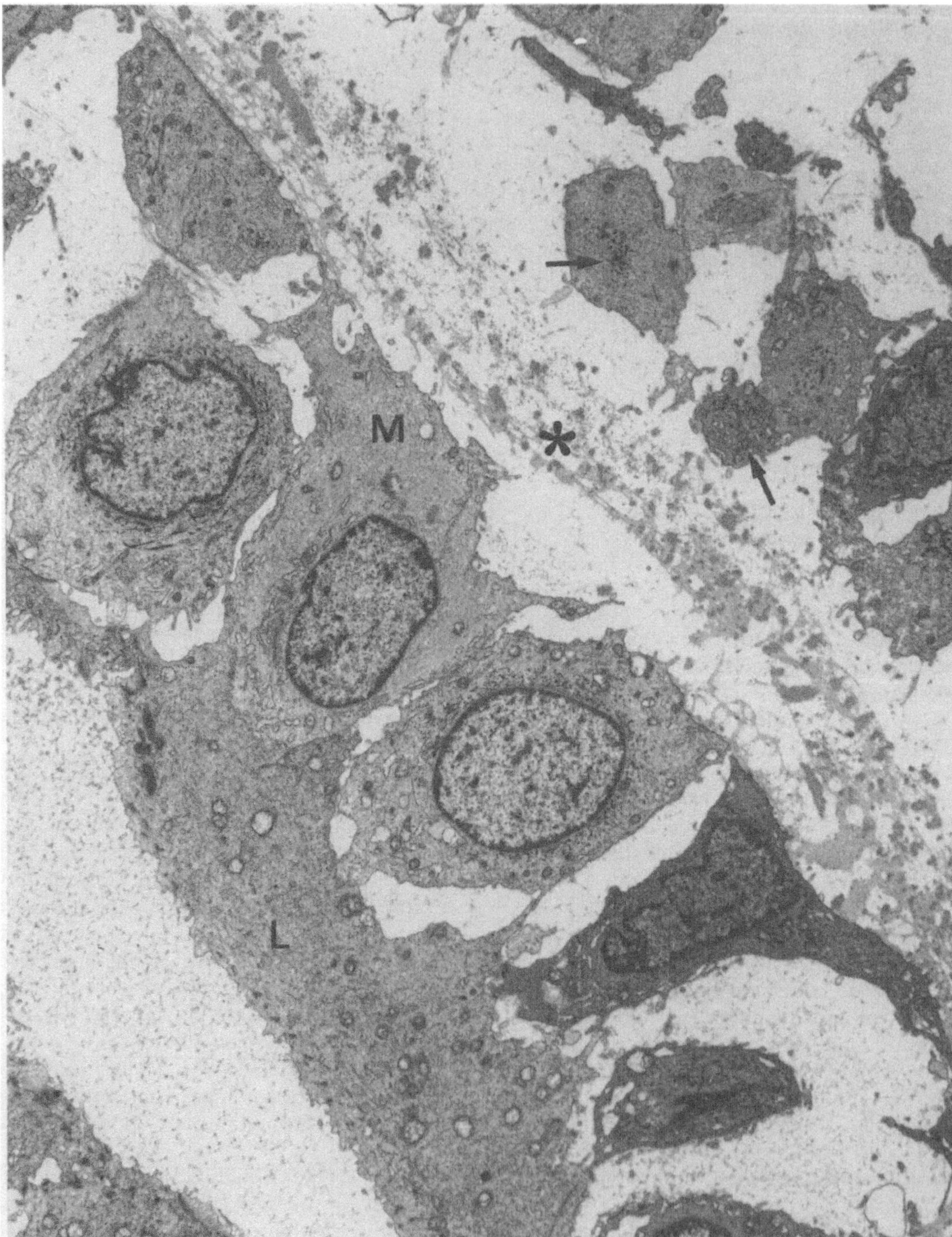

Fig. 5-11. Pleomorphic adenoma, parapharyngeal area. Ultrastructurally, the glands are formed of two cell types; closely associated, microvillus-bearing inner cells (L) and an outer layer of polygonal or irregular shaped tumor cells (M) separated by widened intercellular spaces. A layer of reduplicated basal lamina (asterisk) and proteoglycans separates the glandular component from loosely arranged tumor cells containing tonofilament aggregates (arrows). Uranyl acetate and lead citrate. x 5,800.

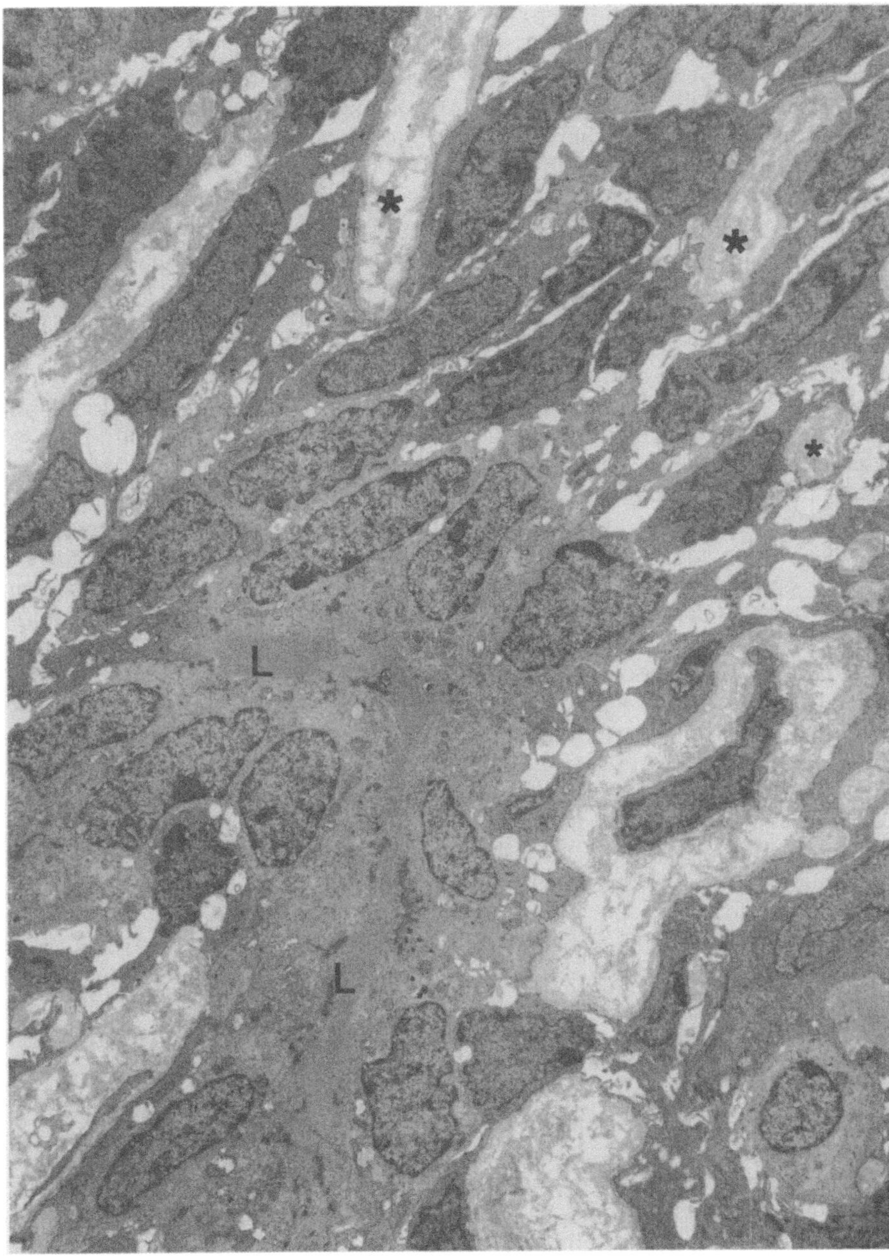

Fig. 5-12. Monomorphic adenoma, parotid gland. Survey-type electron micrograph reveals that at least some tumors in this class ae also composed of cells enclosing a lumen (L). Angular tumor cells, separated by widened intercellular spaces, are also associated with basal lamina-lined intercellular spaces of varying sizes (asterisks). Uranyl acetate and lead citrate. x 3,800.

ferentiation in the monomorphic adenomas is further illustrated by a recent ultrastructural and immunohistochemical survey of five examples of an unusual variant [66]. In this subtype of basal cell and tubulo-trabecular adenomas, there was a separate population of myoepithelial-like cells interspersed within the stromal-type regions between the compact nests and cords of epithelial cells composed of luminal and modified myoepithelials cells [66].

7. Adenoid cystic carcinoma

Ultrastructural studies have confirmed the presence of small glandular or ductlike lumens lined by closely apposed luminal epithelial cells, bearing microvilli on the apical surfaces, surrounded by the more numerous angular and variably shaped modified myoepithelial cells [22, 23, 25, 67]. Between the myoepithelial cell component, usually, there are many widened intercellular spaces interrupted by broad desmosome-bearing cytoplasmic processes projecting from adjacent cells. Both cell types may display tonofilament bundles, but these are usually more obvious in the luminal cells [24, 25]. The development of myofilaments in the modified myoepithelial cells of adenoid cystic carcinomas is variable, both in the extent and number of tumor cells involved in each case, but it is a feature reported in a number of studies [22–25, 40, 67, 68]. The electron microscopic counterpart of the pseudolumens (pseudocystic spaces or spheroidal bodies) are basal-lamina-lined intercellular spaces arising in relation to the myoepithelial cell component. Such spaces either contain an excess of basal lamina and glycosaminoglycans or admixtures of glycosaminoglycans, collagen, and elastic tissue [22, 23, 25]. Recently, the ultrastructural features of tyrosine-containing crystalloids within pseudocystic spaces of a palatal adenoid cystic carcinoma were described by Gould et al. [69].

The complexity of cellular differentiation in adenoid cystic carcinoma is evident in the presence of occasional neoplastic myoepithelial cells abutting on both the basal-lamina-lined pseudolumen and the true lumen, with the latter cellular surface bearing microvilli [14]. This feature has also been reported in a salivary gland monomorphic adenoma [16]. A gradual transition of ultrastructural cellular features from myoepithelial cells to luminal cells is also apparent in adenoid cystic carcinoma [25, 70]. Based on morphological ultrastructural evidence, Chaudhry and associates [40] have suggested that adenoid cystic carcinomas are formed by intercalated duct, myoepithelial, secretory, and pluripotential reserve/stem cells. In the case of the latter cell type, the authors have failed to provide a rationale for labeling them as *stem* cells or evidence that the described cells directly differentiate as intercalated duct and myoepithelial cells. Histogenetic implications and proliferative capacities can not be easily inferred from ultrastructural observations. However, by making possible the observation of cell-to-cell relationships and specific synthetic activities of tumor cells in adenoid cystic carcinoma, electron microscopy has improved the understanding of the development of the varied histological patterns in this tumor, i.e., solid, tubular, and cribriform types. In addition, the basic similarities of cell types, organizational aspects, and synthetic processes between pleomorphic adenoma and adenoid cystic carcinoma is evident in survey-type electron micrographs from the two lesions [14].

Figure 5-13 shows the placement of luminal cells and modified myoepithelium and the formation of basal-lamina-lined intercellular spaces in a cellular zone between pseudocysts of a typical cribriform nest in a submandibular gland, adenoid cystic carcinoma. By comparing Fig. 5-13 with Fig. 5-3, it is readily apparent that basic structural and functional aspects of adenoid cystic carcinoma and pleomorphic adenoma are identical.

8. Acinic cell carcinoma

Recent ultrastructural studies of acinic cell carcinoma [40,43] have revealed a greater degree of tumor cell differentiation and organizational complexity in this neoplasm than had been acknowledged in some earlier ultrastructural studies [71–73]. Mainly secretory granule morphology in acinar cells had been described in these studies. Although differentiation of intercalated duct cells has been apparent at both the light [74–76] and electron [67, 77–81] microscopic levels, we have

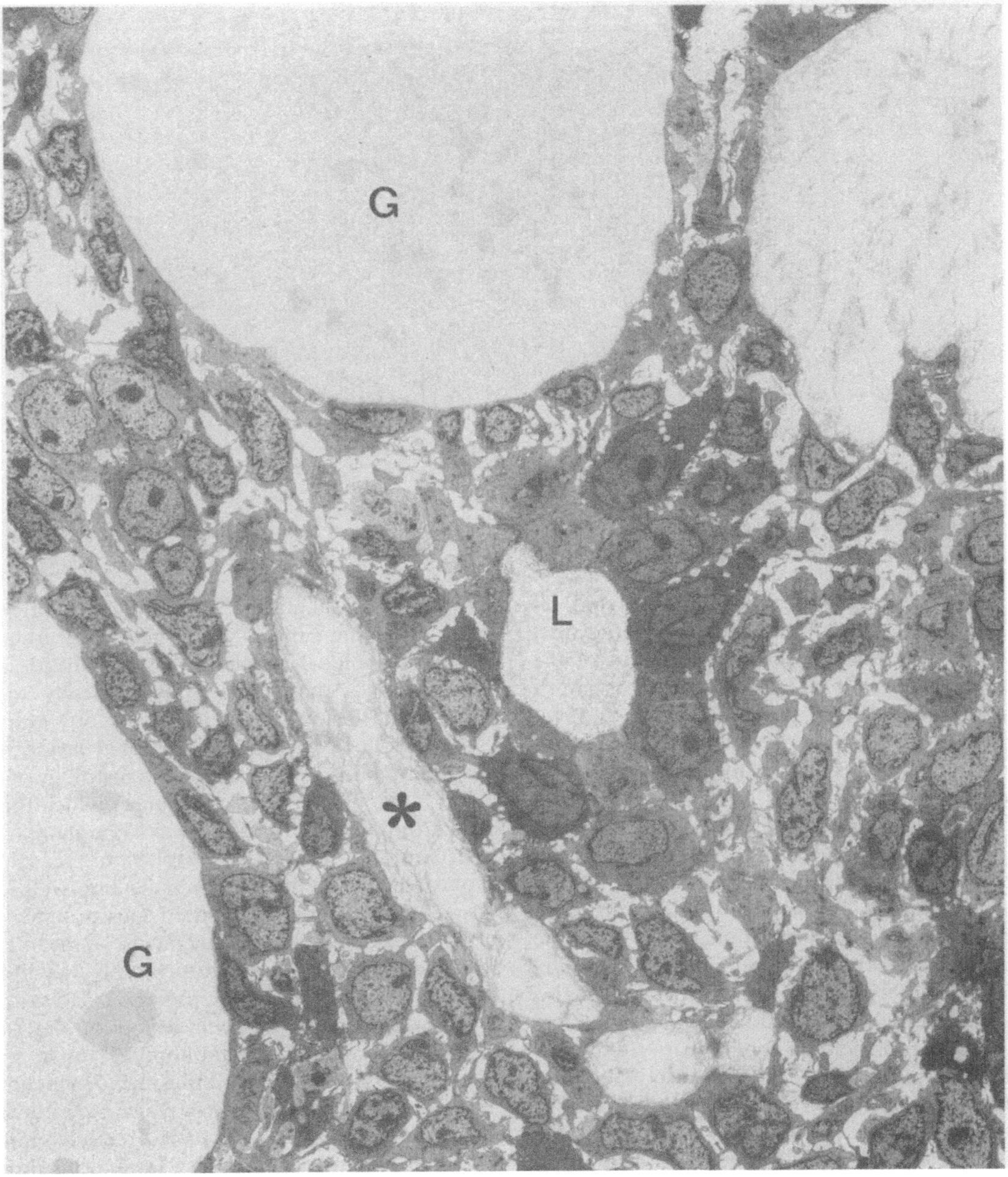

Fig. 5-13. Adenoid cystic carcinoma, submandibular gland. In a cellular area of a cribriform region of this tumor, epithelial cells form a lumen (L) complete with microvilli on the apical surfaces. The surrounding loosely arranged tumor cells, with many cytoplasmic processes, surround intercellular spaces, some of which contain reduplicated basal lamina (asterisk), while others contain excessive glycosaminoglycans (G), the so-called pseudocysts or pseudolumens. Uranyl acetate and lead citrate. x 2,600.

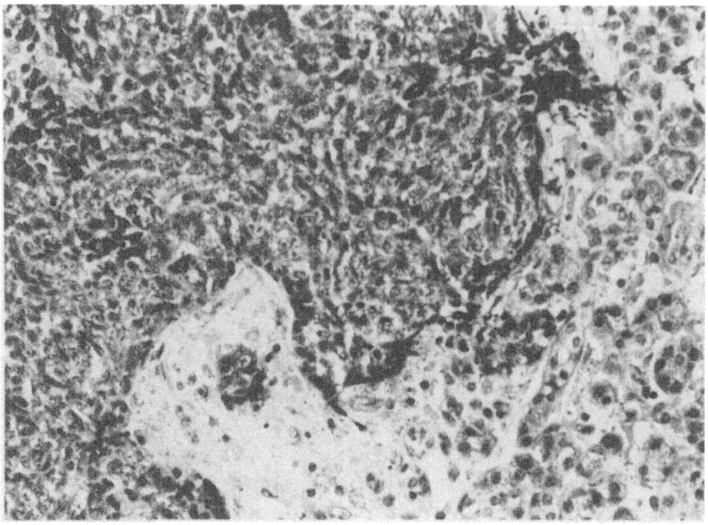

Fig. 5-14. Small-cell epidermoid carcinoma, parotid gland. A sheet of small tumor cells, exhibiting considerable crush artifact, invade adjacent degenerating salivary gland parenchyma (right). Hematoxylin and eosin. x 250.

provided electron microscopic evidence that dilated lumens lined by flattened intercalated duct cells are responsible for the development of the microcystic pattern in acinic cell carcinoma [82]. Ultrastructurally, there is no evidence that vacuolization of tumor cell cytoplasm, due to degenerative changes, is the mechanism for this histological feature, as had been suggested previously [74, 75]. Well-organized, small ductal structures are also evident coursing between the secretory-granule-bearing acinar cells [14, 82]. In addition, in two series myoepithelial cells have now been reported in 4 of 19 examples of acinic cell carcinoma examined by electron microscopy [81, 82]. Chaudhry and associates [81] have also identified a cell type in acinic cell carcinoma that they indicate represents a progenitor (stem) cell. Unfortunately, this conclusion is based solely on morphological features of what is interpreted as an "undifferentiated" cell in this tumor. No direct, or even indirect, evidence is offered to support claims that this type of cell is part of the dividing cell pool or that its presence bolsters the reserve cell hypothesis [3, 4].

9. Small cell tumors

In salivary gland lesions, electron microscopy has played a dual role: one of these has been the differential diagnosis of small cell tumors and the other the subtyping of small cell carcinomas.

In the elderly patient with a parotid salivary gland mass that proves histologically to be a small cell lesion, electron microscopy may allow separation of a primary epithelial lesion from a lymphoma. Fig. 5-14 shows the histology of such a lesion from a 76-year-old man in which the clinicopathological impression was a lymphoma. Ultrastructurally, however, despite the scanty cytoplasmic organelles, the small, but well-formed and tonofilament-associated desmosomes (Fig. 5-15) established the diagnosis of a small cell carcinoma. Neurosecretory-type cytoplasmic granules were absent in this case. However, the most common small cell lesion of the salivary gland is a malignant lymphoma. Lymphoma cells will also have few cytoplasmic organelles, but usually have more irregularly contoured nuclear profiles and an absence of intercellular junctions and basement membranes. Immunophenotyping will confirm the B- or T-cell nature of such a lesion.

The example illustrated in Figs. 5-14 and 5-15 reflects the features of one subtype of primary small cell carcinoma of the salivary gland, the epidermoid or ductal form, in which tonofilaments and desmosomes are the principal diagnostic feature. The ultrastructural aspects of a small number of such cases have been described

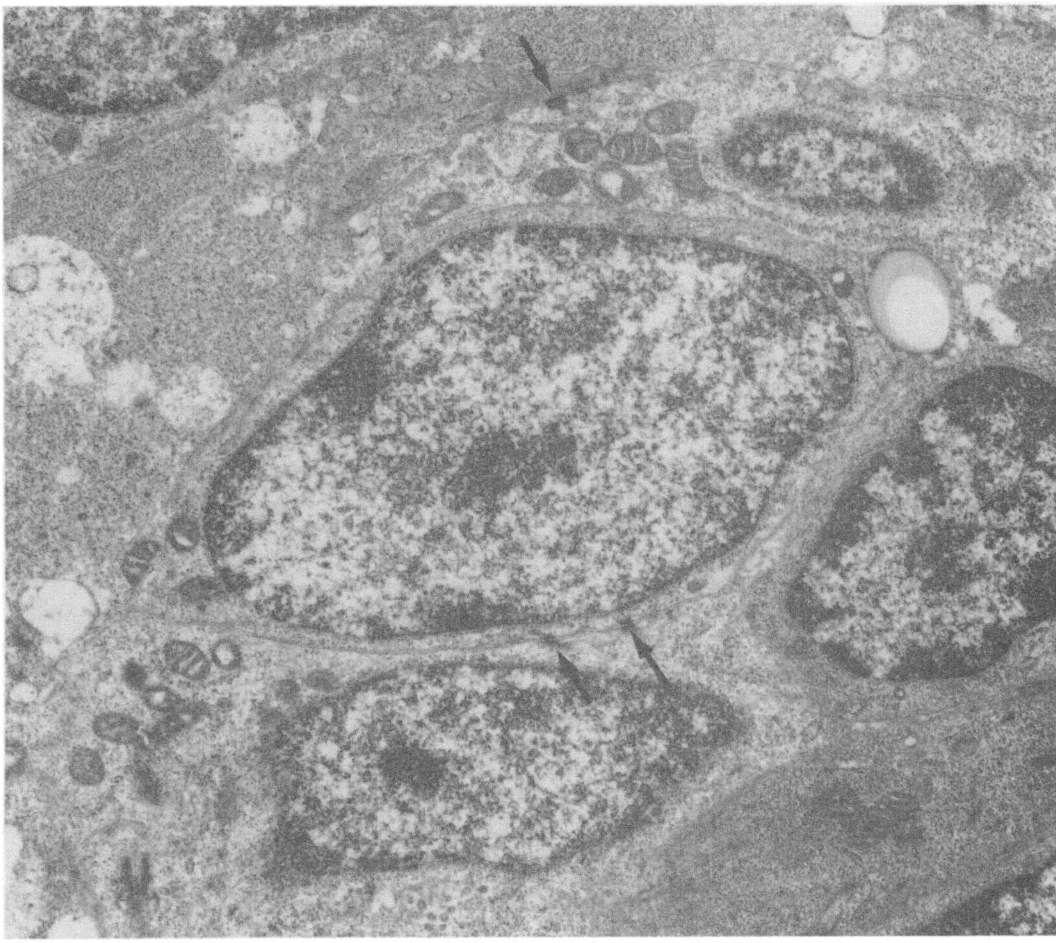

Fig. 5-15. Small-cell epidermoid carcinoma, parotid gland. Ultrastructurally, the tumor cells have a high nuclear to cytoplasmic ratio and scanty cytoplasmic organelles, features suggesting a lymphoma. However, the presence of small desmosomes (arrows) with associated tonofilaments establishes the lesion as a carcinoma. Uranyl acetate and lead citrate. x 15,400.

in the literature [83–87]. In the second type, small-cell neuroendocrine carcinoma, membrane-bound neurosecretory granules are present within the cytoplasm, particularly in blunt cytoplasmic processes between adjacent tumor cells [86,87]. In most reported cases, the number of neurosecretory granules has been limited, and these usually range in size from 80 nm to 240 nm in diameter [43,84,85]. Although this is the size range of neurosecretory granules, immunohistochemical studies corroborating the neuropeptide content of these granules are required. Since neuroendocrine cells have not been reliably identified in the human salivary gland by electron microscopy or immunohistochemistry, the presence of neurosecretory granules may simply represent the morphological result of abnormal gene expression in some small cell and anaplastic carcinomas. What influences this will have on prognostic aspects in such carcinomas will await reporting of additional cases with ultrastructural and, hopefully, immunohistochemical confirmation. Interestingly, of the two subtypes, ultrastructural studies to date reveal that the neuroendocrine form is the least common [86]. As a group, small cell carcinomas of the salivary gland appear to have a more favorable clinical course than morphologically comparable carcinomas of the lung [86].

In terms of the potential avenues of differentiation in small cell carcinomas of the salivary gland, it is noteworthy that a component of myoepithelial cells has been described by electron microscopy of three samples [83, 84, 87].

10. Experimental studies of salivary gland tumors

Electron microscopy and immunohistochemistry have played a limited role in experimental studies of salivary gland tumors, and such studies have largely been confined to pleomorphic adenomas [52, 88, 89]. Takeuchi and associates [52] performed histochemical and ultrastructural studies on cell lines isolated from five pleomorphic adenomas, three of which had cellular growth patterns and two of which were primarily myxoid. The cellular pleomorphic adenomas contained limited amounts of glycosaminoglycans, and in culture the tumor cells were epithelial-like and developed a glandular growth pattern, as sèen in electron micrographs. In contrast, the myxoid tumors contained isolated, round or spindle cells and produced extensive amounts of chondroitin sulfates, hyaluronic acid, and heparan sulfate [52]. When cultured in flasks, tumor cells from this type of pleomorphic adenoma had nonepithelial features, but culture in a sponge matrix or inoculation of tumor cells into mice produced a typical epithelial growth pattern. Review of an electron micrograph of one of the myxoid lesions (their Fig. 10 of case 4) reveals a growth pattern surrounding basal-lamina-lined and glycosaminoglycan-containing microcystic spaces [52]. These features indicate that the lesion contained a significant component of modified myoepithelial cells [14, 17, 18] and that mesenchymal-like cells in pleomorphic adenoma are of epithelial origin. In a subsequent autoradiographic study, Takeuchi et al. [90] suggested that the tumor cells in pleomorphic adenoma were responsible for the development of the myxoid and chondroid zones in this lesion. Using ultrastructural histochemical techniques, Lam [50] has shown that the modified myoepithelial cells in pleomorphic adenoma are the principal source for the proteoglycan-rich tissues so characteristic of this tumor.

Electron microscopy of in-vitro cultures established from a minor salivary gland pleomorphic adenoma has confirmed the myoepithelial cell differentiation of resulting cloned cell lines [88]. Implantation of such cultured cell lines into nude mice produced myoepitheliomas [88]. These cultured human neoplastic myoepithelial cells have been further characterized immunohistochemically (S100 protein and myosin positive; secretory component, lactoferrin, and carcinoembryonic antigen negative) in a comparative study [91] with a duct epithelial cell line established from a resected human submandibular gland that had been irradiated earlier during a course of radiotherapy for a carcinoma of the floor of the mouth [92]. Not only did the cultured cells have markers for duct-type cells [91], but an adenocarcinoma developed when cloned cells were injected into nude mice [92]. This human salivary duct cell line, grown as tumor nodules in nude mice, has recently been shown to be capable of differentiating into myoepithelial cells (induction of S100 protein and myosin positivity, displayed with appropriate antibodies, and typical myoepithelial cell characteristics ultrastructurally) following injection of dibutyryl cyclic AMP into the tumor nodules [93]. Such findings have important implications for salivary gland histogenesis by suggesting that neoplastic induction, even of a differentiated duct luminal cell, might result in a benign or malignant tumor exhibiting two cell types organized in a caricature of the normal ductal or acinar unit, i.e., luminal or acinar cells and myoepithelial cells.

A few studies have indicated the relative ease with which human pleomorphic adenomas can be established in nude mice [58, 89], despite the small proportion of the tumor cell population that is actually cycling [58, 89]. The observation that labeling indices were similar over solid epithelial and myxoid or chondromyxoid regions is also important [58]. Histological features of the transplanted tumors reflected the original pleomorphic adenoma, and immunohistochemical staining patterns were also maintained during growth in the nude mouse [89].

One model system for the development of pleomorphic adenomas following subcutaneous injection of polyoma virus into the neck of newborn mice has been reported [94]. Resulting tumors have histological features resembling human pleomorphic adenomas. Ductal and myoepithelial cells were evident in cellular and myxoid regions and, ultrastructurally, the myoepithelial cell component is described as having classical features. This aspect is difficult to discern in the electron micrographs provided in the report [94].

Ultrastructural examinations have been part of experimental studies to establish cultures from outgrowths of fragments of human submandibular gland [95] and enzyme-dissociated rat submandibular gland [96]. Despite the success of such techniques, as well as the culture of fragments of ductal tissue from the human breast [97], these systems have not been employed to study the development of tumors in the salivary gland. Ultrastructural studies of the culture of human-breast duct fragments as organoids in collagen gels reveals the continued proliferation of myoepithelial and luminal epithelial cells organized as in the original tissue [97].

A recent ultrastructural and histochemical study has clarified the nature of the crystalloids that occasionally occur in salivary gland tumors [98]. While one case contained tyrosine-rich crystalloids, six others had crystalloids formed by numerous collagen fibers radiating out from a clear, central zone [98]. Intraluminal, laminated bodies (corpora amylacea) have been described in a Warthin's tumor [99] and a low-grade adenocarcinoma of the parotid gland [100]. In the latter, electron microscopy revealed the matted, 10-nm wide fibers typical of amyloid [100].

11. Conclusions and future directions

The studies and findings reviewed in this chapter emphasize the major role of electron microscopy in establishing some of the mechanisms underlying the development of the diverse morphological patterns within any one subtype of salivary gland neoplasm and the wide and unique range of tumor histomorphology in salivary gland tumors generally. Such information will assist in clarifying diagnostic problems in this tumor class and should eventually help reorganize classification schemes for salivary gland neoplasms.

Except in a few situations, such as small cell and spindle cell tumors of the salivary gland, electron microscopy has a minimal role in diagnosis. However, by combining immunocytochemical and ultrastructural characteristics of salivary gland tumors, considerable new information has been and should continue to be generated about the vital role of the myoepithelium in such tumors and the range of structural modification that this cell type is capable of undergoing. We have attempted to organize the key aspects of this new ultrastructural data in order to highlight the main advances that have been achieved in the area of salivary gland pathology. A better understanding of the organization and functional relationships of the various types of cells in normal human salivary gland is needed. Dynamic studies using immunological techniques and autoradiography at the ultrastructural level could have a major application in such studies and could then be applied equally in similar studies in salivary gland tumors. Hopefully, the recent resurgence of interest in the pathology of this organ will continue, as these lesions offer a unique opportunity to acquire information concerning cell biological aspects of tumor cell differentiation and tumor morphogenesis generally.

Acknowledgement

Funded by a generous gift from the Moe Levin Family Foundation, Montreal, Canada. Drs. R. Byard and B. Renick provided valuable research assistance. The technical skills of D. Hoppe, M. Boivin, S. Cavell, P. Rippstein, and B. Parks are appreciated, along with the secretarial assistance provided by M. McDougall.

References

1. van Nostrand AWP: The quiet revolution: techniques of cell biology applied to tissue diagnosis. *Ann Otol Rhinol Laryngol* 95: 601–602, 1986.
2. Eversole LR: Histogenetic classifications of salivary gland tumours. *Arch Pathol* 92: 433–443, 1971.
3. Regezi JA, Batsakis JG: Histogenesis of salivary gland neoplasms. *Otolaryngol Clin North Am* 10: 297–307, 1977.
4. Batsakis JG: Salivary gland neoplasia: an outcome of modified morphogenesis and cytodifferentiation. *Oral Surg* 49: 229–232, 1980.
5. Barka T: Induced cell proliferation: the effect of isoproterenol. *Exp Cell Res* 37: 662–679, 1965.
6. Schneyer CA: β-adrenergic effects of autonomic agents

on mitosis and hypertrophy in rat parotid gland. *Proc Soc Exp Biol Med* 131: 71–75, 1969.
7. Cutler LS, Chaudhry AP: Differentiation of the myoepithelial cells of the rat submandibular gland in vivo and in vitro: an ultrastructural study. *J Morphol* 140: 343–354, 1974.
8. Taga R, Sesso A: Ultrastructural studies on developing parotid gland of the rat at early post-natal periods. *Arch Histol Jpn* 42: 427–444, 1979.
9. Ferguson DJP: Ultrastructural characterization of the proliferative (stem ?) cells within the parenchyma of the normal "resting" breast. *Virchows Arch* (Pathol Anat) 407: 379–385, 1985.
10. Smith CA, Monaghan P, Neville AM: Basal clear cells of the normal human breast. *Virchows Arch* (Pathol Anat) 402: 319–329, 1984.
11. Leeson CR, Jacoby F: An electron microscopic study of the rat submaxillary gland during its post-natal development and in the adult. *J Anat* 93: 287–295, 1959.
12. Jacoby F, Leeson CR: The post-natal development of the rat submaxillary gland. *J Anat* 93: 201–216, 1959.
13. Young JA, van Lennep EW: The Morphology of Salivary Glands. London: Academic Press, 35–44, 1978.
14. Dardick I, van Nostrand AWP: Morphogenesis of salivary gland tumors: a prerequisite to improving classification. *Pathol Annu* 22 (pt 1): 1–53, 1987.
15. Batsakis JG, Kraemer B, Sciubba J: The pathology of head and neck tumours: the myoepithelial cell and its participation in salivary gland neoplasia, part 17. *Head Neck Surg* 5: 222–233, 1983.
16. Dardick I, van Nostrand AWP: Myoepithelial cells in salivary gland tumours — Revisited. *Head Neck Surg* 7: 395–408, 1985.
17. Dardick I, van Nostrand AWP, Jeans MTD, Rippstein, P, Edwards V: Pleomorphic adenoma. I. Ultrastructural organization of "epithelial" regions. *Hum Pathol* 14: 780–797, 1983.
18. Dardick I, van Nostrand AWP, Jeans MTD, Rippstein, P, Edwards V: Pleomorphic adenoma. II. Ultrastructural organization of "stromal" regions. *Hum Pathol* 14: 798–809, 1983.
19. Dardick I, Kahn HJ, van Nostrand AWP, Baumal R: Salivary gland monomorphic adenoma: ultrastructural, immunoperoxidase and histogenetic aspects. *Am J Pathol* 115: 334–348, 1984.
20. Erlandson RA, Cardon-Cardo C, Higgins PJ: Histogenesis of benign pleomorphic adenoma (mixed tumour) of the major salivary glands: an ultrastructural and immunohistochemical study. *Am J Surg Pathol* 8: 803–820, 1984.
21. Chaudhry AP, Satchidanand S, Peer R, Culter LS: Myoepithelial cell adenoma of the parotid gland: a light and ultrastructural study. *Cancer* 49: 288–293, 1982.
22. Hoshino M, Yamamoto I: Ultrastructure of adenoid cystic carcinoma. *Cancer* 25: 186–198, 1970.
23. Tandler B: Ultrastructure of adenoid cystic carcinoma of salivary gland origin. *Lab Invest* 24: 504–512, 1971.
24. Lawrence JB, Mazur MT: Adenoid cystic carcinoma: a comparative pathologic study of tumors in salivary gland, breast, lung and cervix. *Hum Pathol* 13: 916–924, 1982.
25. Orenstein JM, Dardick I, van Nostrand AWP: Ultrastructural similarities of adenoid cystic carcinoma and pleomorphic adenoma. *Histopathology* 9: 623–638, 1985.
26. Donath K, Seifert G, Schmitz R: Zur Diagnose und Ultrastruktur des tubularen Speichelgangcarcinoma: Epithelial-myoepitheliales Schaltstuckcarcinom. *Virchows Arch* (Pathol Anat) 356: 16–31, 1972.
27. Corio RL, Sciubba JJ, Brannon RB, Batsakis JG: Epithelial-myoepithelial carcinoma of intercalated duct origin. *Oral Surg* 53: 280–287, 1982.
28. Daley TD, Wysocki GP, Smout MS, Slinger RP: Epithelial-myoepithelial carcinoma of salivary glands. *Oral Surg* 57: 512–519, 1984.
29. Luna MA, Ordonez NG, Mackay B, Batsakis JG, Guillamondegui O: Salivary epithelial-myoepithelial carcinomas of intercalated ducts: a clinical, electron microscopic, and immunocytochemical study. *Oral Surg* 59: 482–490, 1985.
30. Dardick I, Daya D, Hardie J, van Nostrand AWP: Mucoepidermoid carcinoma: ultrastructural and histogenetic aspects. *J Oral Pathol* 13: 342–358, 1984.
31. Shirasuna K, Watatani K, Miyazaki T: Ultrastructure of a sialadenoma papilliferum. *Cancer* 53: 468–474, 1984.
32. Dardick I, Jeans MTD, Sinnott NM, Wittkuhn JF, Kahn HJ, Baumal R: Salivary gland components involved in the formation of squamous metaplasia. *Am J Pathol* 119: 33–43, 1985.
33. Dardick I, van Nostrand AWP, Phillips MJ: Histogenesis of salivary gland pleomorphic adenoma (mixed tumor) with an evaluation of the role of the myoepithelial cell. *Hum Pathol* 13: 62–75, 1982.
34. Dairkee SH, Blayney C, Smith HS, Hackett AJ: Monoclonal antibody that identifies human myoepithelium. *Proc Natl Acad Sci USA* 82: 7409–7413, 1985.
35. Dairkee SH, Blayney-Moore CM, Smith HS, Hackett AJ: Concurrent expression of basal and luminal epithelial markers in cultures of normal human breast analyzed using monoclonal antibodies. *Differentiation* 32: 93–100, 1986.
36. Kahn LB: Benign lymphoepithelial lesion (Mikulicz's disease) of the salivary gland: an ultrastructural study. *Hum Pathol* 10: 99–104, 1979.
37. Boquist J, Kumlien A, Ostberg Y: Ultrastructural findings in a case of benign lymphoepithelial lesion (Sjögren's syndrome). *Acta Otolaryng* 70: 216–226, 1970.
38. Kjorell U, Ostberg Y: Distribution of intermediate filaments and actin microfilaments in parotid autoimmune sialoadenitis of Sjögren syndrome. *Histopathology* 8: 991–1011, 1984.
39. Palmer RM, Eveson JW, Gusterson BA: 'Epimyoepithelial' islands in lymphoepithelial lesions. *Virchows Arch* (Pathol Anat) 408: 603–609, 1986.
40. Chaudhry AP, Yamane GM, Satchidanand S, Labay G, Sunderraj M: Light and ultrastructural features of lymphoepithelial lesions of the salivary glands in Mikulicz's disease. *J Pathol* 146: 239–250, 1986.
41. Donath K, Seifert G: Ultrastruktur und Pathogenese der myoepithelialen Sialadenitis: uber das vorkommen von Myoepithelzellen bei der benignen lymphoepithelialen

Lesion. *Virchows Arch Abt A Path Anat* 356: 315–329, 1972.
42. Caselitz J, Osborn M, Wustrow J, Seifert G, Weber K: Immunohistochemical investigations on the epimyoepithelial islands in lymphoepithelial lesions: Use of monoclonal keratin antibodies. *Lab Invest* 55: 427–432, 1986.
43. Dardick I, van Nostrand AWP, Rippstein P, Skimming L, Hoppe D, Dairkee SH: Characterization of epimyoepithelial islands in benign lymphoepithelial lesions of major salivary gland: an immunohistochemical and ultrastructural study. *Head Neck Surg*, 10:168–178, 1988.
44. Sciubba JJ, Brannon RB: Myoepithelioma of salivary glands: report of 23 Cases. *Cancer* 49: 562–572, 1982.
45. Sciubba JJ: Diagnostic electron microscopy of salivary gland tumors. *The Cancer Bull* (MD Anderson Hosp and Tumor Inst) 36: 95–105, 1984.
46. Lomax-Smith JD, Azzopardi JG: The hyaline cell: a distinctive feature of "mixed" salivary gland tumours. *Histopathology* 2: 77–92, 1978.
47. Luna MA, Mackay B, Gamez-Araujo J: Myoepithelioma of the palate: review of a case with histochemical and electron microscopic observations. *Cancer* 32: 1429–1435, 1973.
48. Tanimura A, Nakamura Y, Nagayama K, Tanaka S, Hachisuka H: Myoepithelioma of the parotid gland: report of two cases with immunohistochemical technique for S-100 protein and electron microscopic observation. *Acta Pathol Jpn* 35: 409–417, 1985.
49. Kahn LB, Schoub L: Myoepithelioma of the palate: histochemical and ultrastructural observations. *Arch Pathol* 95: 209–212, 1973.
50. Lam RMY: An electron microscopic histochemical study of the histogenesis of major salivary gland pleomorphic adenoma. *Ultrastruct Pathol* 8: 207–223, 1985.
51. Mills SE, Cooper PH: An ultrastructural study of cartilaginous zones and surrounding epithelium in mixed tumors of salivary glands and skin. *Lab Invest* 44: 6–12, 1981.
52. Takeuchi J, Sobue M, Yoshida M, Esaki T, Katoh Y: Pleomorphic adenoma of the salivary gland with special reference to histochemical and electron microscopic studies and biochemical analysis of glycosaminoglycans in vivo and in vitro. *Cancer* 36: 1771–1789, 1975.
53. Chisholm DM, Waterhouse JP, Kraucunas JP, Sciubba JJ: A quantitative ultrastructural study of the pleomorphic adenoma (mixed tumor) of human minor salivary glands. *Cancer* 34: 1631–1641, 1974.
54. Palmer RM, Lucas RB, Langdon JD: Ultrastructural analysis of salivary gland pleomorphic adenoma, with particular reference to myoepithelial cells. *Histopathology* 9: 1061–1076, 1985.
55. Evans RW, Cruickshank AH: Epithelial tumours of the salivary gland. Major Problems in Pathology. JL Bennington (ed), Philadelphia: WB Saunders, 167–225, 1970.
56. Caselitz J, Loning T, Staquet MJ, Seifert G, Thivolet J: Immunohistochemical demonstration of filamentous structures in the parotid gland: occurrence of keratin and actin in normal and tumoral parotid gland with special respect to the myoepithelial cells. *J Cancer Res Clin Oncol* 100: 59–68, 1981.
57. Krepler R, Denk H, Artilier U, Moll R: Immunohistochemistry of intermediate filament proteins present in pleomorphic adenomas of the human parotid gland: characterization of different cell types in the same tumor. Differentiation 21: 191–199, 1982.
58. Barfoed C, Graem N, Bretlau P, Rygaard J: Human pleomorphic adenomas transplanted to nude mice. *Arch Otolaryngol Head Neck Surg* 112: 946–948, 1986.
59. Batsakis JG, Brannon RB, Sciubba JJ: Monomorphic adenomas of major salivary glands: a histologic study of 96 tumours. *Clin Otolaryngol* 6: 129–143, 1981.
60. Gardner DG, Daley TD: The use of the terms monomorphic adenoma, basal cell adenoma and canalicular adenoma as applied to salivary gland tumors. *Oral Surg* 56: 608–615, 1981.
61. Chen S-Y, Miller AS: Canalicular adenoma of the upper lip. An electron microscopic study. *Cancer* 46: 552–556, 1980.
62. Nelson JF, Jacoway JR: Monomorphic adenoma (canalicular type): report of 29 cases. *Cancer* 31: 1511–1513, 1973.
63. Daley TD, Gardner DG, Smout MS: Canalicular adenoma: not a basal cell adenoma. *Oral Surg* 57: 181–188, 1984.
64. Jao W, Keh PC, Swerdlow MA: Ultrastructure of the basal cell adenoma of parotid gland. *Cancer* 37: 1322–1333, 1976.
65. Suzuki J: Basal cell adenoma with acinic differentiation. *Acta Pathol Jpn* 32: 1085–1092, 1982.
66. Dardick I, Daley TD, van Nostrand AWP: Basal cell adenoma with myoepithelial cell-derived "stroma": a new major salivary gland tumor entity. *Head Neck Surg* 8: 257–267, 1986.
67. Hubner G, Klein HJ, Kleinsasser O: Zur Feinstrucktur und Genese der Azinuzelltumoren der Glandula parotis. *Virchows Arch Pathol* 345: 1–14, 1968.
68. Gould VE, Miller J, Wellington J: Ultrastructure of medullary, intraductal, tubular, and adenocystic breast cancers. Comparative patterns of myoepithelial differentiation and basal lamina deposition. *Am J Pathol* 78: 401–416, 1975.
69. Gould AR, van Arsdall LR, Hinkle SJ, Harris WR: Tyrosine-rich crystalloids in adenoid cystic carcinoma: histochemical and ultrastructural observations. *J Oral Pathol* 12: 478–490, 1983.
70. Zaloudek C, Oertel YC, Orenstein JM: Adenoid cystic carcinoma of breast. *Am J Clin Pathol* 81: 297–307, 1984.
71. Kay S, Schatzki PF: Ultrastructure of acinic cell carcinoma of the parotid salivary gland. *Cancer* 29: 235–244, 1972.
72. Martin AG, Palacios JJN, Sanchez LH: Acinar cell carcinoma of minor salivary glands. Light and ultrastructural study of two cases. *Morfologia Normal y Pathologica* 4: 441–450, 1980.
73. Perzin KH, Cantor JO, Johannessen JV: Acinic cell carcinoma arising in nasal cavity: report of a case with

ultrastructural observations. *Cancer* 47: 1818–1822, 1981.

74. Abrams AM, Cornyn J, Scofield HH, Hansen LS: Acinic cell adenocarcinoma of the major salivary glands: a clinicopathologic study of 77 cases. *Cancer* 18: 1145–1162, 1965.
75. Abrams AM, Melrose RJ: Acinic cell tumors of minor salivary gland origin. *Oral Surg* 46: 220–233, 1978.
76. Ellis GL, Corio RL: Acinic cell carcinoma: a clinicopathologic analysis of 294 cases. *Cancer* 52: 542–549, 1983.
77. Hirtzler R, Oberman B, Kulis M, Ljubesic N: Acinuszelladenocarcinome der Speicheldrusen. Bericht uber 4 Falle mit elektronmikropischer Bearbeitung von 2 Geschwulsten. *Arch Klin Exp Ohr Nas Kehlkopfheilk* 195: 68–80, 1969.
78. Echevarria RA: Ultrastructure of the acinic cell carcinoma and clear cell carcinoma of the parotid gland. *Cancer* 20: 563–571, 1967.
79. Erlandson RA, Tandler B: Ultrastructure of acinic cell carcinoma of the parotid gland. *Arch Pathol* 93: 130–140, 1972.
80. Batsakis JG, Wozniak KJ, Regezi JA: Acinous cell carcinoma: a histogenetic hypothesis. *J Oral Surg* 35: 904–906, 1977.
81. Chaudhry AP, Cutler LS, Leifer C, Satchidanand S, Labay G, Yamane G: Histogenesis of acinic cell carcinoma of the major and minor salivary glands and ultrastructural study. *J Pathol* 148: 307–320, 1986.
82. Dardick I, George D, Jeans MTD, Wittkuhn JF, Skimming L, Rippstein P, van Nostrand AWP: Ultrastructural morphology and cellular differentiation in acinic cell carcinoma. *Oral Surg* 63: 325–334, 1987.
83. Wirman JA, Battifora HA: Small cell undifferentiated carcinoma of salivary gland origin: an ultrastructural study. *Cancer* 37: 1840–1848, 1976.
84. Yaku Y, Kanda T, Yoshihara T, Kanekot T, Nagao K: Undifferentiated carcinoma of the parotid gland. *Virchows Arch* (Pathol Anat) 401: 89–97, 1983.
85. Leipzig B, Gonzales-Vitale JC: Small cell epidermoid carcinoma of salivary glands: "pseudo"-oat cell carcinoma. *Arch Otolaryngol* 108: 511–514, 1982.
86. Gnepp DR, Corio RL, Brannon RB: Small cell carcinoma of the major salivary glands. *Cancer* 58: 705–714, 1986.
87. Kraemer BB, Mackay B, Batsakis JG: Small cell carcinomas of the parotid gland: a clinicopathologic study of three cases. *Cancer* 52: 2115–2121, 1983.
88. Shirasuna K, Sato M, Miyazaki T: A myoepithelial cell line established from a human pleomorphic adenoma arising in minor salivary gland. *Cancer* 45: 297–305, 1980.
89. Caselitz J, Lunau U, Hamper K, Walther B, Schmiegelow P: The pleomorphic adenoma of salivary glands transplanted on athymic mice. *Virchows Arch* (Pathol Anat) 408: 191–209, 1985.
90. Takeuchi J, Sobue M, Yoshida M, Sato E: Glycosaminoglycan synthetic activity of pleomorphic adenoma, adenoid cystic carcinoma and nonneoplastic tubuloacinar cells of the salivary gland. *Cancer* 42: 202–208, 1978.
91. Sato M, Hayashi Y, Yoshida H, Yanagawa T, Yura Y, Nitta T: Search for specific markers of neoplastic epithelial duct and myoepithelial cell lines established from human salivary gland and characterization of their growth in vitro. *Cancer* 54: 2959–2967, 1984.
92. Shirasuna K, Sato M, Miyasaki T: A neoplastic epithelial duct cell line established from an irradiated human salivary gland. *Cancer* 48: 745–752, 1981.
93. Hayashi Y, Yanagawa T, Azuma M, Yura Y, Yoshida H, Sato M: Induction of cells with a myoepithelial cell phenotype by treatment with dibutyryl cyclic AMP in human salivary adenocarcinoma cells grown in athymic nude mice. *virchows Arch* (Cell Pathol) 50: 1–11, 1985.
94. Lamey P-J, Waterhouse JP, Ferguson MM: Pleomorphic salivary adenoma: virally induced pleomorphic salivary adenoma in the CFLP mouse. *Am J Pathol* 109: 129–132, 1982.
95. Sens DA, Hintz DS, Rudisill MT, Sens MA, Spicer SS: Explant culture of human submandibular gland epithelial cells : evidence for ductal origin. *Lab Invest* 52: 559–567, 1975.
96. Kanamura S, Barka T: Short term culture of dissociated rat submandibular gland cells. *Lab Invest* 32: 366–372, 1975.
97. Foster CS, Smith CA, Dinsdale EA, Monoghan P, Neville AM: Human mammary gland morphogenesis *in vitro*: the growth and differentiation of normal breast epithelium in collagen gel cultures defined by electron microscopy, monoclonal antibodies and autoradiography. *Dev Biol* 96: 197–216, 1983.
98. Campbell Jr WG, Priest RE, Weathers DR: Characterization of two types of crystalloids in pleomorphic adenomas of minor salivary glands. *Am J Pathol* 118: 194–202, 1985.
99. David R, Buchner A, Corpora amylacea in adenolymphoma (Warthin's tumor). *Am J Clin Pathol* 69: 173–175, 1978.
100. David R, Kim KM: β-fibrillary bodies in low-grade adenocarcinoma of parotid gland: a histochemical and ultrastructural study. *Hum Pathol* 13: 1028–1038, 1982.

Author's address:
Dr. Irving Dardick
Department of Pathology
Banting Institute
100 College Street
Toronto, Ontario
M5G 1L5, Canada

CHAPTER 6

Development and comparative aspects of the pancreas

YASUO UCHIYAMA & MASAHIKO WATANABE

1. Introduction

The pancreas in most vertebrates consists of the exocrine and endocrine portions. The exocrine portion occupies large parts of the pancreas, while the endocrine portion, the islets of Langerhans, is scattered throughout the exocrine portion. The exocrine pancreas is a compound, serous, tubulo-acinar gland whose lobules are bound together by loose connective tissue. The blood and lymphatic vessels, nerves, and excretory ducts run through the interlobular connective tissue. The acini are various in shape: round or short tubular. They are composed of a single layer of pyramidal epithelial cells that encircle a central lumen and are lined by the basement membrane. The acinar cells produce the exocrine secretory product. In general, fine structures of acinar cells show well-developed microvilli on the free surface, desmosomes, a juxtaluminal junctional complex (tight junction, desmosome, and gap junction) on the lateral surface, and a lining of basement membrane on the basal surface. Zymogen granules accumulate in the apical cytoplasm, the Golgi complex is seen in the supranuclear region, and rough endoplasmic reticulum (RER) occupies the basal region. Fine structures of acinar cells, however, greatly vary depending on the circadian stages [1]. Intercalated ducts that enter into the acini (centroacinar cells) are continuous with intralobular and interlobular ducts. The larger ducts consist of columnar epithelial cells, while the smaller, distal ducts consist of cuboidal epithelial cells.

We have previously reported that the rat pancreatic acinar cells exhibit strong circadian variations in morphology and function [1, 2]. The fact that the mature acinar cells vary morphologically and functionally over 24 hours is very important when investigating differentiation of the pancreatic acinar cells; that is, fine structures of the rat pancreatic acinar cells at 7 or 14 days after birth are similar to those of the mature acinar cells. (This will be described in a later section.) However, when examining 24-hour variations in morphological quantities of the acinar cells obtained from rats during the weaning period, their subcellular structures show bimodal changes, which clearly differ from the variations in those of the mature acinar cells. Moreover, the morphological changes in the acinar cells over 24 hours do not exhibit the same patterns as in mature acinar cells until 6 weeks of age. Those results are in good agreement with those reported by Oates and Morgan [3], who have clarified that the nuclear number and DNA content in rat pancreatic acinar cells become constant by 56 days after birth. Thus developmental analysis should reveal not only prenatal changes in morphology and function of the cells, but also their postnatal changes.

In cyclostomes and several bony fishes, no distinct, massive pancreas develops, while microscopic bits of pancreatic tissue are located in the walls of the gut. In those animals, the endocrine islet tissue has no connection with the acinar pancreatic parenchyma [4–6]. As shown in Fig. 6-1, the exocrine pancreatic tissue of the goldfish

Riva, A. and Motta, P.M., (eds.), Ultrastructure of the Extraparietal Glands of the Digestive Tract.
ISBN 978-1-4612-8211-2.

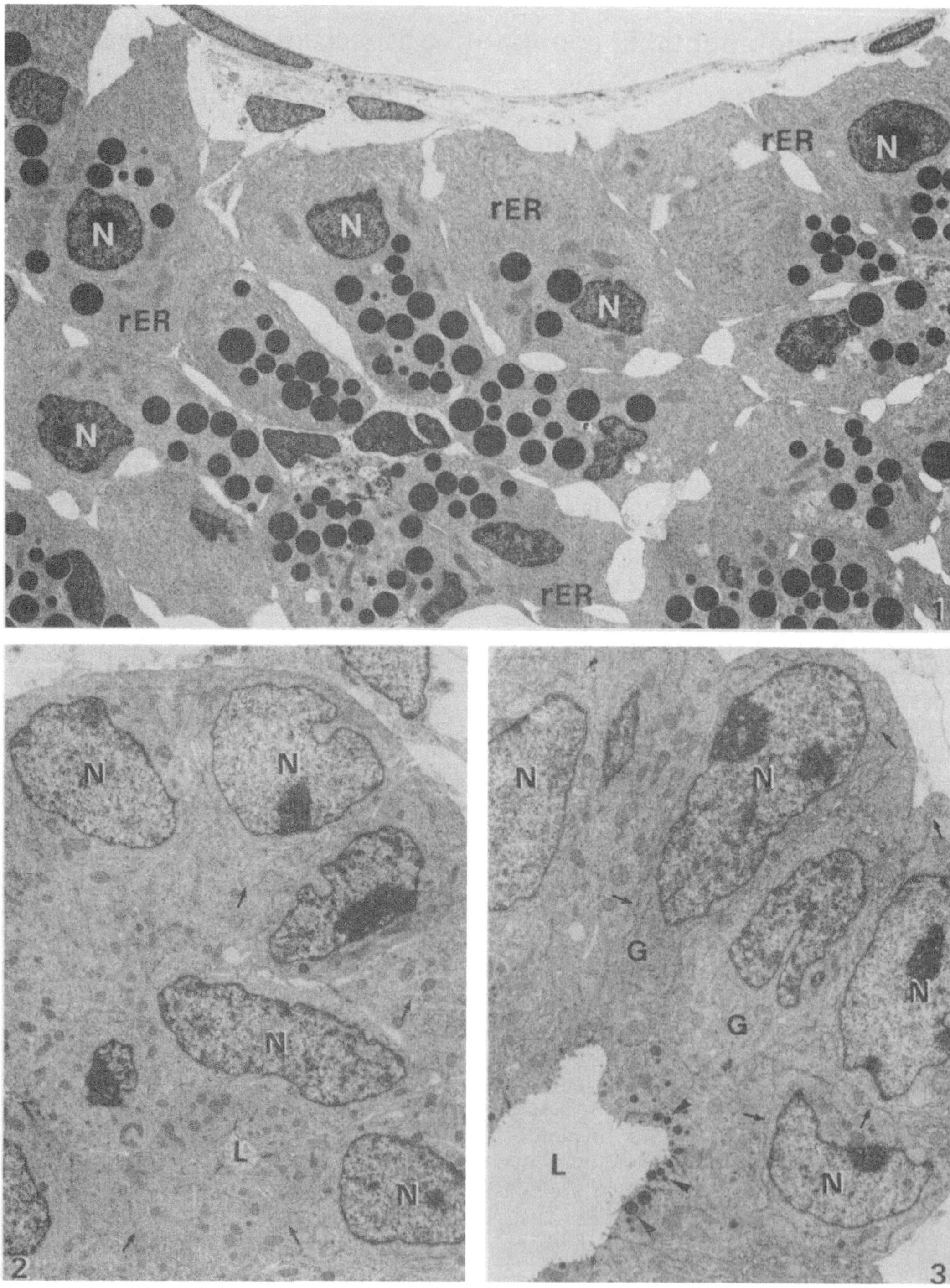
rER
N
rER
N
N
rER
N
N
rER
1
N
N
N
N
L
N
2
N
N
G
G
N
N
L
3

is seen in the dorsal mesentery close to the gut, although almost no pancreatic mass can be detected without a binocular or light microscope. In this bony fish, the endocrine portion can be seen in the exocrine portion. The acinar cells contain large electron-dense zymogen granules in the apical cytoplasm. The cisternal profiles of the RER are arranged in parallel and are densely located in the lateral and basal parts of the cytoplasm. The fine structures of the acinar cells in the goldfish pancreas appear to be similar to those in higher vertebrates. In higher vertebrates, the pancreas is either a massive or compact tissue and is located either in the dorsal mesentery or in the retroperitoneal portion. The fine structures of the pancreatic acinar cells are similar in all vertebrates, as stated above.

The pancreas is formed by two separate primordia of endodermal epithelium just caudal to the presumptive stomach: the dorsal and ventral pancreatic buds. In human embryos, the dorsal pancreatic bud appears earlier on the dorsal wall of the duodenum, while the ventral bud arises later near the hepatic rudiment. At about the seventh week of gestation, the two pancreatic primordia fuse to form a single organ [7]. Falin [8] has reported that secretory granules appear in the acinar cells during the fifth month of fetal life, and, further, that the release of pancreatic juice into the intestinal lumen occurs at this time.

The prenatal development of the exocrine pancreas has been precisely investigated using rat and mouse embryos [9–20]. Structural alterations in developing mouse acinar cells have been described by Munger [12]. From careful comparison of pancreatic histogenesis and cytodifferentiation of acinar cells between in-vivo and in-vitro conditions using rat embryos, Parsa et al. and Pictet et al. [13–15] found no difference in development between the two conditions. In the early phase of rat pancreatic development, the epithelial cells rapidly proliferate but do not distinctly differentiate (into acinar or ductal cells); low levels of special exocrine proteins are dispersed in many cells and are not distributed in a few fully differentiated cells [15]. In the later phase, the specialization of epithelial cells occurs rapidly; the RER and zymogen granules increase in the cytoplasm of the acinar cells. This biphasic pattern of pancreatic differentiation has been considered as two regulatory transitions: the protodifferentiated and cytodifferentiated states [18].

Morphometry is available for the analysis of distinct cellular and subcellular alterations. The point-counting method of Weibel [21] is useful for the estimation of relative and absolute amounts of various cytoplasmic organelles; the relative amounts are expressed as volume, surface and numerical densities (volumes, surface areas, and numbers of objective structures per unit volume, e.g., $\mu m^3/\mu m^3$, $\mu m^2/\mu m^3$, $1/\mu m^3$). These morphometric techniques have been applied to the clarification of rapid and drastic changes in subcellular structures of the pancreatic acinar cells during prenatal and postnatal life [19, 22, 23] and of 24-hour variations in subcellular structures of the mature acinar cells [1, 24]. Using those results, the precise course of cytodifferentiation of rat pancreatic acinar cells and their postnatal development will be explained in the following sections.

2. Cytodifferentiation of rat pancreatic acinar cells during fetal life

The pancreatic diverticulum appears in the gut at 9.5 days of gestation in the mouse and at 11 days in the rat. In the early phase of the rat pan-

←

Fig. 6-1. The acinar cells of the goldfish pancreas. Acinar cells possess numerous profiles of the RER with a parallel arrangement and highly electron-dense zymogen granules. N = nucleus. x 3,000.

Fig. 6-2. At 15 days' gestation. The acinar epithelium consists of several layers of acinar cells, which possess large nuclei (N) with homogeneous karyoplasm. The acinar lumen (L) is still small, and the acinar cell plasmalemma facing the lumen is provided with a few microvilli. Cytoplasmic organelles are poorly developed in the cytoplasm, where free polysomes are densely distributed. In particular, the cisternal profiles of the RER are short and single layered (arrows). x 2,900.

Fig. 6-3. At 17 days' gestation. Acinar cells possess slightly long cisternal profiles of the RER (arrows) and small, round zymogen granules (arrowheads), which chiefly lie beneath the apical plasmalemma. The Golgi complex (G) is seen in the supranuclear region. N = nucleus; L = lumen. x 2,900.

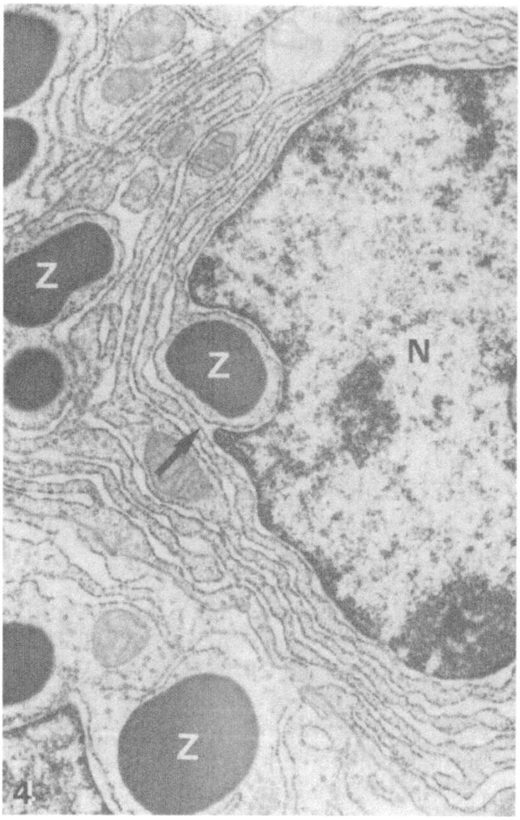
Z
Z
N
Z
4

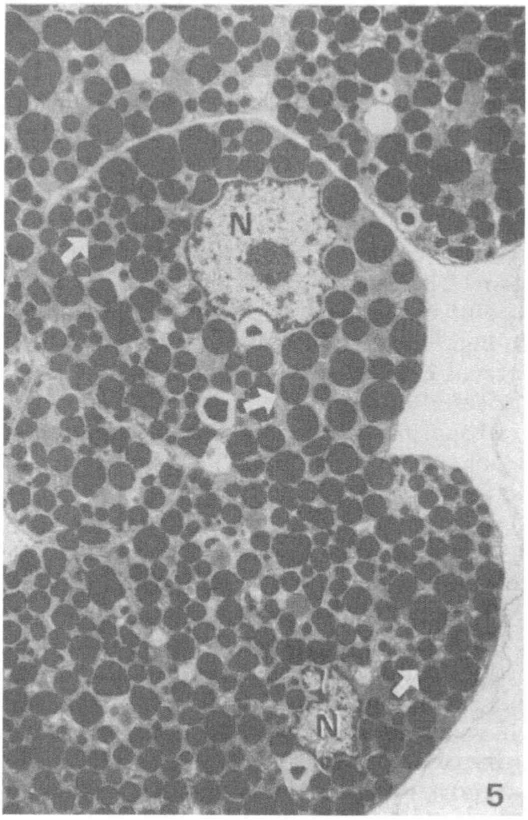
N
N
5

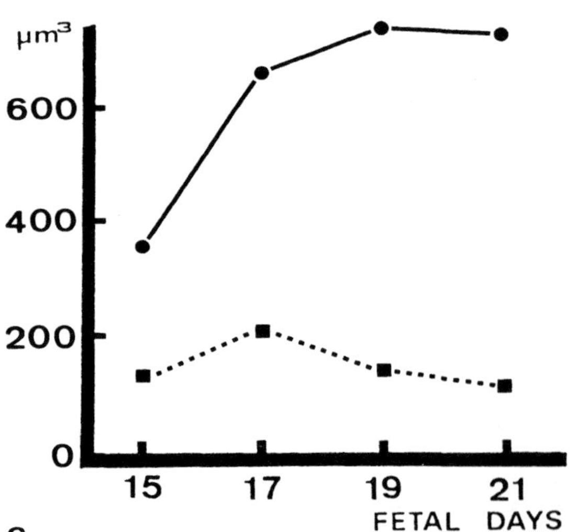
μm³
600
400
200
0
15
17
19
21
FETAL DAYS
6

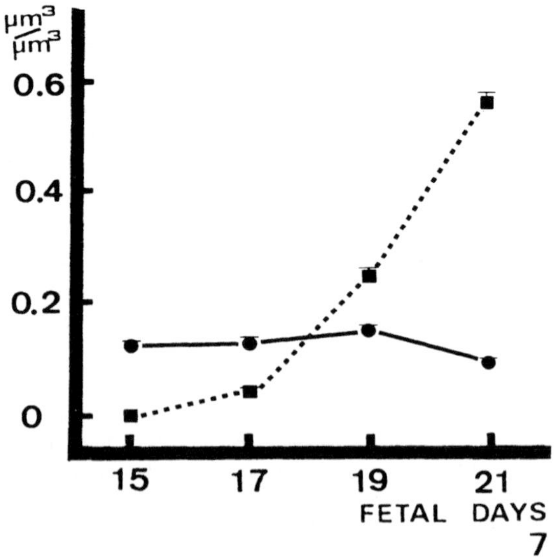
μm³/μm³
0.6
0.4
0.2
0
15
17
19
21
FETAL DAYS
7

creatic development, most epithelial cells have large nuclei with a homogeneous karyoplasm, and free polysomes are densely distributed in the cytoplasm, where other cytoplasmic organelles are poorly developed. On day 15 in rat fetuses, the acinar epithelium is composed of several cellular layers (Fig. 6-2). The acinar lumen is narrow and the acinar epithelial cells which are cuboidal in shape, are provided with a few small microvilli on the luminal surface. Profiles of the RER in the acinar cells are short tubular, and the RER is loosely distributed in the cytoplasm. Ribosomes are loosely attached to the cisternal membranes of the RER, so that the tubular RER often resembles smooth ER. The Golgi complex is small and compact, and its vacuoles and vesicles are less in number. After the 15th day of gestation, cytodifferentiation of the acinar cells rapidly occurs. On day 17, the acinar epithelium consists of single-layered cells. The acinar cells, having a columnar shape, distinctly contain small zymogen granules (Fig. 6-3). In our observation of developing acinar cells, zymogen granules appear exactly on day 17, although small secretory granules sometimes appear near the Golgi complex of the acinar cells on day 16, as Pictet et al. have reported [15]. Cisternal profiles of the RER, whose cisternal membranes are densely attached by ribosomes, become longer. The Golgi complex is clearly larger than that at the previous stage and is provided with condensing vacuoles and small, dense zymogen granules. On day 19, zymogen granules are distinctly larger in size and number, while their profiles are not only round, but often irregular. The zymogen granules are mainly located in the apical and supranuclear parts of the cells, but they often appear in the basal part. Long cisternal RER is arranged in parallel and is densely distributed in the basal part of the cells. It is interesting that the membrane continuity of the RER with the nuclear envelope is often seen at this stage (Fig. 6-4). The greatly expanded Golgi complex is located in the supranuclear region and is provided with large condensing vacuoles. On day 21, most acinar cells are occupied with zymogen granules (Fig. 6-5). The stacked RER completely disappears from the cytoplasm, and the single-layered RER is located around zymogen granules. Even at this stage, membrane continuity of the RER with the nuclear envelope is occasionally seen. The Golgi complex often exists as vacuolated cisternae among zymogen granules.

As shown in Fig. 6-6, the average cell volumes of the acinar cells rapidly increase from day 15 to day 17, peaking on day 19. The cell volume on day 19 is twice as large as on day 15. The cell volume on day 19 does not significantly differ from that on day 21. The average nuclear volume peaks on day 17, differing from the change in the cell volume during the prenatal life. Corresponding to the increase in quantities of zymogen granules, the nuclear volumes decrease from day 17 to day 21. The volume densities of the RER elevate from day 17 to day 19, while they decrease strongly on day 21 (Fig. 6-7). The volume density of the RER on day 19 is 1.25 times larger than on day 15 and 1.58 times larger than on day 21. The increase in the volume densities of zymogen granules during fetal life is tremendously rapid and large compared to that in the RER; that is, the volume densities of zymogen granules increase rapidly from day 17 to day 21. The surface densities of the RER and zymogen granules change in a manner that is similar to the changes in volume densities, respectively. Moreover, the numerical densities of zymogen granules vary concomitantly with the volume and surface densities.

With regard to the secretory activity of the

←

Fig. 6-4. At 19 days' gestation. Stacks of the RER are well developed in the cytoplasm. The membrane of the RER is connected directly with the nuclear envelope (arrow). Zymogen granules (Z) are not only round but often are irregular in shape. N = nucleus. x 11,600.

Fig. 6-5. At 21 days' gestation. Acinar cells are filled with zymogen granules. Single-layered cisternae of the RER encircle zymogen granules (arrows). Nuclei (N) of the acinar cells are irregular in shape x 2,900.

Fig. 6-6. Changes in average cell and nuclear volumes of the acinar cells during fetal life. Solid line: average cell volume; dotted line: average nuclear volume.

Fig. 6-7. Changes in volume densities of the RER and zymogen granules in the acinar cells during fetal life. Vertical bar ± S.E.; solid line: RER; dotted line: zymogen granules. n = 20 at each sampling time.

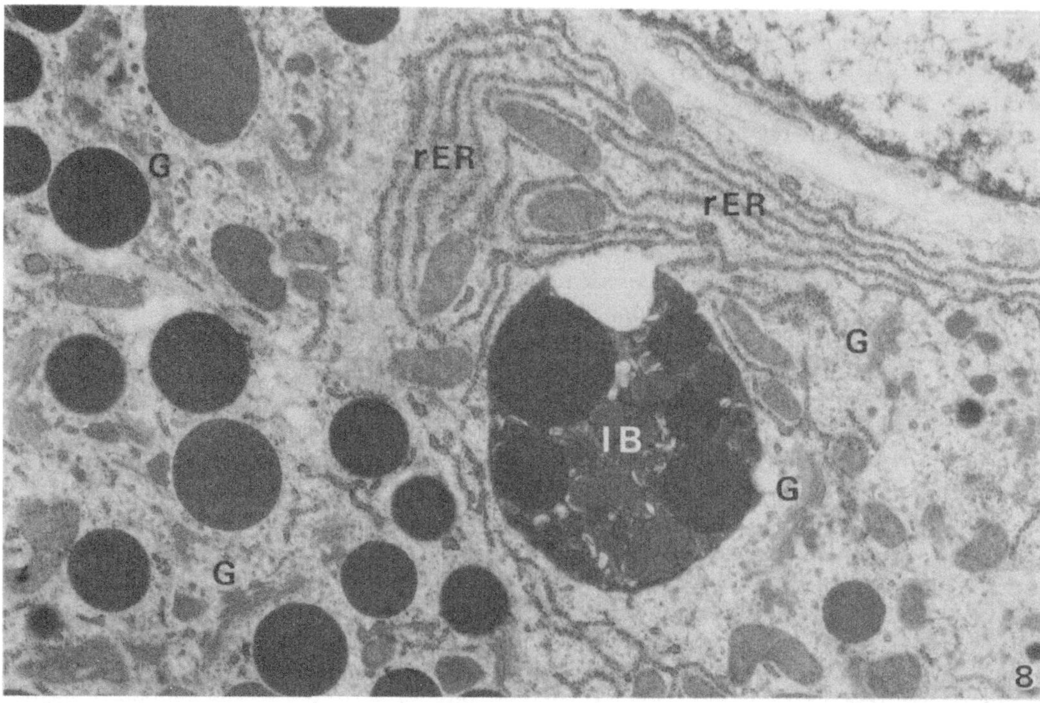
G
rER
rER
G
IB
G
G
8

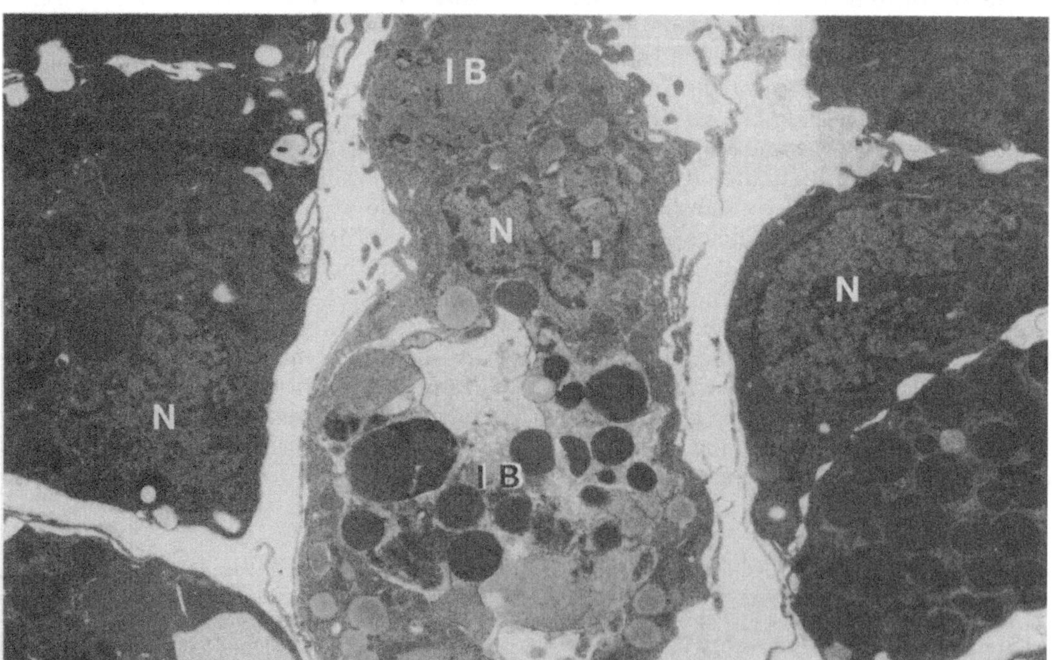
IB
N
N
N
IB
9

acinar cells during prenatal life, no structural evidence showing secretory processes of zymogen granules from the cells into the lumen is demonstrated at any of the fetal stages examined. This observation agrees with that by Pictet et al. [15]. This tendency of the developing acinar cells may be important when considering the membrane dynamics of the cells. Corresponding to the marked increase in zymogen granules on day 19, the volume and surface densities of the RER peak, and the membrane of the RER is connected with the nuclear envelope. As far as we have observed in pancreatic acinar cells from the embryonic to mature stages, no direct membrane connection between the RER and nuclear envelope is seen, except in the acinar cells on day 19 and sometimes on day 21. This indicates that the membrane of the RER may be produced by the RER itself (and free polysomes) in the early stage of cytodifferentiation of the acinar cells. On day 19, however, the membrane of the RER may be generated from the nuclear envelope, corresponding to the rapid increase in protein synthesis. Moreover, the surface areas of the RER and zymogen granules calculated from their surface densities, and from the cytoplasmic volumes of the acinar cells, increase on day 19. On day 21, the surface area of the RER decreases strongly, although that of zymogen granules increases. Furthermore, it is also important that almost no lysosomal structure, which may be concerned with membrane degradation, is found in acinar cells examined during fetal life. From those results, it is likely that the membrane generated by the RER or nuclear envelope is transported into that of zymogen granules via the Golgi complex. Of course, the maturation process of the membrane may occur in the Golgi lamellae. It remains unknown whether the membrane dynamics occurring in developing acinar cells are consistent with those of mature acinar cells. However, it is beyond doubt that the membrane dynamics in fetal acinar cells are important for understanding membrane recycling in mature acinar cells.

Acinar cells in the adult exocrine pancreas possess a clear-cut polarity, zymogen granules being located in the apical part of the cytoplasm. Even in the starved state, zymogen granules are never seen in the basal part. We have previously suggested that protein transport between the RER and Golgi complex of the pancreatic acinar cells in the adult rat is regulated on a circadian basis [1]. In the fetal pancreas, however, zymogen granules are often located in the basal part of the cells. In particular, secretory protein seems to be abundantly produced on day 21, zymogen granules being densely distributed throughout the cytoplasm. Thus, fetal acinar cells appear to have an immature regulatory mechanism for synthesis and intracellular transport of secretory proteins.

3. Early postnatal development of rat pancreatic acinar cells

Early postnatal development of pancreatic acinar cells has been poorly studied. Pharmacological approaches, however, have contributed to the clarification of the maturation process of stimulus-secretion coupling [11, 25, 26]. As stated above, rat pancreatic acinar cells do not secrete during fetal life. The onset of secretion may be the most important event after birth; that is, the pancreatic acinar cells differ in their activity between prenatal and postnatal stages. The excellent finding on the difference in the activity of the acinar cells between prenatal and postnatal stages has been morphologically verified by Munger [12]. He has reported a rapid decrease in zymogen granules of mouse pancreatic acinar cells during the 2 days after birth and has also reported cytoplasmic changes suggesting degeneration. We have analyzed, using ultrastructural and morphometric techniques, the reorganization of rat pancreatic acinar cells from the initiation of secretion.

←

Fig. 6-8. At birth. Acinar cells possess an inclusion-body-like structure (IB) containing zymogen granules and degenerating organelles. Stacked RER and its short cisternal profiles are seen in the cytoplasm. The Golgi complex (G) is small and compact, its cisternae being highly electron dense. x 11,000.

Fig. 6-9. At birth. An interstitial cell having inclusion-body-like structures (IB) is located near acinar cells, which are highly electron dense. N = nucleus. x 5,100.

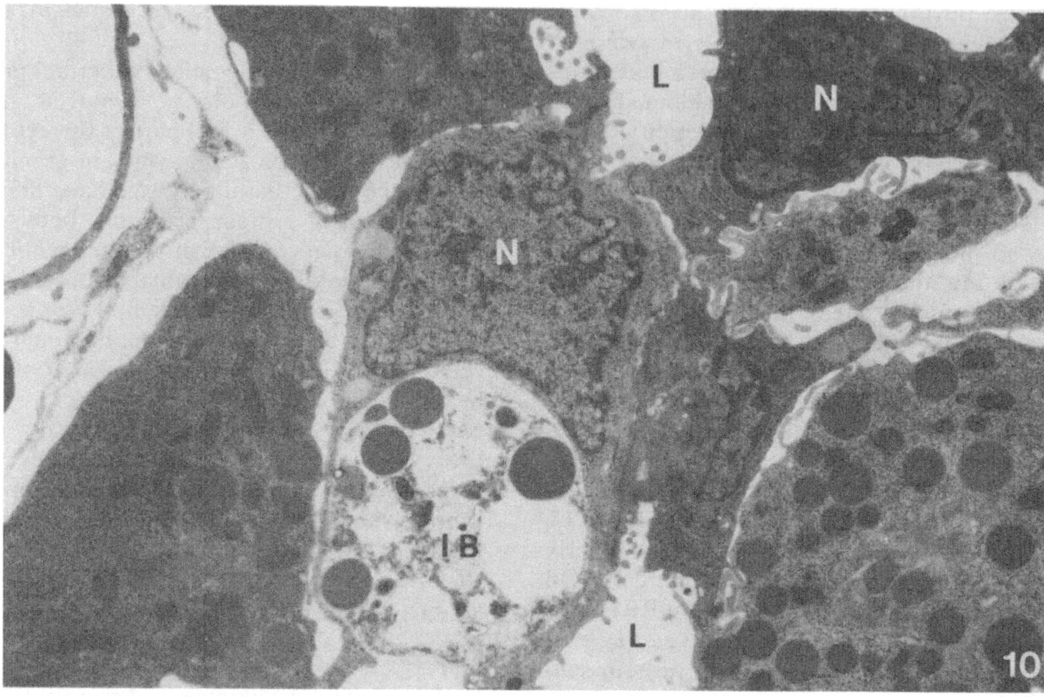
L
N
N
IB
L
10

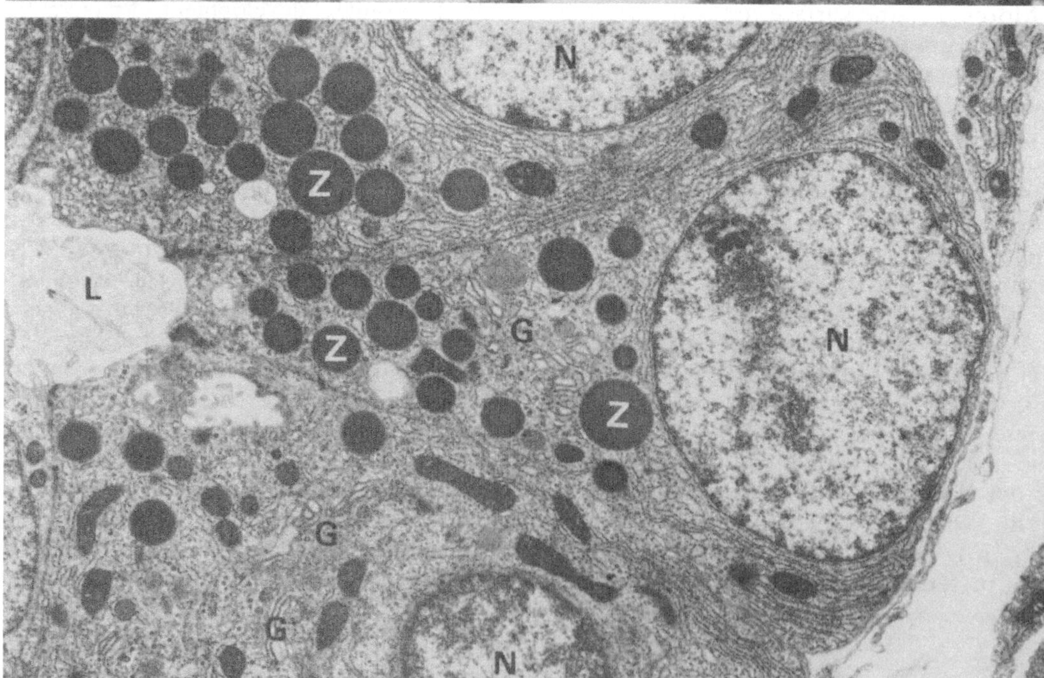
N
Z
L
Z
G
N
Z
G
G
N
11

At birth, some acinar cells are filled with zymogen granules, resembling those at 21 days of gestation. Zymogen granules in these cells are large and round, but are sometimes irregular in shape. Their electron density is high, although some zymogen granules possess dense cores only. In other cells, zymogen granules are often found in aggregations, and aggregated zymogen granules resemble inclusion bodies (Fig. 6-8). Some of the those inclusion-body-like structures contain amorphous structures or degenerating cytoplasmic organelles. The acinar cells containing the inclusion-body-like structures also possess small numbers of zymogen granules and RER in parallel arrangements. Sometimes the acinar cells are highly electron dense, containing numerous zymogen granules (Fig. 6-9). Inclusion-body-like structures are also frequently found in interstitial cells located near the acinar cells and in cells in the intercalated portion between the acinus and duct (Figs. 6-9 and 6-10). Moreover, degenerating acinar cells, cell debris, and inclusion-body-like structures are seen in the acinar or ductal lumen.

In general, the acinar cells at 2 days of age are similar to those at birth. However, the electron-dense acinar cells decrease slightly in number. The acinar cells often contain fewer zymogen granules than those at birth. Profiles of the RER are short cisternal, but in some cases stacked RER cisternae are seen in the basal part of the cytoplasm. The inclusion-body-like structures are still present in acinar cells, interstitial cells, and intercalated cells between the acinus and duct, although their occurrence in these cells decreases slightly. At 3 days, the number of zymogen granules is considerably reduced in acinar cells. Most zymogen granules, which are round and highly electron dense, are located only in the apical part of the cells. The relatively long cisternal profiles of the RER are often parallel in arrangement in the basal and lateral parts of the cells. The Golgi complex and condensing vacuoles are seen in the supranuclear region. The inclusion-body-like structures showing aggregated zymogen granules occasionally appear in acinar cells and intercalated cells between the acinus and duct. At 5 days, the general appearance of acinar cells is similar to that of mature acinar cells: long cisternal RER with a parallel arrangement in the lateral and basal parts of the cytoplasm, a Golgi complex with condensing vacuoles in the supranuclear region, and zymogen granules in the apical part. Inclusion-body-like structures containing zymogen granules and debris of cytoplasmic organelles are seen only rarely in acinar cells, but not in interstitial cells and intercalated cells between the acinus and duct. No prominent difference in morphological features of the acinar cells is found between 7 or 14 days of age and the adult stage (Fig. 6-11). Moreover, binucleate acinar cells are not discerned from birth to 14 days of age.

Serum α-amylase levels are examined at 21 days of gestation; at birth; and on 2, 3, 7, 14, and 21 days after birth. The α-amylase activity is low at 21 days of gestation. Its activity is largely augmented at birth, and the activity level is 1503 ± 113.4 IU (n = 5). The serum α-amylase level is low from 2 days until 21 days of age.

The cell and nuclear volumes of acinar cells gradually increase from birth until 7 days of age, while they decrease at 14 days of age (Fig. 6-12). The volume density of the RER is lowest at birth (Fig. 6-13). Then, the volume densities increase rapidly during the next 2 days. At 5 days, the volume density is slightly lower than that at 3 days. The volume densities of the RER increase gradually at 7 and 14 days, the density at 14 days being 2.4 times higher than that at birth. The volume densities of the Golgi complex and condensing vacuoles also show the lowest values at birth, while they increase rapidly during the next 4 days, peaking at 5 days. Then the volume densities decrease slightly at 7 and 14 days. As for the volume density of zymogen granules, it

←

Fig. 6-10. At birth. An intercalated cell between the acinus and duct possesses an inclusion-body-like structure (IB) containing zymogen granules and degenerating organelles. L = lumen; N = nucleus. x 7,700.

Fig. 6-11. At 7 days after birth. Acinar cells resembling a pyramidal profile contain zymogen granules (Z) in the apical region, the Golgi complex (G) in the supranuclear region, and the stacked RER in the lateral and basal regions. The nuclei (N) of the cells with a round profile are located in the basal part. L = lumen. x 7, 900.

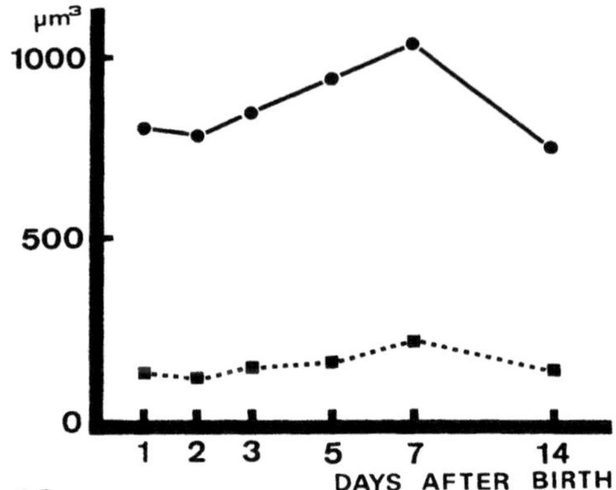

12

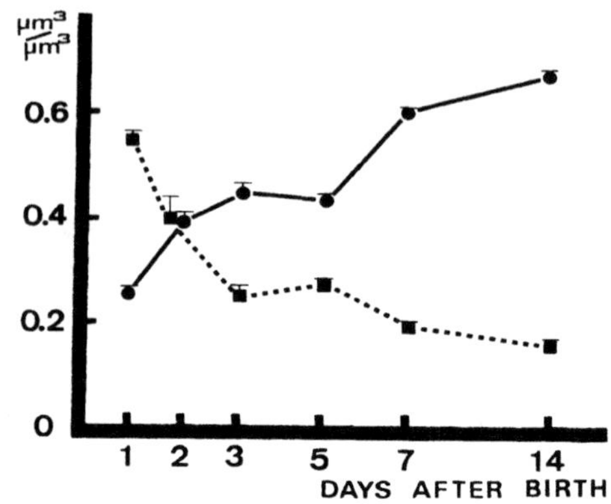

13

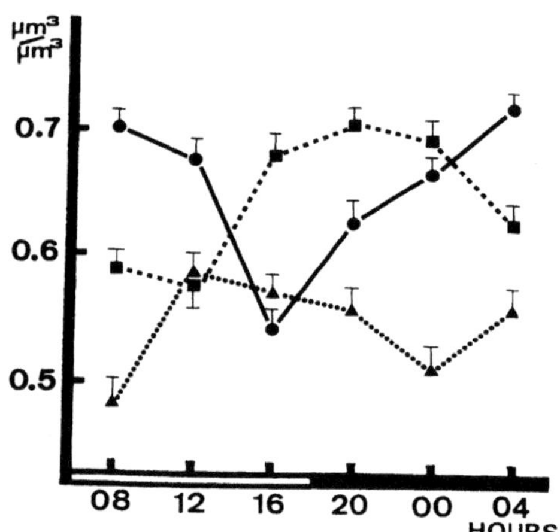

14

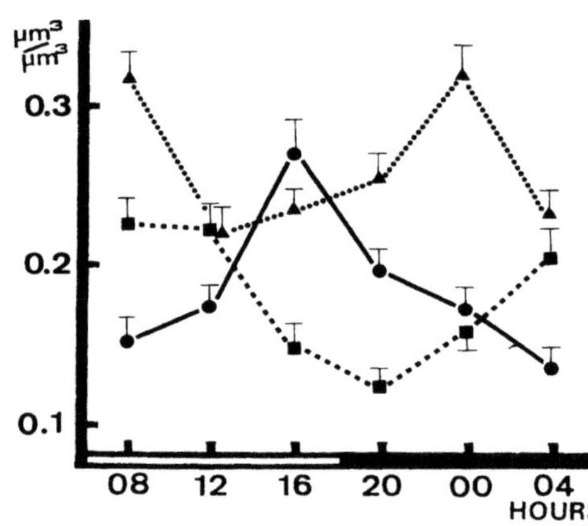

15

Fig. 6-12. Changes in the average cell and nuclear volumes of the acinar cells during early postnatal life. Solid line: average cell volume; dotted line: average nuclear volume.

Fig. 6-13. Changes in the volume densities of the RER and zymogen granules of the acinar cells during early postnatal life. Vertical bar: ± S.E.; solid line: RER; dotted line: zymogen granules. n = 48 at each sampling time point.

Fig. 6-14. Twenty-four-hour variations in the volume densities of the RER in the acinar cells of rats obtained at 21 days (before weaning), 31 days (after weaning), and 42 days of age (after weaning). Th daily mean volume densities are 0.543 ± 0.005 (± S.E., $\mu m^3/\mu m^3$) at 21 days, 0.645 ± 0.004 (± S.E., $\mu m^3/\mu m^3$) at 31 days, and 0.655 ± 0.004 ($\mu m^3/\mu m^3$) at 42 days. White bar: light phase; black bar: dark phase; vertical bar: ± S.E.; solid line: 42 days; dotted (square dot) line: 31 days; dotted (round dot) line: 21 days. n = 48 at each sampling time point.

Fig. 6-15. Twenty-four-hour variations in the volume densities of zymogen granules in the acinar cells of rats obtained at 21 days (before weaning), 31 days (after weaning), and 42 days of age (after weaning). The daily mean volume densities are 0.262 ± 0.004 (± S.E., $\mu m^3/\mu m^3$) at 21 days, 0.180 ± 0.004 (± S.E., $\mu m^3/\mu m^3$) at 31 days, and 0.183 ± 0.004 (± S.E., $\mu m^3/\mu m^3$) at 42 days. White bar: light phase; black bar: dark phase; vertical bar: ± S.E.; solid line: 42 days; dotted (square dot) line: 31 days; dotted (round dot) line: 21 days. n = 48 at each sampling time point.

completely changes in a manner that is exactly opposite the change in volume density of the RER from birth to 14 days of age (Fig. 6-13). At first, 2 days after birth, the volume densities of zymogen granules are high, in particular at birth. Then, the density at 3 days decreases rapidly, and at 5 days it shows almost the same value. At 7 and 14 days, the densities are diminished, and the density at birth is 3.4 times higher than that at 14 days. The volume density of inclusion-body-like structures is found to be 0.020 ± 0.006 (±S.E., $\mu m^3/\mu m^3$) at birth, whereas it decreases to 0.001 ± 0.001 (±S.E., $\mu m^3/\mu m^3$) at 3 days.

The most notable event in rat pancreatic acinar cells during early postnatal life is the beginning of the secretion of zymogen granules. As stated above, exquisite changes in morphology and morphometry of acinar cells occur around the period of the onset of the secretory activity. Before birth, acinar cells overproduce zymogen granules, which are not secreted and completely occupy the cytoplasm. This is due to the fact that stimulus-secretion coupling is not formed in the acinar cells. In such states of the acinar cells, we can find inclusion-body-like structures containing zymogen granules and debris of degenerating organelles. The reason for the appearance of inclusion-body-like structures in the acinar cells remains unknown. However, it is reasonable to suppose that inclusion-body-like structures are formed in acinar cells to degrade or extrude unsecreted zymogen granules. Therefore, extruded inclusion-body-like structures are found in the luminal space of the acinus and duct, the intercalated cells between the acinus and duct, and the interstitial cells.

Thus, it should be taken into consideration which surface of the acinar cells the inclusion-body-like structures are extruded from. It is certain that zymogen granules are secreted in the luminal surface of the mature acinar cells. However, judging from the region where the inclusion-body-like structures are found, they are extruded not only from the luminal surface of the acinar cells, but also from the basolateral surface. The extruded material located inside the basement membrane of the acinar epithelium is taken into the intercalated cells between the acinus and duct. This indicates that the cells located in the intercalated portion between the acinus and duct have a capacity for phagocytosis and act as scavengers in the ductal or acinar epithelial layer. Moreover, some material may be secreted into the interstitial space and engulfed by the interstitial cells. This is consistent with the fact that the blood α-amylase level is extremely high at birth. As α-amylase activity in salivary glands at birth is very low and morphological features of their acinar cells are not fully developed [27], the high α-amylase level must be derived from pancreatic acinar cells. In addition, some acinar cells become electron dense, thus suggesting degeneration. Some of these highly electron-dense cells may dissociate from the acinar epithelium to produce a holocrine secretion; degenerating acinar cells, cell debris, and inclusion-body-like structures containing zymogen granules and degenerating cytoplasmic organelles are found in the acinar and ductal lumina. It is possible that the degeneration of acinar cells is attributed to over production of secretory proteins. Indeeed, the regulatory system for protein synthesis, intracellular transport of synthesized proteins, and their secretion is not fully developed.

As for stimulus-secretion coupling, Larose and Morisset [11] have reported that the pancreas of fetal and newborn rats does not respond to 10^{-5} M urecholin, whereas 3-day- to 15-day-old pancreas increasingly responds to this dose. Doyle and Jamieson [25] have suggested that the response to hormonal stimuli in pancreatic acinar cells first occurs at birth. Moreover, Werlin and Grand [26] have found that amylase secretion from the pancreas in response to carbachol is comparable in 1-day-old, 8-day-old, and adult rats, although fetal pancreas shows no response to it. Such pharmacological results are compatible with the morphological findings seen during the first 2 or 3 days after birth and, in particular, at birth.

Previously, we have shown that the volume densities of the RER and zymogen granules in pancreatic acinar cells of adult rats vary in a reciprocal manner over 24 hours [1]. In the acinar cells from 1 to 14 days after birth, the volume densities of the RER change in a manner that is reciprocal to those of zymogen granules; the increase in the volume densities of the RER coincides with the decrease in those of zymogen granules. In the acinar cells, no direct connection of the membrane of the RER with the nuclear envelope is observed after birth. Therefore, it seems likely that the increase in the membrane of the RER after birth is dependent upon the retrieval of membranes from the plasmalemma, following the secretion of zymogen granules.

4. Twenty-four hour variations in subcellular structures of rat pancreatic acinar cells during the periweaning period

Recent findings of morphological, morphometric, biochemical and pharmacological analyses on developing acinar cells have played an important role in ascertaining the maturity of these cells [3, 9, 11, 15, 22, 23, 25, 26, 28–30]. As stated above the acinar cells at about 1 week after birth morphologically appear to be similar to those from the exocrine pancreas of adult rats, while stimulus-secretion coupling is formed soon after birth. Sesso et al. [28] and de Carvalho et al. [9] have reported that the morphofunctional pattern of rat pancreatic acinar cells remains constant after 30–40 days of age. Moreover, Oates and Morgan [3] have examined the incidence of binucleation and the nuclear DNA content, using isolated rat pancreatic acinar cells of various ages, and have shown that those parameters become mature at 56 days of age; 64% of cells are binucleate, with a diploid DNA content in each nucleus.

We have examined variations in subcellular structures of the pancreatic acinar cells of adult rats over 24 hours [1]. Housing conditions of the rats used were constant (temperature 22 ± 1°C; relative humidity 55% to 65%; 12:12 hour light-dark cycle, light period 06.00 h – 18.00 h). Variation patterns of the cytoplasmic organelles suggest an intracellular route for secretory proteins over 24 hours. From rhythmic changes in subcellular structures of the acinar cells, it is possible to divide the 24-hour period into three stages: 1) the resting or protein synthetic stage (from the mid-dark to the early-light period): the volume density of the RER is strongly increased and that of zymogen granules is clearly decreased; 2) the granule accumulation stage (from the early-light to the late-light period): the volume density of the RER is markedly decreased and that of zymogen granules is increased; 3) the secretion stage (from the late-light to the mid-dark period): as a result of the release of zymogen granules from the acinar cells, the volume density of zymogen granules decreases, and that of the RER increases. Moreover, the relationship between the volume densities of the RER and zymogen granules varies distinctly in a reciprocal manner. Thus, the 24-hour variations in subcellular structures of the acinar cells at various stages after birth can be one of the important parameters for ascertaining the maturity of acinar cells.

We have morphometrically analyzed 24-hour variations in subcellular structures of the acinar cells obtained from young Wistar rats aged 21, 31, and 42 days. Until day 21, these animals were housed with mother rats (eight offspring with one mother rat). Two experimental groups of 48 rats were weaned at 21 days (four rats to one plastic cage). The first experimental group of 24 rats was sacrificed at 21 days (before weaning). Animals used for the experiments were housed in well-ventilated chambers (temperature 24 ± 1°C, relative humidity 55% to 65%) and were fed ad libitum. The lighting schedule was a 12:12 hour light-dark cycle (light period: 06.00 h – 18.00 h).

Ultrastructures of the acinar cells show no difference between 21, 31, and 42 days, being similar to the mature acinar cells. The interstitial space in the pancreas is clearly wider at 21 days of age than at 31 and 42 days, so acini are still sparse at 21 days.

The daily mean volumes of acinar cell are smallest at 21 days, no difference being seen in the volumes between 31 and 42 days (1295.5 ± 94.3 μm^3 at 21 days, 1562.2 ± 73.5 μm^3 at 31 days, 1542.8 ± 75.0 μm^3 at 42 days). The variation curves of the cell volumes at 31 and 42 days are similar to one another, although those of the nuclear volumes differ from each other. The daily mean nuclear volumes are large at 42 days, compared to those at 21 and 31 days (173.5 ± 4.7 μm^3 at 21 days, 186.0 ± 5.6 μm^3 at 31 days, 194.4 ± 7.4 μm^3 at 42 days). At 21 days, binucleate acinar cells appeared at each time that they were examined. To clarify the incidence of binucleate acinar cells at each stage, we counted 250 acinar cells per rat, using semithin sections stained with toluidine blue. The daily mean incidence of binucleate acinar cells markedly increased with age, being 1.5 times greater at 31 days and 2.9 times greater at 42 days than at 21 days (5.54 ± 0.43% at 21 days, 8.27 ± 0.43% at 31 days, 15.8 ± 0.67% at 42 days). The variation curves of the incidence of the binucleate acinar cells over 24 hours are prominent at each age but are not synchronized with each other.

The daily mean volume densities of the RER are significantly lower at 21 days than at 31 and 42 days, while they are almost similar at 31 and 42 days. The volume density of the RER at 21 days varies bimodally over 24 hours, peaking at the mid-light (12.00 h) and the late-dark times (04.00 h) and decreasing at the early-light (08.00 h) and the mid-dark times (00.00 h) (Fig. 6-14). The maximal-minimal difference in the curve (amplitude) is 21.4% ($p < 0.001$). At 31 days, the volume density fluctuates unimodally over 24 hours, the peak appearing at 20.00 h and the trough at 12.00 h (amplitude: 22.8%; $p < 0.001$) (Fig. 6-14). At 42 days, the variation curve of the density is also unimodal, with a peak at 04.00 h and a trough at 16.00 h (Fig. 6-14). The amplitude of the variation is 32.7% ($p < 0.001$). Comparing the curves at 31 and 42 days, the maximum at 42 days occurs 8 hours later than at 31 days, while the minimum at 42 days occurs 4 hours later.

The daily mean volume densities of the Golgi complex do not show any differences among the three stages examined. The fluctuation of the volume density of the Golgi complex is bimodal at 21 days (amplitude: 71.7%; $p < 0.001$), while it is unimodal at 31 and 42 days (amplitudes: 30.2% at 31 days, 42.1% at 42 days; $p < 0.001$ for both). Condensing vacuoles located near the Golgi complex exhibit a lower electron density than zymogen granules. The daily mean volume densities of condensing vacuoles are highest at 21 days and lowest at 31 days. The variation in the volume density of condensing vacuoles is not significant at 31 days. The density at 21 days changes significantly over 24 hours, peaking at 08.00 h and decreasing at 00.00 h and 04.00 h (amplitude: 56.0%; $p < 0.05$). The clear-cut unimodal variation is demonstrated in the volume density of condensing vacuoles at 42 days, exhibiting a maximum at 20.00 h and a minimum at 04.00 h (amplitude: 90.5%; $p < 0.01$).

The daily mean volume densities of zymogen granules are highest at 21 days, while no difference is seen in the mean volume densities between 31 and 42 days. At each stage, the volume density of zymogen granules varies significantly over 24 hours (Fig. 6-15). The volume density at 21 days peaks at 08.00 h and 00.00 h, while it is decreased at 12.00 h and 04.00 h (maximal-minimal difference: 45.0%; $p < 0.001$). At 31 days, the variation curve of the volume density is unimodal, the maximum appearing at 08.00 h and the minimum at 20.00 h (amplitude: 82.9%; $p < 0.001$). At 42 days, the curve of the volume density shifts 8 hours from that at 31 days: a maximum at 16.00 h and a minimum at 04.00 h (amplitude: 100.7%; $p < 0.001$).

By comparing the variation patterns of subcellular structures of acinar cells over 24 hours at each stage, one finds the following characteristics: 1) Most parameters examined differ distinctly between 21 days of age (before weaning) and 31 and 42 days (after weaning). The differences are found in the daily means of cell and nuclear volumes and in the volume densities of the RER and zymogen granules. Moreover, variation curves of the volume densities of the RER and zymogen granules are bimodal at 21 days, while they are unimodal at 31 and 42 days. 2) Some parameters measured also differ between 31 and 42 days. The prominent difference is an 8-hour phase shift in the variation curves of the volume densities of RER and zymogen granules between the two stages. In addition, it is interesting that the daily means of the incidence of binucleate acinar cells at 31 days shows a transitional value between 21 and 42 days.

Kachar et al. [29] have reported that mitosis actively occurs in acinar cells during the third and fourth week after birth, although these cells are in an immature developmental state at that time. This correlates well with the fact that the distribution of acinar cells in the pancreatic tissue at 21 days is less dense than that at 31 and 42 days and that the size of acinar cells increases greatly during the mitotic period (from 21 to 31 days). Moreover, feeding and housing conditions before weaning differ from those after weaning. It seems likely that those differences in cellular state and housing conditions are responsible for variations in the parameters measured during the periweaning period.

Sesso et al. [28] and de Carvalho et al. [9] have described that from 30 to 40 days of age onwards, pancreatic acinar cells exhibit a more highly developed and stable morphofunctional pattern. This agrees with our results that the daily mean volume densities of the RER and zymogen granules are rather similar between 31 and 42 days, respectively. However, with respect to the variation curves of the volume densities of the RER and zymogen granules over 24 hours, a distinct 8-hour phase shift is present in each parameter between 31 and 42 days. As we have previously clarified, pancreatic acinar cells obtained from adult rats housed under the same conditions as in the experiment mentioned above can be divided into three stages: the protein synthesis or morphologically resting stage (00.00 h–04.00 h), the zymogen granule accumulation stage (08.00 h–16.00 h), and the secretion stage (16.00 h–00.00 h) [1]. From the variations in the volume densities of the RER and zymogen granules over 24 hours, the acinar cells, both at 31 and 42 days of age, can also be divided into these three stages. However, only at 42 days are the variation patterns of the RER and zymogen granules compatible with those of adult rats [1]. The acquisition of adult variation patterns of subcellular structures in rat pancreatic acinar cells thus takes 3 weeks to develop after weaning.

5. Concluding remarks

Rat pancreatic acinar cells during prenatal and postnatal life were intensively studied using fine structural and morphometric techniques. The results can be summarized as follows.

During prenatal life, the volume density of the rough endoplasmic reticulum increases, peaking on day 19, while it decreases on day 21. The volume density of zymogen granules, which appear on day 16, increases from days 17 to 21. Zymogen granules are not secreted during fetal life, occupying the cytoplasm on day 21. These results suggest that membrane transport involved with intracellular movement of newly synthesized proteins from the RER via the Golgi complex to zymogen granules occurs in one direction and lacks regulation. The reorganization of the acinar cells corresponds to the onset of secretion, especially during the first 3 days after birth. The dramatic findings during this period are the appearance of degenerating acinar cells and of inclusion-body-like structures containing zymogen granules and degenerating organelles in the acinar, intercalated, and interstitial cells. Ultrastructures of the acinar cells at 7 days resemble those at the adult stage. In comparing circadian variations in the RER and zymogen granules of the acinar cells among days 21, 31, and 42, only at 42 days were their variation patterns compatible with those of adult rats. This indicates that circadian studies using morphometry may be used for the determination of developmental stages.

References

1. Uchiyama Y, Saito K: A morphometric study of 24-hour variations in subcellular structures of the rat pancreatic acinar cells. *Cell Tissue Res* 226: 609–620, 1982.
2. Uchiyama Y: A histochemical study of variations in the localization of 5′-nucleotidase activity in the acinar cell of the rat exocrine pancreas over the twenty-four hour period. *Cell Tissue Res* 230: 411–420, 1983.
3. Oates PS, Morgan RGH: Changes in pancreatic acinar cell nuclear number and DNA content during aging in the rat. *Am J Anat* 177: 547–554, 1986.
4. Schirner H: Das Pankreas von *myxine glutinosa* und *bdellostoma stouti*. *Nytt Mag Zool* 11: 5–18, 1963.
5. Schreiner KE: The pancreas-like organ of *Myxine glutinosa*. Avhandl Norske Vidensk-Akad., Oslo, *I Mat-Naturv Kl* 1: 1–19, 1957.
6. Falkmer S, Patent GJ: Comparative and embryological aspects of the pancreatic islets. In: Handbook of Physiology, Section 7: Endocrinology, Vol. I. Endocrine Pancreas. 1–23, 1972.
7. Hamilton WJ, Mossman HW: Human Embryology, 4th ed. Baltimore: Williams & Wilkins, 1972.
8. Falin LI: The development and cytodifferentiation of the islets of Langerhans in human embryos and foetuses. *Acta Anat* 68: 147–168, 1967.
9. de Carvalho AF, Laurindo FRM, Taga R, Sesso A: Ultrastructural morphometric study on developing acinar cells of the rat pancreas and parotid gland. *Acta Anat* 101: 234–244, 1978.
10. Kallman F, Grobstein C: Fine structure of differentiating mouse pancreatic exocrine cells in transfilter culture. *J Cell Biol* 20: 399–413, 1964.
11. Larose L, Morisset J: Acinar cell responsiveness to urecholine in the rat pancreas during fetal and early postnatal growth. *Gastroenterology* 73: 530–533, 1977.
12. Munger BL: A phase and electron microscopic study of cellular differentiation in pancreatic acinar cells of the mouse. *Am J Anat* 103: 1–33, 1958.
13. Parsa I, Marsh WH, Fitzgerald PJ: Pancreas acinar cell differentiation. I. Morphological and enzymatic comparisons of embryonic rat pancreas and pancreatic anlage grown in organ culture. *Am J Pathol* 57: 457–487, 1969.
14. Parsa I, Marsh WH, Fitzgerald PJ: Pancreas acinar cell differentiation. II. Comparative DNA and protein synthesis of the embryonic rat pancreas and the pancreatic anlage grown in organ culture. *Am J Pathol* 57: 489–521, 1969.

15. Pictet LR, Clark WR, Williams RH, Rutter WJ: An ultrastructural analysis of the developing embryonic pancreas. *Dev Biol* 29: 436–467, 1972.
16. Rutter WJ, Weber CS: Specific proteins in cytodifferentiation. In: Developmental and Metabolic Control Mechanisms and Neoplasia. DL Ward (ed), Baltimore: Williams & Wilkins, 195–218, 1965.
17. Rutter WJ, Wessels NK, Grobstein C: Control of specific synthesis in the developing pancreas. *J Nat Cancer Inst Monogr* 13: 51–65, 1964.
18. Rutter WJ, Kemp JD, Bradshaw WS, Clark WR, Ronzio RA, Sanders RG: Regulation of specific protein synthesis in cytodifferentiation. *J Cell Physiol* 72 (Suppl 1, 2): 1–18, 1968.
19. Uchiyama Y, Watanabe M: A morphometric study of developing pancreatic acinar cells of rats during prenatal life. *Cell Tissue Res* 237: 117–122, 1984.
20. Wessels NK, Cohen IH: Early pancreas organogenesis: Morphogenesis, tissue interactions and mass effects. *Develop Biol* 15: 237–270, 1967.
21. Weibel ER: Stereological Methods. Practical Methods for Biological Morphometry. London: Academic Press, 1979.
22. Uchiyama Y, Watanabe M: Morphometric and fine-structural studies of rat pancreatic acinar cells during early postnatal life. *Cell Tissue Res* 237: 123–129, 1984.
23. Uchiyama Y, Watanabe M: A morphometric study of the 24-hour variations in subcellular structures of rat pancreatic acinar cells during the periweaning period. *Cell Tissue Res* 237: 131–138, 1984.
24. Uchiyama Y, Watanabe M: Correlation of rhythmic variations in subcellular structures of rat hepatocytes, pancreatic acinar cells, and islet B cells. In: Advances in Chronobiology, Part A. JE Pauly, LE Scheving (eds), New York: Alan R. Liss, 153–162, 1987.
25. Doyle CM, Jamieson JD: Development of secretagogue response in rat pancreatic acinar cells. *Dev Biol* 65: 11–27, 1978.
26. Werlin SL, Grand RJ: Development of secretory mechanisms in rat pancreas. *Am J Physiol* 236: E446–E450, 1979.
27. Redman RS, Sreebny LM: Proliferative behavior of differentiating cells in the developing rat parotid gland. *J Cell Biol* 46: 81–87, 1970.
28. Sesso A, Carneiro J, Cruz AR, de Leite JBA: Biochemical, cytochcmical and electron microscopic observations on the enhancement of the pancreatic acinar cell secretory activity in the rat during early postnatal growth. *Arch Histol Jpn* 35: 343–364, 1973.
29. Kachar B, Taga R, Kniebel GA, Sesso A: Morphometric evaluation of the number of pancreatic exocrine cells during early postnatal growth in the rat. *Acta Anat* 103: 11–15, 1979.
30. Ermak TH, Rothmn SS: Large decrease in zymogen granule size in the postnatal rat pancreas. *J Ultrastr Res* 70: 242–256, 1980.

Author's address:
Yasuo Uchiyama, M.D.
Department of Anatomy
Institute of Basic Medical Sciences
University of Tsukuba
Tsukuba, Ibaraki-Ken 305
Japan

CHAPTER 7

The human exocrine pancreas in health and disease

GUIDO ADLER & HORST F. KERN

1. Normal human pancreas

The structure-function relationship of the exocrine pancreas in various mammals, especially laboratory animals, has been summarized extensively in recent years [1–7]. The purpose of this chapter is to describe the fine structure of the exocrine pancreas in the human, relate this to recent biochemical findings concerning intracellular aspects of the secretory pathway, and then apply this knowledge to the pathophysiology of the major pancreatic diseases.

1.1. Fine structure of acinar cells

From its macroscopic and microscopic appearance, the exocrine pancreas has traditionally been compared to a bunch of grapes in which the enzyme-producing acinar cells (from latin *acinus* = berry or grape) are grouped together in round or oval structures and are tagged to the duct system. From early developmental studies [8], it is known that the secretory end-pieces (comprised of enzyme-producing cells and duct cells) divide dichotomously and finally form branched structures. After retrograde injection of silicone rubber into this terminal part of the human pancreas, an anastomosing tubular arrangement of acinar cells has been postulated [9, 10]. This has recently been questioned, at least for the rat pancreas [11].

For conventional reasons we use the term *acini* or *acinar cells* as a general denomination for enzyme-producing cells in the pancreas. Electrophysiological studies, combined with injection of fluorescent dyes in acinar cells [12], have demonstrated that several hundred acinar cells are electrically coupled via gap junctions to act in synchrony upon stimulation, also indicating functional complexity of the acinus.

Sections through the acini of the human pancreas at the electron microscope commonly reveal oblong luminal spaces or two acinar lumina in close proximity (Fig. 7-1A) underlying its composite architecture. The acinar cells comprise 80% to 90% of the total cell volume in the pancreas, and they have served as a model system for the study of the cellular process of secretion. More than a hundred years ago, distinct subcellular compartments were first described by Heidenhain [13], and he postulated that secretion is not a passive diffusion process from the blood across secretory epithelia but a specific cellular activity of secretory cells themselves [14]. This paradigm has been verified over the past 30 years by G.E. Palade [15] using electron microscopy, tissue autoradiography, and cell fractionation.

The acinar cells usually are pyramid shaped and show a polarized distribution of their cytoplasmic membrane systems: the infranuclear and paranuclear regions of the cytoplasm are occupied mainly by the membrane system of the rough endoplasmic reticulum (RER), while the Golgi complex and the zymogen granules are located predominantly at the apical pole of the cell (Fig. 7-1A). At the cell basis, the membrane is surrounded by structures of the extracellular matrix (basal lamina) and occasional attachment of nerve

Riva, A. and Motta, P.M., (eds.), Ultrastructure of the Extraparietal Glands of the Digestive Tract.
ISBN 978-1-4612-8211-2.

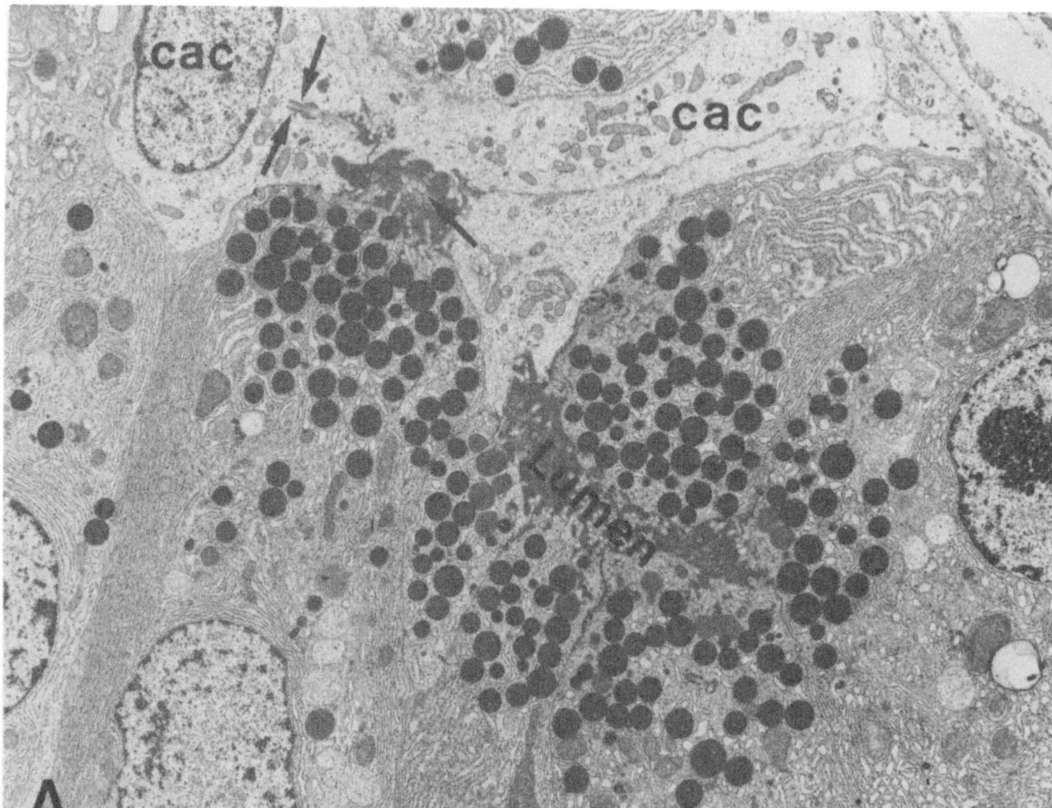
cac
cac
Lumen
A
Sy
bl
B
EX
z.o.
C
D

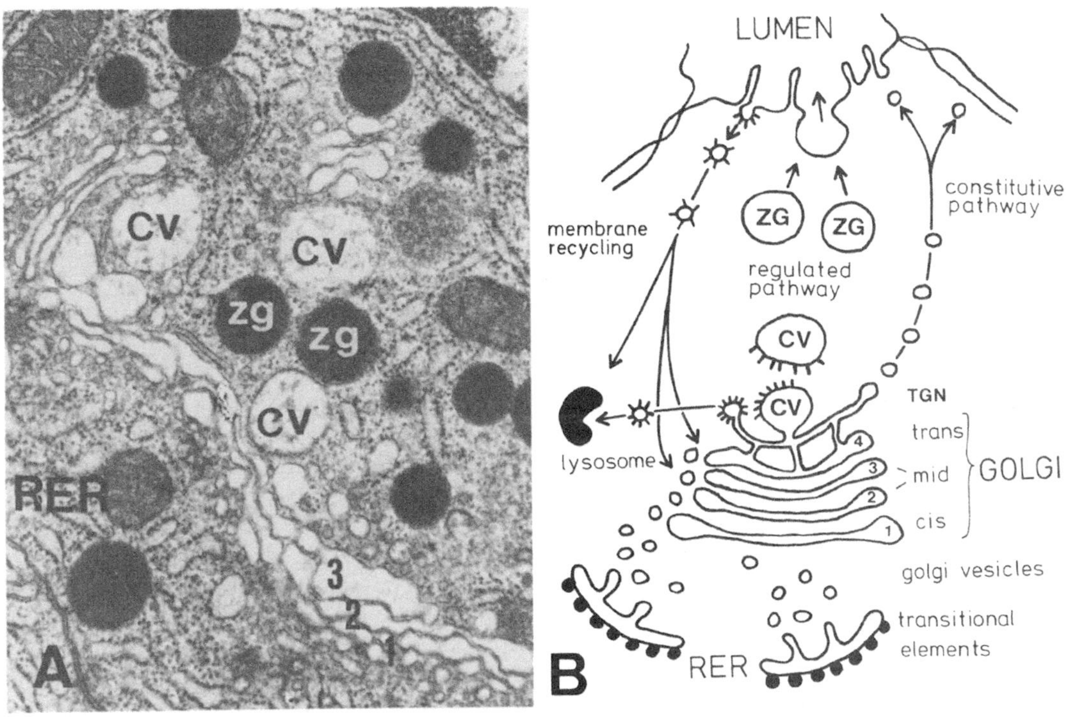

Fig. 7-2. Structure and function of the Golgi complex in the human acinar cell.
A: Rough endoplasmic reticulum (RER), Golgi cisternae (1, 2, 3) and granules (CV = condensing vacuole; zg = zymogen granule) are organized in a polarized fashion. x 24,000. **B**: central role of Golgi complex as sorting compartment for secretory, lysosomal, and membrane proteins. The dominant pathway in the acinar cell of protein processing and sorting is represented by the secretagogue-regulated pathway. After exocytosis of zymogen granules (ZG), membrane material is recycled, to a large extent, to the Golgi complex, where it is reutilized. Targeting of membrane proteins to polar surfaces of the cell occurs via a constitutive pathway. TGN = trans-Golgi network.

endings is observed (Fig. 7-1B). These are often associated with local interdigitations of the basal plasma membrane and contain vesicles of various sizes and densities, indicating possible storage of neurotransmitters and neuropeptides [16].

The Golgi complex of acinar cells appears as a horseshoelike structure composed of three to five stacks of flattened membrane-bound cisternae, numerous small vesicles, and various stages in the maturation of zymogen granules (Fig. 7-2A). The cisternae are arranged in a polar fashion. They face on one side (*cis* or forming face) transitional elements of the RER and are oriented on the other side (*trans* or condensing face) towards secretory granules (Figs. 7-2A and 7-2B). Mature zymogen granules are stored in the apical portion of the acinar cell and accumulate beneath the luminal plasma membrane (Fig. 7-1A). Numer-

←
Fig. 7-1. Fine structure of acinar unit in the human pancreas. **A**: low-power view of central segment showing polarized arrangements of cellular organelles and concentration of zymogen granules around the acinar lumen. Centro-acinar cells (cac) containing kinocilia (arrows) are also shown. x 4,900. **B**: nerve ending (Sy) at basal plasma membrane of an acinar cell, which shows local membrane interdigitations (arrows) and is surrounded by the basal lamina (bl) of the extracellular matrix. x 24,150. **C**: release

ous microvilli project from this surface into the luminal space, containing bundles of microfilaments made up of actin and associated proteins [17].

Upon nerval and/or hormonal stimulation, the zymogen granules move towards the apical plasma membrane and intermittently fuse with it to release their content via exocytosis (Fig. 7-1C). This step, in addition to complex events in stimulus-secretion coupling [18], also involves a specific recognition of the granule membrane and the apical plasma membrane and their transient fusion. The membrane material of the granule that is incorporated into the apical plasma membrane is subsequently removed by endocytosis of coated vesicles, which are taken up into the cytoplasm and recirculate to the Golgi complex [19]. The acinar lumen is sealed off from the intercellular space by elaborate junctional complexes, which are arranged in tandem and consist of occluding junctions (zonula occludens) close to the luminal space (Fig. 7-1C), belt desmosomes (zonula adhaerens) at the level of the terminal web, and spot desmosomes (macula adhaerens) along the lower parts of the lateral plasma membrane (for details see Chapter 2).

1.2. Secretory pathway in acinar cells

The classical studies of G.E. Palade [15] have demonstrated that the secretory process in pancreatic acinar cells occurs in time sequence along well-defined cellular compartments. The biosynthesis of about 15 different enzyme and isoenzyme proteins [20] takes place at ribosomes attached to the RER. Experimental studies have indicated that segregation of newly synthesized proteins into the cisternal space of the RER is closely coupled to biosynthesis. Blobel and Dobberstein [21] have demonstrated that the process of segregation is determined by specific sequences at the level of each mRNA: mRNAs coding for secretory proteins (including lysosomal enzymes) and for integral membrane proteins contain at the 5' end following the start codon a sequence of 45 to 90 nucleotides that code for mostly hydrophobic amino acids (signal sequence). At the beginning of biosynthesis, these signal codons are translated, first resulting in the formation of a signal peptide, which emerges from the secretory pole of the free ribosome [21]. This newly synthesized signal peptide is recognized by a nucleoprotein particle (signal recognition particle, SRP) in the cytoplasm, which by binding to the signal peptide mediates the binding of the free ribosome to receptor proteins in the membrane of the RER [22]. Until the binding of the ribosome to the RER is complete, the association of SRP with the signal peptide results in the arrest of protein synthesis, which only continues after the RER binding has been completed and the SRP has been released from the ribosome [22]. The binding of the ribosome to the membrane of the RER leads to formation of a protein pore complex (translocon), which then mediates the translocation of the nascent polypeptide chain into the cisternal space of the RER. Upon entry into the cisternal space, the signal peptide is cleaved from the nascent polypeptide chain, which during elongation forms its secondary and tertiary structure. After translation of the coding sequence of the mRNA up to the 3' end, it is released from the ribosome, which by itself detaches from the RER membrane and the same cycle can start from the beginning.

Acinar cells in the pancreas with a continuously high rate of biosynthesis of exportable proteins therefore have a well-developed RER, which is densely studded with ribosomes. Internally the RER represents a continuous space that is filled with a composite mixture of various types of newly synthesized proteins. Autoradiographic studies by Jamieson and Palade [23] have demonstrated that this mixture is transported in a directional fashion within 20 minutes towards the Golgi region and is translocated into the Golgi complex. This represents a second membrane-bound compartment within the cell and serves as a collecting and processing organelle for various types of proteins [24, 25]. The transfer of newly synthesized proteins from the cisternal space of the RER into the Golgi complex is mediated by transporting vesicles, which bud off from ribosome-free segments of the RER (transitional elements), migrate to the Golgi complex, and fuse with the *cis* face (Fig. 7-2B). Further processing occurs within the Golgi cisternae in which proteolytic cleavage and terminal glycosylation of proteins takes place, followed by sorting of the different proteins into functional groups, i.e., lysosomal proteins into lysosomes, secretory proteins into secretion graıules, and membrane proteins into membrane-lestined vesicles (for details see Fig. 7-2B). The unctional processing and sorting of the different ırotein species seems to be associated with disinct Golgi cisternae and involves "recognition equences" contained in the respective polypepide chains [24, 25]. Sorting of the proteins into lifferent classes finally involves packaging into nembrane-bound vesicles or granules, which are lerived from a specific trans-Golgi cisterna, the rans-Golgi network (Fig. 7-2B) [26, 27]. Within ecretory granules a further maturation process, ;enerally referred to as *condensation,* takes place, he details of which are not fully understood 28, 29]. In the structural and functional sequence ıf the secretory pathway, the secretion granules herefore represent the third independent comıartment, which is involved in storage and release ıf secretory proteins.

1.3. Centroacinar and intralobular duct cells

The initial part of the duct system draining the acinar lumen is formed by the intercalated ducts (ductules), which are composed of flat to low cuboidal cells. Since they penetrate into the acinar unit, sections through the center of acini usually reveal cells of the ductules, also referred to as *centro-acinar* and *intralobular* duct cells (Figs. 7-1A, 7-3A). Whereas the luminal surface of acinar cells reveals numerous microvilli projecting into the acinar lumina, the corresponding surface of intercalated and intralobular duct cells is smooth (Fig. 7-3A). The cytoplasm is less electron dense than in acinar cells, mainly due to the lack of profiles of the RER. Ribosomes are mainly found free in the cytoplasm, together with a large number of mitochondria, which are generally concentrated around the Golgi complex. Compared

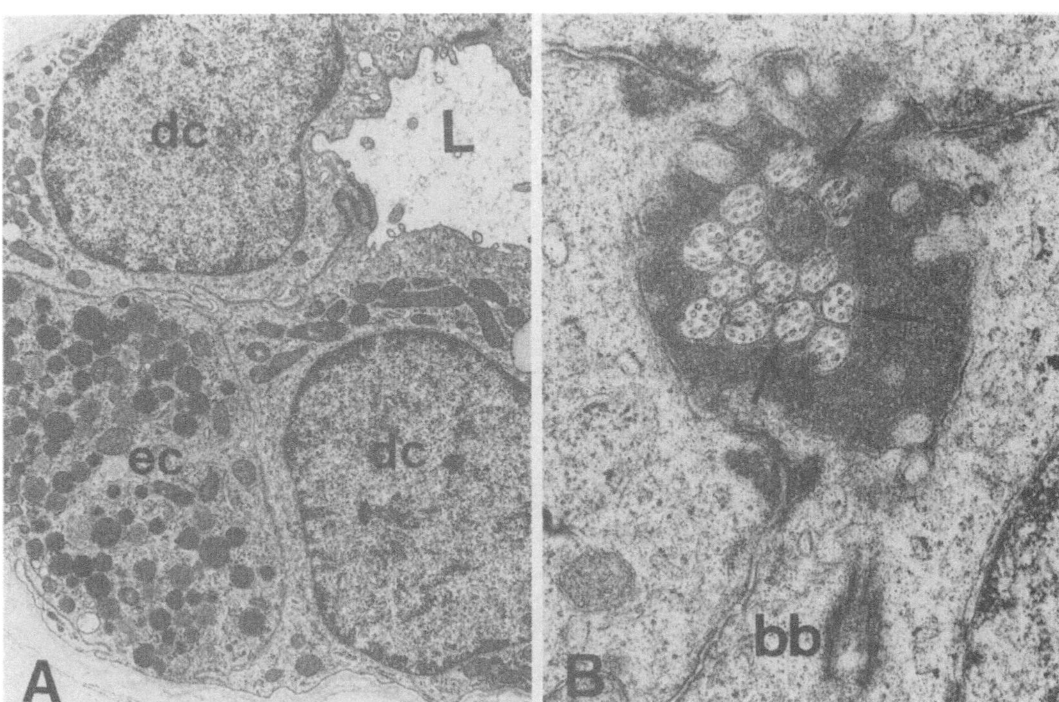

Fig. 7-3. **A**: fine structure of intralobular duct cells (dc) and an endocrine cell within the duct epithelium (ec) in the human pancreas. L = lumen. x 7,200. **B**: numerous cross sections of kinocilia in the duct lumen (arrows) and basal body (bb) in the cytoplasm of an intralobular duct cell. x 14,800.

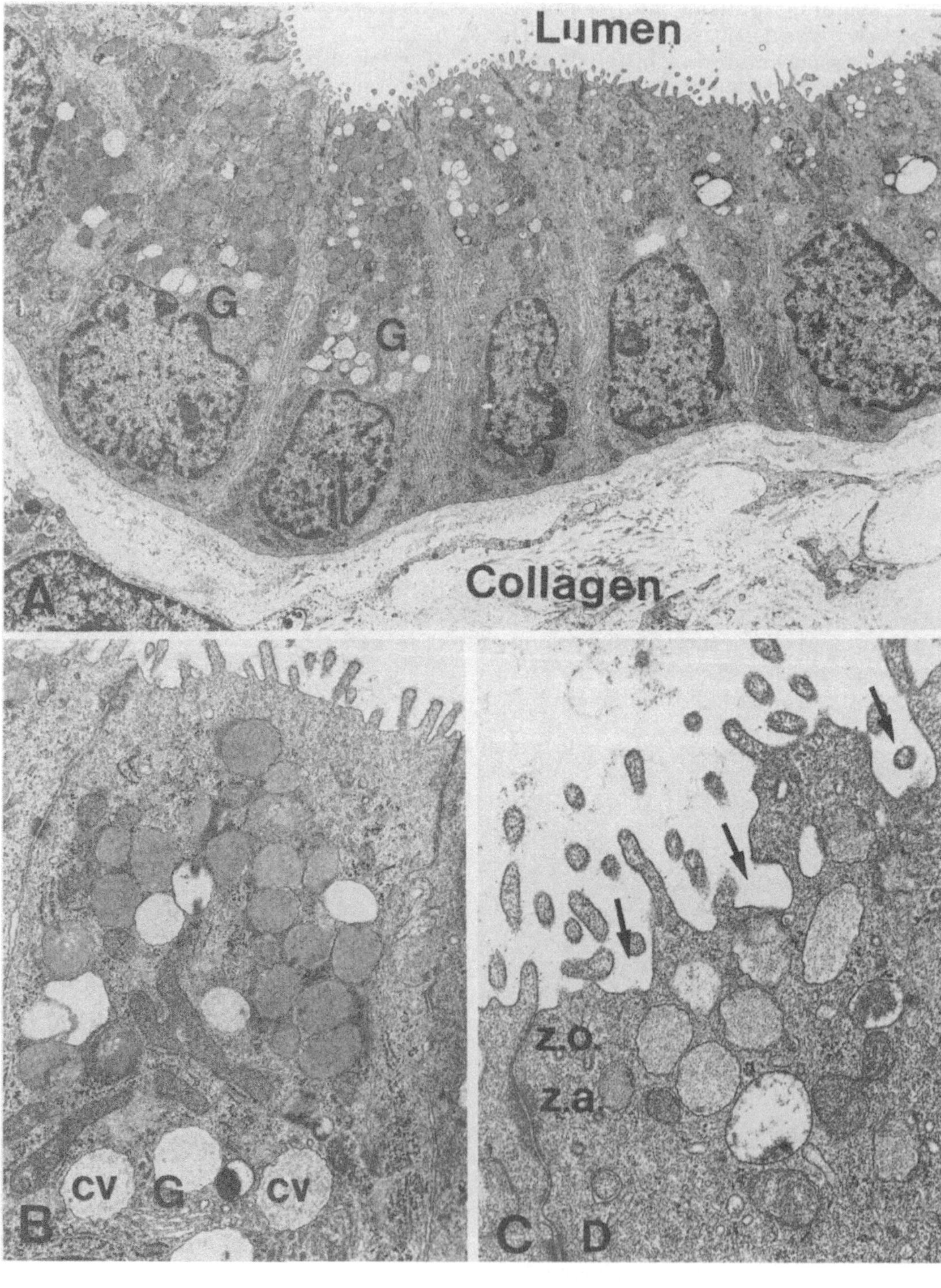
Lumen
G
G
Collagen
A
CV
G
CV
B
z.o.
z.a.
C
D

to acinar cells, the Golgi complex in intralobular duct cells is small and lacks typical secretion granules. Instead, numerous smooth vesicles are found in the Golgi area and in close proximity to both the apical and lateral plasma membrane. A characteristic feature of intralobular duct cells is the presence of elaborate interdigitations at the lateral and basal plasma membrane [4]. In this structural feature they resemble the cells of the striated ducts in salivary glands or the proximal convoluted tubules of the kidney. The conclusion that they are engaged in the transport of fluid and ions [7] is further supported by the immunocytochemical demonstration of carbonic anhydrase mainly in the intralobular ducts of the human pancreas [30].

Another characteristic feature of centro-acinar and intralobular duct cells are solitary kinocilia, which are derived from a basal body in the supranuclear region (Figs. 7-1A, 7-3B) and which project into the luminal space.

Kinocilia derived from six to ten adjacent ductular cells are observed to course in parallel for a distance of 20 to 30 μm in the direction of the secretory flow [4] and can also be identified in cross sections (Fig. 7-3B). It has been hypothesized that kinocilia in the initial portion of the duct system could play a role in both the correct mixing of fluid and proteins in the acinar lumen and in the propulsion of pancreatic juice [31].

1.4. Interlobular and main pancreatic duct cells

The wall of the interlobular and the main pancreatic ducts in the human pancreas is formed by cuboidal to pyramid-shaped cells, which project short microvilli from their luminal surface and contain abundant secretory granules in their apical cytoplasm (Fig. 7-4A). The basal and paranuclear region of the cytoplasm is filled with elements of the RER and mitochondria, while the Golgi complex and secretory granules, in various stages of maturation, are localized in the supranuclear cytoplasm (Fig. 7-4B). Elaborate interdigitations and numerous spot desmosomes are observed along the lateral plasma membrane. The cytoplasm contains a dense network of intermediate filaments that terminate in the desmosomal plaques (Fig. 7-4C).

The specific secretory products in interlobular ducts and in the main pancreatic duct in humans have only been poorly analyzed. Histochemical studies indicate the production and secretion of four types of mucins, which differ in their staining characteristics [32]. At the ultrastructural level, the content of secretory granules appears as a fine fibrillar material of low electron density. Exocytotic discharge of the granules is observed, and frequently a flocculent material of similar composition is found at the surface of duct cells (Figs. 7-4B and 7-4C). The regulation of mucin secretion from the human pancreatic ducts has not been studied so far.

On the basis of in-vivo duct perfusion studies carried out in cats [33], the presence of an epithelial barrier that prevents the diffusion of larger molecules from the luminal space across the duct was postulated. The structural basis for this barrier could be in specialized intercellular junctions (Fig. 7-4C) and in the secretion of mucins that remain associated with the luminal surface as a protective coat. There is no significant structural difference between interlobular duct cells and those of the main pancreatic duct. One special feature concerns the high frequency of cell exfoliation and replacement by mitosis, which is evident even at the level of the light microscope. Such exfoliating cells show ultrastructural signs of degenerated organelles, pycnotic nuclei, and loss of intercellular connections (Fig. 7-5A). Exfoliating cells are replaced by new cells that appear in the basal region of the epithelium (Fig. 7-5A) and by mitotic division of cells within the epithelium (Fig. 7-5B). It can only be hypothesized that this higher rate of cell renewal renders the terminal part of the duct system especially susceptible to noxious events, which eventually could lead to cell transformation and tumor formation.

←

Fig. 7-4. Fine structure of interlobular duct in human pancreas. For details see text. G = Golgi complex; CV = condensing vacuole; z.o. = zonula occludens; z.a. = zonula adhaerens; D = desmosome. The arrows in C mark exocytotic figures. A: x 4,750; B: x 13,600; C: x 24,000.

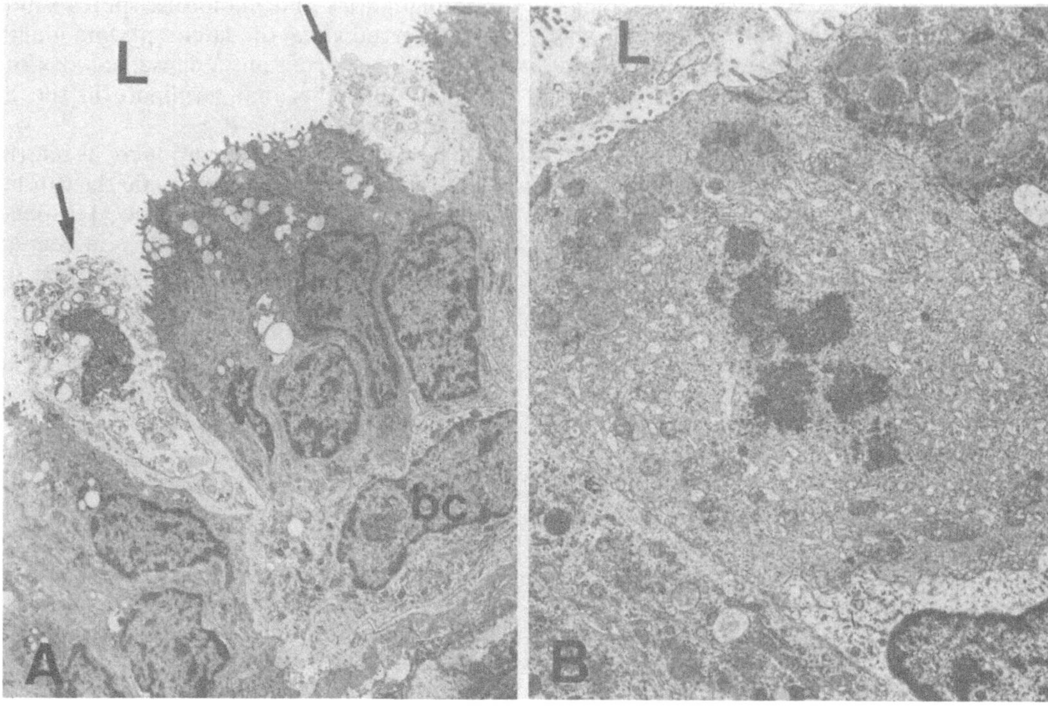

Fig. 7-5. **A**: exfoliation of degenerated cells (arrows). **B**: mitosis in the main pancreatic duct. bc = basal cell for replacement; L = lumen A: x 3,800; B: x 5,700.

2. Acute pancreatitis: Clinical and morphological definition

Clinically, acute pancreatitis is characterized by acute abdominal pain and is associated with elevated pancreatic enzymes in the blood and urine [34, 35]. Acute pancreatitis occurs as a single episode or as recurrent exacerbations of chronic pancreatitis. Most attacks of acute pancreatitis show a benign course and require little treatment. Severe attacks, however, may lead to multiorgan failure and pancreatic complications. They are fatal in 10% to 20% of the patients despite intense therapeutic measures [36]. The possible transition from a mild to a severe form can not be predicted. Generally, recurrent attacks of chronic pancreatitis are less severe. In acute pancreatitis, endocrine and exocrine pancreatic functions are impaired for a variable duration, but are restored to normal if the cause and complications of acute pancreatitis are eliminated [37].

Biliary tract disease is the most frequent cause (33% to 65%) of acute pancreatitis in Europe and in the United States [38, 39]. Recurrent attacks of acute biliary pancreatitis occur until the biliary tract is cleared from stones. Other extrapancreatic causes of acute pancreatitis are surgery, trauma, drugs, hyperparathyroidism, hyperlipidemia, and infection. Alcohol abuse is the cause of acute pancreatitis in 14% to 51% of patients, depending on the geographic area [38, 39]. It is not known whether acute alcoholic pancreatitis always manifests chronic pancreatitis or if in certain cases no progression to chronic disease occurs.

As can be expected from the varying clinical picture, there is no sharp morphological distinction between mild and severe pancreatitis [40]. The mild form is characterized by edema, cellular infiltration, and peripancreatic and intrapancreatic fat necrosis. Parenchymal necroses are rare.

In cases with severe necrotizing pancreatitis, large and confluent foci of peripancreatic fat necrosis appear. They extend into the parenchyma along the interstitial space and are associated with perilobular necrosis of acinar tissue. Thrombosis and rupture of veins and arteries lead to hemorrhage and focal ischemic necrosis of pancreatic lobules. As a rule, the lesions are irregularly distributed over the organ. Therefore, normal tissue and edematous lesions can be found adjacent to necrotic areas.

Ultrastructural studies of the pancreas in acute pancreatitis have mainly been performed on surgical specimens [41–47]. At present, ultrasound-guided biopsy of the pancreas is not a commonly used procedure. In one ultrastructural study, the pancreas from patients dying from shock was rapidly sampled within minutes of somatic death [48]. Since surgery is generally not performed in the beginning of the disease, it is expected that only little information about early changes in pancreatitis can be obtained from surgical material. However, when specimens are taken from different regions of the gland, all stages of cell injury, as defined by Trump and coworkers, are found along with normal acinar tissue [49].

2.1. Acinar cells

Although acinar cells from areas neighbouring fat necrosis at the light microscopic level are without structural alterations, at the ultrastructural level different alterations become evident (Fig. 7-6A). The polarized organization of the acinar cell prevails, the cisternae of the RER are slightly distended, and whorl-like configurations are focally revealed. In all specimens, mitochondrial changes precede other structural alterations (Fig. 7-6B). Mitochondria are enlarged, condensed, or exhibit ruptures of inner and outer membranes. In some acinar cell nuclei, the chromatin is slightly clumped.

The zymogen granules are concentrated mainly in the apical part of the cytoplasm. Their number, however, is less, and their size is decreased when compared to the normal pancreas. This has been described by Bockman et al. [45] and was recently quantitated by Willemer [Willemer et al., unpublished]. In another study, an increase in the size and number of zymogen granules was observed [42]. Focally, in some acinar cells the zymogen granules are concentrated along the basolateral, instead of the apical, cell membrane (Fig. 7-6C) [46]. This finding could indicate a rerouting of the normal secretory pathway and a release of enzymes from the basal part of the cell. In general, the zymogen granules are well preserved and reveal an intact membrane enclosing a homogeneous electron-dense core, even in stages of severe cell injury. The integrity of the zymogen granules is further documented when secretory proteins are labeled using the protein A gold antitrypsin complex. The colloidal gold particles are seen almost exclusively over zymogen granules or condensing vacuoles and not in the cytoplasm of injured cells (Fig. 7-7A).

In early stages of cell injury, the Golgi complex is well preserved, revealing slightly dilated cisternae. Small bundles of filaments are located in the vicinity of the Golgi complex (Fig. 7-6C). They belong to the 10-nm intermediate filaments of the cytoskeleton [50].

In various stages of cell injury, large aggregates of closely packed filaments are found that are partly surrounded by a single membrane (Fig. 7-7B). It is not clear whether they represent accumulations of intermediate filaments or fibrillar residual bodies.

Large autophagic vacuoles are found frequently. They are filled with cell organelles, including fragments of RER, mitochondria, and zymogen granules, and are limited by a single membrane (Fig. 7-7C). Their contents reveal marked variability, ranging from fine floccular material to membrane remnants, as a result of lysosomal degradation. In a few cases, fusion of zymogen granules among each other is found (Fig. 7-6B), as has been described in shock patients [48]. This could lead to the formation of vacuoles that are larger than zymogen granules but contain electron-dense material similar in texture to zymogen granules. In areas of apparent fusion, a breakdown of zymogen granules occurs (Fig. 7-7D). The electron-dense core of the granules is now irregular and is surrounded by an electron-translucent halo. Finally, when cellular membranes are digested within autophagic vacuoles, large residual bodies remain that are composed of translucent and granular areas (Fig. 7-7D). More severe stages of cellular injury are found in closer proximity to areas of necrosis. The severity of structural alterations is not uniform and varies within one acinus. Thus necrotic cells are found in close proximity to acinar cells, which only reveal slight dilatation of the RER. Further destruction

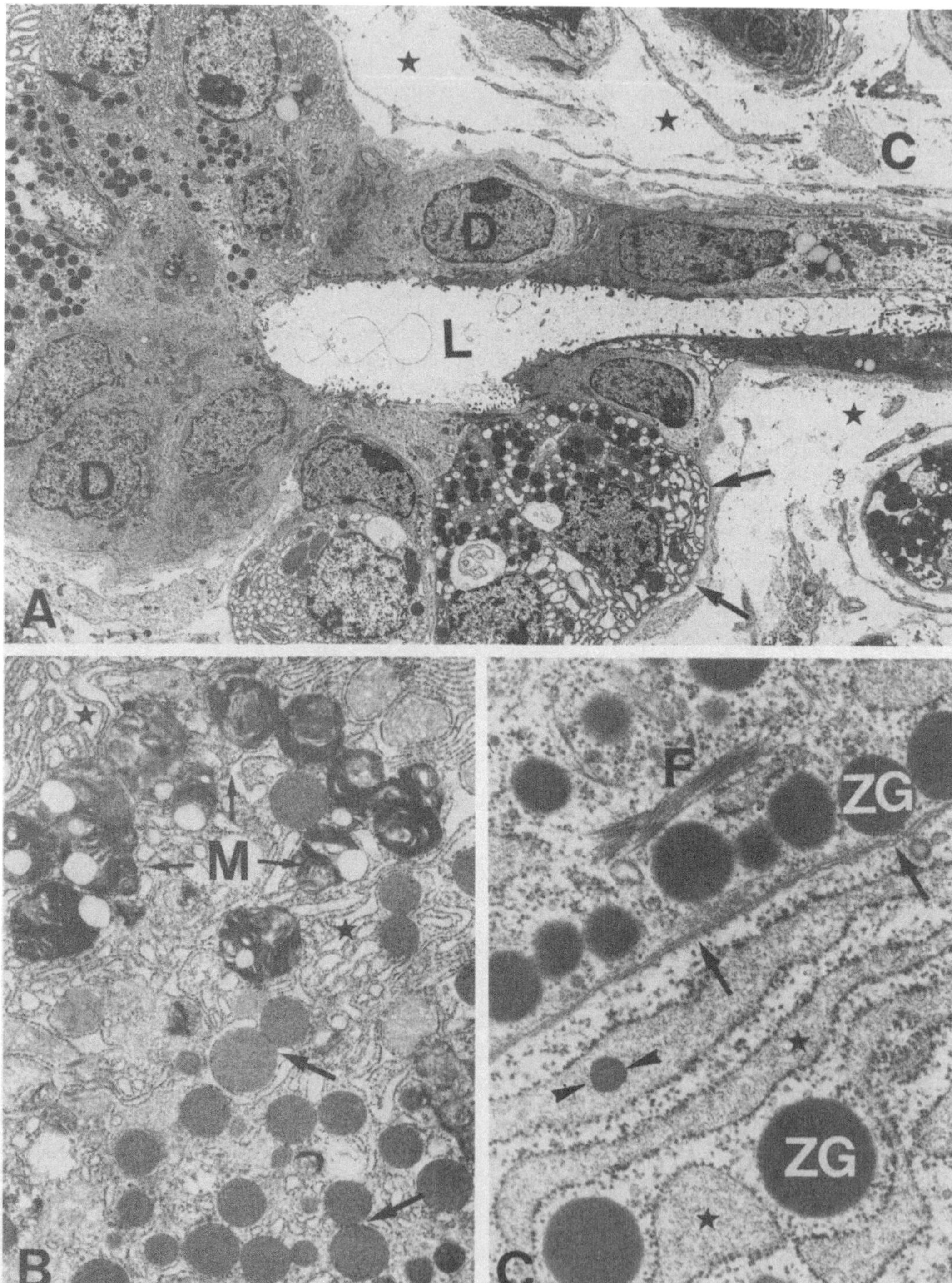

Fig. 7-6. Fine structure of exocrine pancreas in acute pancreatitis (tissue from surgery 10 days after onset of the disease). **A**: coexistence of intact acinar cells and acinar cells with dilated RER and cytoplasmic vacuoles (arrows). Normal fine structure of duct cells (D). Membrane remnants within the duct lumen (L). Edema (asterisks) and collagen fibers (C) in the interstice. x 2,700. **B**: marked condensation of mitochondria (M). Slight dilatation of cisternae of RER (asterisks). Fusion of zymogen granules (arrows). x 11,400. **C**: irregular distribution of zymogen granules (ZG) along the basolateral plasma membrane (arrows). Intracisternal granule (arrowheads) within dilated cisternae of the RER (asterisks). Bundles of filaments (F). x 26,400.

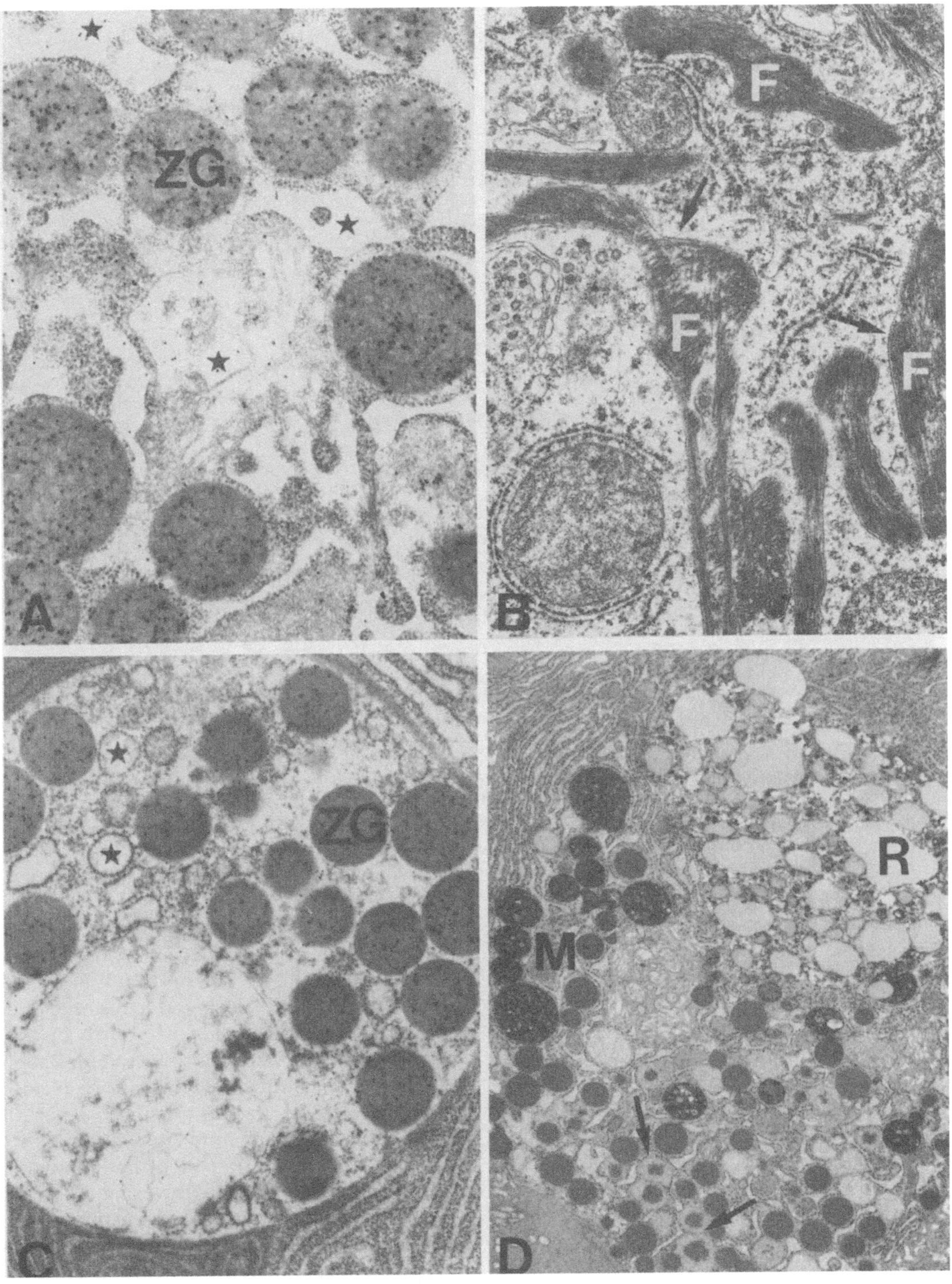

Fig. 7-7. Fine structure of acinar cells in human acute pancreatitis (tissue from surgery 3 and 10 days after onset of disease). **A**: intact zymogen granules in injured acinar cell with massive dilatation and fragmentation of RER (asterisks). The electron-dense gold particles are coupled with trypsin antibody and are mainly found over zymogen granules (ZG). x 28,800. (Courtesy of S. Willemer). **B**: bundles of closely packed filaments (F) that are partly surrounded by a limiting membrane (arrows). x 21,600. **C**: autophagic vacuole containing zymogen granules (ZG), fragments of RER (asterisks), and membrane remnants. The gold-antitrypsin label is restricted to the zymogen granules. x 19,600. (Courtesy of S. Willemer). **D**: focal breakdown and fusion of zymogen granules (arrows). M = condensed mitochondria; R = residual body. x 9,100.

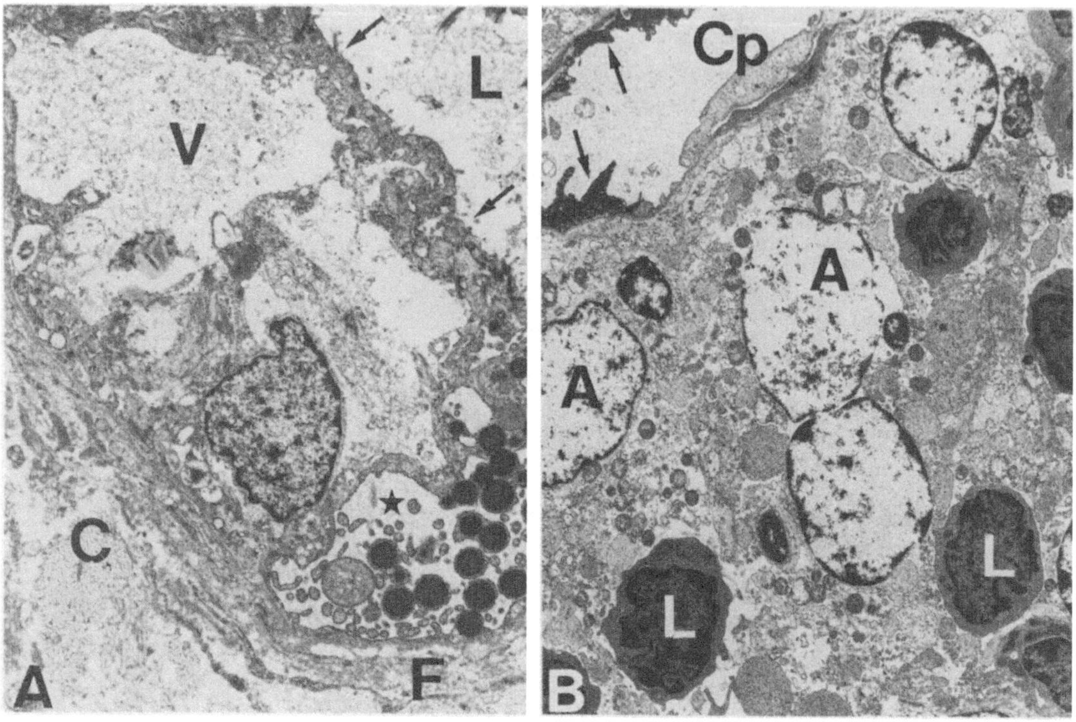

Fig. 7-8. Destruction of acinar cells in human acute pancreatitis (material from surgery 10 days after onset of disease). **A**: large cytoplasmic vacuoles (V) filled with granular material occupy most of the cytoplasm. Loss of microvilli (arrows) at the luminal membrane. L = acinar lumen; F = fibrin deposits along the basal lamina; C = collagen fibers. Note fragmentation of RER (asterisk) and structurally intact zymogen granules in neighboring acinar cell. x 5,800. **B**: remnants of necrotic acinar cells (A). Infiltration of lymphocytes (L). Necrotic endothelium (arrows) of a capillary (Cp). x 5,100.

results in acinar cells consisting of a small rim of cytoplasm surrounding a large, central space filled with granular and fibrillar material (Fig. 7-8A). At this stage, the plasma membrane and the basal lamina appear intact. In the most advanced stages, complete fragmentation of membrane systems and karyolysis are found (Fig. 7-8B). Even in those injured cells that are morphologically intact, zymogen granules are observed. Finally, necrotic remnants of cellular organelles are intermingled with lymphocytes, macrophages, plasma cells, erythrocytes, and activated fibroblasts (Fig. 7-8B).

2.2. Acinar lumen

Loss of microvilli, deposition of fibrillar material, dilatation of acinar lumina, and formation of tubular complexes are the main alterations of the acinar lumen in acute pancreatitis. An irregular loss of microvilli is often observed, even when acinar cell injury is not severe. The remaining microvilli are mostly plump [46, 47]. Some acinar lumina are filled with aggregates that are composed of membrane remnants, partly digested cellular organelles, and amorphous material [42, 45]. At higher magnification, the latter reveals a network of densely packed fibrils that vary in diameter from 10 to 30 nm (Fig. 7-9A).

Focally, a dilatation of acinar lumina is seen that is paralleled by a lowering of acinar cell height and invaginations of the luminal plasma membrane (Fig. 7-9B). A progressive lowering of the acinar cell leads to formation of tubular complexes that occur in acute pancreatitis [51]. Their formation has been observed in various pancreatic diseases [52, 53] and in experimental models of pancreatic injury [54]. Bockman and colleagues have shown that retrogressive changes in acinar cells lead to those ductlike structures [9].

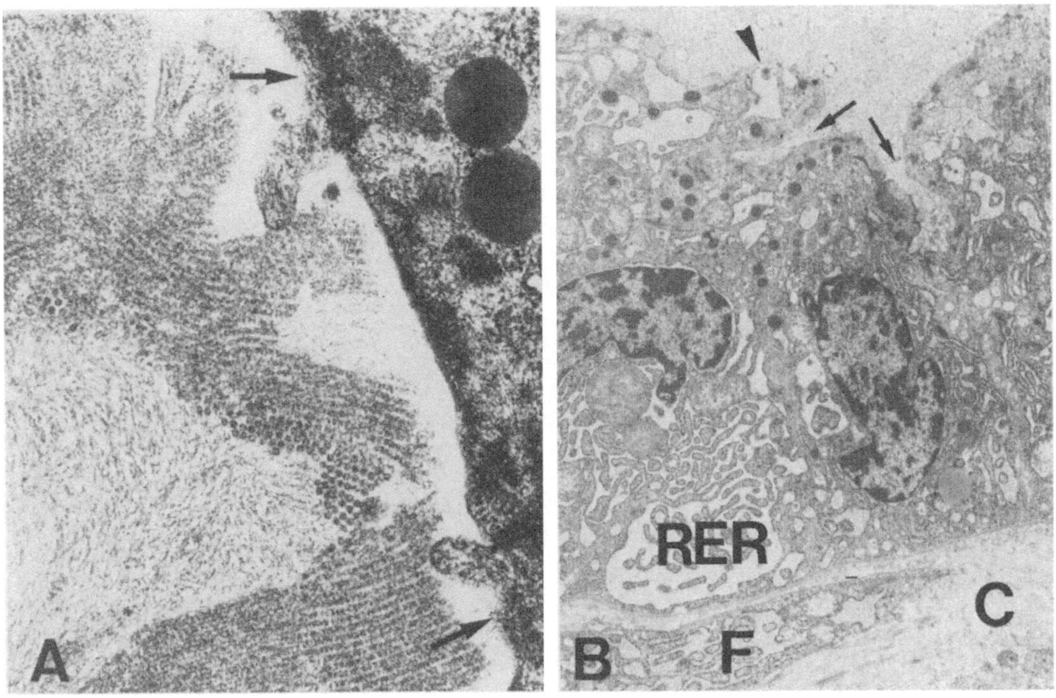

Fig. 7-9. Acinar lumen in human acute pancreatitis (material from surgery 5 days after onset of disease). **A**: densely packed network of fibrils of different diameter. Loss of microvilli at the luminal plasma membrane (arrows). x 36,000. **B**: formation of tubular complex. Lowering of acinar cell height. Invagination of luminal plasma membrane (arrows). Protrusion of vacuolated cytoplasm (arrowhead). RER = fragmentation of RER; F = activated fibroblasts; C = collagen fiber. x 5,750.

Immunocytochemical studies demonstrated a loss of pancreatic enzymes and a reduced lectin binding for L-fucose and N-acetylgalactosamine in acinar cells during formation of tubular complexes [51]. From the careful reconstruction studies performed by Bockman et al., two possibilites for the development of tubular complexes are suggested [9]: a flattening of acinar cells could occur when single cells replace necrotic cells. Second, the lowering of acinar cells could result from autophagocytosis and release of cellular debris into the acinar lumen. No mitotic activity was found in tubular complexes during experimental pancreatitis [54]. In the rat, tubular complexes are finally digested by macrophages [54].

2.3. *Duct system*

In general, the epithelial lining of the duct system is preserved, even in regions of damaged acinar and interstitial tissue (Fig. 7-6A). In the luminal space of almost all ducts, fibrillar material is noticed, and sometimes intact zymogen granules are found in the duct lumen. Centroacinar cells in close proximity to necrotic acinar cells retain their structural integrity (Fig.7-10A). In some cases, swelling of mitochondria and focal cytoplasmic degeneration of intralobular duct cells occurs. The intercellular spaces of duct cells are narrow, and no destruction of intercellular connections is seen. The interlobular duct system is well preserved.

2.4. *Interstitial space*

Interstitial edema is a common finding in areas close to fat necrosis. Focally, fibrin deposits are attached to the basal lamina (Fig. 7-10B). A variable number of polymorphonuclear leucocytes, lymphocytes, and macrophages is present [45]. Inflammatory cells are found invading the basal lamina and protruding between individual acinar cells (Fig. 7-10C). An electron-translucent space

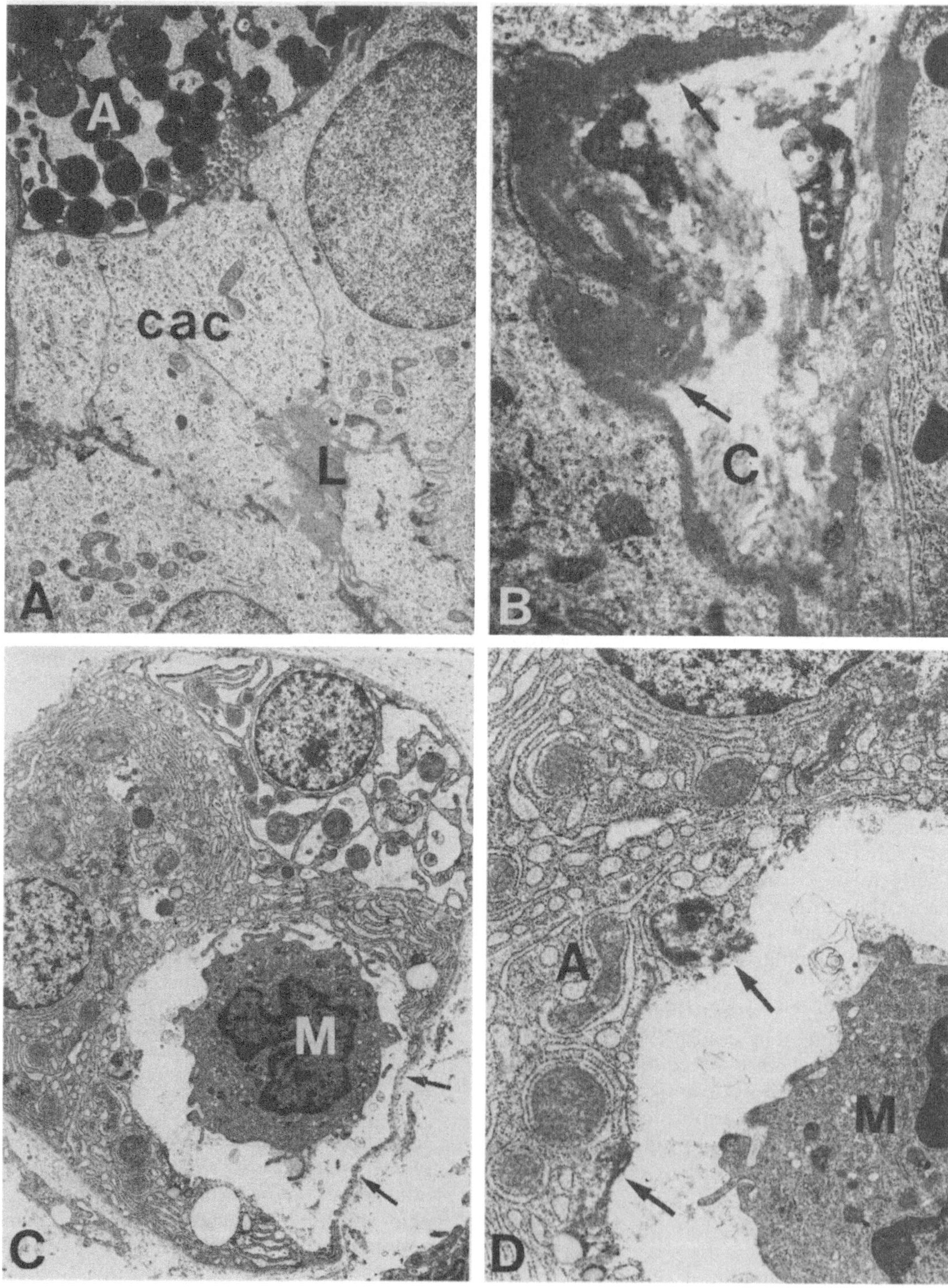
A
cac
L
A
C
B
M
C
A
M
D

separates the inflammatory cells from the acinar cells and contains membrane remnants (Fig. 7-10D). At the edge of this space the plasma membrane and the cytoplasm of the acinar cells are destroyed. Cell necrosis progresses concomitant with the inflammatory reaction until only necrotic remnants of acinar cells can be identified between interstitial cells (Fig. 7-8B). The macrophages often contain large amounts of lipid droplets, which are partially degraded by lysosomes.

In all these stages a significant proliferation of fibroblasts accompanies the cellular infiltration. Some fibroblasts are spindle shaped without well-developed cellular organelles. Others contain enlarged Golgi complexes and prominent cisternae of RER. Broad bands of collagen fibers are located in the interlobular and intralobular interstitial space. This is observed even in areas of cellular infiltration without acinar necrosis.

Capillary thrombosis and hemostasis has been described by Bockman et al. [45]. Structural alterations of capillaries comprise swelling, disruption, and necrosis of endothelial cells (Fig. 7-8B). Focal destruction of Schwann cells and of nerve fibers are also observed. Within the pancreatic islets the B cells are generally more altered compared to A cells and D cells. Residual bodies are a common finding in B cells.

2.5. *Pathogenesis of acute pancreatitis*

Only little information is available explaining the cellular and molecular mechanisms by which biliary stones, alcoholism, duct adenocarcinoma, and other factors induce acute pancreatitis [44]. Furthermore, there is no clear distinction between factors that initiate pancreatic injury and those that occur as a consequence of injury and play a role in the progression of the disease. Finally, it is still a matter of controversy whether pancreatitis is initiated in acinar cells or in the interstitial space [55, 56].

The different etiological factors of human acute pancreatitis induce similar morphological changes in all ultrastructural studies and in several necropsy cases (40, 45, 48, 57, 58). In a recent review of 31 necropsy cases of acute pancreatitis, Foulis described an association of etiological factors with histological changes at the light microscopic level [59]. In patients with biliary and alcoholic pancreatitis, he noticed periductal necrosis, while perilobular necrosis was observed in patients dying from shock. In contrary, Jones et al. showed that ultrastructural changes correlated with the degree and duration of shock and are similar to those seen in other forms of pancreatitis [48]. An interstitial edema represents the initial consequence of pancreatic injury [56]. The edema contains pancreatic secretory proteins, plasma proteins, and mediators of inflammatory cells, all of which govern the progression and severity of pancreatic disease. The initiation and amplification of acute pancreatitis is a result of several pathobiological processes: obstruction of the normal transport of secretory proteins, alteration in compartmentation, activation of pancreatic enzymes, and disturbance of the microvasculature [56].

Obstruction in the normal flow of secretory proteins along the pancreatic duct and into the duodenum may occur by a variety of clinical conditions, including biliary stones, pancreatic carcinoma, and spasm and edema of the sphincter. It is, however, not well defined how stones in the distal bile duct induce pancreatitis. In patients with biliary disease, the passage of small gallstones through the ampulla could induce transient obstruction [60]. In experimental models, obstruction of pancreatic flow, as well as the reflux of bile and duodenal contents, lead to a dramatic loss of compartmentation between the pancreatic ductal lumen and the interstitial space, which results in dissolution of intercellular junctions and destruction of duct cells [56]. The extravasation of pancreatic juice into the interstitial space will rapidly lead to edema and inflammation. This is

←

Fig. 7-10. Fine structure of human acute pancreatitis (surgical material). **A**: normal structure of centroacinar cells (cac) in close proximity to severely injured acinar cells (A). L = duct lumen. x 6,000. **B**: fibrin deposits (arrows) along the basolateral plasma membrane. C = collagen fibers. x 24,000. **C**: invasion of a macrophage (M) into acinar unit. Basal lamina (arrows). x 4,180. **D**: detail from Fig. 7-10 C. Membrane remnants in the electron-lucent space between the macrophage (M) and the acinar cell (A). destruction of cytoplasm and cellular organelles (arrows). x 9,600.

caused by vascular changes that allow the passage of cellular and soluble factors from the blood circulation into the interstitial space. Structural and functional studies in experimental models and in human acute pancreatitis revealed significant alterations of vascular permeability [45,61,62]. Thus, factors that are normally confined to the vascular system and potential hydrolases, which are largely confined to the ductal space, are brought together within one compartment. Although the biochemical events that lead to the initial activation of pancreatic enzymes within the interstitial space are poorly understood, the consequences are disastrous.

According to the above concept of extravasation of pancreatic enzymes from the duct system, pancreatitis starts within the interstitial space. There is, however, ample evidence from experimental and human studies that acute pancreatitis is initiated by events occuring within the acinar cells [55,56]. This concept has emerged from two noninvasive models of experimental pancreatitis. Lombardi and coworkers induced hemorrhagic pancreatitis in female mice by feeding them a choline-deficient diet supplemented with ethionine [63]. Lampel and Kern induced edematous pancreatitis by infusion of the CCK analogue, caerulein, in excess of the amount that stimulates pancreatic secretion [64]. Under both experimental conditions, intracellular transport and secretion of pancreatic enzymes was inhibited. The reaction of the acinar cells to such changes involved activation of lysosomes and degradation of sequestered cellular organelles by autophagocytosis [65,66]. Additionally, the process that normally separates secretory proteins from lysosomal enzymes in the Golgi region was disturbed [55]. This led to the formation of large vacuoles where lysosomal enzymes and digestive enzymes are colocalized. This represents a hazardous condition, since one lysosomal enzyme, cathepsin B, is able to activate trypsinogen. The activation of enzymes could induce leakage and disruption of lysosomal structures and could thus start intracellular autodigestion [67].

Using the caerulein model, an additional pathobiologic process was defined by which the normal compartmentation of secretory proteins is lost [65]. During supramaximal stimulation with caerulein, secretion via the duct system is markedly inhibited and discharge is redirected to the basolateral plasma membrane. Ectopic discharge in this manner delivers large quantities of secretory proteins to the interstitial space. Even in this concept, changes of vascular permeability could then aggravate organ destruction.

Similar structural changes have been observed in human acute pancreatitis by immunohistochemical and ultrastructural methods. In the vicinity of fat necrosis phospholipase A_2, trypsin, lipase, and other enzymes were identified in the interstitial space, while their cellular content was decreased [68,69]. Furthermore, autophagic vacuoles and fusion of zymogen granules have been demonstrated by several authors [44,48]. The acinar cells contained less zymogen granules, which in some sections were localized along the basolateral instead of the apical plasma membrane [46]. These data support the concept of an intracellular start of pancreatitis. This could be induced by alteration of cellular metabolism by different factors such as alcohol, several drugs, and disturbance of the blood supply.

Both concepts involve the alteration of compartmentation, which leads to the appearance of pancreatic enzymes in the interstitial space. This results in edema and cellular infiltration as a consequence of changes in vascular permeability [62] and in adipose tissue necrosis by lipolytic enzymes [70]. At this stage only little damage of acinar cells is found. The further progression of the disease not only depends on the dominance of the etiological factor, but also on the extent of the inflammatory reaction. Careful morphometrical and immunohistochemical studies at both the ultrastructural and light-microscopic level, as well as scanning electron microscopic studies, are needed to further clarify the pathogenesis of acute pancreatitis. Only recently, the loss of microvilli and the preservation of zymogen granules in acute pancreatitis was shown by scanning electron microscopy [71].

3. Chronic pancreatitis: Clinical and morphologic definition

Chronic pancreatitis is characterized clinically by recurrent or persisting abdominal pain in most patients. Pain may not be present or may be reduced at the onset of the disease [72]. Depending on the duration of chronic pancreatitis and the extent of the destruction of pancreatic tissue, exocrine and endocrine insufficiency occur [34, 35, 37]. The course of chronic pancreatitis is often marked by recurrent episodes of acute pancreatic inflammation. These attacks clinically present as acute pancreatitis. They become less severe the longer the underlying chronic disease lasts.

According to the 1984 Marseille classification, chronic pancreatitis is morphologically characterized by destruction of acinar parenchyma, which is paralleled by segmental or diffuse fibrosis, and might be accompanied by focal necrosis and calculi [35]. Based mainly on the classical work of Henri Sarles, two distinct forms of chronic pancreatitis are classified [73]. Chronic calcifying pancreatitis is characterized by the presence of protein precipitates or calcified calculi in the acinar and duct lumen, lobular distribution of lesions, and dilatation of interlobular ducts [74–77]. In addition to these chronic alterations, focal tissue necrosis and pseudocyst formation occur. Probable etiologic factors of this disease are chronic alcohol consumption, a high-caloric protein and fat diet, hypercalcemia, and malnutrition [73]. Over time, 76% to 85% of the patients will develop intraductal calcification that is visible on plain x-ray analysis of the abdomen [78]. The less frequent form is called *chronic obstructive pancreatitis*. This is characterized by dilatation of the ductal system proximal to occlusion of one of the major ducts by a tumor or a scar, diffuse atrophy of acinar parenchyma, and diffuse fibrosis. In this form of chronic pancreatitis calculi are uncommon.

The development of chronic pancreatitis is best followed by studies that compare the clinical findings with the histology of pancreatic biopsy. On the basis of such correlation, Howard and Nedwich described three forms of chronic pancreatitis [79]. In minimal chronic pancreatitis, an increase in perilobular fibrosis and little inflammatory reaction is seen in patients with recurrent abdominal pain but no other signs of pancreatic disease. The pancreatic duct system is almost normal and the acini are well preserved. Focally irregular scarring occured. These changes are classified as stage I of chronic pancreatitis by Stolte [80].

Moderate pancreatitis in patients with more frequent episodes of pain and increased serum amylase levels is characterized by irregular loss of acini and an increase in the connective tissue in the interstitial space. The lining of the ducts is irregular and the media of arterioles is thickened. The number of islets appear increased, some islets being hypertrophic. These changes correlate with stage II–III of chronic pancreatitis.

A pancreas from patients with continuous abdominal pain and exocrine insufficiency reveals the advanced stage of chronic pancreatitis (stage IV). The almost complete loss of acini is compensated by large amounts of fibrous tissue. The ducts are irregularly dilated and are lined by cuboidal to cylindric cells that produce large amounts of mucus. The walls of the blood vessels are thickened. The inflammatory infiltrate is patchy in distribution and is composed mainly of lymphocytes.

In surgical material from patients with chronic pancreatitis (23 cases of chronic alcoholic pancreatitis; 14 cases of nonalcoholic pancreatitis) almost always acute and chronic structural changes are seen side by side at the electron microscopic level [Adler et al., unpublished]. When morphometry is performed on semithin sections, alcoholic pancreatitis is separated from nonalcoholic pancreatitis by dilatation of ducts, epithelial lesions, and ductal stones. At the ultrastructural level, in all studies no significant difference was observed between the various etiological forms of chronic pancreatitis [52, 81–85]. Atrophy of acini and separation of acini by broad bands of collagen fibers are the most characteristic features in chronic pancreatitis.

3.1. Acinar cells

In advanced stages of chronic pancreatitis, the acini are no longer composed of pyramid-shaped cells that are located around a small acinar lumen. Instead, the cells are cuboidal or flattened, while the diameter of the acinar lumen is enlarged (Fig. 7-11A). The luminal membrane of the acinar cells is almost devoid of microvilli. Most acinar cells have lost their polarized organization and contain only few zymogen granules, which show a marked heterogeneity in size. The cisternae of the RER are distended and reveal annular figures.

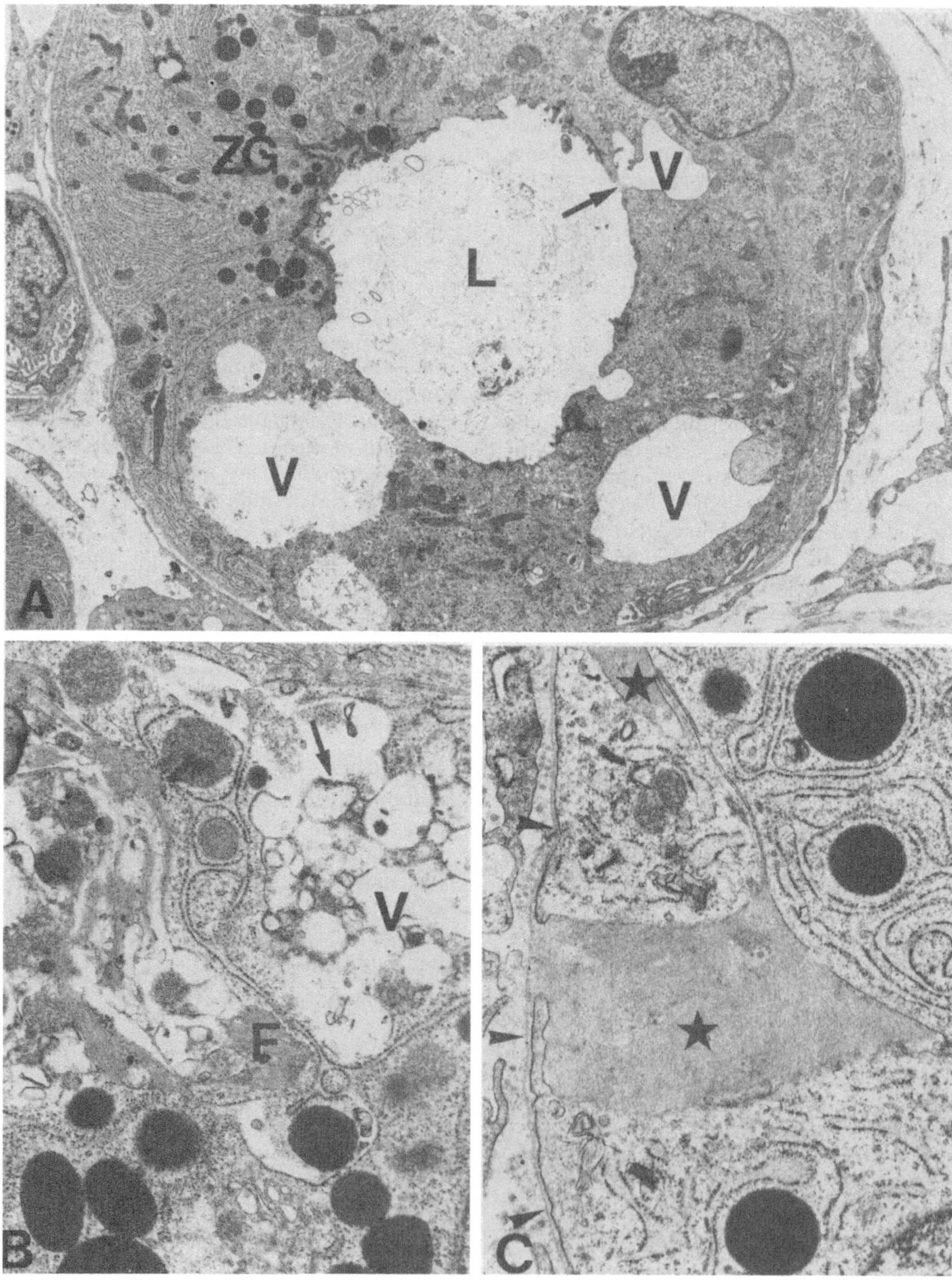
ZG
V
L
V
V
A
V
F
B
C

Varying amounts of residual bodies are found. Focal cytoplasmic degeneration induces formation of large autophagic vacuoles containing fibrillar material and membrane remnants (Fig. 7-11B). These intracellular vacuoles often come in contact with the acinar lumen (Fig. 7-11A). The lowering of acinar cell height could result from an increased autophagocytosis and releases of the digested cellular material into the acinar lumen. These retrogressive changes lead to the formation of acini that are composed of some typical acinar cells containing zymogen granules and also of flattened cells with ductular characteristics. As discussed earlier (2.2), those tubular complexes are a common finding in all diseases of the pancreas but occur most frequently in chronic pancreatitis [10, 51, 52]. The interstitial space between acinar cells is narrow and the intercellular juntions appear intact. In some cases of chronic alcoholic pancreatitis, the intercellular space is focally enlarged and is filled with a fine fibrillar material (Fig. 7-11C). This material separates the acinar cells mainly in the basal region and comes in direct contact with the basal lamina. It is suggested that alterations of the normal secretory process lead to discharge of proteins into the interstitial space, where they form fibrillar precipitations.

3.2. *Duct system*

The enlarged luminal spaces of acini, intercalated ducts, and small intralobular ducts are often completely filled with precipitations that are tightly apposed to the apical surface of the bordering cells (Fig. 7-12A). The precipitations are formed by a network of densely packed fibrils that vary in diameter from 10 nm to 30 nm. The fibrils do not reveal a substructure such as regular spacing or cross-striation. Focally, a homogenous, electrondense material is found that packs several fibrils into a homogenous mass. In chronic pancreatitis, eosinophilic precipitates in pancreatic ducts are composed of concentric laminar layers of proteins [76]. Similar precipitates appear in pancreatic juice from the main pancreatic duct from alcoholic subjects and from patients with chronic pancreatitis [86]. It is conceivable that the electron-dense condensations in the luminal space represent focal points of calcium carbonate crystallization within the protein matrix and therefore are precursors of calcified stones [87]. The intralobular ducts are often dilated and lined by a flattened epithelium. In some instances, the centroacinar and the intralobular cells are hypertrophied and protrude into the duct lumen (Fig. 7-12B). The microvilli at the luminal membrane are replaced by irregular cytoplasmic protrusions. The intracellular organization of the hypertrophied duct cells is not different from healthy controls, except that some cells contain increased amounts of 10 nm intermediate filaments. In the region of hypertrophied duct cells, the duct lumen is drastically reduced (Fig. 7-12B). Close to this constriction, the duct lumen is filled with densely packed cellular debris. The interlobular ducts reveal similar dilatation and flattening of lining cells. In addition, different dysplastic changes of duct cells are found. As shown in Fig. 7-12C, some cells of the single-layered epithelium are hypertrophied. Their apical part is domed, protrudes into the duct lumen, and is not connected by intercellular junctions to the neighboring cells. Similar hypertrophic changes have been observed using scanning electron microscopy [88]. In addition to hypertrophy, hyperplasia, mucoid transformation, and squamous metaplasia is found in many cases of chronic pancreatitis (Fig. 7-12D). Stolte has shown that the occurence of dysplastic alterations of duct cells parallels the severity of fibrosis and acinar cell destruction of the pancreas [89].

3.3. *Pathogenesis of chronic pancreatitis*

The pathogenesis of chronic pancreatitis is still a matter of controversy. The main question is whether or not the formation of protein plugs and

←

Fig. 7-11. Fine structure of exocrine pancreas in chronic pancreatitis. **A**: formation of tubular complex by lowering of acinar cells. Note the reduced content of zymogen granules (ZG). V = intracellular vacuoles. Continuity of a vacuole (arrow) with the acinar lumen (L). x 4,900. **B**: large autophagic vacuole (V) including membrane remnants (arrow) and fibrillar material (F). x 16,800. **C**: fibrillar material (asterisks) in the enlarged interstitial space between acinar cells. Basal lamina (arrowheads). x 19,200.

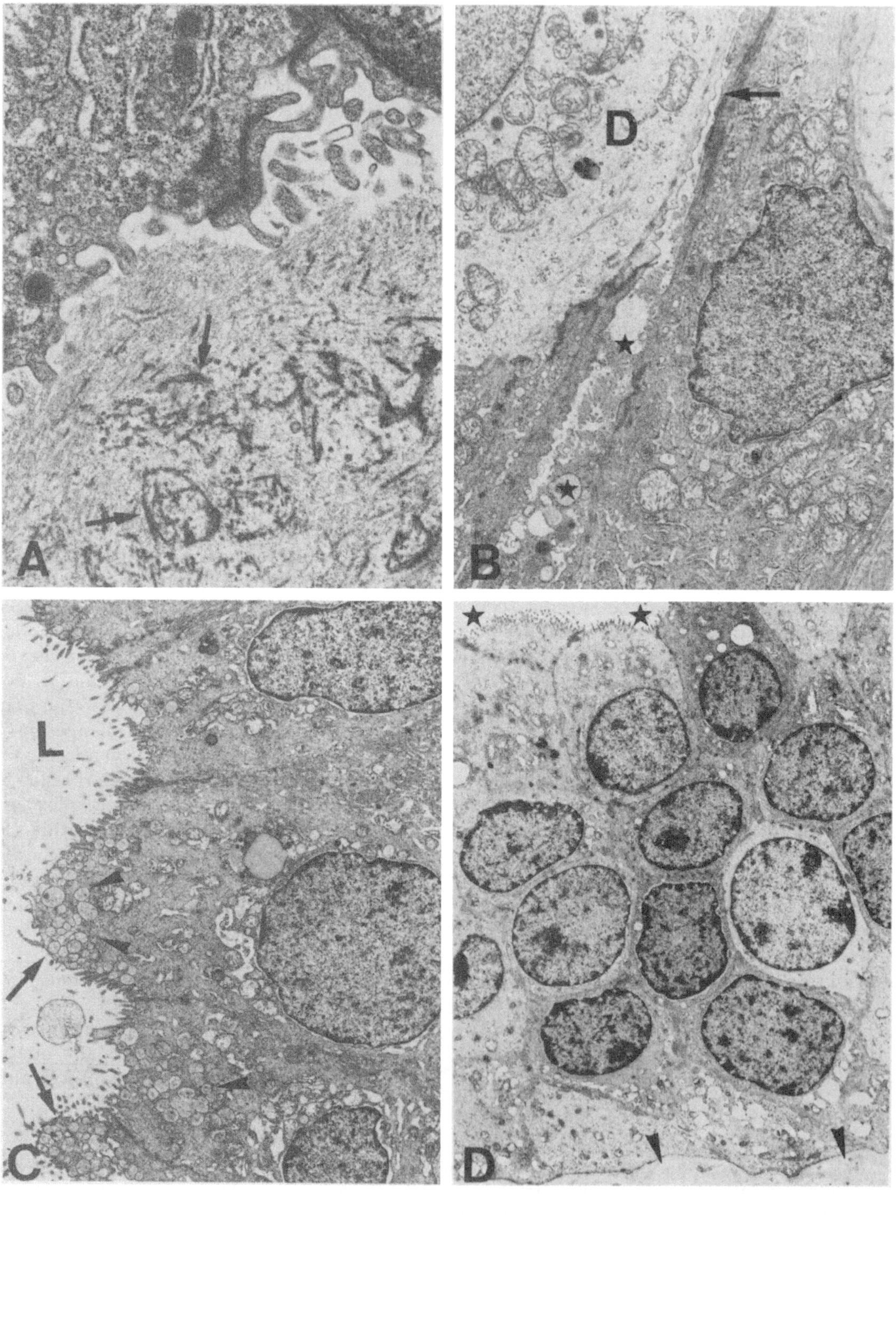

D
A
B
L
C
D

calcification are the primary lesions of chronic pancreatitis.

From extensive clinical, experimental, and biochemical work, Sarles and his coworkers suggested that the initial lesion in chronic pancreatitis of alcoholics is the precipitation of protein plugs that later become calcified and then obstruct the duct lumen and damage the ductal lining cells [73–77]. This concept is supported by the occurrence of plugs even in the early stages of chronic pancreatitis and the transition to pancreatic stones. Their formation in the pancreatic duct system is a consequence of different factors. Pancreatic juice is normally supersaturated in calcium ions. In normal subjects, the formation of calcium salt crystals is prevented by a stabilizer, the pancreatic stone protein (PSP), which is secreted parallel to the other exocrine proteins (90–91). Therefore, a decrease in PSP could support stone formation. A low PSP was found in patients with chronic calcifying pancreatitis of alcoholic or other origins. Furthermore, in alcoholic subjects the protein concentration of pancreatic juice is increased. This could alter the viscosity of the juice and decrease washout of calcium crystals. Protein plugs and calcium crystals then destroy the duct epithelium. Desquamated cells and debris, however, provide further nucleation centers for stone formation. Finally, the epithelium is replaced by scars that enhance the obstruction of the ductal lumen. At this time, at least at the ultrastructural level, no difference exists between chronic calcifying pancreatitis and chronic obstructive pancreatitis.

Although the concept of calcium precipitation suficiently explains the formation of pancreatic stones, the biochemical mechanisms, and especially the metabolic disturbances due to alcohol consumption, are not clarified.

An alternative pathogenetic view of chronic pancreatitis is proposed by several pathologists [47, 88, 89, 92]. Their concept centers around focal autodigestive tissue necrosis, which leads to perilobular scarring. Repeated necrosis could aggravate fibrosis and cause duct strictures. This may slow down the secretory flow, increase viscosity, and by this support the formation of protein plugs. According to this concept, the pathogenesis of chronic pancreatitis is linked more directly to repeated attacks of acute pancreatitis.

3.4. Cystic fibrosis

Cystic fibrosis is an autosomal recessive homozygous disease that leads to exocrine pancreatic insufficiency in 85% of cases. Pancreatic secretions from cystic fibrosis patients contain significantly higher concentrations of protein than those from controls, while fluid secretion is decreased [93]. The limited flow of a protein-rich juice predisposes patients to precipitations and ductal obstruction. Therefore, the pathogenesis of cystic fibrosis in the pancreas resembles that of chronic obstructive pancreatitis. At the ultrasturctural level, protein precipitates and cellular debris are found in the acinar and duct lumen [94]. As in chronic pancreatitis of other etiologies, the precipitates are composed of fibrillar material. The epithelium of acinar and duct cells reveal all stages of cell injury. The interstitial space is widened and edematous, and the basal membrane is disrupted. Finally, the acini are replaced by dilated cysts and fibrosis.

4. Fine structure of pancreatic adenocarcinoma

Among the epithelial neoplasms arising in the pancreas, so-called ductal adenocarcinoma represents the dominant type, comprising more than 80% of all tumors. In their histological, fine-structural, and immunocytochemical (CEA, CA 19–9, cytokeratins) characteristics, these neo-

←

Fig. 7-12. Fine structure of exocrine pancreas in chronic pancreatitis. **A**: densely packed fibrils and condensed, electron-dense material (arrows) in the acinar lumen. x 20,000. **B**: narrowing of duct lumen (arrow) by hyperplastic intralobular duct cell (D). Cellular debris (asterisks). x 10,000. **C**: hyperplastic cell of interlobular duct. Domed protrusions of apical cytoplasm (arrows) into the lumen (L). Mucin granules (arrowheads). x 7,020. **D**: squamous metaplasia of the epithelium of an interlobular duct. Acinar lumen (asterisks); basal lamina (arrowheads). x 3,420

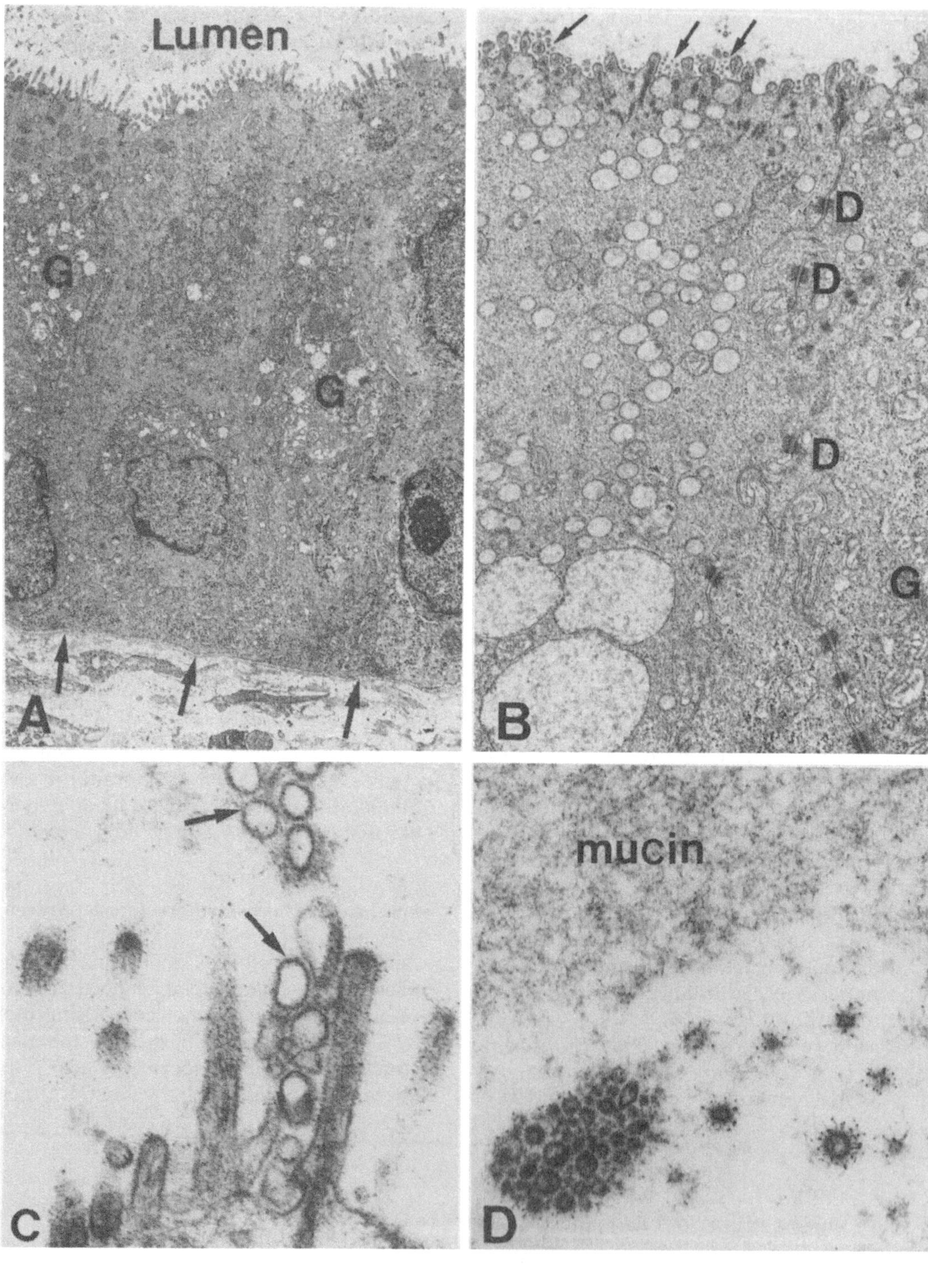
Lumen
G
G
A
D
D
D
G
B
C
mucin
D

plasms closely resemble the normal duct system [95–97]. Tumors are preferentially localized in the head of the pancreas, in an area behind the papilla of Vater or in proximity to the common bile duct, which runs through the dorsal portion of the pancreatic head [95]. In spite of detailed systems for classification and staging of these tumors [98, 99], our knowledge of the biology of the dominant tumor cell type is limited. For a better understanding of the differences observed in tumor growth, survival time, and response to chemotherapy, a grading system has recently been introduced in which a large group of ductal adenocarcinoma has been subdivided into three major grades of malignancy [100]. Glandular differentiation, nuclear size and anaplasia, and mitotic activity were used as histologic parameters.

The same grading system has been applied at the fine structural level [101] to analyze in more detail

1. the heterogeneity of cell types within the same tumor and among tumors of different grades,
2. the fine structure of cellular constituents and indications for specific cellular functions, and
3. details of intercellular connections and the relation of tumor cells to elements of the extracellular matrix.

4.1. Grade I tumors

Tumors of the highest grade of differentiation (grade I), which made up roughly 42% of a series of 51 tumors, are composed of well-differentiated ductlike structures, reminiscent of the previously described larger interlobular ducts (Fig. 7-13A). The tumor cells are pyramid shaped and are strictly polarized, resting with their basal plasma membrane on a basal lamina of the extracellular matrix and having their apical plasma membrane exposed to a common luminal space. Neighboring tumor cells form regular junctional complexes, and their epithelial contact is enforced by numerous spot desmosomes (Fig. 7-13B). As in normal pancreatic duct cells, elaborate networks of intermediate filaments terminate in the desmosomal plaques, also in tumor cells. Different size classes of granules measuring 0.4 μm to 2.0 μm in diameter, are found in the apical cytoplasm. The fine structure of the granule content varies among cells in the same tumor and between different tumors (Fig. 7-13B), but the granules largely correspond to the secretion granules in the normal interlobular duct. In grade I tumors, discharge of secretion granules occurs exclusively at the apical plasma membrane (monotopic discharge) via exocytosis. From previous histochemical studies [32], it is known that the secretory products in tumor cells are predominantly mucins, which also appear in the epithelial cells of normal interlobular ducts. A finding distinctly different from the secretory products in the normal duct system concerns the regular occurrence of small vesicles, measuring 50 nm to 100 nm in diameter, which are found in close proximity to the apical plasma membrane and in association with exocytotic figures (Fig. 7-13B). In the luminal space, aggregates of such vesicles are observed embedded in the electron-dense material of the mucins (Fig. 7-13D). Detailed analysis of the cell surface of a large number of tumors revealed the formation of these vesicles from the luminal plasma membrane, predominantly by pinching off from the tip of microvilli or by fragmentation of whole microvilli (Fig. 7-13C). It has therefore been concluded that the secretory material from pancreatic adenocarcinoma represents a mixture of mucins and of membrane material [101].

4.2. Grade II tumors

About 47% of pancreatic adenocarcinomas are of an intermediate grade of differentiation (grade II) in which the ductlike tubular structures, as de-

←

Fig. 7-13. Fine structure of grade I pancreatic adenocarcinoma. **A**: polarized arrangement of tumor cells with connection to the basal lamina (arrows). G = Golgi complex. x 3,800. **B**: various size classes of secretion granules in the cytoplasm. In the apical part, tumor cells are connected by junctional complexes and by membrane interdigitations combined with spot demosomes (D). The arrows point to vesicles in the luminal space. x 12,000. **C**: formation of vesicles by fragmentation of surface microvilli (arrows). x 43,800. **D**: fine structure of luminal content. x 43,800.

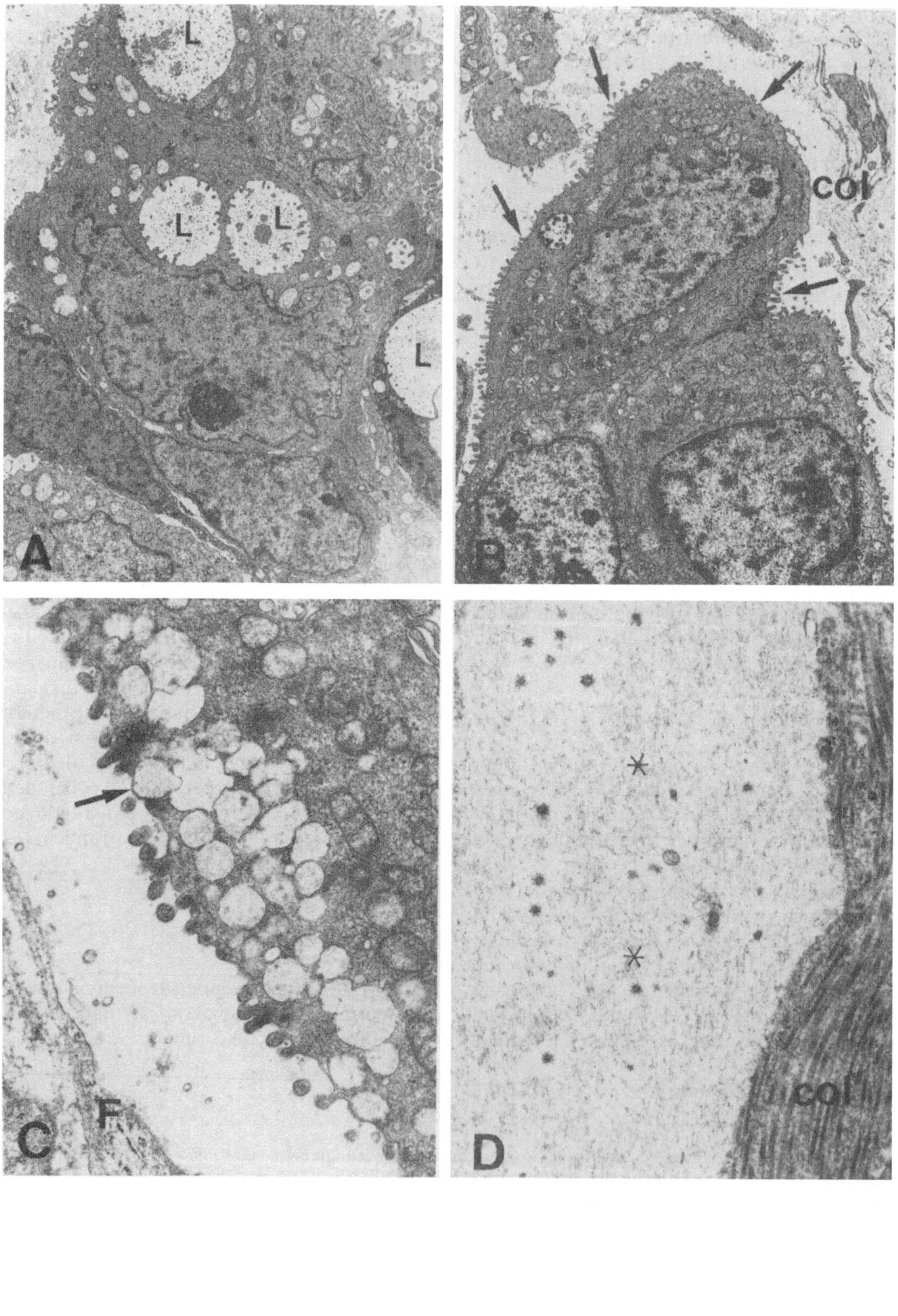
L
L
L
L
A
col
B
C
F
D
col

scribed in the previous grade, are combined with solid-growing buds of tumor cells. Nuclear size and anaplasia is more variable, and mitotic activity is more pronounced [100]. At the fine structural level, the ductlike structures in grade II tumors are lined by irregular-shaped cells that closely resemble those in grade I, except with a greater variability in their equipment with cell organelles such as RER, the Golgi complex, and secretion granules. The most pronounced change is the *loss of cell polarity* and the formation of multiple bizarre lumina within the cytoplasm (Fig. 7-14A). Microvilli are observed as a lining of the cytoplasmic lumina, as well as on several surfaces of the tumor cells (Fig. 7-14B). Frequently the luminal space of the ductlike tubular structures is not closed but communicates with the interstitial connective tissue. The solid-growing buds or cords of tumor cells also reveal microvilli on the surface, which are oriented towards the surrounding connective tissue (Fig. 7-14B). Because of the loss of cell polarity in grade II tumors, secretion granules are randomly distributed within the cytoplasm and beneath several membrane surfaces (Fig. 7-14C). Discharge of secretion granules occurs into the various luminal spaces and into the interstitial connective tissue (polytopic discharge), which frequently is displaced from the tumor cell surface by the accumulation of a mixture of electron-opaque material and of vesicles (Fig. 7-14C and 7-14D). This is similar to the secretory material in cytoplasmic lumina (Fig. 7-14A) and to the luminal content observed in grade I tumors (Fig. 7-13D). Therefore, the loss of cell polarity in grade II tumors results in polytopic discharge of mucins and membrane material into the interstitial compartment, from where it may be absorbed into the circulation.

4.3. *Grade III tumors*

About 10% of pancreatic adenocarcinomas are composed of irregular aggregates of large polymorphous cells, which show little tendency to tubular or epithelial formation (grade III). The nuclei in these tumor cells are large and irregular, sometimes highly segmented (Fig. 7-15A). The cytoplasm contains predominantly free ribosomes and only occasional elements of the RER. The sizes and forms of Golgi complexes are variable but usually are smaller than in previous grades and signs of granule formation are rare or absent (Fig. 7-15B). Formation of membrane junctions and desmosomes is rare, therefore the intercellular space is usually distended and contains highly variable microvilli (Fig. 7-15C). During the search for secretory activity from grade III tumors, instead of typical secretion granules, a high incidence of coated pits and both coated and uncoated vesicles were observed along the whole membrane surface (Fig. 7-15C). Their occurrence in the vicinity of the Golgi complex and in fusion with the lateral rims of individual Golgi cisternae indicated endocytotic activity of tumor cells, which could be verified using electron-opaque tracers (cationized ferritin) in cell culture experiments (Fig. 7-15D).

4.4. *Fine structure of the basal lamina in pancreatic adeno-carcinoma*

It has previously been demonstrated in a variety of carcinomas, including pancreatic adenocarcinoma, that malignant tumors in general lack an intact basal lamina, while benign lesions in the same organs have their basal lamina intact [102]. It is also known that a variety of differentiated functions of epithelial cells, including the development and preservation of polarity, depend on a specific interaction between components of the extracellular matrix and the basal plasma membrane of the respective epithelial cell [103]. In tumor cells, it was hypothesized that basal lamina components might be involved in the migration and resettlement of tumor cells during metastasis [104, 105]. Using affinity-purified antibodies to major components of the extracellular matrix

←

Fig. 7-14. Fine structural characteristics of grade II tumor cells. **A**: formation of numerous luminal spaces (L) in the cytoplasm of the tumor cells. x 3,900. **B**: loss of cell polarity with orientation of surface microvilli (arrows) towards interstitial connective tissue (col). x 4,370. **C**: accumulation of secretion granules at basal part of cytoplasm. The arrow points to a granule prior to release. F = fibrocytic cell. x 18,400. **D**: secretory products (asterisks) accumulated in interstitial connective tissue (col). x 24,000.

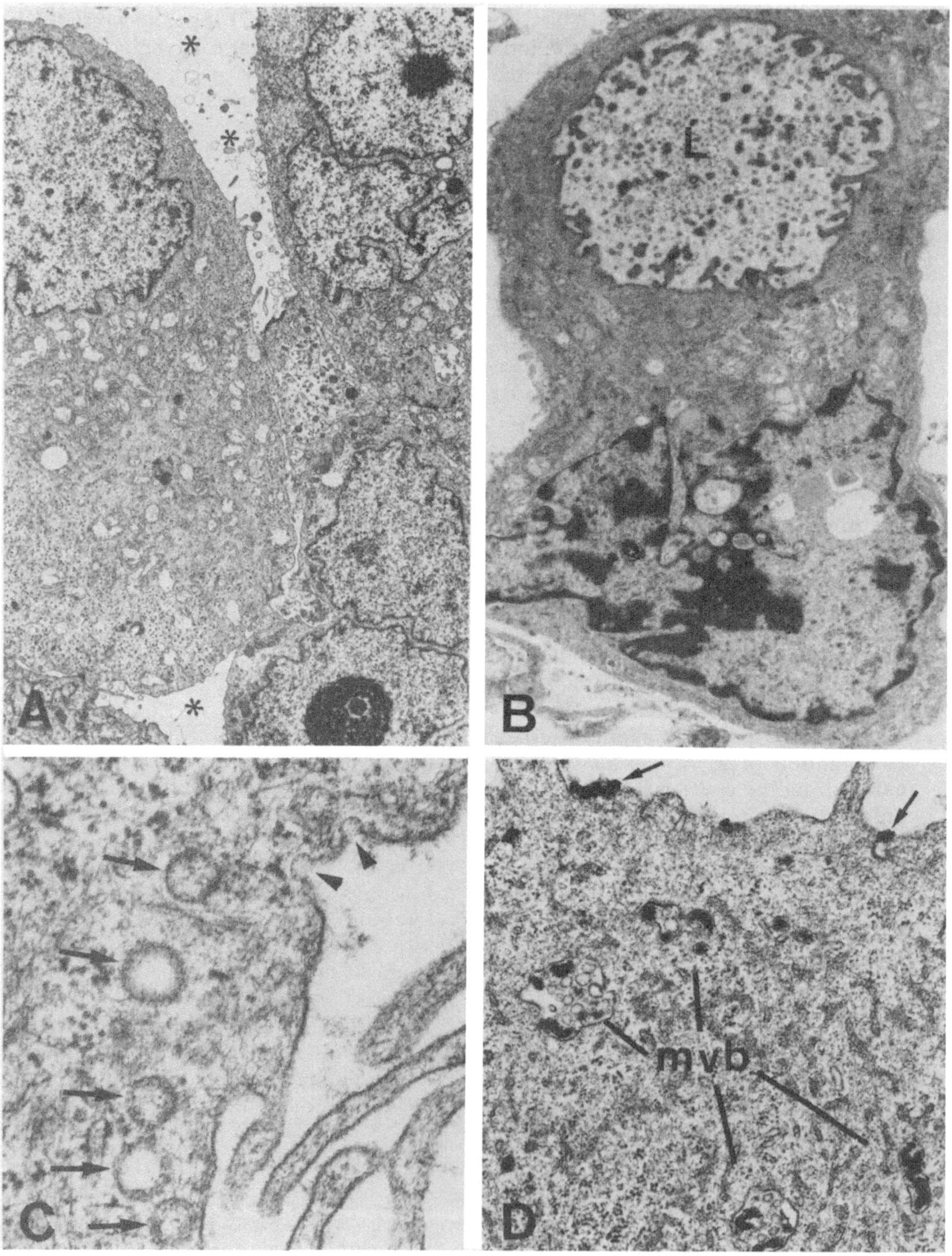
A
B
L
C
D
mvb

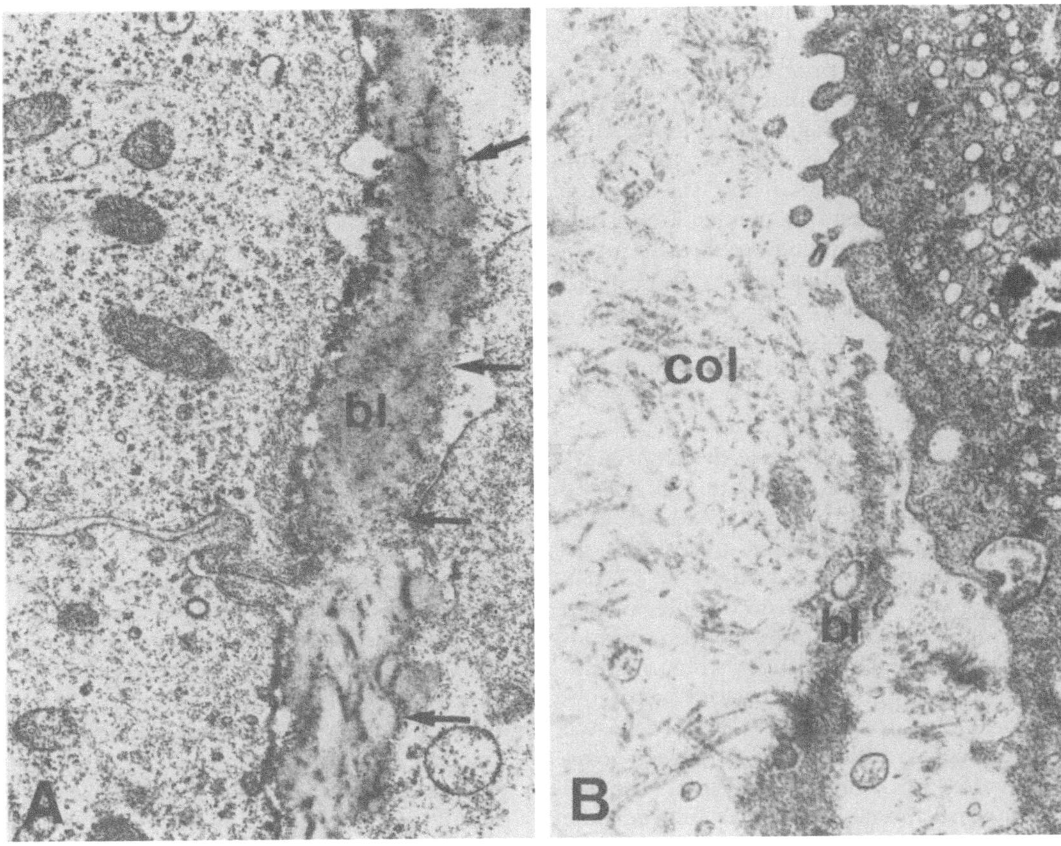

Fig. 7-16. Fine structure of basal lamina (bl) in grade I (A) and grade II (B) tumors. col = collagen fibers. A: x 20,300; B: x 20,300.

(fibronectin, collagen type I) and of the basal lamina (laminin) and electron microscopy, we have recently analyzed their distribution in the three major grades of malignancy of pancreatic adenocarcinoma [106]. Intact basement membranes, also with structural irregularities, were only present in the highest grade of differentiation where tumor cells grow as tubular structures. The basal lamina of tumor cells, in comparison to normal duct cells, however, shows irregularities in thickness and occasional interruptions (Fig. 7–16A). In grade II tumors, especially in areas of solid epithelial growth, the sheathlike deposition of basal lamina is greatly reduced. Instead, electron-dense material reminiscent of the basal lamina is randomly distributed at tumor cell surfaces and is often deposited in patches (Fig. 7–16B). In grade III tumors,no definable structures comparable to the basal lamina were observed. Instead collagen fibers with typical cross striations are deposited around the tumor cells and often come in contact with their plasma mem-

←

Fig. 7-15. Fine structure of grade III tumor cells. **A**: lack of intercellular junctions and widening of intercellular space (asterisks). x 4,370. **B**: luminal space (L) in the cytoplasm of a grade III tumor cell. x 8,800. **C**: multiple smooth and coated vesicles beneath the plasma membrane (arrows) as a sign of endocytotic activity. The arrowheads point to invaginations of the the plasma membrane. x 60,500. **D**: endocytosis of cationized ferritin in an established tumor cell line, showing association of the tracer with coated pits (arrows) and uptake into multivesicular bodies (mvb). x 12,800.

branes [106]. These findings indicate a disturbance of the macromolecular organization of the basal lamina that seems to be progressive in the three grades of tumor cell differentiation. On the other hand, the interaction of malignant cells with the components of the basal lamina seems to be important for the complex event of metastasis. There is increasing evidence that the production of lytic enzymes, including plasminogen activator, lysosomal hydrolases, and collagenases, plays a key role in tissue invasion [107, 108].

4.5. Monocyte-related functions in pancreatic carcinoma cell lines

The observed endocytotic activity in primary tumors of lower grade differentiation suggested active uptake of substances from the extracellular milieu, which had to be further quantitated. In addition, one monoclonal antibody (BW 227/19) produced against pancreatic tumor cells of low differentiation, besides binding to primary pancreatic carcinoma, reacted with 80% blood monocytes and with macrophages in a variety of tissues in cytofluorometric studies [109]. This suggested common antigenic determinants between pancreatic carcinoma cells and monocytes. In order to also study functional similarities between the two cell types, we have measured typical monocyte-related functions such as endocytosis, lysosomal enzyme secretion, and superoxide anion production in five cell lines of pancreatic carcinoma and in five cell lines from tumors of other origins (two from small lung-cell carcinoma, two from adenocarcinoma of the colon, and one from squamous-cell carcinoma of the lung). The functional response in tumor cells was compared to isolated human monocytes and was studied under both basal conditions (PBS) and after exposure to zymosan or immune complexes as functional stimuli [110, 111]. As in primary tumors, the established cell lines grow as epithelial sheets and form luminal spaces that are sealed by junctional complexes. Numerous vesicles were observed along the plasma membrane. Exposure of the tumor cells to cationized ferritin as an electron-dense tracer for 10 seconds demonstrated its association with numerous coated pits along the membrane surface (Fig. 7-15D). These were internalized within 5 to 10 minutes as coated vesicles and transferred into multivesicular bodies (Fig. 7-15D). Within 30 to 60 minutes, these multivesicular bodies were progressively filled with tracer and were clustered in the area of the Golgi complex. Quantitation of endocytotic activity by measuring the uptake of radiolabeled collodial gold (^{198}Au) demonstrated that 4 out of 5 pancreatic carcinoma cell lines had a basal endocytotic activity of 50% to 60% of normal human monocytes, while the other tumor cell lines showed only 20% to 25% of this activity. Exposure to zymosan or immune complexes as stimuli of phagocytosis in monocytes increased the uptake of collodial gold by a factor of two, and a similar increase was observed in pancreatic carcinoma cell lines. No change, however, was seen in all other tumor cell lines tested [110]. A similar and even more pronounced functional response was seen when the discharge of lysosomal enzymes was measured under the same conditions. Both the cell lines from pancreatic tumors and from other tumors showed the same secretory activity as human monocytes under basal conditions (10% to 15% release of total content in 24 hours). Exposure to either zymosan or immune complex had no effect on cell lines derived from lung or colon carcinomas but increased enzyme release from pancreatic cancer cell lines to the same extent as from monocytes. This stimulation of lysosomal enzyme secretion was not due to destructive effects, since the release of lactate dehydrogenase is not altered under these conditions (for details see [110]). These findings support the hypothesis that the invasive and/or metastatic potential of tumor cells is associated with the production and secretion of lytic enzymes, which are able to disrupt structural barriers, such as the basal lamina, and to facilitate the penetration of tumor cells into adjacent normal tissue. It remains to be shown that pancreatic cancer cells also secrete proteases, e.g., collagenases that, together with lysosomal enzymes, would enhance their lytic potential during tissue invasion.

5. Concluding remarks

Considerable progress has been made in recent years in understanding the basic cell biology of acinar cells in the pancreas. Major questions concern the handling of proteins within compartmental spaces, such as the forces governing the directional movement of the protein mixture in the cisternal space of the RER and their correct sorting within Golgi cisternae. Still a greater enigma concerns the recognition processes of the various membrane carriers during intracellular transport and during the final event of exocytosis.

Little progress has been made in understanding the pathogenesis of inflammatory pancreatic diseases, both acute and chronic. The cellular and molecular mechanisms by which the major etiological factors (alcohol and biliary stones) induce pancreatitis are not known. Our understanding is mainly based on experimental animal models, which indicate disturbances in the sorting process of digestive enzymes from lysosomal enzymes and alterations of the specific membrane fusion process of zymogen granules with the plasma membrane as early pathogenetic events. Additional factors that lead to aggravation and progression of acinar cell distruction, such as disturbances in pancreatic microcirculation, have to be studied in the future. One problem remains, namely, direct study of the human situation, which is only possible on selected cases of surgically obtained biopsies. So far, all functional and morphological methods give little information about regeneration of the pancreas from acute damage. It is of great interest to know whether pancreatic necrosis causes fibroblast proliferation and transient collagen deposition, and how this is replaced by regenerating exocrine tissue. An imbalance between tissue regeneration and collagen deposition could be the key event in the development of chronic pancreatitis.

The most pressing clinical problems are encountered in pancreatic carcinoma. Our basic biological knowledge is minimal, thus preventing new strategies for the early diagnosis and treatment of this fatal disease. It seems clear that the tumors arise from the cells of the larger ducts, which have to be studied with the same scrutiny as were the acinar cells in the past 20 to 30 years.

Acknowledgements

The authors thank Prof. Dr. H. D. Röher, Prof, Dr. D. Maroske, Prof, Dr. G. Klöppel, and Dr. S. Willemer for collaboration. The technical and editorial help of Brigitte Agricola, Roswitha Naumann, and Volkwin Kramer is gratefully acknowledged.

References

1. Case RM: Synthesis, intracellular transport and discharge of exportable proteins in the pancreatic acinar cell and other cells. *Biol Rev* 53: 211–354, 1978.
2. Dixon JS: Histology and ultrastructure. In: The Exocrine Pancreas. HT Howat, H Sarles (eds), London: WB Saunders, 31–49, 1979.
3. Gorelick FS, Jamieson JD: Structure function relationship of the pancreas. In: Physiology of the Gastrointestinal Tract. LR Johnson (ed), New York: Raven Press, 773–794, 1981.
4. Kern HF: Fine structure of the human exocrine pancreas. In: The Exocrine Pancreas: Biology, Pathobiology, and Diseases. VLW Go et al. (eds), New York: Raven Press, 9–19, 1986.
5. Scheele GA, Kern HF: The exocrine pancreas. In: Molecular and Cellular Basis of Digestion. P Desnuelle, H Sjöström, O Norén (eds), Amsterdam: Elsevier, 173–194, 1986.
6. Scheele GA, Kern HF: Cellular compartmentation and protein processing in the exocrine pancreas. In: Handbook of Physiology. The Gastrointestinal System. SG Schultz (ed), Washington DC: American Physiological Society, in press, 1989.
7. Case RM, Argent BE: Pancreatic secretion of electrolyte and water. In: Handbook of Physiology. The Gastrointestinal System. SG Schultz (ed), Washington DC: American Physiological Society, in press.
8. Heidenhain M: Über die teilungsfähigen Drüseneinheiten oder Adenomeren, sowie über die Grundbegriffe der morphologischen Systemlehre. *Arch Entw Mech Org* 49: 1–178, 1921.
9. Bockman DE, Boydston WR, Parsa J: Architecture of human pancreas: implications for early changes in pancreatic disease. *Gastroenterology* 83: 55–61, 1983.
10. Akao S, Bockman DE, Lechene de la Porte P, Sarles H: Three-dimensional pattern of ductulo-acinar associations in normal and pathological human pancreas. *Gastroenterology* 90: 661–668, 1986.
11. Takahashi H: Scanning electron microscopy of the rat exocrine pancreas. *Arch Histol Jpn* 47: 387–404, 1984.
12. Iwatsuki N, Petersen OH: Direct visualization of cell to cell coupling: transfer of fluorescent probes in living mammalian pancreatic acini. *Pflügers Arch* 380: 227–281, 1979.
13. Heidenhain R: Beiträge zur Kenntnis des Pankreas. *Pflügers Arch Physiol* 10: 557–632, 1875.
14. Heidenhain R: Absonderungsvorgänge. In: Handbuch der Physiologie, Vol 6. L Hermann (ed) Leipzig: FCW Vogel, 173–208, 1883.
15. Palade GE: Intracellular aspects of the process of pro-

tein secretion. *Science* 189: 347–358, 1975.
16. Holst JJ: Neural regulation of pancreatic exocrine function. In: The Exocrine Pancreas: Biology, Pathobiology, and Diseases. VLW Go et al. (eds), New York: Raven Press, 287–300, 1986.
17. Drenckhahn D, Mannherz HG: Distribution of actin and actin-associated proteins myosin, tropomyosin, alpha-actinin, vinculin, and villin in rat and bovine exocrine glands. *Eur J Cell Biol* 30: 167–176, 1983.
18. Williams JA, Hootman SR: Stimulus-secretion coupling in pancreatic acinar cells. In: The Exocrine Pancreas: Biology, Pathobiology, and Diseases. VLW Go et al. (eds), New York: Raven Press, 123–139, 1986.
19. Herzog V, Farquhar MG: Luminal membrane retrieved after exocytosis reaches most Golgi cisternae in secretory cells. *Proc Natl Acad Sci USA* 74: 5073–5077, 1977.
20. Scheele GA, Bartelt D, Bieger W: Characterization of human exocrine pancreatic proteins by two-dimensional isoelectric focusing/sodium dodecylsulfate gel electrophoresis. *Gastroenterology* 80: 461–473, 1981.
21. Blobel G, Dobberstein B: Transfer of proteins across membranes. Presence of proteolytically processed and unprocessed nascent immunoglobulin light chains on membrane bound ribosomes of murine myeloma. *J Cell Biol* 67: 835–851, 1975.
22. Walter P, Lingappa VR: Mechanism of protein translocation across the endoplasmic reticulum membrane. *Ann Rev Cell Biol* 2: 499–516, 1986.
23. Jamieson JD, Palade GE: Intracellular transport of secretory proteins in the pancreatic exocrine cell. I. Role of peripheral elements of the Golgi complex. *J Cell Biol* 34: 577–596, 1967.
24. Farquhar MG: Progress in unraveling pathways of Golgi traffic. *Ann Rev Cell Biol* 1: 477–488, 1985.
25. Rothman JE: Protein sorting by selective retention in the endoplasmic reticulum and Golgi stack. *Cell* 50: 521–522, 1987.
26. Griffith G, Simons K: The *trans* Golgi network: sorting at the exit site of the Golgi complex. *Science* 234: 438–443, 1986.
27. Orci L, Ravazzola M, Amherdt M, Perrélet A, Powell SK, Quinn DL, Moore HPH: The *trans*-most cisternae of the Golgi complex: a compartment for sorting of secretory and plasma membrane proteins. *Cell* 51: 1039–1051, 1987.
28. Mellman J, Fuchs R, Helenius A: Acidification of the endocytic and exocytic pathways. *Ann Rev Biochem* 55: 663–700, 1986.
29. Orci L, Ravazzola M, Anderson GW: The condensing vacuole of exocrine cells is more acidic than the mature secretory vesicle. *Nature* 326: 77–79, 1987.
30. Spicer SS, Sens MA, Tashian RE: Immunocytochemical demonstration of carbonic anhydrase in human epithelial cells. *J Histochem Cytochem* 30: 864–873, 1982.
31. Kern HF, Ferner H: Die Feinstruktur des exokrinen Pankreasgewebes vom Menschen. *Z Zellforsh* 113: 322–343, 1971.
32. Roberts PF, Burns J: A histochemical study of mucins in normal and neoplastic human pancreatic tissue. *J Pathol* 107: 87–94, 1972.
33. Reber HA, Adler G, Wedgwood KR: Studies in the perfused pancreatic duct in the cat. In: The Exocrine Pancreas: Biology, Pathobiology, and Diseases. VLW Go et al. (eds), New York: Raven Press, 255–273, 1986.
34. Sarner M, Cotton PB: Classification of pancreatitis. *Gut* 25: 756–759, 1984.
35. Gyr K, Singer MV, Sarles H (eds): Pancreatitis Concepts and Classification. Amsterdam: Elsevier, 1984.
36. Howard JM: Treatment of acute pancreatitis. In: Surgical Diseases of the Pancreas. JM Howard, GL Jordan Jr, HA Reber (eds), Philadelphia: Lea & Febiger, 426–439, 1987.
37. Sarner M: Pancreatitis: definitions and classification. In: The Exocrine Pancreas: Biology, Pathobiology, and Diseases. VLW Go et al. (eds), New York: Raven Press, 459–464, 1986.
38. Howard JM: Pancreatitis in the United States of America. In: Surgical Diseases of the Pancreas. JM Howard, GL Jordan Jr, HA Reber (eds), Philadelphia: Lea & Febiger, 231–233, 1987.
39. Cuilleret J, Guillemin G: Pancreatitis on the continent of Europe. In: Surgical Diseases of the Pancreas. JM Howard, GL Jordan Jr, HA Reber (eds), Philadelphia, Lea & Febiger, 241–249, 1987.
40. Austin JL, Reber HA: Pathologic features of acute pancreatitis. In: Surgical Diseases of the Pancreas. JW Howard, GL Jordan Jr, HA Reber, (eds), Philadelphia: Lea & Febiger. 355–370, 1987.
41. Taura S, Tsunoda T, Yoshino R, et al.: Ultrastructural studies of human acute pancreatitis. *Gastroent Jpn* 10: 132–140, 1975.
42. Helin H, Mero M, Markkula H, Helin M: Pancreatic acinar ultrastructure in human acute pancreatitis. *Virchows Arch* (Pathol Anat) 387: 259–270, 1980.
43. Aho HJ, Nevalainen TJ, Havia VT, Heinonen RJ, Aho AJ: Human acute pancreatitis. *Acta Path Microbiol Immunol Scand* Sect A 90: 367–373, 1982.
44. Adler G, Kern HF: Fine structural and biochemical studies in human acute pancreatitis. In: Pancreatitis — Concepts and Classification. K Gyr, MV Singer, H Sarles (eds), Amsterdam: Elsevier, 37–42, 1984.
45. Bockman DE, Büchler M, Beger HG: Ultrastructure of human acute pancreatitis. *Int J Pancreatol* 1: 141–153, 1986.
46. Klöppel G, Dreyer T, Willemer S, Kern HF, Adler G: Human acute pancreatitis: its pathogenesis in the light of immunohistochemical and ultrastructural findings in acinar cells. *Virchows Arch* (Pathol Anat) 409: 791–803, 1986.
47. Klöppel G, Adler G, Kern HF: Pathomorphology of acute pancreatitis in relation to its clinical course and pathogenesis. In: Diagnostic Procedures in Pancreatic Disease. P Malfertheiner, H Ditschuneit, (eds), Berlin: Springer Verlag, 11–18, 1986.
48. Jones RT, Garcia JH, Mergner WJ, Pendergrass RE, Valigorshy JM, Trump BF: Effects of shock on the pancreatic acinar cell. *Arch Pathol* 99: 634–644, 1975.
49. Arstila AU, Hirsimaki P, Trump BF: Studies on the subcellular pathophysiology of sublethal chronic cell injury. *Beitr Pathol* 152: 211–242, 1974.

50. Goldmann RD, Milsted A, Schloss JA, Starger J, Yerna MJ: Cytoplasmic fibers in mammalian cells: cytoskeletal and contractile elements. *Ann Rev Physiol* 41: 703–722, 1979.
51. Willemer S, Adler G: Histochemical and ultrastructural characteristics of tubular complexes in human acute pancreatitis. *Dig Dis Sci*, in press, 1987.
52. Bockman DE, Boydston WR, Anderson MC: Origin of tubular complexes in human chronic pancreatitis. *Am J Surg* 144: 243–249, 1982.
53. Cubilla AL, Fitzgerald PJ: Morphological patterns of primary nonendocrine human pancreas carcinoma. *Cancer Res* 35: 2234–2248, 1975.
54. Willemer S, Elsässer HP, Kern HF, Adler G: Tubular complexes in cerulein- and oleic acid-induced pancreatitis: glycoconjugate pattern, immunocytochemical and ultrastructural findings. *Pancreas* 2: 669–675, 1987.
55. Steer ML, Meldolesi J: The cell biology of experimental pancreatitis. *N Engl J Med* 316: 144–150, 1987.
56. Adler G, Kern HF, Scheele GA: Experimental models and concepts in acute pancreatitis. In: The Exocrine Pancreas: Biology, Pathobiology and Diseases. VLW Go et al. (eds), New York: Raven Press, 407–421, 1986.
57. Schmitz-Moormann P: Comparative radiological and morphological study of the human pancreas. *Path Res Pract* 163: 93–108, 1981.
58. Klöppel G, von Gerkan R, Dreyer T: Pathomorphology of acute pancreatitis. Analysis of 367 autopsy cases and 3 surgical specimens. In: Pancreatitis — Concepts and Classification. KE Gyr, MV Singer, H Sarles (eds), Amsterdam: Elsevier, 29–35, 1984.
59. Foulis AK: Histological evidence of initiating factors in acute necrotising pancreatitis in man. *J Clin Pathol* 33: 1125–1131, 1980.
60. Acosta JM, Ledesma C: Gallstone migration as cause of acute pancreatitis. *N Engl J Med* 290: 484–487, 1974.
61. Rao SS, Watt IA, Donaldson LA, Crocket A, Joffee SN: A serial histologic study of the development and progression of acute pancreatitis in the rat. *Am J Pathol* 103: 39–46, 1981.
62. Nuutinen P, Kivisaari L, Standertskjöld-Nordenstam CG, Lempinen M, Schröder T: Microangiography of the pancreas in experimental oedemic and hemorrhagic pancreatitis. *Scand J Gastroenterol* 21 (Suppl 126): 12–17, 1986.
63. Lombardi B, Estes L, Longnecker DS: Acute hemorrhagic pancreatitis (massive necrosis) with fat necrosis induced in mice by DL-ethionine fed with a choline-deficient diet. *Am J Pathol* 79: 465–480, 1975.
64. Lampel M, Kern HF: Acute interstitial pancreatitic in the rat induced by excessive doses of a pancreatic secretagogue. *Virchows Arch* (Pathol Anat) 373: 97–117, 1977.
65. Adler G, Rohr G, Kern HF: Alteration of membrane fusion as a cause of acute pancreatitis in the rat. *Dig Dis Sci* 27: 993–1002, 1982.
66. Adler G, Hahn C, Kern HF, Rao KN: Cerulein-induced pancreatitis in the rat: increased lysosomal enzyme activity and autophagocytosis. *Digestion* 32: 10–18, 1985.
67. Scheele GA, Adler G, Kern HF: Role of lysosomes in the development of acute pancreatitis. In: Pancreatitis — Concepts and Classification. KE Gyr, MV Singer, H Sarles (eds). Amsterdam: Elsevier, 17–23, 1984.
68. Aho H, Putzke HP, Nevalainen TJ, et al.: Immunohistochemical localization of trypsinogen and trypsin in acute and chronic pancreatitis. *Digestion* 27: 21–28, 1983.
69. Aho HJ, Sternby B, Nevalainen TJ: Fat necrosis in human acute pancreatitis. *Acta Path Microbiol Immunol Scand*. Sect A 94; 101–105, 1986.
70. Schmitz-Moormann P, Böger A: Tissue damage by fatty acids released by lipolysis. *Path Res Pract* 171, 303–313, 1981.
71. Morgenroth K, Kozuschek W: Scanning electron microscopic observations in acute pancreatitis. *Verh Dtsch Ges Path* 71: 237–241, 1987.
72. Ammann RW, Akovbiantz A, Largiader F, Schueler G: Course and outcome of chronic pancreatitis. *Gastroenterology* 86: 820–828, 1984.
73. Sarles H: Etiopathogenesis and definition of chronic pancreatitis. *Dig Dis Sci* 31: (Suppl): 91S–107S, 1986.
74. Sarles H, Sahel J: Pathology of chronic calcifying pancreatitis. *Am J Gastroenterol* 66: 117–139, 1976.
75. Payan H, Sarles H, Demirdjian M, Gauthier AP, Cros RC, Durbec JP: Study of the histological features of chronic pancreatitis by correspondance analysis. Identification of chronic calcifying pancreatitis as an entity. *Rev Europ Etudes Clin Biol* 17: 663–670, 1972.
76. Nakamura K, Sarles H, Payan H: Three-dimensional reconstruction of the pancreatic ducts in chronic pancreatitis. *Gastroenterology* 62: 942–949, 1972.
77. Sahel J, Cros RC, Durbec JP, et al.: Multicenter pathological study of chronic pancreatitis. *Pancreas* 1: 471–477, 1986.
78. Ammann RW, Buehler H, Muench R, Freiburghaus AW, Siegenthaler W: Differences in the natural history of idiopathic (nonalcoholic) and alcoholic chronic pancreatitis. A comparative long-term study of 287 patients. *Pancreas* 2: 368–377, 1987.
79. Howard JM, Nedwich A: Correlation of the histologic observations and operative findings in patients with chronic pancreatitis. *Surg Gynec Obstet* 132: 387–395, 1971.
80. Stolte M: Chronic pancreatitis. *Verh Dtsch Ges Path* 71: 175–186, 1987.
81. Ribet A, Fedou R: Les Pancréatites Chroniques. Paris: Masson et Cie, 1965.
82. Tasso F: Etude ultrastructurale du pancréas dans 8 cas de pancréatites chroniques. *Ann Anat Pathol* 12: 343–372, 1967.
83. Tasso F, Stemmelin N, Sarles H: Comparative morphometric study of the human pancreas in its normal state and in primary chronic calcifying pancreatitis. *Biomedicine* 18: 134–144, 1973.
84. Noronha M, Salgadinho A, Ferreira De Almeida MJ, Dreiling DA, Bordalo O: Alcohol and the pancreas. I. Clinical associations and histopathology of minimal pancreatic inflammation. *Am J Gastroent* 76: 114–119, 1981.
85. Noronha M, Bordalo O, Dreiling DA: Alcohol and the pancreas. II. Pancreatic morphology of advanced al-

coholic pancreatitis. *Am J Gastroent* 76: 120–124, 1981.
86. Guy O, Roblez-Diaz G, Adrich Z, Sahel J, Sarles H: Protein content of precipitates present in pancreatic juice of alcoholic subjects and patients with chronic calcifying pancreatitis. *Gastroenterology* 84: 102–107, 1983.
87. Kern HF, Warshaw AL, Scheele GA: Fine structure of protein precipitations in acinar lumina of the normal human pancreas and in chronic pancreatitis. In: Pancreatitis — Concepts and Classification. KE Gyr, MV Singer, H Sarles (eds), Amsterdam: Elsevier, 101–105, 1984.
88. Lederer P, Stolte M, Tulusan H: Wirsung's duct of the normal pancreas and chronic pancreatitis. *Virchows Arch* (Pathol Anat) 372: 109–121, 1976.
89. Stolte M: Chronic pancreatitis. *Verh Dtsch Ges Path* 71: 175–186, 1987.
90. Giorgi D, Bernard JP, De Caro A et al.: Pancreatic stone protein. I. Evidence that it is encoded by a pancreatic messenger ribonucleic acid. *Gastroenterology* 89: 381–386, 1985.
91. Multigner L, Sarles H, Lombardo D, De Caro A: Pancreatic stone protein. II. Implication in stone formation during the course of chronic calcifying pancreatitis. *Gastroenterology* 89: 387–391, 1985.
92. Bagenstoss AH: Pathology of pancreatitis. In: Pancreatitis. EE Gambill (ed), Saint Louis: C Mosby, 179–212, 1973.
93. Kopelman H, Durie P, Gaskin K, Weizman Z, Forstner G: Pancreatic fluid secretion and protein hyperconcentration in cystic fibrosis. *N Engl J Med* 312: 329–334, 1985.
94. Lebenthal E, Lerner A, Heitlinger L: The pancreas in cystic fibrosis. In: The Exocrine Pancreas: Biology, Pathobiology, and Diseases. VLW Go et al. (eds), New York: Raven Press, 783–817, 1986.
95. Klöppel G, Fitzgerald PJ: Pathology of nonendocrine pancreatic tumors. In: The Exocrine Pancreas: Biology, Pathobiology, and Diseases. VLW Go et al. (eds), New York: Raven Press, 649–674, 1986.
96. Osborne M, van Essen G, Weber K, Klöppel G, Altmannsberger M: Differential diagnosis of gastrointestinal carcinomas by using monoclonal antibodies specific for individual keratin polypeptides. *Lab Invest* 55: 497–504, 1986.
97. Bätge B, Bosslet K, Sedlacek HH, Kern HF, Klöppel G: Monoclonal antibodies against CEA-related components discriminate between pancreatic duct-type carcinomas and nonneoplastic duct lesions as well as nonduct type neoplasias. *Virchows Arch* (Pathol Anat) 408: 361–374, 1986.
98. Gall FP, Zirngibl: Maligne Tumoren des Pankreas und der periampullären Region. In: Chirurgische Onkologie. Histologie und Stadiengerechte Therapie Maligner Tumoren. FP Gall, P Hermaneck, J Tonak (eds), Berlin: Springer Verlag, 234–250, 1986.
99. Klapdor R: TNM — Klassifikation, Staging und Prognose des Pankreaskarzinoms. *Dtsch Med Wschr* 111: 229–233, 1986.
100. Klöppel G, Lingenthal G, von Bülow M, Kern HF: Histological and fine structural features of pancreatic ductal adenocarcinomas in relation to growth and prognosis: studies in xenografted tumors and clinicohistopathological correlation in a series of 75 cases. *Histopathology* 9: 841–856, 1985.
101. Kern HF, Röher HD, von Bülow M, Klöppel G: Fine structure of three major grades of malignancy of human pancreatic adenocarcinoma. *Pancreas* 2: 2–13, 1987.
102. Barsky SA, Siegal GP, Janotta F, Liotta LA: Loss of basement membrane components by invasive tumors but not by benign counterparts. *Lab Invest* 49: 140–147, 1983.
103. Ekblom P, Vestweber D, Kemler R: Cell-matrix interactions and cell adhesion during development. *Ann Rev Cell Biol* 2: 27–47, 1986.
104. Liotta LA, Rao CN, Barsky SH: Tumor invasion and the extracellular matrix. *Lab Invest* 49: 636–649, 1983.
105. Liotta LA, Rao CN, Wewer UM: Biochemical interactions of tumor cells with the basement membrane. *Ann Rev Biochem* 55: 1037–1057, 1986.
106. Mollenhauer J, Roether J, Kern HF: Distribution of extracellular matrix proteins in pancreatic ductal adenocarcinoma and its influence on tumor cell proliferation in vitro. *Pancreas* 2: 14–24, 1987.
107. Mullins DE, Rohrlich ST: The role of proteinases in cellular invasiveness. *Biochim Biophys Acta* 695: 177–214, 1983.
108. Mignatti P, Robbins E, Rifkin DB: Tumor invasion through the human amniotic membrane: requirement for a proteinase cascade. *Cell* 47: 487–498, 1986.
109. Bosslet K, Kern HF, von Bülow M, Röher HD, Klöppel G, Schorlemmer HU, Kurrle R, Sedlacek HH: A human monocyte cell surface antigen highly expressed on an established pancreatic carcinoma cell line (PaTu II). *Dig Dis Sci* 28: 928, 1983.
110. Kern HF, Bosslet K, Mollenhauer J, Sedlacek HH, Schorlemmer HU: Monocyte-related functions expressed in cell lines established from human pancreatic adenocarcinoma. I. Comparative analysis of endocytotic activity, lysosomal enzyme secretion, and superoxide anion production. *Pancreas* 2: 212–221, 1987.
111. Kern HF, Bosslet K, Sedlacek HH, Schorlemmer HU: Monocyte-related functions expressed in cell lines established from human pancreatic adenocarcinoma. II. Inhibition of stimulated activity by monoclonal antibodies reacting with surface antigens on tumor cells. *Pancreas* 2: 627–639, 1987.

Author's address:
Prof. Dr. Horst F. Kern
Institut für Zytobiologie und Zytopathologie
Robert-Koch-Strasse 5
D-3550 Marburg
Federal Republic of Germany

CHAPTER 8

Three-dimensional features of pancreatic cells

TOMONORI NAGURO & AKIHIRO IINO

1. Introduction

1.1. Advances in the SEM study of the pancreas

Almost 20 years have passed since the scanning electron microscope (SEM) was first applied to the biological field. In the beginning of the last decade, studies on the pancreatic cells with this instrument were limited to observing luminal and basal cell surfaces, since no feasible method to reveal internal cell structures existed. Observation of internal cell structures subsequently commenced with the development of various cracking methods [1–4]. However, progress in SEM in this field was delayed until quite recently due to difficulty in obtaining three-dimensional information on cell organelles prepared solely by the cracking method. Moreover, conventional SEM did not have enough resolving power to give distinct images of the fine details of cell organelles. As a result, there was little new discovery beyond the information already obtained through transmission electron microscopy (TEM).

With the advent of field emission SEM, with its high resolving power, the problem of resolution was practically solved. As to the problem of the specimen preparation technique, remarkable progress has been made in this field in recent years with the development of new preparation methods. Nowadays SEM has come to play an important role in the three-dimensional study of cell structures, that is generally difficult to achieve by TEM. In the present report, the authors use SEM to demonstrate the three-dimensional fine structure of pancreatic cells.

2. General aspects of the pancreas

As is well known, the pancreas has no definite fibrous capsule on its outer limit but is covered by a thin layer of loose connective tissue from which thin septa extend to divide the pancreas into obvious lobules. Blood and lymphatic vessels, nerves, and excretory ducts run in these connective tissue septa. Fine delicate reticular fibers surround individual acini (Figs. 8-1 and 8-2). The pancreas is both an exocrine and an endocrine gland, and these two functions are carried out by distinctly different groups of cells.

3. The exocrine pancreas

The exocrine pancreas, which comprises about 90% of the pancreas, is classified as a compound tubulo-acinar gland. The pancreatic acinar cells and the centro-acinar cells form the terminal portion of this gland (Fig. 8-3). The acini are composed of a single row of pyramidal cells arranged around a small central lumen, with their broad bases resting upon a basal lamina. Since the acinar cells show obvious polarity, the cell surface can be divided into three parts:

1. the *basal surface*, facing the basal lamina,
2. the *lateral surface*, facing the adjacent cell, and
3. the *luminal surface*, facing the central lumen.

Riva, A. and Motta, P.M., (eds.), Ultrastructure of the Extraparietal Glands of the Digestive Tract.
ISBN 978-1-4612-8211-2.

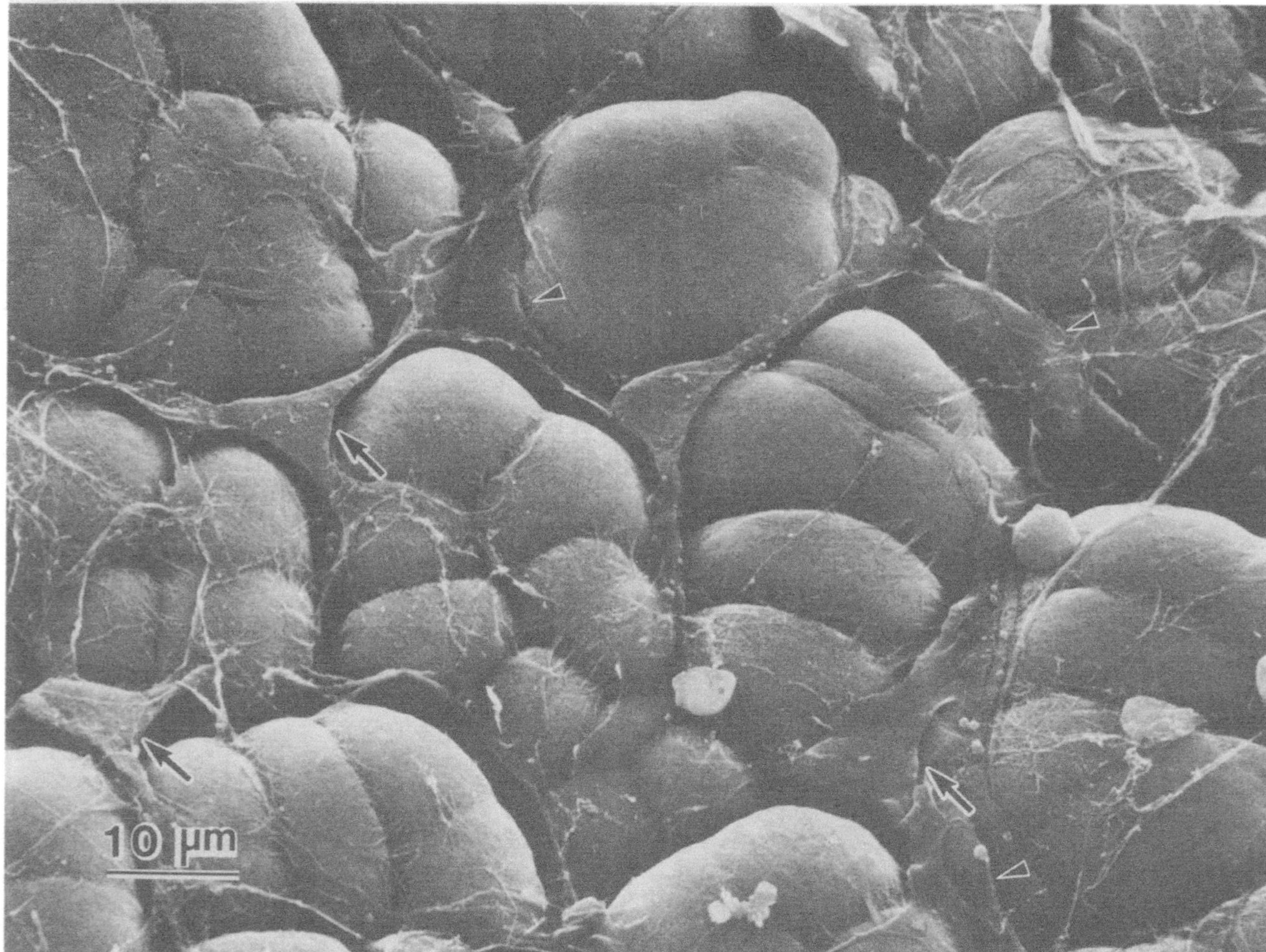

Fig. 8-1. External feature of a pancreatic lobule in the rat. Capillaries (arrowheads) and fibroblasts (arrows) are visible between the acini. The fibroblasts extend their long processes and seem to be linked together to form a kind of network.

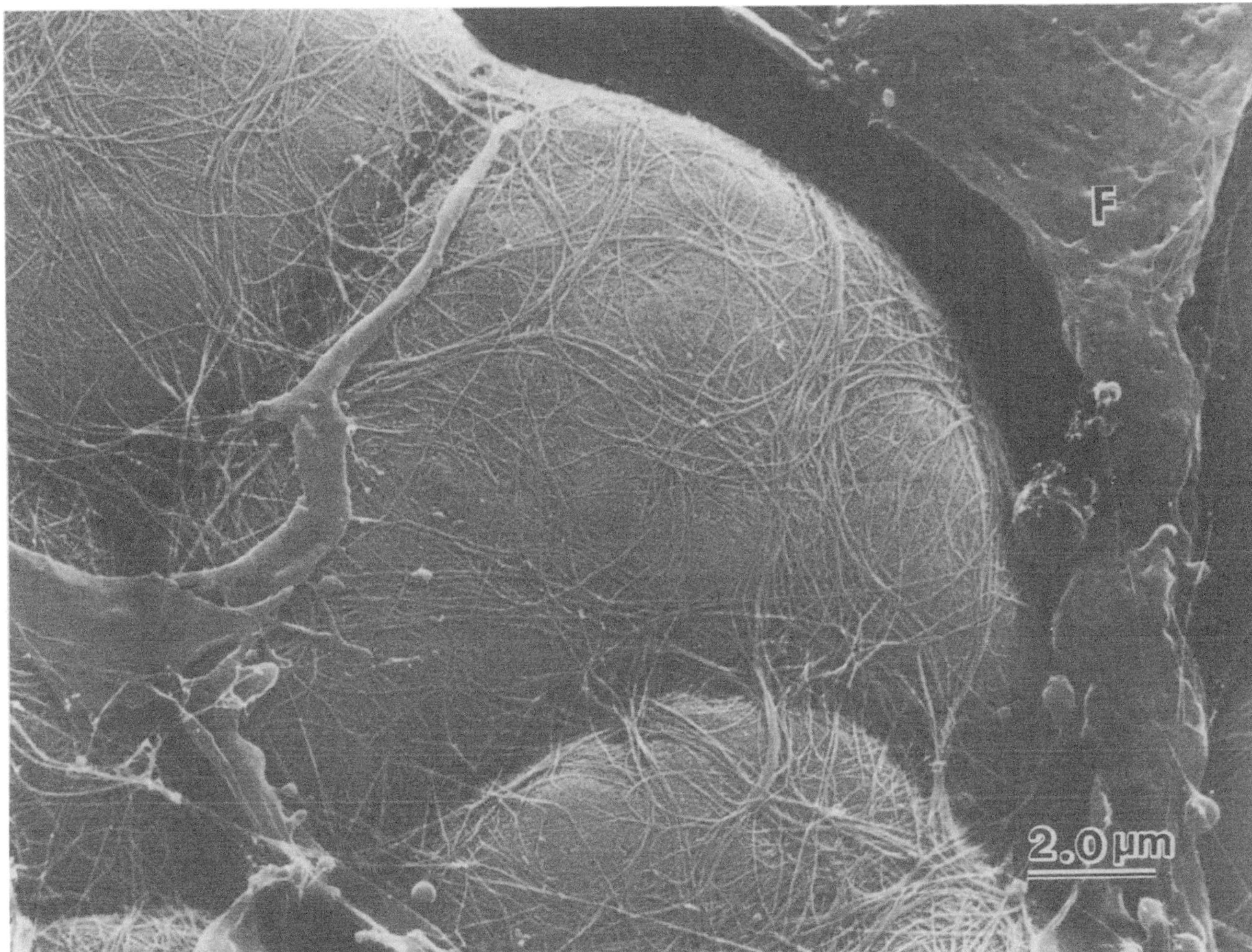

Fig. 8-2. Basal surface of acinar cells in the rat pancreas. Fine reticular fibers surround the acini in the shape of a basketwork. F-fibroblast.

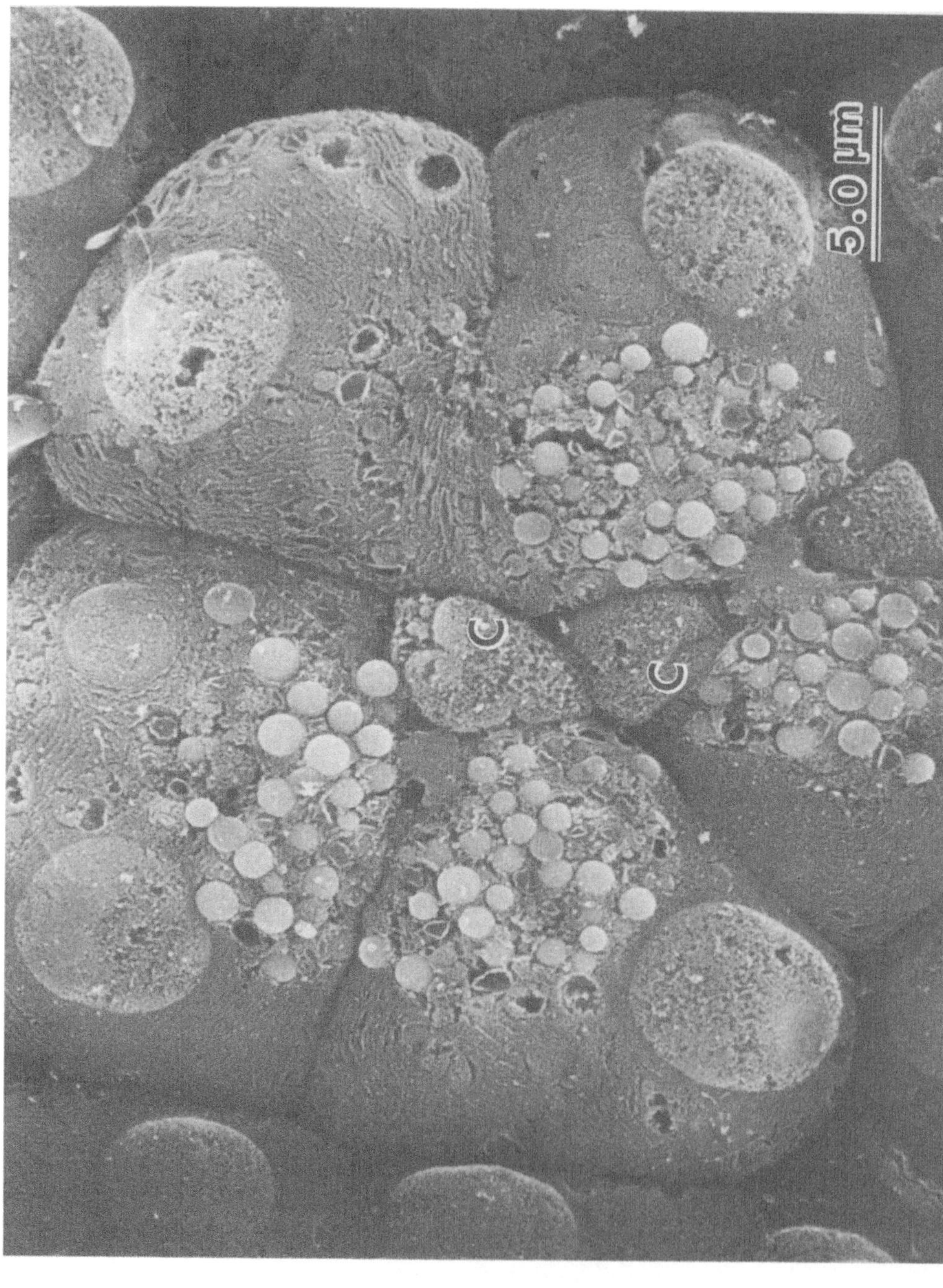

Fig. 8-3. A cracked acinus in the rat pancreas. The acinus is composed of a single row of pyramidal acinar cells arranged around a small central lumen. C-centroacinar cells.

Besides the cell surfaces described above, the intracellular organization of the acinar cells is also an important object of SEM observations.

3.1. External features of acinar cells

3.1.1. Observations of the basal surface.

A section of a pancreatic lobule observed by SEM is shown in Fig. 8-1. This sample was prepared routinely by double fixation using glutaraldehyde and osmium, without any special process of specimen preparation. Acini are covered with a basement membrane and packed together within the lobulus in an irregular fashion. Capillaries, collagen fibers, and fibroblasts are seen between the acini, but myoepithelial cells are not present. The fibroblasts extend their processes around the acini. They seem to be linked together to form a kind of network. The outlines of individual acinar cells can be discerned through the basement membrane. It has been shown clearly by TEM that the basement membrane is composed of three layers, namely, the lamina lucida, the densa (basal lamina), and the fibroreticularis. SEM observations on the basement membrane give a better understanding of the relationship between the basal lamina and the lamina fibroreticularis. Reticular fibers (unit fibers of collagen), which are found in close association with the basal lamina, surround the acini and give it a basket-woven appearance. (Fig. 8-2). A bundle of collagen fibers rarely takes part in the formation of the basketwork. In order to observe the basal cell membrane, the basement membrane must be completely removed. For this purpose, the HCl digestion method devised by Evan et al. [5] has been widely used up to the present. Takahashi [6] extended the application of this method to the pancreas and demonstrated the basal surface aspects of rat pancreatic acini. This method is very effective for removing the basement membrane, although the fine structures of the cell membrane tend to be preserved poorly because of the high concentration of HCl employed in this procedure.

3.1.2. Observations of the lateral surface of acinar cells

In order to expose the lateral surface, which is normally concealed by adjacent cells, one feasible technique is to mechanically split adjacent cells from an acinus using a scalpel or forceps, after initial fixation with glutaraldehyde or after critical-point drying. When this method is employed the initial fixation must be performed carefully. Neither hypotonic fixative nor short-time fixation produces the desired effect, since the fracture tends to occur not along the cell boundaries, but at the interior of the cells. The lateral surface of acinar cells, as observed by SEM using this technique, is shown in Fig. 8-4. Many small, irregularly shaped processes are distributed over almost the entire lateral surface, as opposed to the traditional notion, based on TEM studies, that the lateral surface is evenly flat due to the lack of interdigitation. Takahashi [6] reported a detailed study on intercellular canaliculi by the use of the HCl digestion method.

3.1.3. Observations of the luminal surface of acinar cells

The above-mentioned methods for observing the lateral surface of acinar cells can also be used to expose the luminal surface, although it would be more advisable to adopt the cracking methods described in the introduction. The luminal surface of acinar cells usually bears a few irregularly oriented, short, fingerlike microvilli although its appearance is easily influenced by the functional condition (Fig. 8-5). Openings of about 1 μm in diameter, which seem to be created during secretion by the fusion of the limiting membrane of the zymogen granules and the cell surface membrane, are occasionally found on the luminal surface. The acinar lumen is usually filled with secretory products, which may often obstruct the view of the luminal surface under SEM. In this case, the maceration procedure described in the following section would be recommended.

3.2. Internal features of the acinar cells

Both the osmium-DMSO-osmium method (the ODO method), devised by Tanaka and Naguro [7], and a revised ODO method [8] are widely used for observing intracellular structures by SEM. These methods are characterized by the maceration procedure, by which the excess cytoplasmic matrix is removed from the cracked sur-

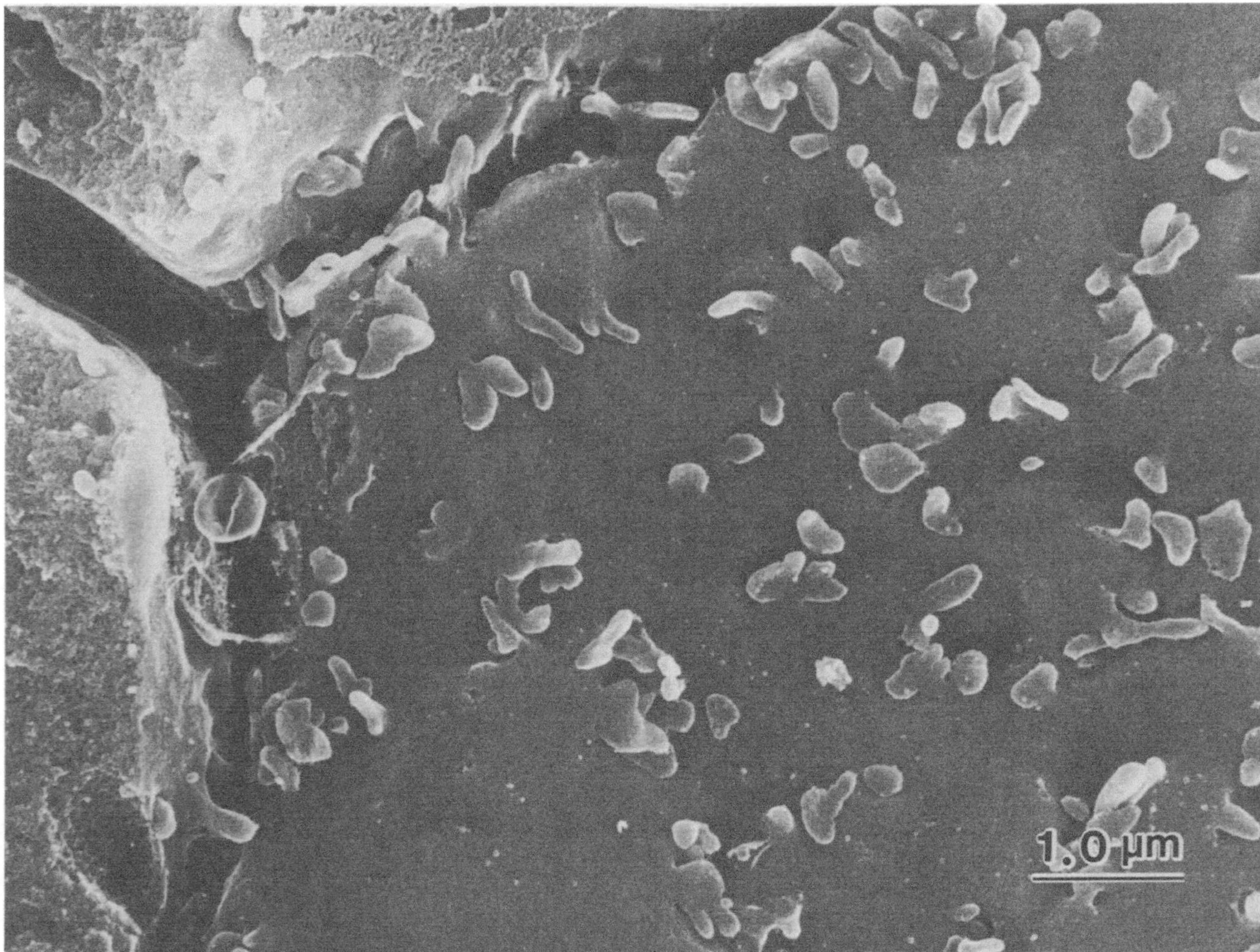

Fig. 8-4. Lateral surface of an acinar cell in the rat pancreas. Many small irregularly shaped processes are distributed over almost the entire lateral surface.

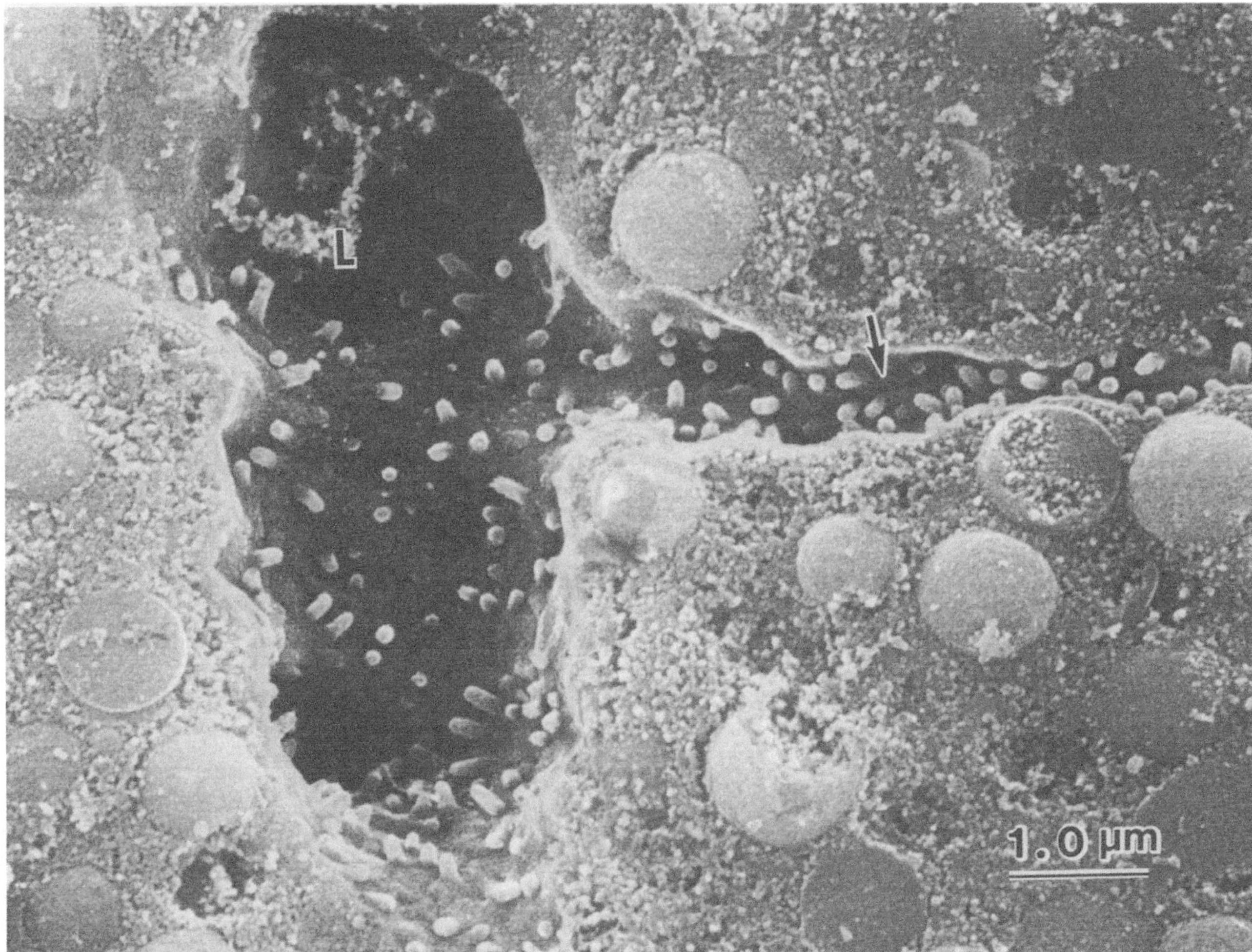

Fig. 8-5. Luminal surface of acinar cells in the rat pancreas. The lumen (L) is surrounded by acinar cells. Irregularly oriented fingerlike microvilli are found on the luminal surface of acinar cells and also on the surface of the intercellular canaliculus (arrow).

face of the cells. An outline of the ODO method follows.

3.2.1. The ODO method

1. Fixation: small blocks of pancreatic tissue (about 1 × 1 × 5 mm) are removed from the animals and fixed, while fresh, with 1% osmium tetroxide in M/15 phosphate buffer solution at pH 7.4 at 4°C for 1–2 hours.
2. DMSO treatment: after rinsing with buffer solution, the specimens are successively immersed in 15%, 30%, and 50% dimethyl sulfoxide (DMSO) in water for 30 minutes each in order to prevent ice-crystal damage caused by the freezing in the next step.
3. Freeze cracking: the specimens are frozen on metal plate and cracked in two with a razor blade and hammer.
4. Thawing: the cracked pieces are immediately placed in 50% DMSO aqueous solution.
5. Maceration: after rinsing with buffer solution, the specimens are immersed in 0.1% osmium tetroxide in buffer solution and are left standing at 20°C for 1–3 days.
6. Post fixation: the specimens are fixed again, for 1 hour, in buffered 1% osmium solution.
7. Conductive staining: the specimens are treated with 2% tannic acid aqueous solution for 1 hour, 1% osmium aqueous solution for 1 hour [9].
8. Dehydration: the specimens are dehydrated in a graded ethanol series.
9. Drying: after treatment with isoamyl acetate, the specimens are dried by the critical-point drying method.
10. Metal coating: the dried specimens are coated with about 3 nm of platinum in an ion coater equipped with a rotary tilt stage.

The maceration procedure, which involves the exposure of intracellular structures, requires considerable care. During this process some cell components, such as the cytoskeleton, zymogen granules, and nuclear and cytoplasmic matrices, tend to disappear from the exposed surface very easily. The length of the maceration time must therefore be controlled appropriatly so that these cell components are preserved in a good state.

A common feature of pancreatic acinar cells prepared by the ODO method is shown in Fig. 8-6. This typical serous cell is obviously polarized in the arrangement of its organelles. The spherical nucleus is situated near the center of the basal region. The granular endoplasmic reticulum is particularly prominent in the basal region but also extends into the perinuclear region. The Golgi apparatus is located in the supranuclear region, and the apical region is filled with numerous zymogen granules. A detailed description of the fine structure of acinar cell organelles, as observed under SEM, follows.

3.2.2. Nucleus

An acinar cell generally has one nucleus but often has two. The frequency of occurrence of the binucleated cell is about 10% of the total acinar cells in the rat [10]. Its shape is usually spherical, about 6–8 μm in diameter, but varies with the cell's physiologic activity. The nucleus of a fasting animal, which is thought to be in an inactive state, shows a more irregular shape and a smaller volume, as compared with that of a feeding animal. The nucleus encloses numerous chromatin fibers having a diameter of about 15–25 nm (Fig. 8-7) and contains one or more prominent nucleoli. The outer limit of the nucleus is clearly demarcated by two membranes that are known as the *nuclear envelope*. The cytoplasmic side of the nuclear envelope is encrusted with numerous ribosomes about 20 nm in diameter (Fig. 8-8). Nuclear pores are also plentiful and are fairly evenly distributed, showing no special regularity in their arrangement. Each nuclear pore is rimmed with eight small granules of about 40 nm in diameter (see inset, Fig. 8-8). These granules seem to form the so-called annulus, which is known to be a main component of the nuclear pore complex. These granules tend to disappear very easily during the maceration procedure, while the ribosomes are well preserved. In this case, nuclear pores of about 70 nm that lack annuli can be observed. They are larger than the pores with annuli which have an inside diameter of about 50 nm.

3.2.3. Granular endoplasmic reticulum

Rough-surfaced or granular endoplasmic reticulum (GER) is easily noted in SEM photographs

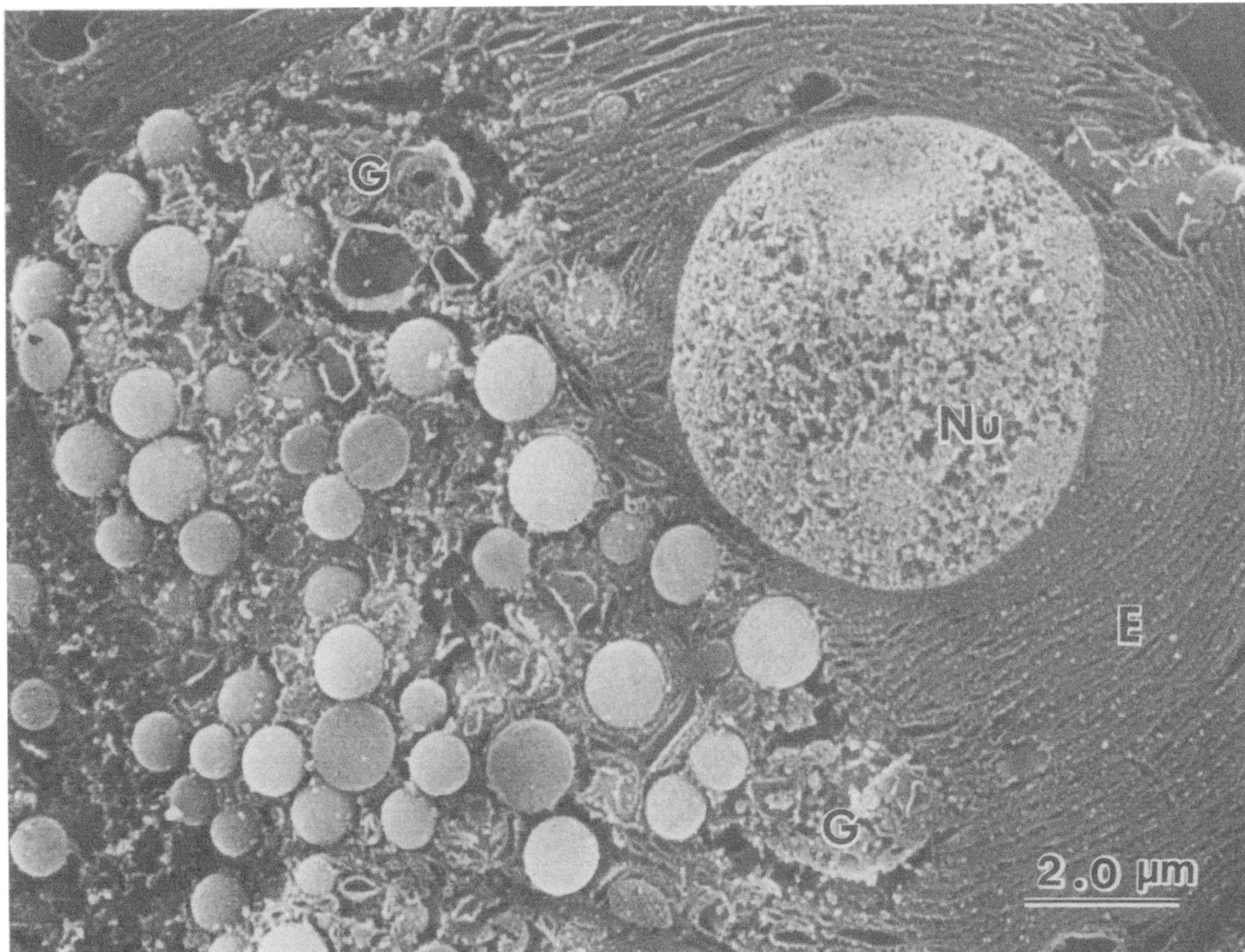

Fig. 8-6. Internal features of an acinar cell in the rat pancreas. Granular endoplasmic reticulum (E) is prominent in the basal region of the cell. The apical region is filled with numerous zymogen granules. The Gogli apparatus (G) is located above the nucleus (Nu) and occupies a position between granular endoplasmic reticulum and zymogen granules.

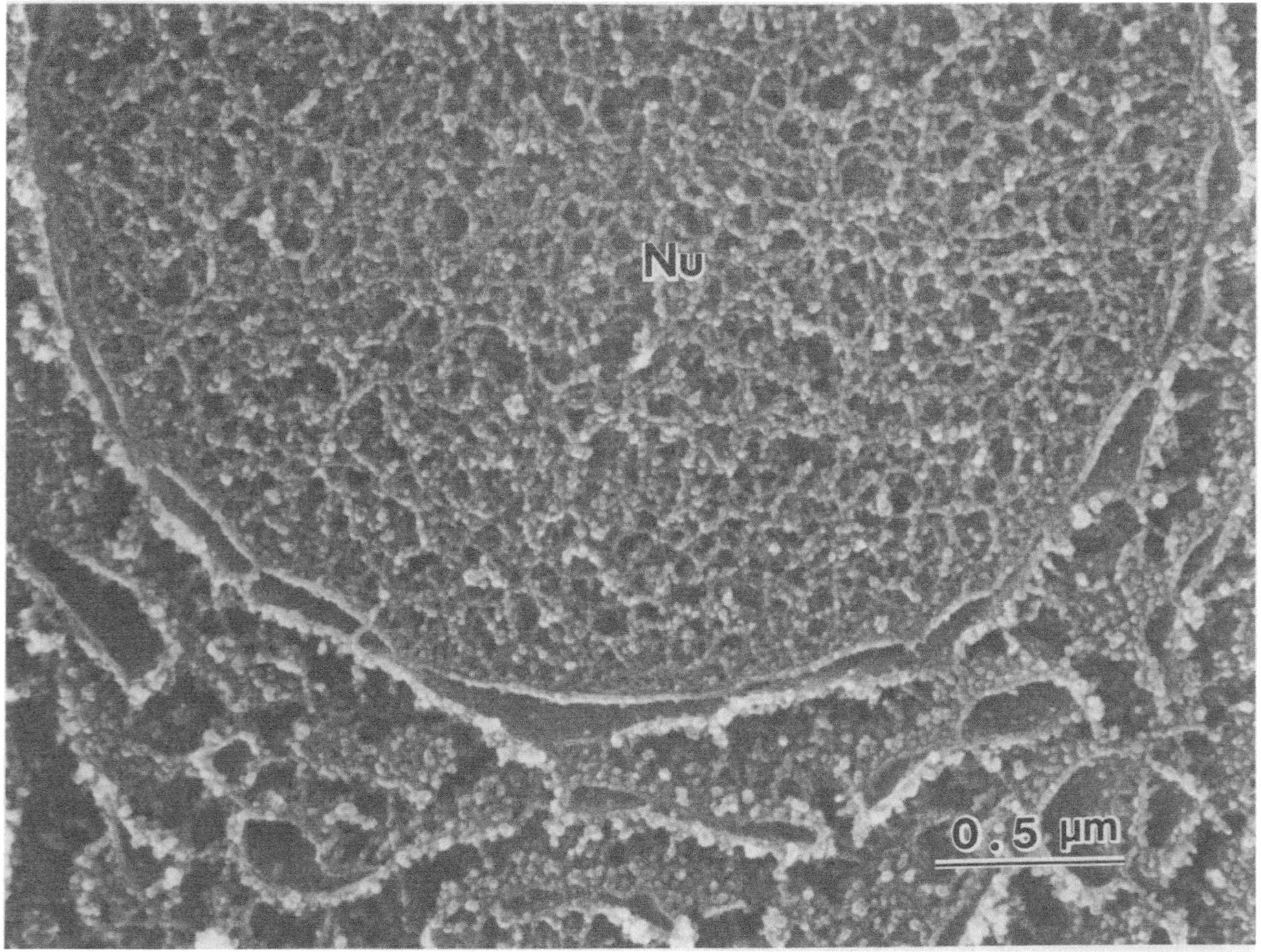

Fig. 8-7. A cracked nucleus of an acinar cell in the rat pancreas. The nucleus (Nu) is occupied by chromatin fibers of about 15-25 nm in diameter. The outer limit of the nucleus is clearly demarcated by the two membranes of the nuclear envelope.

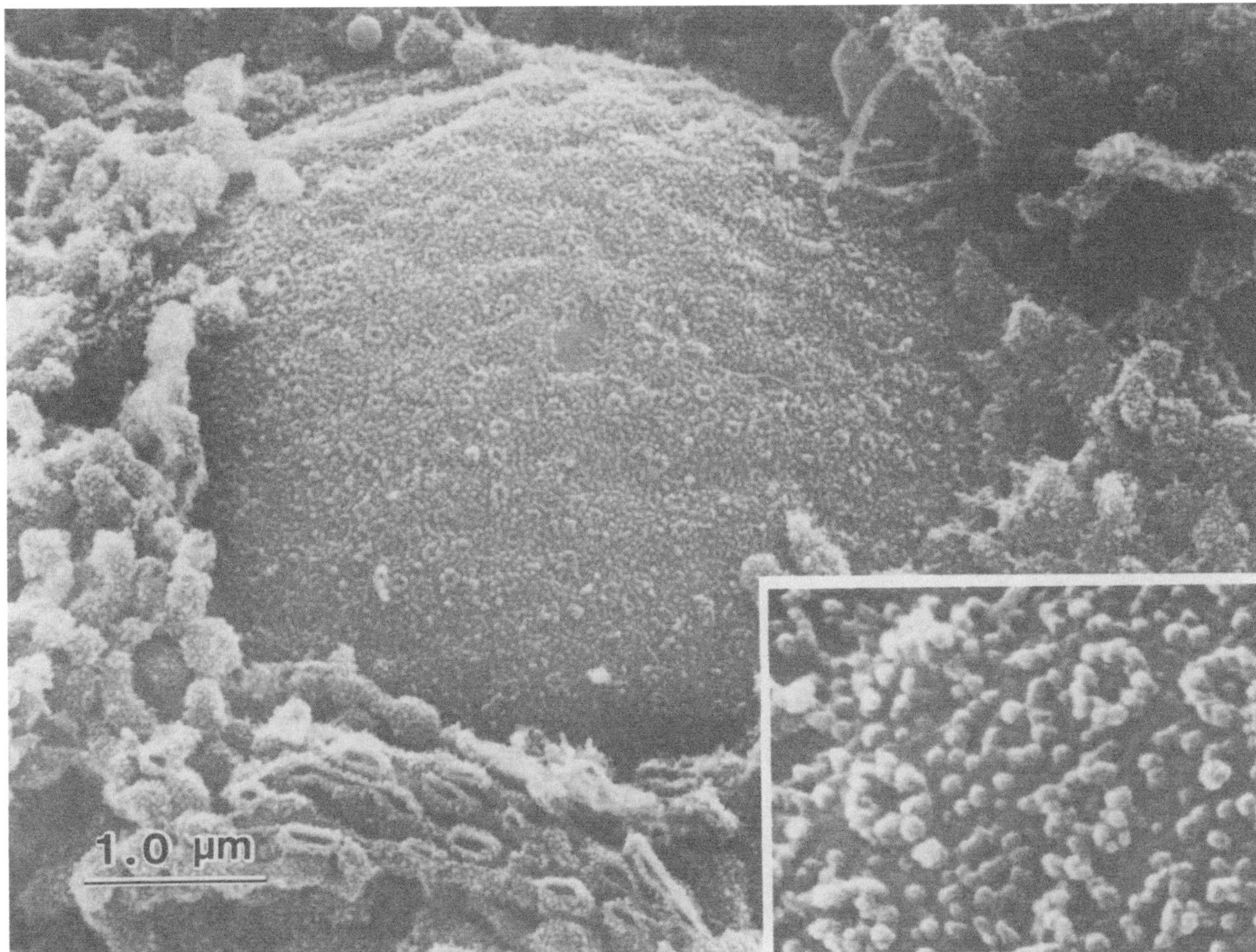

Fig. 8-8. The outer surface of the nuclear envelope in the rat pancreatic acinar cell. Nuclear pores are fairly evenly distributed over the nuclear envelope, showing no special regularity in their arrangement. Each nuclear pore is rimmed with eight small granules of about 40 nm (inset).

as a predominant organelle in the acinar cells, which synthesize and secrete many kinds of digestive enzymes. As shown in Figs. 8-6 and 8-9, the GER appears to be formed from membranous cisternae, the outer surface of which is studded with numerous ribosomes, similar to the outer nuclear membrane. In the basal region, the GER usually occurs in the form of many flattened cisternae lying upon one another in lamellated fashion. In the perinuclear to apical region, it tends to be arranged as meandering tubules or isolated vesicles, though wide structural variations occur in different species and in different conditions of cell activity. When the GER shows a lamellar form, fine threadlike structures interconnecting the lamellae of the GER are often observed (Fig. 8-10). Although the nature of these connecting threads has not yet been determined [11], they seem to play an important role in the maintenance of the GER's lamellar form. The contents of the GER are usually extracted during specimen preparation. Thus, the lumen of the cisternae usually appears empty, except in some animals, such as the dog and guinea pig [12], in which intracisternal granules are often observed.

3.2.4. Golgi apparatus

The appearance of the Golgi apparatus in acinar cells varies greatly, depending upon the cell's activity, although its position remains relatively constant. It is located above the nucleus and occupies a position between the GER and the zymogen granules that accumulate in the apical cytoplasm (Fig. 8-6). Based on TEM studies, the three principal components of the Golgi apparatus are: the Golgi stacks of cisternae (or saccules), Golgi vacuoles, and Golgi vesicles. These three elements are also recognizable by SEM. Three to six cisternae are piled one upon the other in close parallel array forming a Golgi stack (Fig. 8-11). In the thin sections used for TEM, the many Golgi stacks of a given acinar cell may appear to be isolated from one another, but in fact our three-dimensional SEM studies prove that all the Golgi stacks of a cell are connected to each other to form a single Golgi apparatus [13] (Fig. 8-12). The Golgi stack is said to have two different faces; the side facing the GER is called the *forming, immature*, or *cis* face, which is usually convex, and the side facing the zymogen granules is called the *mature* or *trans* face, which is generally concave. In practice, it is difficult to identify the face merely by the morphological pattern of the curved stack, since this pattern is often reversed.

The *cis* face of the Golgi stack shows a curious structure (Fig. 8-13). Numerous small pores of about 30–50 nm are evenly distributed over the entire surface of the *cis*-most cisterna of the Golgi stack. In addition to the presence of the pores, a few large openings of about 100–500 nm in diameter and many small vesicles of about 60–80 nm are also observed on the *cis*-most cisterna. Similarly sized small vesicles are present at the margin of the cisternae of the GER and are thought to be budding vesicles. Occasionally, a slender tubules of about 30 nm in caliber penetrates into the large opening (Fig. 8-11). Usually, the *trans* face of the Golgi stack shows a relatively flat appearance, as compared with the *cis* face, but occasionally it appears as a complicated plexus consisting of anastomotic tubules (Fig. 8-14). The continuity between these tubules and the *trans*-most cisterna of the Golgi stack can be found at several places. Golgi vacuoles of various sizes are also present on the *trans* face.

3.2.5. Mitochondria

Mitochondria of acinar cells are relatively large, as compared with those of other cell types. They may be divided into two groups according to their shape: slender rods and spherical granules. Both types of mitochondria are randomly distributed throughout the cytoplasm. However, the rod-shaped mitochondria are generally situated in the basal cytoplasm, where the lamellar GER is well developed (Fig. 8-15). On the other hand, spherical mitochondria are usually present in the apical cytoplasm and near the periphery of the Golgi apparatus (Fig. 8-16). The spherical mitochondria can hardly be identified merely by their external appearance, but when the mitochondrial cristae are revealed by the ODO method, they are easily distinguished from other granular structures. Cristae of both rod-shaped and spherical mitochondria generally occur in the form of plates or shelves. The mitochondrial matrix is completely

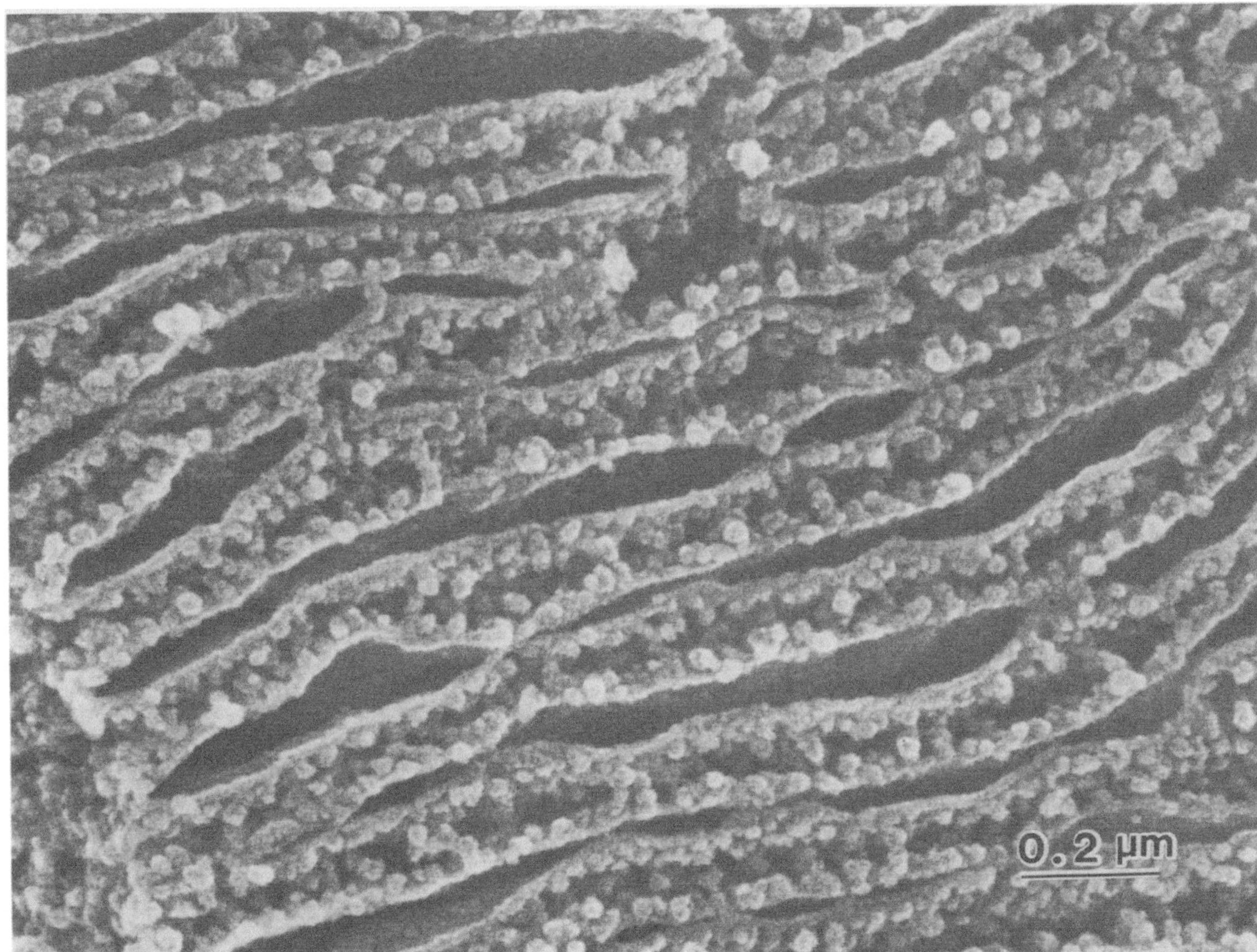

Fig. 8-9. Granular endoplasmic reticulum of an acinar cell in the rat pancreas. It consists of many flattened cisternae lying upon one another in laminated fashion. The lumen of the cisternae appears empty, as the contents of the granular endoplasmic reticulum are extracted during specimen preparation.

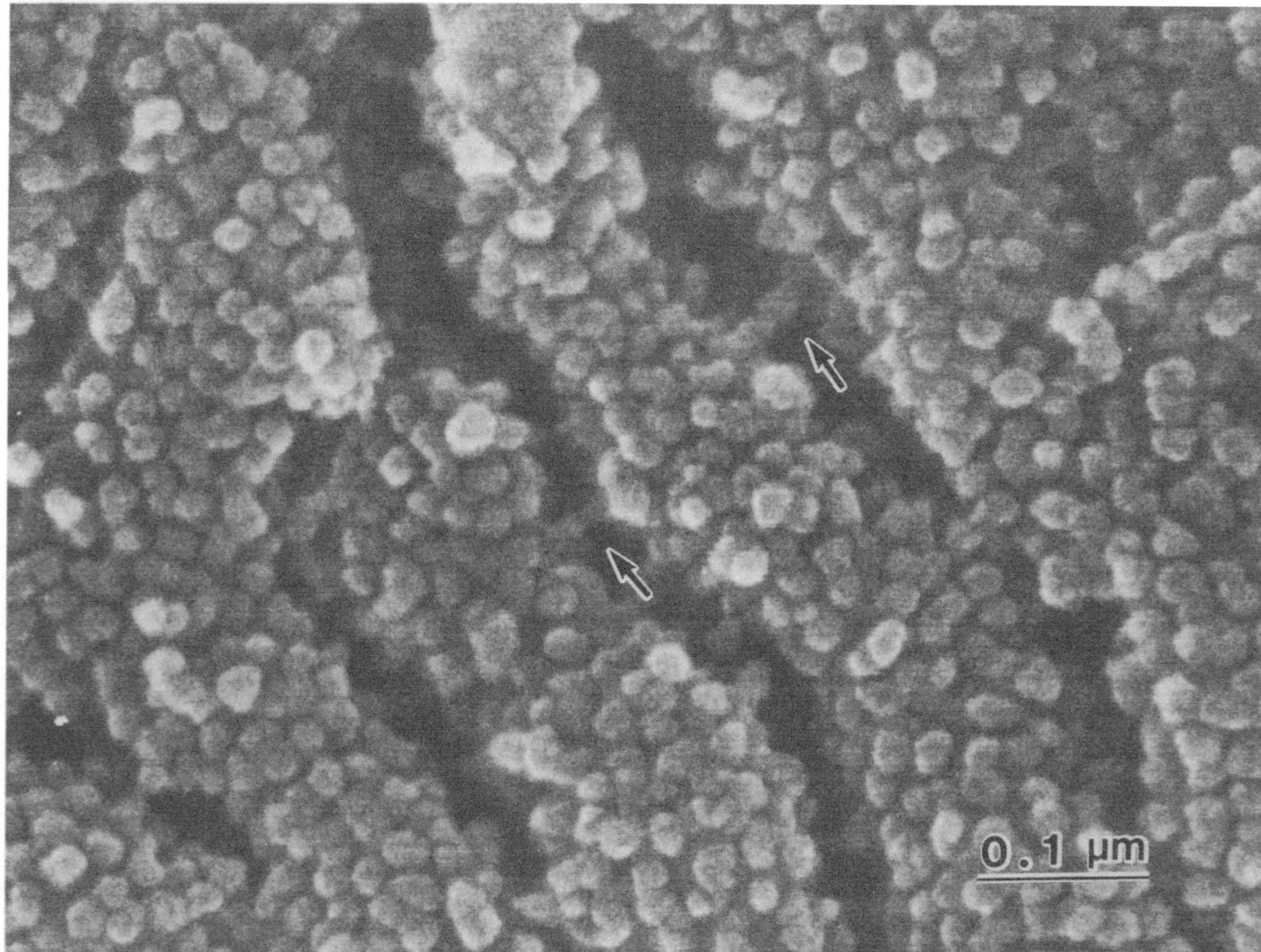

Fig. 8-10. Cytoplasmic side of granular endoplasmic reticulum of rat pancreatic acinar cell. Numerous ribosomes of about 25 nm cover the entire surface of the granular endoplasmic reticulum. A fine threadlike structure interconnects the lamellae of the granular endoplasmic reticulum to each other (arrows).

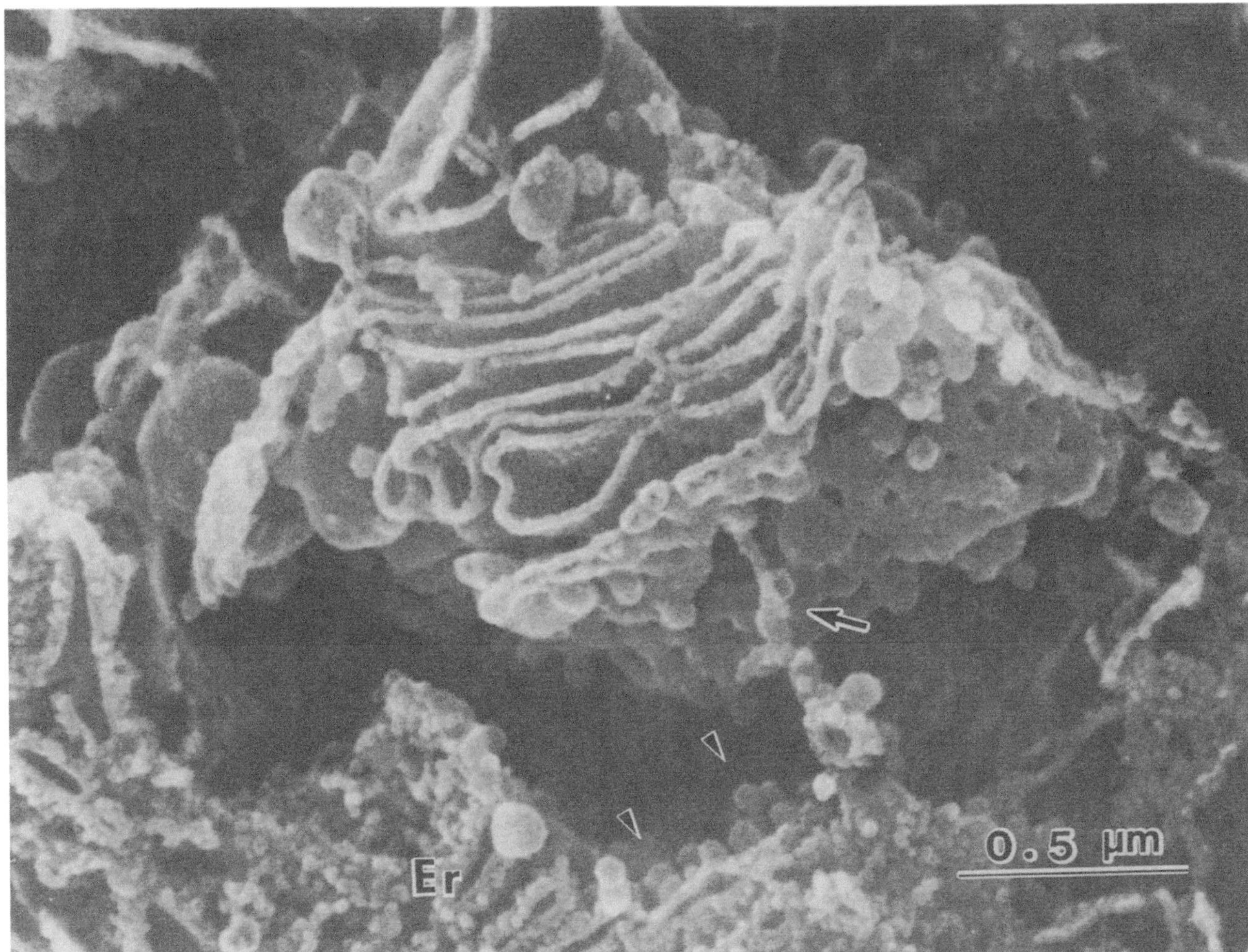

Fig. 8-11. Golgi apparatus of acinar cell in the rat pancreas. Six smooth-surface cisternae are piled one upon the other in a close parallel array, forming a Golgi stack. Small vesicles (arrowheads) of about 80 nm, in diameter are present at the margin of the granular endoplasmic reticulum (Er). A slender tubule of about 30 nm in caliber penetrates into the large opening of the *cis*-most cisterna of the Golgi stack (arrow).

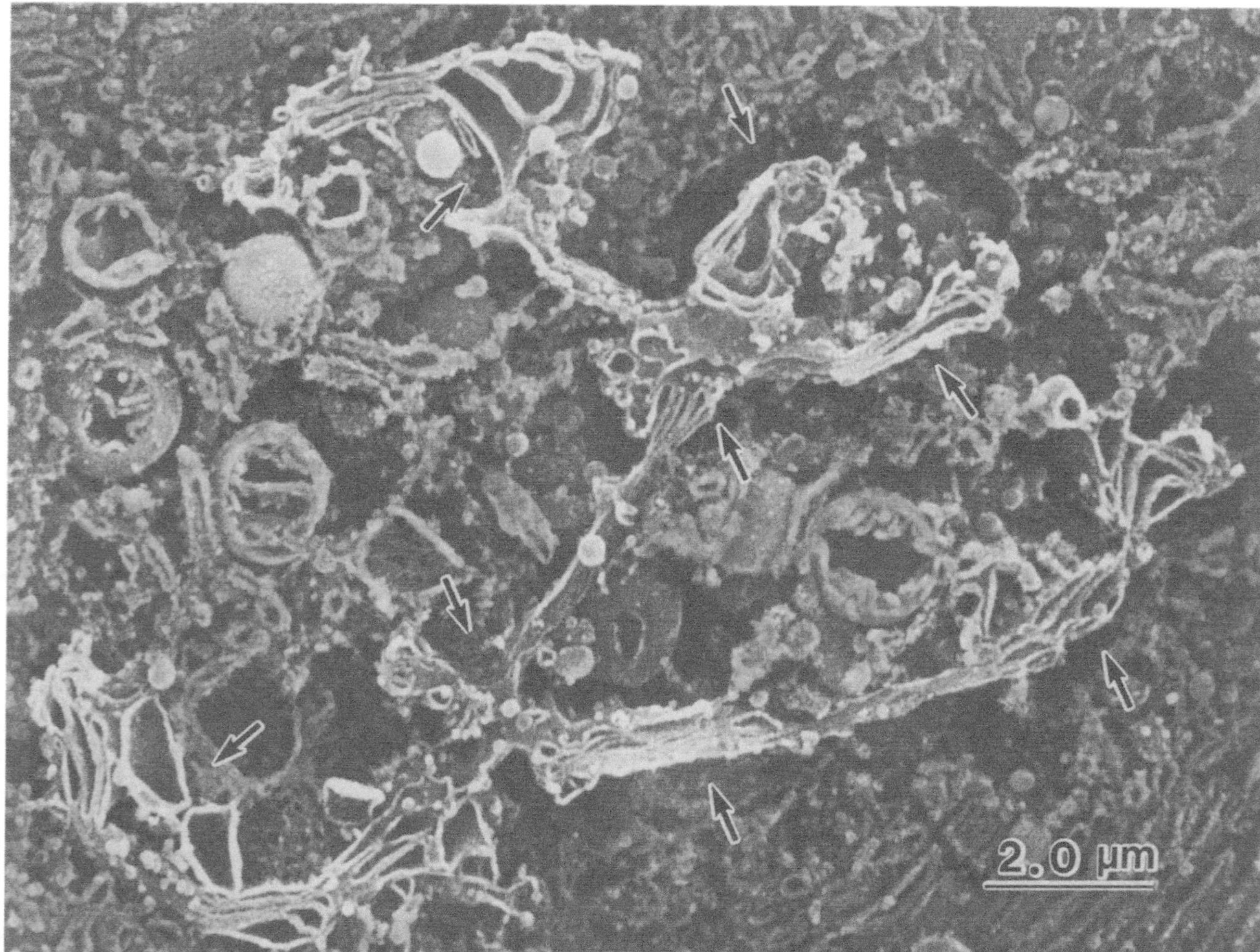

Fig. 8-12. Golgi apparatus of an acinar cell in the rat pancreas. Many Golgi stacks (arrows) in the cell are connected to each other three dimensionally to form a single interconnected Golgi apparatus.

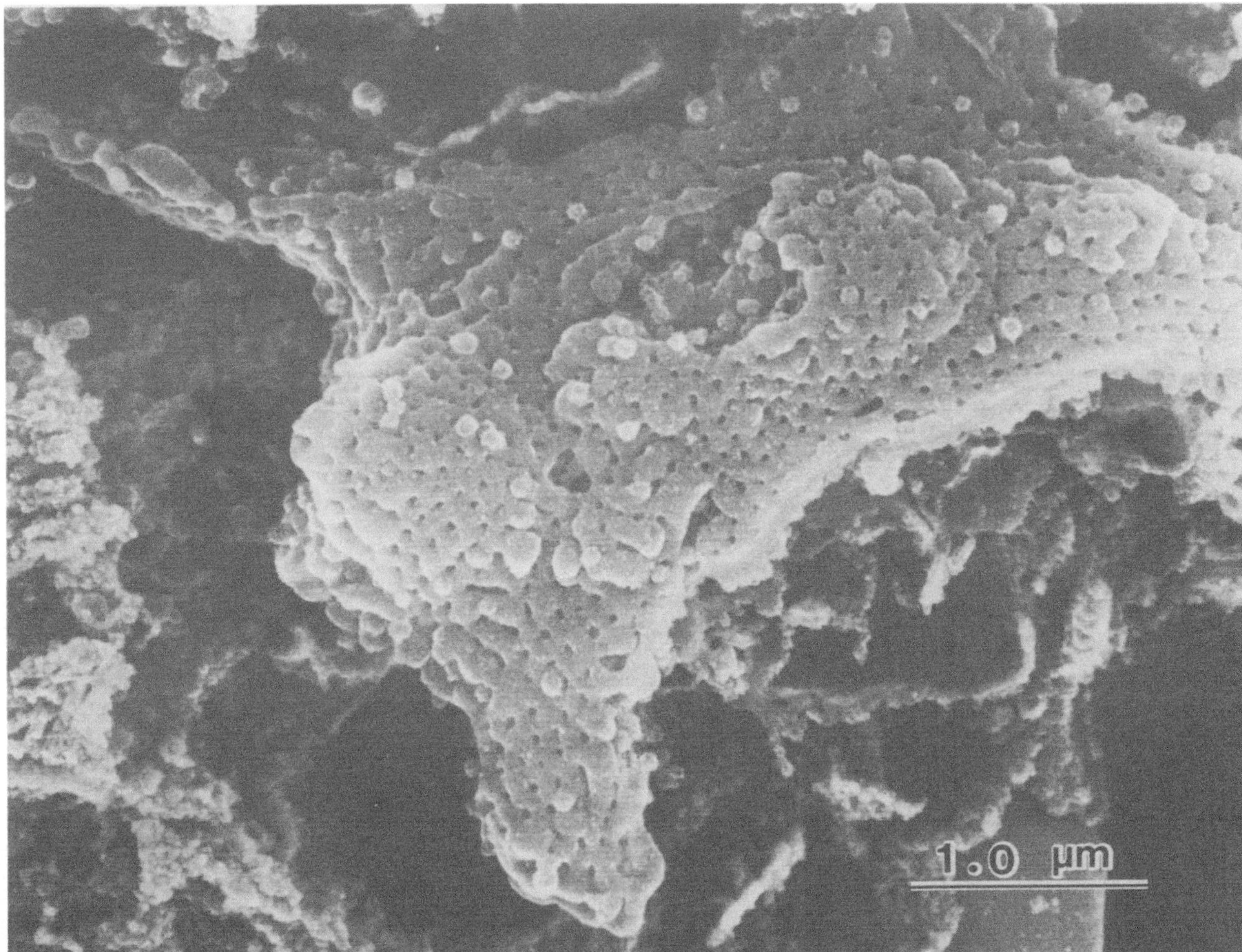

Fig. 8-13. *Cis* face of Golgi stack of acinar cell in the rat pancreas. Small pores of about 30-50 nm are evenly distributed over the entire surface of the cis-most cisterna of the Gogli stack. A few large openings of about 100–500 nm and many small vesicles of about 60–80 nm are also found on the *cis*-most cisterna.

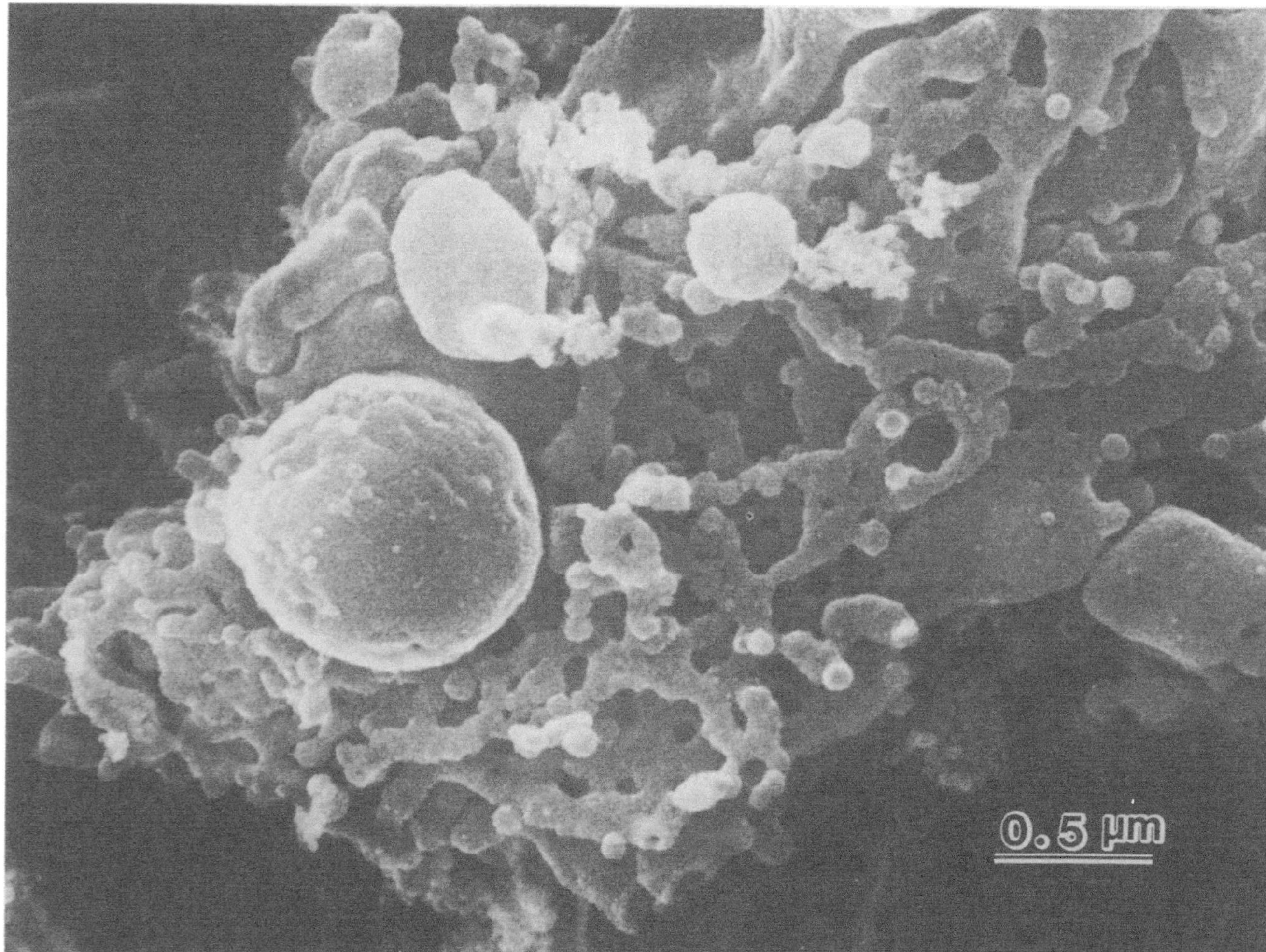

Fig. 8-14. Trans face of Golgi stack of acinar cell in the rat pancreas. It appears as a complicated plexus composed of anastomotic tubules. The continuity between these tubules and the *trans*-most cisterna of the Golgi stack can be found at several places. Golgi vacuoles of various sizes are also present on this face.

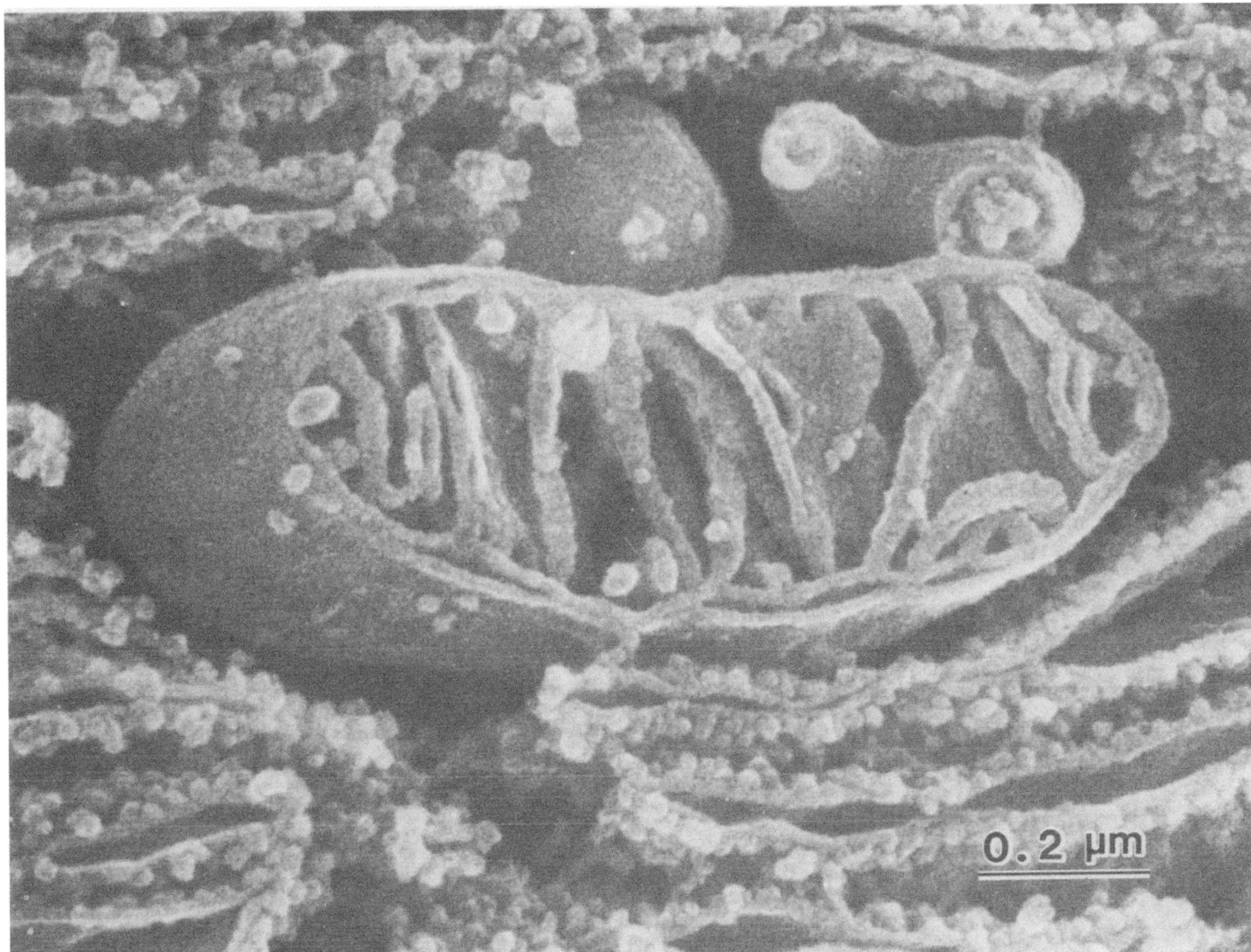

Fig. 8-15. Rod-shaped mitochondria and lamellar granular endoplasmic reticulum in the rat pancreatic acinar cell. Mitochondria of this shape are usually found in the basal cytoplasm, where the lamellar granular endoplasmic reticulum is well developed.

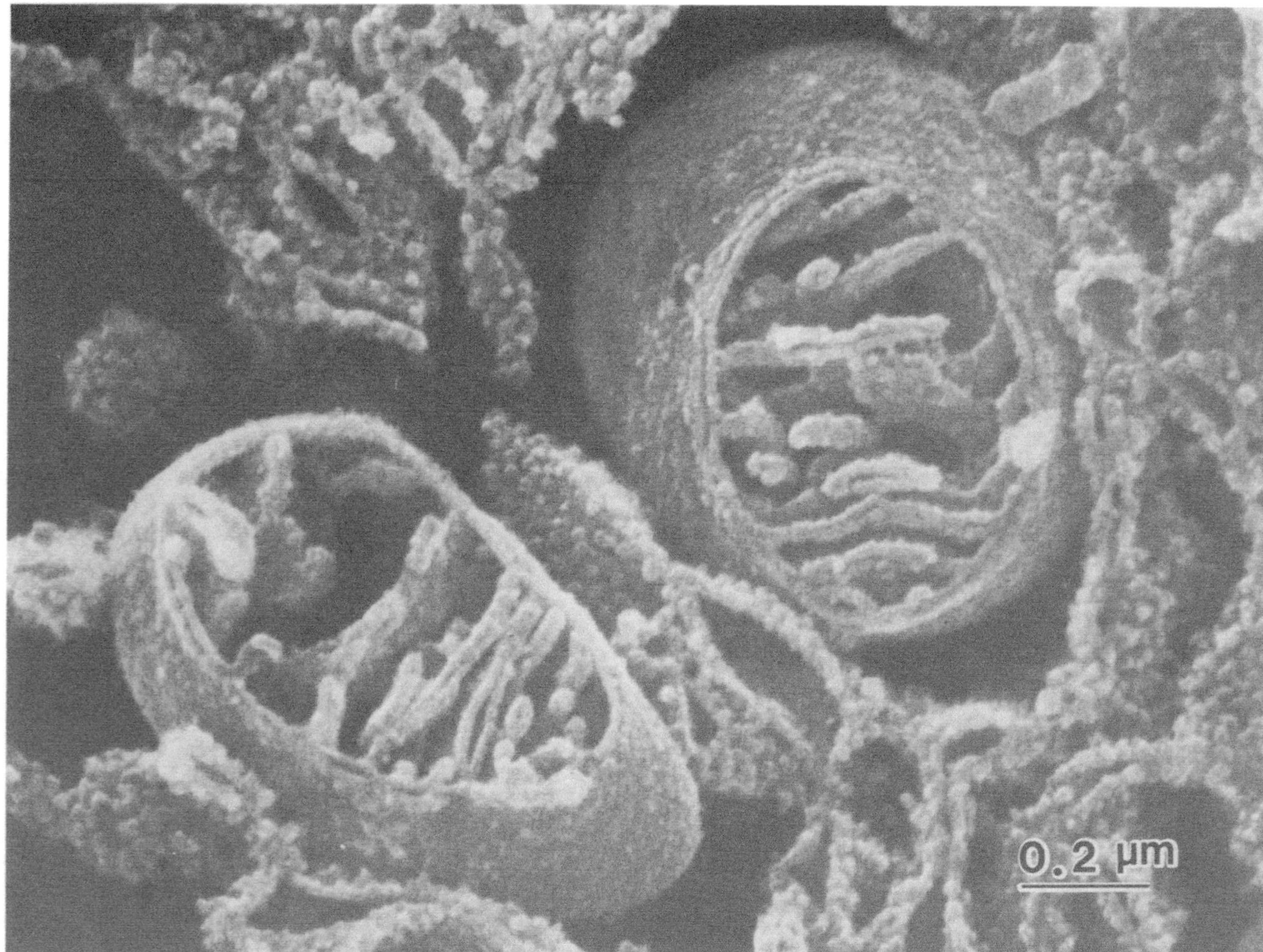

Fig. 8-16. Spherical mitochondria of acinar cell in the rat pancreas. Mitochondria of this shape are generally situated in the apical cytoplasm and in the vicinity of the Golgi apparatus. Mitochondrial cristae are visible and resemble plates or shelves.

extracted from cracked mitochondria during specimen preparation, rendering this technique particularly suitable for the study of the three-dimensional morphology of the cristae. However, the maceration makes it impossible to observe the intramitochondrial granules and other mitochondrial inclusions.

3.2.6. Lysosomes

As shown in Fig. 8-17, a kind of secondary lysosome, which is regarded as an autophagic vacuole, can be observed by SEM. These lysosomes are limited by a single membrane and contain various kinds of cytoplasmic components such as pieces of GER, interspersed vacuoles, and stacks of membranes called *myelin figures*. In the case of the primary lysosome, it may be difficult to distinguish it from other spherical cell components solely by its morphological features. For the identification of lysosomes, a cytochemical technique that tests for acid phosphatase activity was introduced into SEM study [8].

3.2.7. Secretory granules

Zymogen granules of acinar cells, which accumulate in the apical cytoplasm, appear as spherical compact granules about 0.5–1.5 μm in diameter (Fig. 8-6). They vary greatly in size and number depending on the cell's physiological activity [10]. As the contents of zymogen granules tend to be fairly easily extracted during the maceration process, it is better to omit this procedure if one wishes to observe zymogen granules in a good state of preservation. It is generally difficult to recognize the existence of the limiting membrane by SEM, as the zymogen granules are completely filled with compact materials and, consequently, no gap or space is visible between the limiting membrane and their contents. Occasional granules which fuse with each other and seem to be in the process of discharging their contents [14] permit visualization of the limiting membrane [12]. Vacuoles containing materials with a core, that are called *immature secretory granules* or *condensing vacuoles*, are found near the Golgi apparatus and are distinguishable from the mature zymogen granules [12].

3.2.8. Microtubules and filaments

TEM studies have shown that the acinar cells contain microtubules and many kinds of filaments. Such structures are thought to serve a supportive role as a cytoskeleton and may also play an important part in the transport of secretory granules. These filamentous components are completely extracted from the cytoplasm when the maceration procedure is used for a prolonged period of time. If this procedure is not employed at all or if it is done only for a short time, these fine filaments (8–18 nm in thickness) of indefinite length are commonly observed in the cytoplasm by SEM. They are present abundantly around the nucleus and in the vicinity of the Golgi apparatus, although their number varies greatly from cell to cell [12]. In proximity to the zonula adhaerens, many filaments gather together in parallel to form a bundle about 1 μm in width (Fig. 8-18).

4. Endocrine pancreas

The endocrine cells of the pancreas are found in the islets of Langerhans (pancreatic islets), which are highly vascularized by numerous capillaries. Very few SEM studies on the endocrine cells of the pancreatic islets have been performed. [15, 16]. At least two common cell types are distinguishable under SEM by the morphology of their granules. One type, called *alpha* or *A cells*, contains spherical granules, which are relatively uniform in size (Fig. 8-19). The spherical shape of A-cell granules is said to be a constant characteristic of these cells in all mammalian species. Beta or B cells possess granules that have distinct morphological characteristics in different species. As shown in Fig. 8-20, the granules of B cells in the dog are also enclosed in a spherical limiting membrane, but their contents appear as elliptical plates or sometimes as regular hexagonal plates [17]. The remainder of the cytoplasm of endocrine cells, not unlike that of acinar cells, contains the usual organelles, although there are great differences in their size, shape, and quantity. For example, the mitochondria of endocrine cells are distinctly smaller and more slender than those of acinar cells.

One of the most striking findings among the

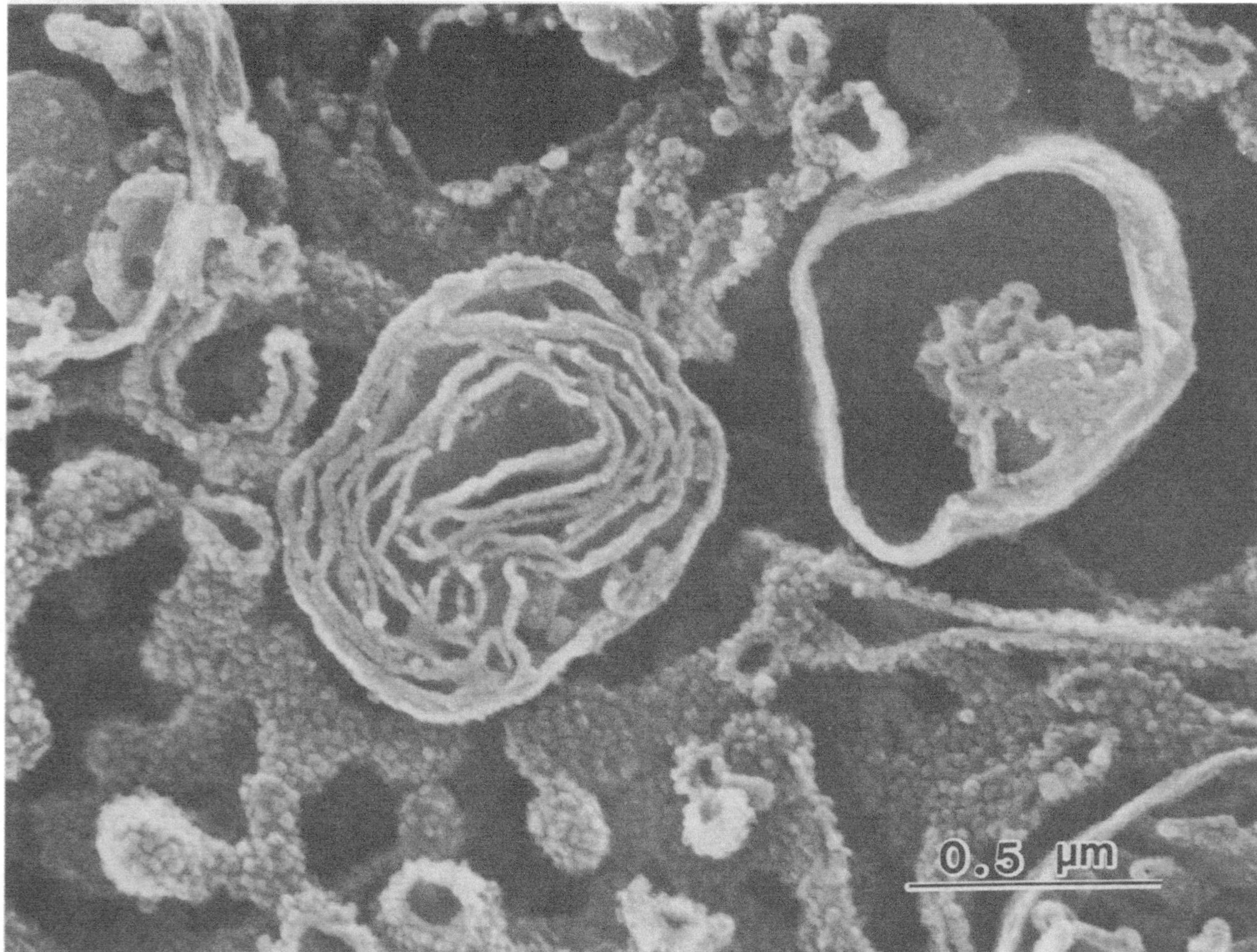

Fig. 8-17. Secondary lysosomes and tubular type of granular endoplasmic reticulum in the rat pancreatic acinar cell. Lysosomes are limited by a single membrane and contain various kinds of cytoplasmic components such as myelin figures (center) and some membranous materials (right).

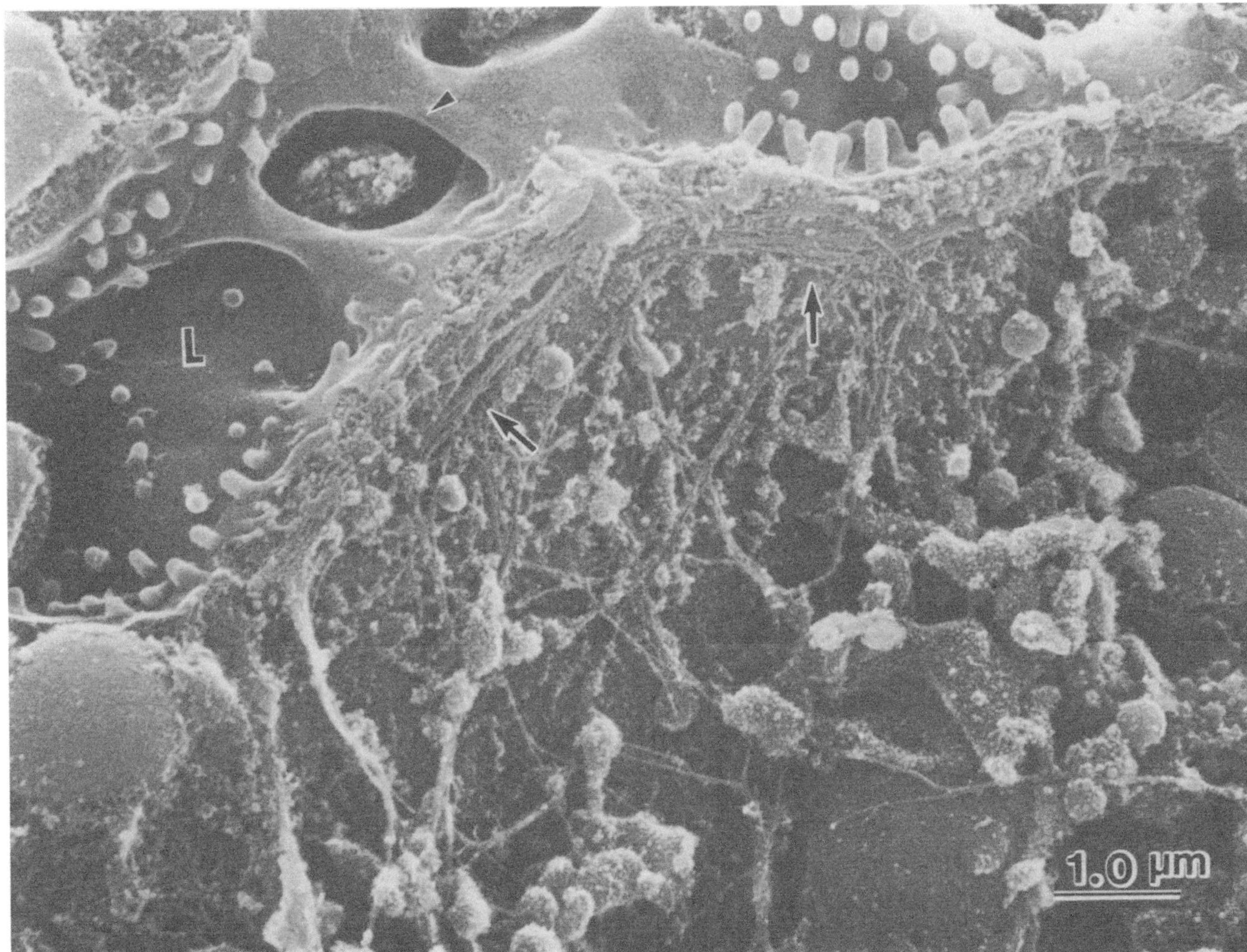

Fig. 8-18. Filaments of acinar cell in the rat pancreas. In the region of the zonula adhaerens many filaments gather together in parallel and form a bundle about 1 µm in width (arrows). An opening (arrowhead) of about 1 µm is found on the luminal surface of acinar cell (L).

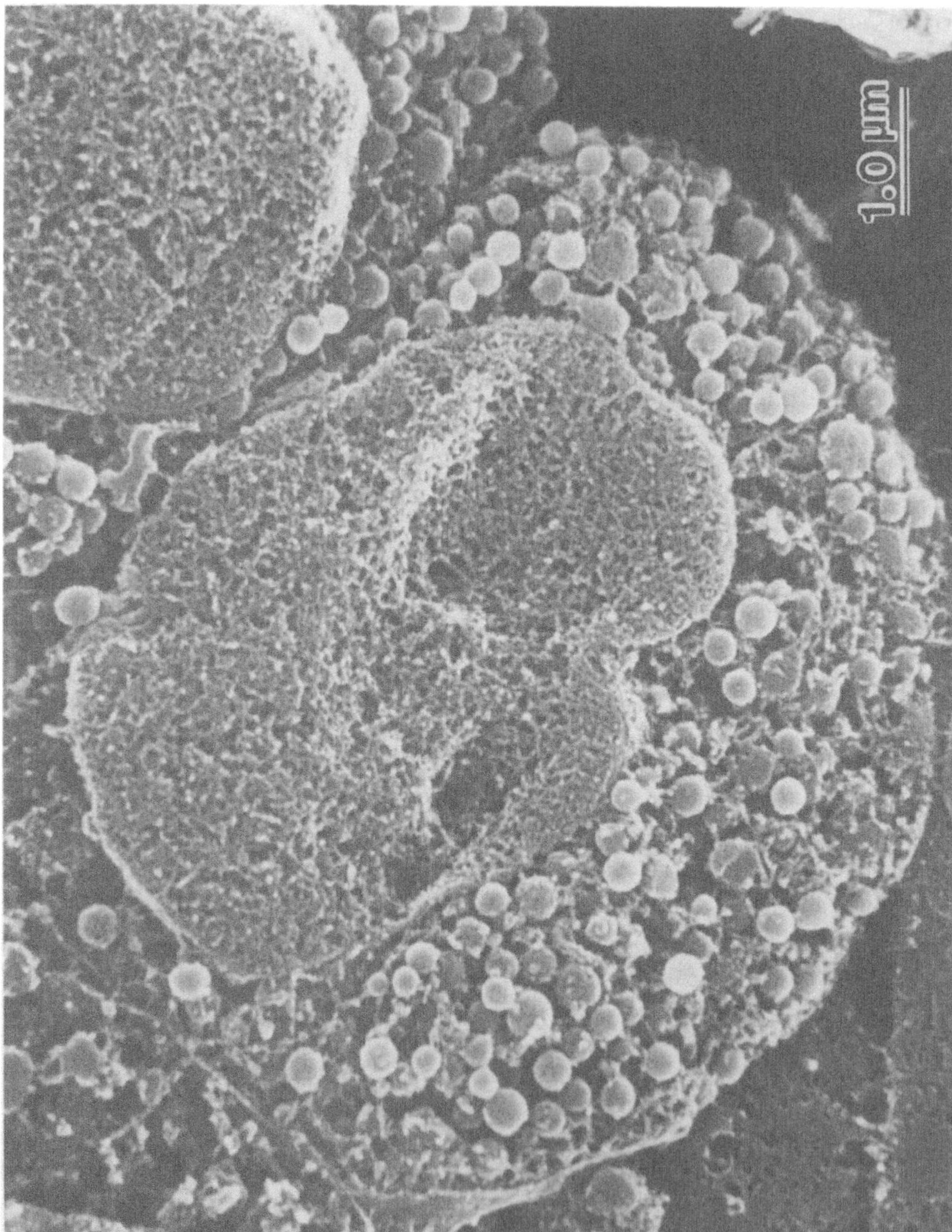

Fig. 8-19. Alpha cells of pancreatic islets in the dog. The cytoplasm is filled with numerous spherical granules.

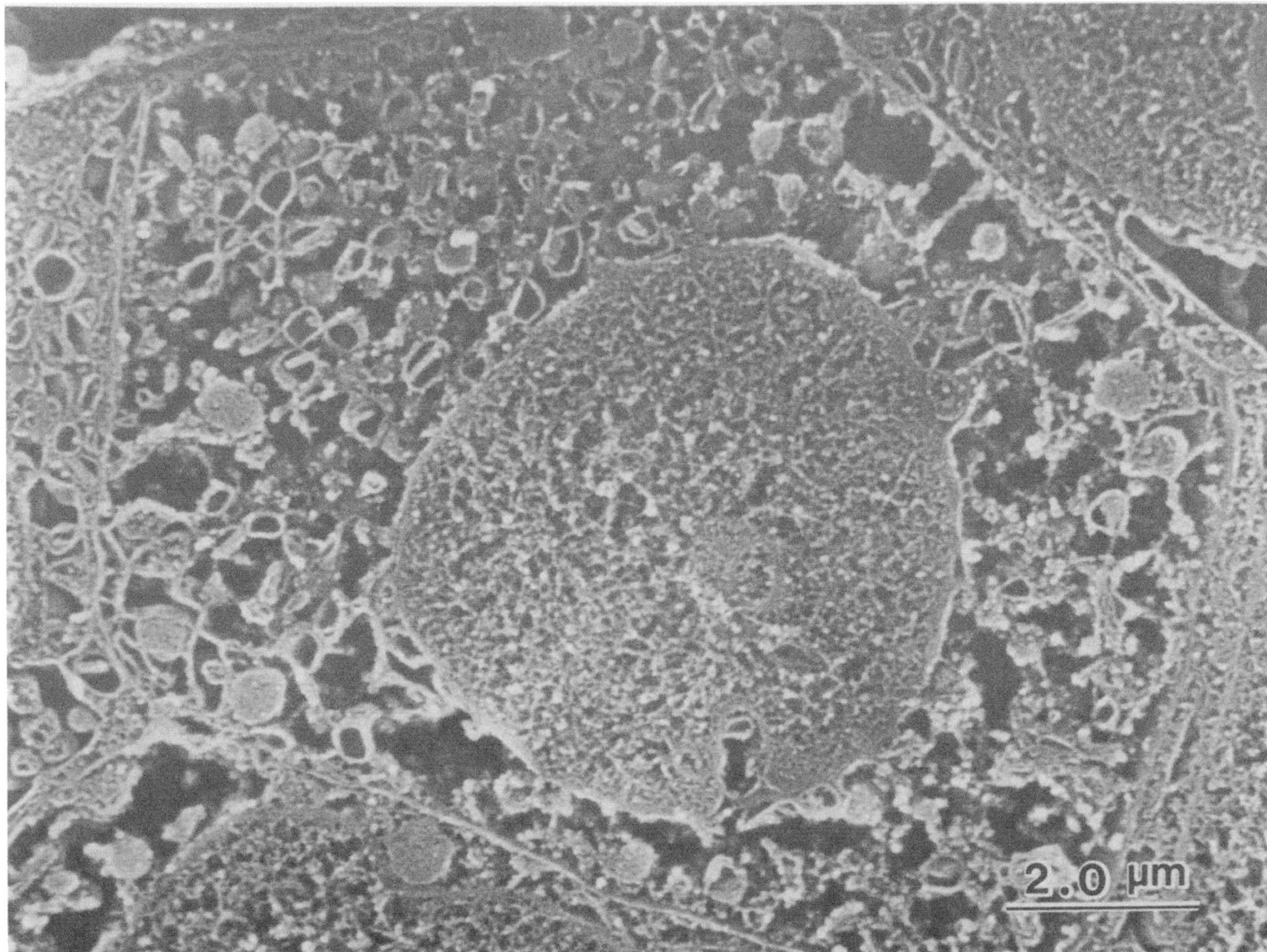

Fig. 8-20. Beta cell of pancreatic islets in the dog. Granules are enclosed in a spherical limiting membrane and their contents are clearly visible.

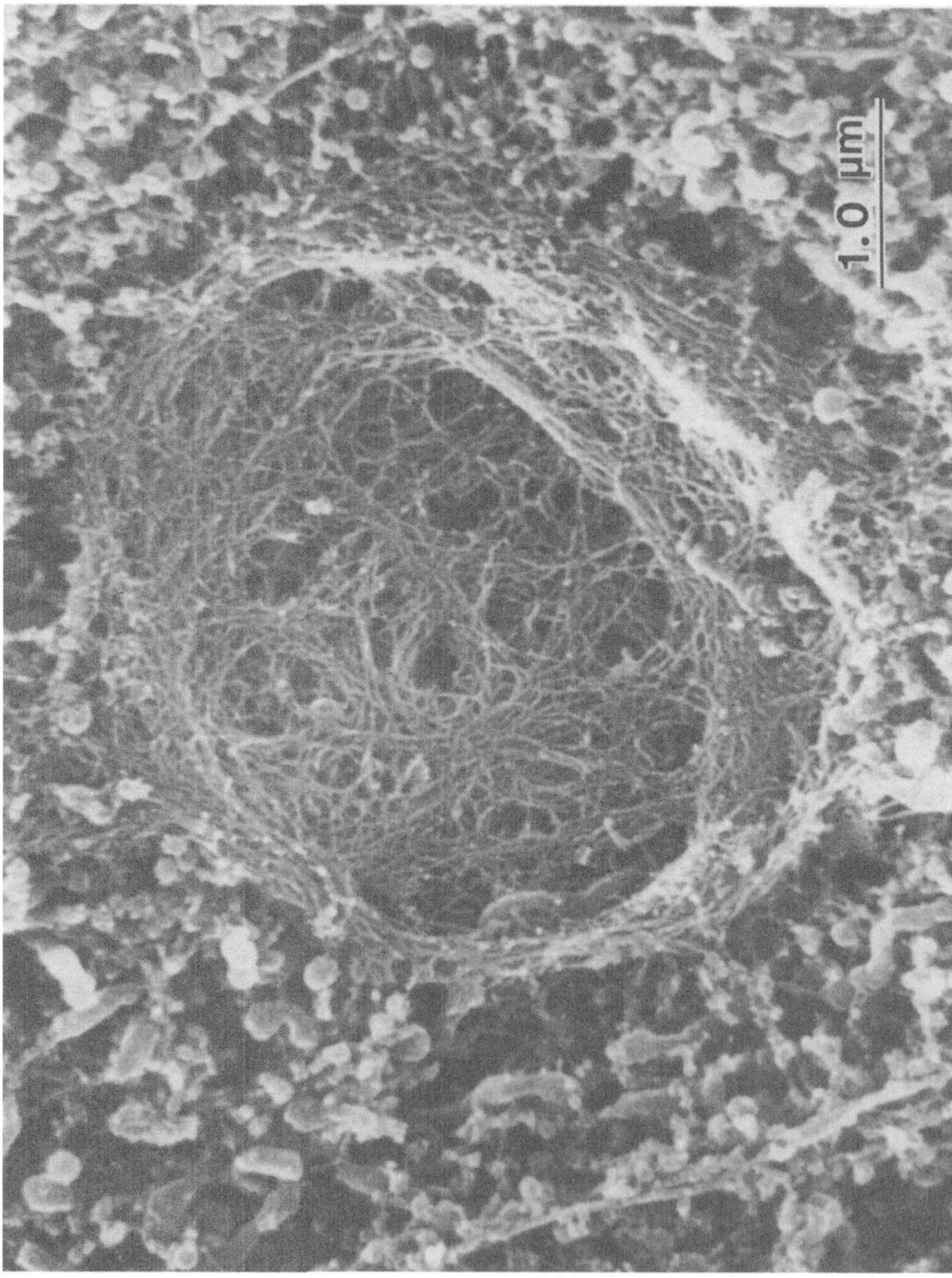

Fig. 8-21. Filaments surrounding the nucleus in a rabbit endocrine cell. Highly developed filaments appear around the nucleus as a round and elaborately knitted mesh.

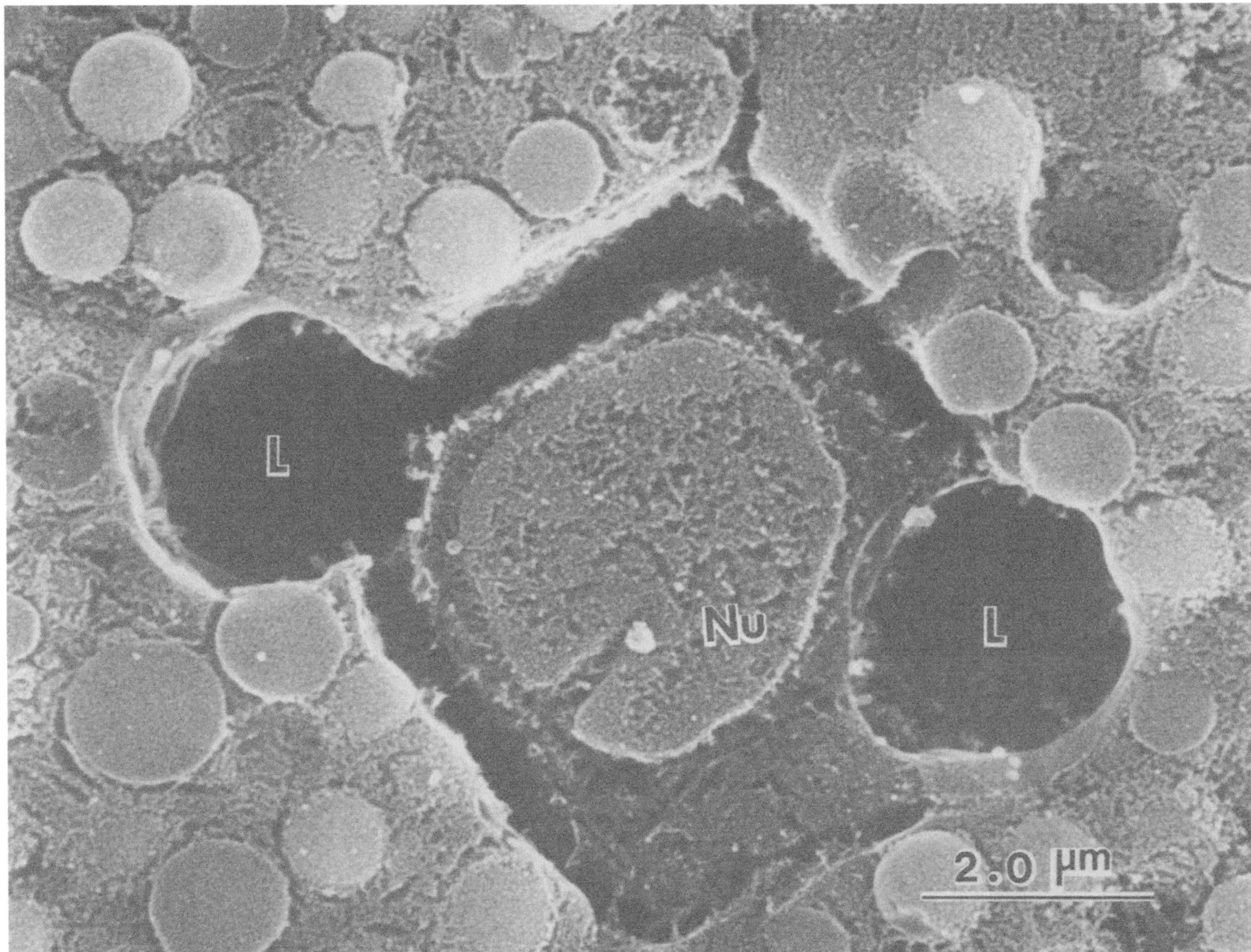

Fig. 8-22. A centroacinar cell in the rat pancreas. The cell appears in the central part of an acinus. In this picture it is sandwiched between two small lumina (L). Zymogen granules accumulate in the apical portion of acinar cells. Nu = nucleus of centroacinar cell.

cell organelles in the endocrine cells is the perinuclear filaments. In the cells of rabbit pancreatic islets, the filaments are highly developed and form a round meshlike structure (Fig. 8-21). This structure, which is situated in the space between the nucleus and the cytoplasm, can be revealed when the nucleus and its envelope are removed during preparation of the specimen [12]. The function and chemical nature of these filaments remain to be determined. Ushiki et al. [18] demonstrated the utility of the backscattered electron image in discerning endocrine cell types immunohistochemically.

5. Duct system

Small cells called *centroacinar cells* can often be seen in the central part of an acinus (Fig. 8-22). The duct system is said to begin with the centroacinar cell and continues with the intercalated, intralobular, interlobular, and main or accessory ducts, respectively. Following this order, the epithelium of the duct system gradually increases in height from squamous through cuboidal to columnar. The fine structure of the centroacinar cells and that of the cells of the intralobular duct are quite similar. The nucleus occupies the greater part of the cell. The very sparse cytoplasm contains few and poorly developed organelles. Scattered short microvilli and an occasional single cilium are found on the luminal surface of the duct cells [19].

6. Conclusions

The three-dimensional features of pancreatic cells were demonstrated by scanning electron microscopy, with special attention given to internal cell structures. The ODO method, devised by Tanaka and Naguro [7], was applied to disclose the intracellular structures such as the nucleus, endoplasmic reticulum, Golgi apparatus, lysosomes, and mitochondria.

Filaments of endocrine cells were seen to form a round meshlike structure surrounding the nucleus.

One of the most interesting findings among the cell organelles was the morphology of the Golgi apparatus in acinar cells:

1. The continuity of the Golgi stack was clearly demonstrated. Many Golgi stacks are virtually connected to each other to form a single interconnected Golgi apparatus.
2. The *cis* face of the Golgi stack showed a peculiar structure. Numerous small pores were evenly distributed over the entire surface of the cisterna of the *cis* face.
3. The *trans* face occasionally showed a complicated plexus, which consisted of anastomotic tubules.

References

1. Haggis GH: Cryofracture of biological material. In: Scanning Electron Microscopy/1970. O Johari (ed) Chicago: IITRI, 99–104, 1970.
2. Tanaka K: Frozen resin cracking method for scanning electron microscopy of biological materials. *Naturwissenshaften* 59: 77, 1972.
3. Tanaka K, Iino A, Naguro T: Styren resin cracking method for observing biological materials by scanning electron microscopy. *J Electron Microsc* 23: 313–315, 1974.
4. Tokunaga J, Edanaga M, Fujita T: Freeze cracking of scanning electron microscope specimens. *Arch Histol Jpn* 37: 165–182, 1974.
5. Evan AP, Dail WG, Dammrose D, Palmer C: Scanning electron microscopy of cell surfaces following removal of extracellular material. *Anat Rec* 185: 443–446, 1976.
6. Takahashi H: Scanning electron microscopy of the rat exocrine pancreas. *Arch Histol Jpn* 47: 387–404, 1984.
7. Tanaka K, Naguro T: High resolution scanning electron microscopy of cell organelles by a new specimen preparation method. *Biomed Res* 2(Suppl): 63–70, 1981.
8. Tanaka K and Mitsushima A: A preparation method for observing intracellular structures by scanning electron microscopy. *J Microscopy* 133: 213–222, 1984.
9. Murakami T: Tannin-osmium conductive staining of biological specimens for non-coated SEM. *Scanning* 1: 127–129, 1978.
10. Uchiyama Y, Saito K: A morphometric study of 24-hour variations in subcellular structures of the rat pancreatic acinar cell. *Cell Tissue Res* 226: 609–620, 1982.
11. Tanaka K: Demonstration of intracellular structures by high resolution scanning electron microscopy. In: Scanning Electron Microscopy/1981. O Johari (ed), Chicago: AMF O' Hare, 1–8, 1981.
12. Naguro T: Three-dimensional view of the intracellular structures of pancreatic cells. *Yonago Acta Med* 24: 81–92, 1980.
13. Noda T, Ogawa K: Golgi apparatus is one continuous organelle in pancreatic exocrine cell of mouse. *Acta His-*

tochem Cytochem 17: 436–451, 1984.
14. Ichikawa A: Fine structural changes in response to hormonal stimulation of the perfused canine pancreas. *J Cell Biol* 24: 369–383, 1965.
15. Krstic R: Scanning electron microscopic aspects of some endocrine glands. *Biomed Res* 2(Suppl): 103–108, 1981.
16. Fujita T, Kobayashi S, Serizawa Y: Intercellular canaliculi system in pancreatic islets. *Biomed Res* 2(Suppl): 115–118, 1981.
17. Tanaka K, Iino A, Naguro T: Scanning electron microscopic observations on intracellular structures of ion-etched materials. *Arch Histol Jpn* 39: 165–175, 1976.
18. Ushiki T, Yonehara K, Iwanaga T, Fujita T: Application of a backscattered electron image to immunocytochemistry in freeze-cracked tissues. *Arch Histol Jpn* 47: 553–557, 1984.
19. Ogrowsky D, Fawcett J, Althoff J, Wilson RB, Pour P: Structure of the pancreas in Syrian hamsters. *Acta Anat* 107: 121–128, 1980.

Author's address:
Dr. Tamonori Naguro
Department of Anatomy
Tottori University
School of Medicine
Yonago, 683
Japan

CHAPTER 9

Cytochemical access to plasma and intracellular membranes of freeze-fractured hepatocytes and salivary gland cells

MARIA ROSARIA TORRISI, PATRIZIA MANCINI, & PEDRO PINTO DA SILVA

1. Introduction and methodology

Cytochemical labeling of intracellular components presupposes direct access to the cell interior. Traditionally, this access is rendered possible through sectioning of tissues or of cell pellets. Sectioning can be performed in fresh tissues, in polymer-embedded specimens, as well as in frozen sections. Alternative, although less conventional, methods do not involve cutting of the tissue but, instead, cleavage or fracture of previously frozen tissues.

Methods that involve the application of classical cytochemical techniques to the characterization of intracellular molecules exposed by the process of fracture were not developed for many years, mainly because it was assumed that irreversible obliteration of intracellular structures would result from the freeze-thaw steps involved in the processing of specimens for freeze-fracture. In large measure, this fear resulted from the assumption of cytoplasm matrices as loose, fragile, relatively open structures.

Over the past 8 years, we have developed a series of cytochemical approaches to characterize the components exposed by the process of fracture [1]. The first approach that we pursued, *fracture label*, involves the cytochemical labeling of the membranous, cytoplasmic, and nucleoplasmic components that are exposed during the course of fracture of a frozen specimen [2–4].

Fracture results from the sudden transfer of kinetic energy from one object to another. The process is complex and, in large measure, stochastic. However, it tends to follow along the path of least resistance. Freezing causes the formation and stabilization of large numbers of hydrogen bonds. In addition, glutaraldehyde, if used, will result in the creation of covalent intermolecular bonds. Entropy considerations, paramount in the definition of living systems, in particular in the organization of biomembranes, become practically irrelevant, due to the loss of long-range mobility of cytoplasmic and membrane components. Other forces, in particular weak bonds of the London/van der Waals type, are much weaker than the hydrogen bonds of ice matrices. In vivo, the bilayer continuum of biological membranes is stabilized mainly because it represents the configuration that allows for maximum overall molecular disorder. In frozen membranes, only intermolecular forces operate, in particular the weak bonds made between apposing lipidic molecules [5, 6].

During freeze fracture, membranes are split along their apolar continuum. The partition of peripheral proteins at the surface is therefore determined by this process. Much less clear is the partition of integral membrane proteins, in particular of transmembrane proteins. Although breakage of covalent bonds by physical means is not impossible (e.g., during the shearing of polymers such as nucleic acids), partition of transmembrane proteins will more likely not involve the breakage of such covalent bonds. Partition of an integral membrane protein with one or the other half of a membrane will probably depend on how relatively strong and cooperative the

Riva, A. and Motta, P.M., (eds.), Ultrastructure of the Extraparietal Glands of the Digestive Tract.
ISBN 978-1-4612-8211-2.

intermolecular bonds that it makes with its immediate neighbors are [7–9].

With fracture label we attempt to label the chemical groups that are exposed by the process of fracture of a frozen specimen. The fracture process tends to follow plasma as well as intracellular membranes, but it also involves exposure by crossfracture of both the cytoplasm and nucleoplasm. To observe the patterns of labeling that are obtained by the labeling of freeze-fractured specimens, we developed two variants of fracture label: *thin-section fracture label* and *critical-point drying fracture label* [2–4]. In the first, after fracture, thawing, and labeling, the specimens are processed for conventional thin-section electron microscopy; in the second, the labeled specimens are dried through the critical point (to avoid deformation) and are replicated by platinum/carbon evaporation. The replicas are observed after digestion of the biological material with acids or bases [2–4].

Here, we review briefly some of the results that we obtained during the development of thin-section fracture label, in particular its application to the labeling of intracellular membranes.

2. Hepatocytes

Thin-section fracture label is particularly suited to study intracellular membrane components: it exposes large areas of the intracellular membranes to the label and provides clear identification of the organelles or cell structures that are labeled. Fracture label, when performed on isolated cells, allows the simultaneous observation and comparison of the pattern of labeling over the unfractured surface and on the fractured intracellular membranes [10].

We used rat hepatocytes to explore the composition of the membrane of intracellular organelles, such as lysosomes, mitochondria, peroxisomes, and lipoprotein secretory granules. Using lectin cytochemistry and the thin-section fracture-label technique, it was possible to investigate the distribution of glycocomponents in intracellular membranes. In particular, the lectin wheat-germ agglutinin (WGA), specific for sialic acid and N-acetyl-D-glucosamine [11], could be used as a marker of terminally glycosylated membrane components, after processing in the Golgi apparatus.

Rat hepatocytes were isolated by digestion with collagenase type IV, fixed in glutaraldehyde, frozen, freeze fractured, and then labeled with WGA indirectly conjugated with colloidal gold. Fig. 9-1a demonstrates, at low magnification, how it is possible to simultaneously examine labeling over fracture faces (F), where intracellular membranes are exposed, and over cell surfaces (S). When the plasma membranes of two adjacent cells are freeze fractured and remain in close proximity, complementary views of protoplasmic and exoplasmic faces of the plasma membranes can be labeled (Fig. 9-1b). Under these circumstances, the partition of transmembrane glycoproteins can be easily assessed. WGA binding sites, corresponding to integral as well as peripheral membrane components, are here almost totally confined to the exoplasmic faces (Figs. 9-1b and 9-1c), as previously observed in pancreatic cells [12]. Labeling of surface antigens and receptors over exoplasmic faces of lectin binding sites results from reorganization of interrupted bilayers, which occurs during thawing [1–7]. The sparse labeling over the protoplasmic faces (Fig. 9-1b), however, may represent transmembrane sialoglycoproteins which partition, during fracture, with the inner leaflet of the membrane, as observed in other cell types [7–9].

WGA fracture labeling of intracellular membranes confirms the existence of two endomembrane compartments [13]: the first is characterized by the absence of WGA labeling over both protoplasmic and exoplasmic faces of the fractured membranes and includes mitochondria (Fig. 9-2c), peroxisomes (Figs. 9-2d and 9-2e), endoplasmic reticulum (Figs. 9-1a and 9-2a), and the nuclear envelope (Fig. 9-1a). The second compartment, characterized by clear WGA labeling, comprises the membranes of lysosomes (Fig. 9-2f) and secretory granules (Fig. 9-2b), as well as plasma membranes. Here, labeling over exoplasmic faces corresponds to lectin binding sites exposed before fracture at the luminal side of the intracellular membrane, in agreement with biochemical evidence [14]. These results show how fracture label can also determine the sideness of the membrane components.

Endoplasmic reticulum membranes are known to lack fully glycosylated components, since the addition of terminal sugars, such as sialic acid,

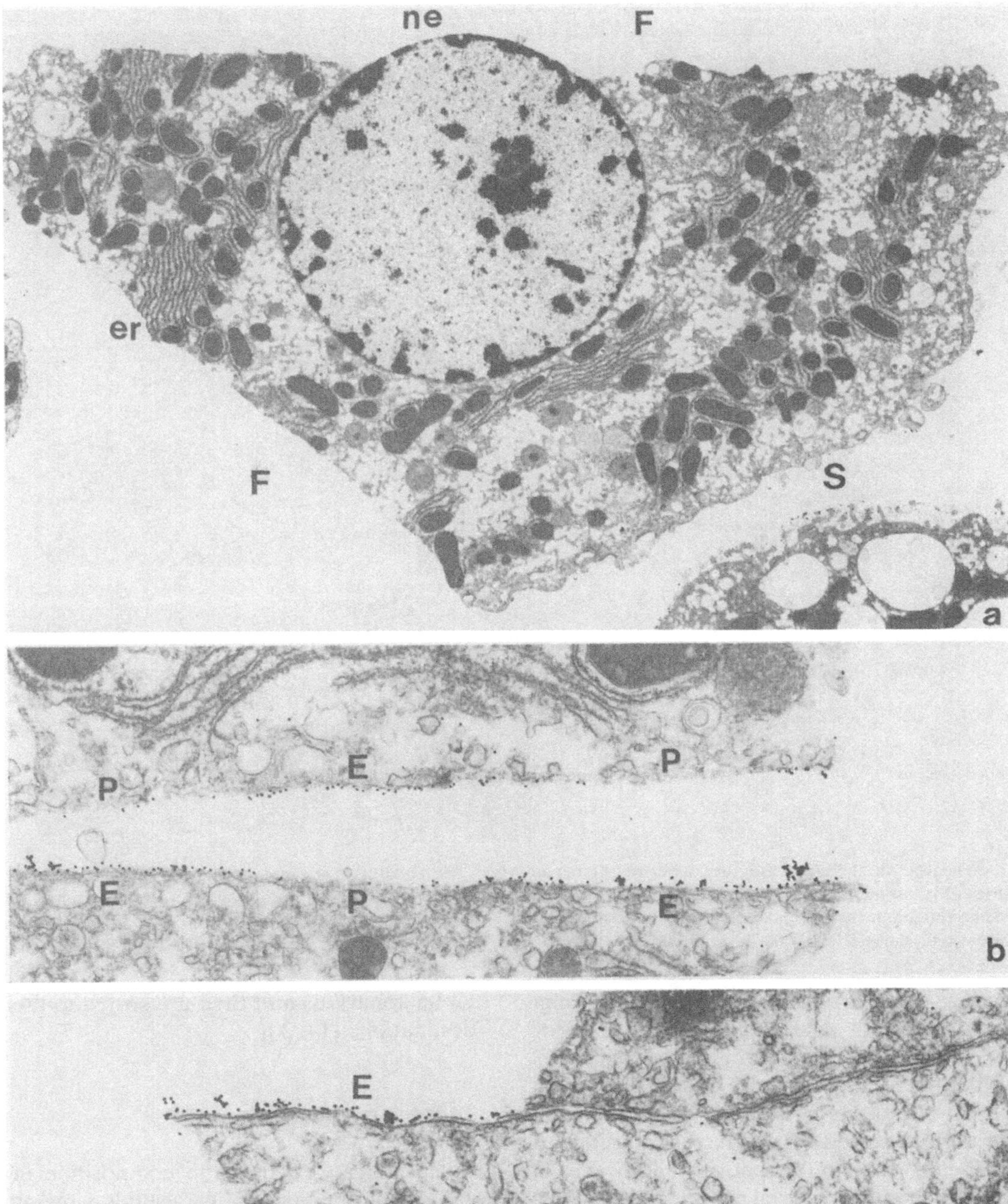

Fig. 9-1. Wheat-germ agglutinin fracture label of rat hepatocytes: **a**: unfractured cell surface (S) and two fractured faces (F) are clearly distinguishable; nuclear envelope (ne), mitochondria (m), and endoplasmic reticulum (er) membranes are exposed by the fracture process; **b**: complementary P and E faces of fracture-labeled plasma membranes of two adjacent hepatocytes show that only the E faces are strongly and uniformly labeled; **c**: E face of a fracture-labeled hepatocyte plasma membrane shows dense labeling. **a**: x 8,000; **b**: x 42,000; **c**: x 45,000.

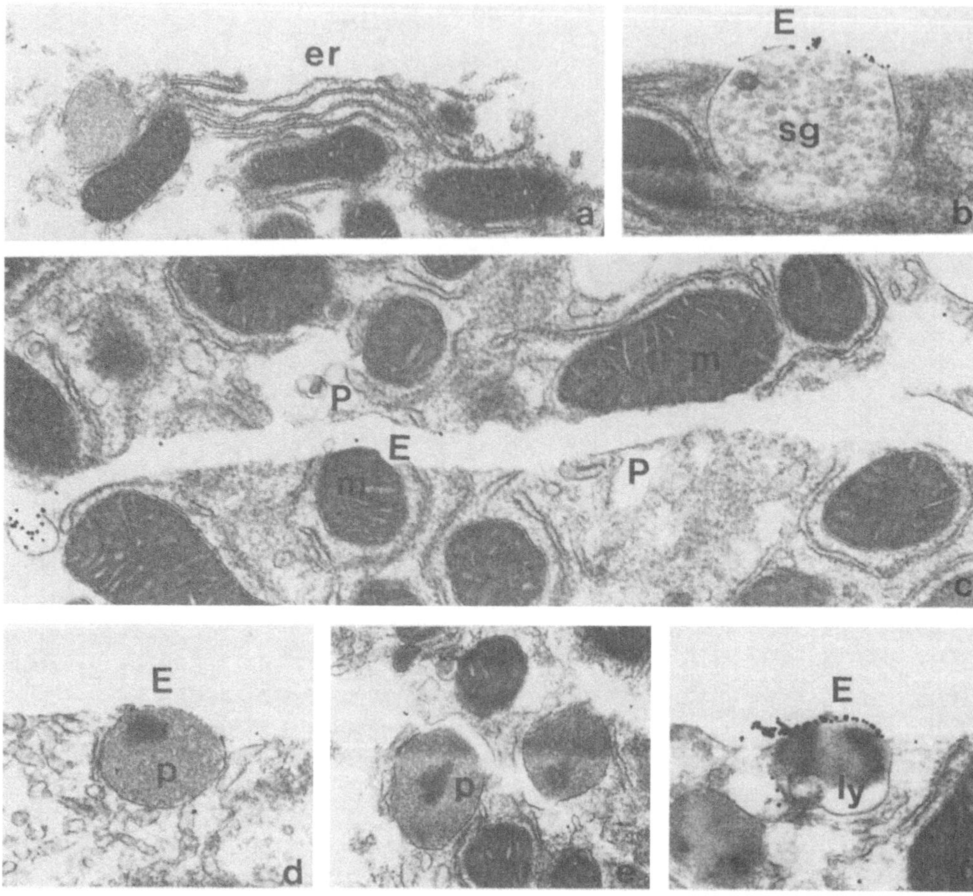

Fig. 9-2. Wheat-germ agglutinin fracture label of rat hepatocytes: The labeling is present only on E faces of freeze-fractured membranes of lysosomes (**f**) and secretory granules (**b**), whereas both P and E faces of mitochondria (**c**), peroxisomes (**d,e**), and endoplasmic reticulum (**a**) are unlabeled. ly = lysosomes; sg = secretory granules; m = mitochondria; p = peroxisomes; er = endoplasmic reticulum; **a**: x 29,000; **b**: x 45,000; **c**: x 50,000; **d**: x 41,000; **e**: x 32,000; **f**: x 46,000.

occurs only during transport through the Golgi apparatus [15–17]. Therefore, membranes of both the endoplasmic reticulum and peroxisomes (thought to derive directly from endoplasmic reticulum) [18] do not contain sialoglycocomponents and can not be labeled by WGA. The absence of WGA labeling over mitochondria membranes is expected: most mitochondrial proteins are synthesized in the cytosol [19] and are not transported through the Golgi complex. In contrast, components of the membranes of lysosomes and secretory granules, as well as lysosomal enzymes and secretory proteins, are terminally glycosilated; after their synthesis in the endoplasmic reticulum, they are transported and processed in the Golgi apparatus and then are sorted to their final destinations [14, 20].

3. Salivary gland cells

To investigate further the composition in sialo-glyco-components of intracellular membranes, particularly the membranes of the Golgi complex and secretory granules, we used collagenase-isolated submaxillary gland cells as an example of typical secretory cells. In these cells, where the Golgi complex is prominent [21], WGA fracture label shows very low labeling over the fractured cisternae, as previously observed in nonsecretory

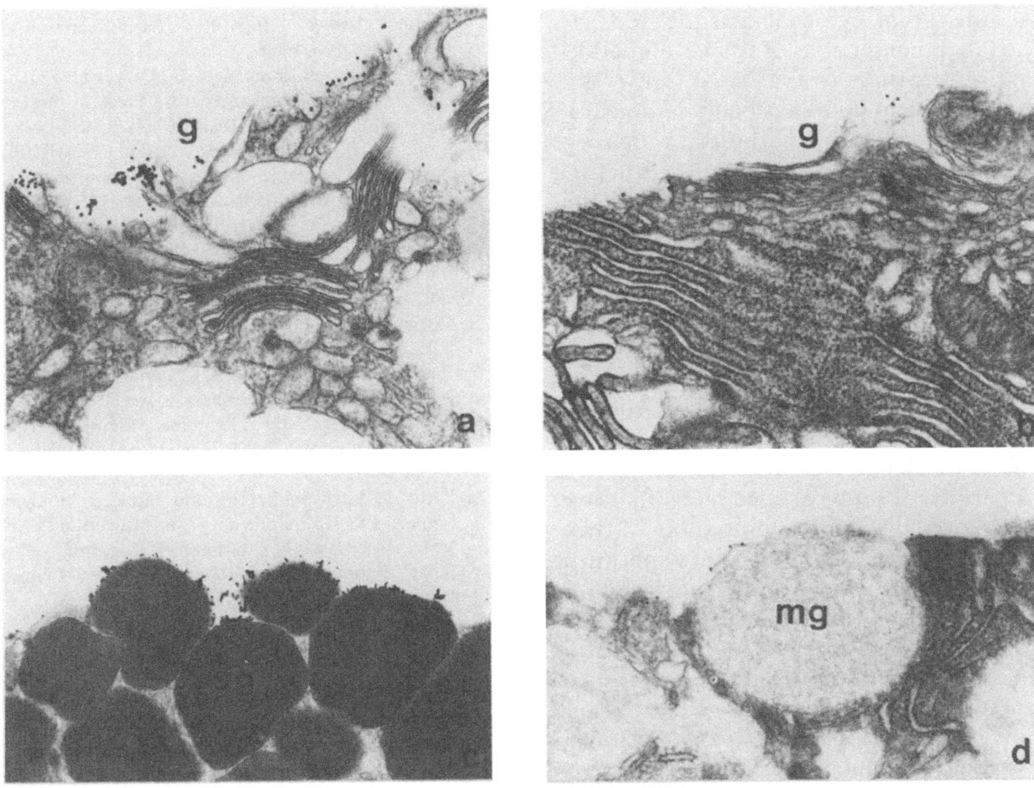

Fig. 9-3. Wheat-germ agglutinin fracture label of rat submaxillory gland cells; labeling is absent or very low over the Golgi fractured cisternae and dense only over the dilated rims (**a,b**). E faces of freeze-fractured membranes of secretory granules are strongly labeled in serous granules (**c**), but only sparsely labeled in mucous granules (**d**). g = Golgi complex; sg = serous granules; mg = mucous granules; **a**: x 45,000; **b**: x 45,000; **c**: x 21,000; **d**: x 36,000.

cells, such as lymphocytes, and in the pancreas [12]. Denser labeling can be seen only over the dilated rims of the complex (Figs. 9-3a and 9-3b). The low labeling over Golgi cisternae can be interpreted, as previously reported [13], as being due to the bridge position of the Golgi apparatus between the two endomembrane compartments described above. No differences in the amount of labeling are observed in typical secretory cells (pancreatic or salivary cells) and in nonsecretory cells (hepatocytes or leukocytes). Membranes of both serous (Fig. 9-3c) and mucous (Fig. 9-3d) secretory granules are labeled over the exoplasmic faces. The higher density of labeling over the serous granules, in comparison with the mucous granules, might be a consequence of a different membrane composition [21].

In fracture label, there is direct cytochemical access of the label to the receptor or antigen that one wishes to label. Steric hindrance may still occur, depending on the size of the labels and the local compactness of the region around the sites. In contrast, in postembedding labeling of polymer-embedded specimens, steric hindrances will occur and may lead to misinterpretation.

4. Concluding remarks

The random course of fracture across tissues and cells leads to direct exposure of extracellular matrices, plasma, and intracellular membranes, as well as cytoplasm and nucleoplasm. With fracture label, the cytochemical access is direct and only local steric considerations may limit the labeling of certain antigens and receptors. This contrasts with recent cytochemical methods developed to label thin sections of tissues embedded in polymerized matrices.

Another important point has been demonstrated during our fracture-label studies, which, we believe, is of particular relevance in ultrastructural pathology. Our studies demonstrated that chemically fixed tissues, after impregnation in a cryoprotectant, can be frozen and thawed without loss of ultrastructural detail or of antigenic activity. Our fracture-label studies show that freezing of chemically fixed, cryoprotected tissues should become the standard method for the storage and preservation of specimens in ultrastructural pathology. Indefinite preservation under optimal conditions renders possible the collection, over extended periods of time, of specimens that may only rarely be obtained. Frozen aliquots of these specimens can be thawed and used in experimental pathology studies when — and only when — representative series of specimens from different patients have been collected. This simultaneous cytochemical examination in a single experiment has a unique practical advantage: it solves one longstanding barrier in the comparative examination of biopsies obtained over months or even years. The reasons for the remarkable preservation of cell ultrastructure, which we observed in frozen/thawed specimens, are not yet clear but they appear to be related to the high compactness of the cytomatrices, as demonstrated by fracture-permeation studies [22, 23].

Acknowledgements

We thank Mr. Giuseppe Lucania for excellent technical assistance and Mr. Alessandro Valia for excellent photographic work. This work was partially supported by grant n. 86. 02397. 44 from Progetto Finalizzato "Oncologia," Consiglio Nazionale delle Ricerche, Italy, and grant n. 0190/87 from Nato.

References

1. Pinto da Silva P: Molecular cytochemistry of freeze-fractured cells: freeze-etching, fracture-label, fracture-permeation and label-fracture. In: Advances in Cell Biology, Vol 1. KR Miller (ed), Geenwich, CT: JAI Press, 157–190, 1986.
2. Pinto da Silva P, Parkinson C, Dwyer N: Fracture-label: cytochemistry of freeze-fracture faces in the erythrocyte membrane. *Proc Natl Acad Sci USA* 78: 343–347, 1981.
3. Pinto da Silva P, Kachar B, Torrisi MR, Brown C, Parkinson C: Freeze-fracture cytochemistry: replicas of critical point-dried cells and tissues after fracture-label. *Science* 213: 230–233, 1981.
4. Pinto da Silva P, Barbosa MLF, Aguas AP: A guide to fracture-label: cytochemical labeling of freeze-fracture cell. In: Advanced Techniques in Biological Microscopy, Vol 3 JK. Koehler (ed), Heidelberg: Spring Verlag, 201–227, 1986.
5. Pinto da Silva P: Topology, dynamics and structure of integral membrane proteins: a freeze-fracture view. In: Electron Microscopy of Proteins. Membranous Structures, Vol 6. JR Harris, RW Horne (eds), Academic Press, 2–38, 1987.
6. Pinto da Silva P: Seeing and thinking biological membranes. In: Biomembranes and Receptor Mechanisms, Vol 7. E Bertoli, D Chapman, A Cambria, U Scapagnini (eds), Padova: Fidia Research Series Liviana Press: Berlin; Springer Verlag, 11–33, 1987.
7. Pinto da Silva P, Torrisi MR: Freeze-fracture cytochemistry: partition of glycophorin in freeze-fractured human erythrocyte membranes. *J Cell Biol* 93: 463–469, 1982.
8. Torrisi MR, Pinto da Silva P: T-lymphocyte heterogeneity: wheat germ agglutinin labeling of transmembrane glycoproteins. *Proc Natl Acad USA* 79: 5671–5674, 1982.
9. Torrisi MR, Bonatti S: Immunocytochemical study of the partition and distribution of Sindbis virus glycoproteins in freeze-fractured membranes of infected baby hamster kidney cells. *J Cell Biol* 101: 1300–1306, 1985.
10. Torrisi MR, Lotti LV, Pavan A, Migliaccio G, Bonatti S: Free diffusion to and from the inner nuclear membrane of newly synthesized plasma membrane glycoproteins. *J Cell Biol* 104: 733–737, 1987.
11. Bhavanandan VP, Kalic AW: The interaction of wheat germ agglutinin with sialoglycoproteins. The role of sialic acid. *J Biol Chem* 254: 4000–4008, 1979.
12. Pinto da Silva P, Torrisi MR, Kachar B: Freeze-fracture cytochemistry: localization of wheat germ agglutinin and concanavalin A binding sites on freeze-fractured pancreatic cells. *J Cell Biol* 91: 361–372, 1981.
13. Torrisi MR, Pinto da Silva P: Compartmentalization of intracellular membrane glycocomponents is revealed by fracture-label. *J Cell Biol* 98: 29–34, 1984.
14. Carey DJ, Hirschberg CB: Topography of sialoglycoproteins and sialyltransferases in mouse and rat liver Golgi. *J Cell Biol* 256: 989–993, 1981.
15. Bretz H, Palade GE: Distribution of terminal glycosyltransferases in hepatic Golgi fractions. *J Cell Biol* 84: 87–101, 1980.
16. Carey DJ, Hirschberg CB: Kinetics of glycosylation and intracellular transport of sialoglycoproteins in mouse liver. *J Biol Chem* 225: 4348–4354, 1980.
17. Roth J: Subcellular organization of glycosylation in mammalian cells. *Biochim Biophys Acta* 906: 405–436, 1987.

18. Tolbert NE, Essner E: Microbodies: peroxisomes and glyoxysomes. *J Cell Biol* 91: 271s–283s, 1981.
19. Lodish HF, Breall WA, Schwartz AL, Strous GJ, Strous AM, Zilberstein A: Synthesis and assembly of membrane and organelle proteins. *Int Rev Cytol* 12: (Suppl) 248–307, 1981.
20. Bennet G, O'Shaughnessy D: The site of incorporation of sialic acid residues into glycoproteins and the subsequent fates of these molecules in various rat and mouse cell types as known by autoradiography after injection of (H^3)-N-acetyl mannosamine: observations in hepatocytes. *J Cell Biol* 88: 1–15, 1981.
21. Pinkstaff CA: The cytology of salivary glands. *Int Rev Cytol* 63: 141–261, 1980.
22. Barbosa MLF, Pinto da Silva P: Ultrastructural patterns of ferritin permeation into glutaraldehyde-fixed sarcomeres characterize stages of contraction in striated muscle. *J Electron Microsc Tech* 4: 329–342, 1986.
23. Barbosa MLF, Pinto da Silva P: Fracture-permeation: a technique to assess cytoplasmic compactness after glutaraldehyde fixation. *J Electron Microsc Tech* 4: 385–397, 1986.

Author's address:
Maria Rosaria Torrisi
Dipartimento di Medicina Sperimentale
Universita' degli Studi di Roma "La Sapienza"
Viale Regina Elena 324
00161 Roma
Italy

CHAPTER 10

The structural basis of mammalian liver function

GUIDO MACCHIARELLI, SAYOKO MAKABE, & PIETRO M. MOTTA

1. Introduction

The mammalian liver is a large exocrine gland connected to the small intestine through the choledochus and to the splanchnic blood circulation by the portal vein and hepatic artery. It is located in the right upper quadrant of the abdomen, just beneath the diaphragm and extends to the hypogastrium and to the left hypochondrium. Topographically, it is divided into two larger (right and left) and two smaller (caudate and quadrate) lobes. Functionally, it can be divided in subsegments (not corresponding to the above-mentioned lobes) according to the intraparenchymal ramification of the blood vessels and biliary tree.

The liver is essential to life, since it exerts many crucial functions. It plays a central role in most of the metabolic processes of the organism, especially in digestion, in the regulation of the splanchnic hemodynamics, in the processes of detoxification, and in immunological defense activities. It also operates as a sort of endocrine organ in that its cells continuously elaborate and secrete into the systemic circulation a variety of substances of general importance (e.g., glucose, proteins, etc.). Furthermore, it is the main hemopoietic organ during prenatal life.

All of these roles are extraordinarily carried out by this gland thanks to a structural organization that is highly complex and heterogeneous in its cellular components. In fact, five principal cell types can be recognized in the hepatic parenchyma: 1) the hepatocyte, 2) the sinusoidal endothelial cell, 3) the Kupffer cell, 4) the perisinusoidal cell, and 5) the pit cell. The latter four are commonly called the *sinusoidal cells*.

The hepatocytes are the most numerous type of liver cells and represent about 60%–70% of the organ's weight [1]. In most mammals, including humans, they are arranged in interconnected unicellular laminae delimiting irregularly shaped vascular lacunae. Here, a complementary series of small and tortuous blood vessels form an extraordinarily complex sinusoidal net of capillaries. The hepatic cells are polyhedral and present six or more facets that are morphofunctionally polarized. The facets in contact with adjacent hepatocytes (the biliary pole) form the bile canaliculi, which are the smallest of the biliary pathways; the other facets (the vascular pole) delimit the interstitial spaces and face the sinusoidal wall. The hepatocytes usually have a single, central nucleus (although two nuclei are not infrequent) and a large cytoplasm with many organelles and inclusions.

The sinusoidal endothelial cell is the principal component of the sinusoidal wall. It has a slightly bulging body containing an elongated, flat nucleus, and large, flat cellular extensions. These are usually in contact with the neighboring cells through junctions with occasionally overlapping borders. The liver endothelial cells are characterized by the absence of a continuous basal lamina and by a typical porosity (the endothelial fenestrations) that permits communication between the vascular and interstitial spaces [2].

The Kupffer cell is a resident macrophage and

Riva, A. and Motta, P.M., (eds.), Ultrastructure of the Extraparietal Glands of the Digestive Tract. © *1990 Kluwer Academic Publishers.*
ISBN 978-1-4612-8211-2.

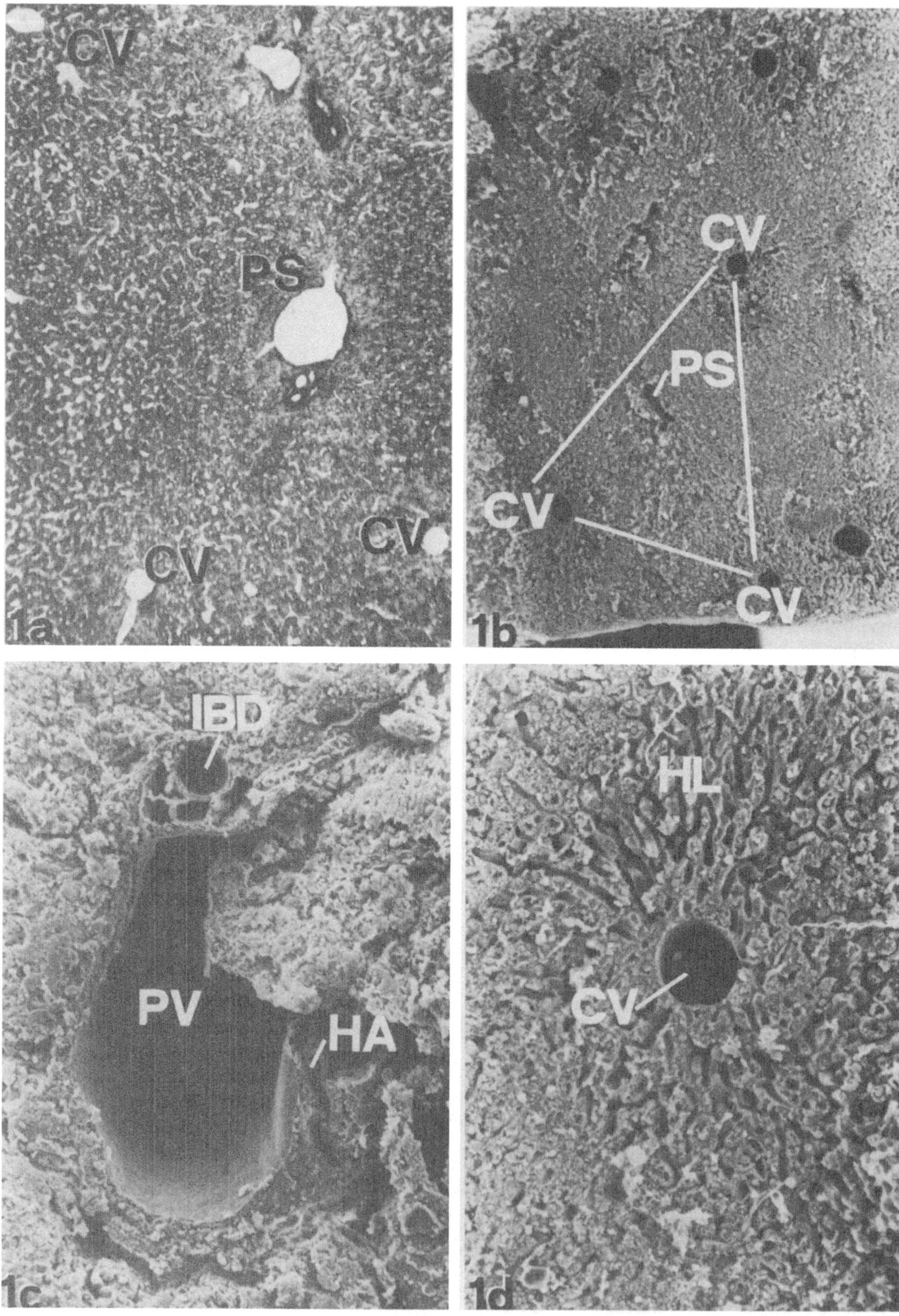
CV
PS
CV
CV
1a
CV
PS
CV
CV
1b
IBD
PV
HA
1c
HL
CV
1d

is characterized by a large body and many surface specializations and prolongations that are anchored to the endothelial wall of the sinusoid. This cell type is strategically embodied in the capillary wall and often is actually located within the sinusoidal lumen, where it can easily exert its immunological role.

The perisinusoidal cell, known as the *fat-storing cell*, *Ito cell*, or *stellate cell* [3], is found in the space of Disse, i.e., the space between the sinusoidal wall and the hepatocyte. It is characterized by an oval cell body containing a nucleus and is provided with many laminar and ramified extensions that surround the sinusoidal wall. The perisinusoidal cell contains typical cytoplasmic lipid inclusions, has vitamin-A storage capability, and, under certain conditions, plays a role in fibrogenesis.

The pit cell, usually located within the sinusoidal lumen in close relation to the sinusoidal endothelium and/or the Kupffer cells (pit cells have also been described in the space of Disse and among hepatocytes) [4], represents the latest discovery among the liver cells [5]. It is characterized by the polar organization of its cytoplasm and by the presence of typical organelles such as electron-dense granules, rod-cored vesicles, and multivesicular bodies. It probably can be considered as a large granular lymphocyte and seems to play a role in the control of neoplastic and other liver diseases [6].

Thus, although the cells composing the liver present different morphological features, they are strategically located within its parenchyma to form a unique structure in which they may exert integrated roles [7–9]. During the last decade, the correlated use of light, transmission, and scanning electron microscopy has offered a better morphological substantiation of such a complementarity of liver cell roles. Further, in allowing a very detailed three-dimensional view of the microtopography of the various liver-cell components, these techniques contributed much towards opening new horizons in both basic and clinical hepatology [10, 11].

In the present chapter, the main fine structural aspects of the liver cells will be described in order to give a comprehensive ultrastructural picture of the morphological basis of liver function.

2. The hepatic microcirculation

2.1. General aspects

The liver presents a particular type of vascularization that differs from that commonly observed in other organs. In fact, it is supplied by two major vessels: the portal vein, which delivers blood that is rich in nutrients absorbed from the digestive system, and the hepatic artery, which carries the oxygen supply.

These two vessels enter the liver at the level of the hepatic hilus and branch within the parenchyma, along with the extensions of the connective tissue capsule (called the *Capsule of Glisson*, which invests the organ almost completely), following retrogradely the intraparenchymal ramifications of the biliary tree. Therefore, the liver assumes a lobular structure (Figs. 10-1a and 10-1b) that is clearly defined in tissue sections by the portobiliary spaces (Fig. 10-1c), which contain a branch of the portal vein, a branch of the hepatic artery, and a bile duct (the so-called portal triad). In addition, these spaces contain lymphatic vessels and nerve fibers. The blood that reaches the sinusoids through the vessels of the portal space is drained by the centrolobular veins, (Fig. 10-1d), which, in tissue sections, can easily be seen at the center of radiating hepatocytic laminae. The centrolobular veins are drained by the sublobular veins and then by larger vessels: the system of hepatic veins, which empty into the inferior vena cava.

2.2. The histofunctional unit of the liver

Such a unique vascular system allowed researchers to easily recognize a tissue unit, the *lobule*, (Figs. 10-1a and 10-1b), which can be considered

←

Fig. 10-1. Liver parenchyma. **a**: light microscopic view of adult monkey liver (H/E, x 150). **b**: SEM view of adult rodent liver, lines delimit a portal lobule. (x 100). **c**: portal area (SEM, x 500). **d**: central area of a classic lobule (SEM x 500). CV = centrolobular vein; PV = portal vein; HA = hepatic artery; IBD = interlobular bile duct; HL hepatic laminae; PS = portal space.

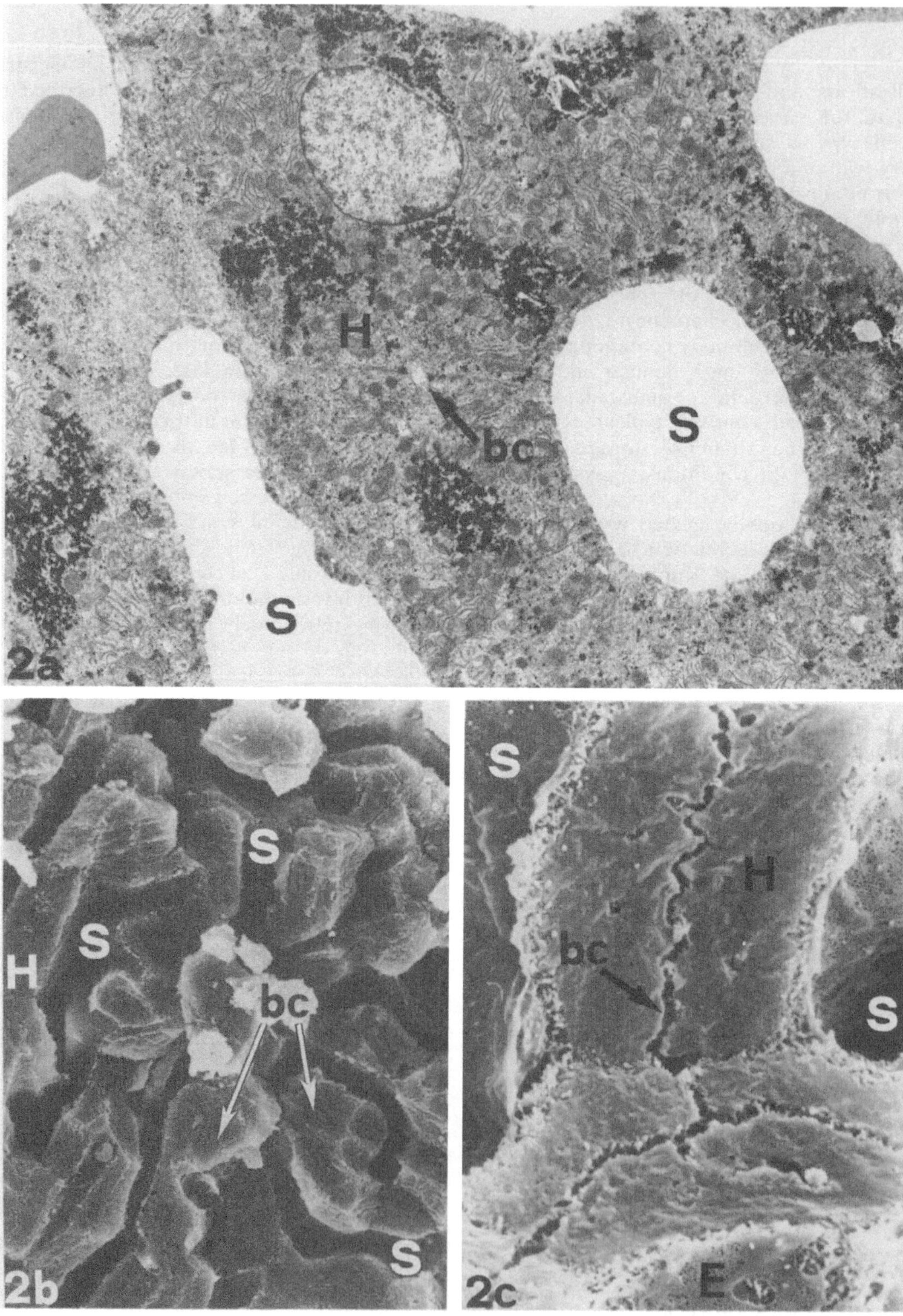
H
bc
S
S
S
2a
S
S
H
bc
S
2b
S
H
bc
S
E
2c

a sort of miniature liver, having all the anatomical and functional characteristics of the entire organ. The first description of the hepatic lobule was given by Kiernan [12], who defined the so-called classic lobule. In histological sections, as well as in scanning electron microscopic (SEM) views, this is a polygonal structure characterized by a small area of liver parenchyma whose center contains a centrolobular vein (Fig. 10-1d) from which the laminae of hepatocytes radiate. The portal spaces represent the angles of this structure. The classic lobule represents a functional point of view of the liver in that the *vascular* roles of the organ are emphasized, since the course of the intrahepatic vessels was considered to define the smallest tissue unit. This is actually the *endocrine* function of the liver. However, as evidenced earlier, the liver is also, and mainly, an exocrine organ, since it secretes the bile through a canalicular and ductular system outside the parenchyma. Therefore, successively Mall [13] tried to identify a histological unit by which the exocrine function of the liver could be better studied. The *lobule of Mall*, or portal lobule, is therefore based in relation to the bile duct present within the portal space. This unit appears as a triangular area, and the centrolobular veins, best called *terminal hepatic veins* in this case, are placed at each angle (Figs. 10-1b and 10-1c). In this morphological system, more emphasis has been given to the *exocrine* role of the liver in that the bile flow converges from the hepatic laminae towards the centrally located bile duct.

More recently, Rappaport et al. [14], in a series of physiological studies, were able to recognize a new liver tissue unit (the *acinus*) that could be better adapted to include all the morphofunctional features of this gland. The *acinus hepaticus* (so-called in honor of M. Malpighi, who in 1666 was the first to describe by microscopy liver structures that he named *acini*) is therefore the first liver unit in which all known hepatic functions are finally acknowledged. It is roughly defined as a mass of liver parenchyma lying between two terminal hepatic veins. In tissue sections, it can be grossly recognized as a rombhoid area encompassing segments of two adjacent classic lobules. The bases of these two triangles are united at the center to form the *axis* of the acinus. The axis of the liver acinus is therefore represented by a small extension of the main portal space, containing a terminal hepatic artery, a terminal portal vein, a bile ductule, lymphatic vessels, and nerve fibers. The peripheral limits of the acinus are delineated by imaginary lines drawn between neighboring terminal hepatic veins. Furthermore, each acinus can be divided in three zones (zone 1, near the axis; zone 2, intermediate; and zone 3, near the terminal hepatic vein), each of which has a different degree of oxygenation and metabolic activity [1]. Thus, from a functional point of view, the acinus hepaticus is better representative of various liver roles, since it stresses the relations between blood supply and cell metabolism. Further, it also helps in interpreting the development of chronic liver disorders (e.g., cirrhosis) [15].

However, due to the many race- and species-dependent histological differences occurring in the liver, both the classic and portal lobule should always be taken into consideration when analyzing tissue preparations in normal and pathological conditions [1, 10, 15]. The classic lobule may serve as a guide for tissue diagnosis, especially when considering pathological processes characterized by centrolobular necrosis. On the other hand, the classic lobule is the most easily recognizable structure in many species (such as the pig or camel) in which the intraparenchymal ramifications of Glisson's capsule are well represented. The portal lobule is comparable to the lobular units of other exocrine glands. Finally, it should also be pointed out that the physiological changes of the pressure gradient along the intraparenchymal vascular ramification confer a certain plasticity and flexibility to the hepatic architecture, which turns in alterations of the spatial location of the hepatic laminae. Consequently, the hepatic laminae may alternatively seem to converge towards the portal triad (seemingly delineating the portal lobule or the acinus) or radiate from the central vein (classic lobule) [10].

←

Fig. 10-2. Hepatic laminae. **a**: TEM of liver parenchyma (rodent x 3,000). **b**: SEM of hepatic laminae (rodent x 1500). **c**: SEM of mammalian hepatocytes: note the close topographic relationship with the sinusoids (rodent, x 7500). H = hepatocyte; S = sinusoid; bc = bile canaliculi; E = endothelium.

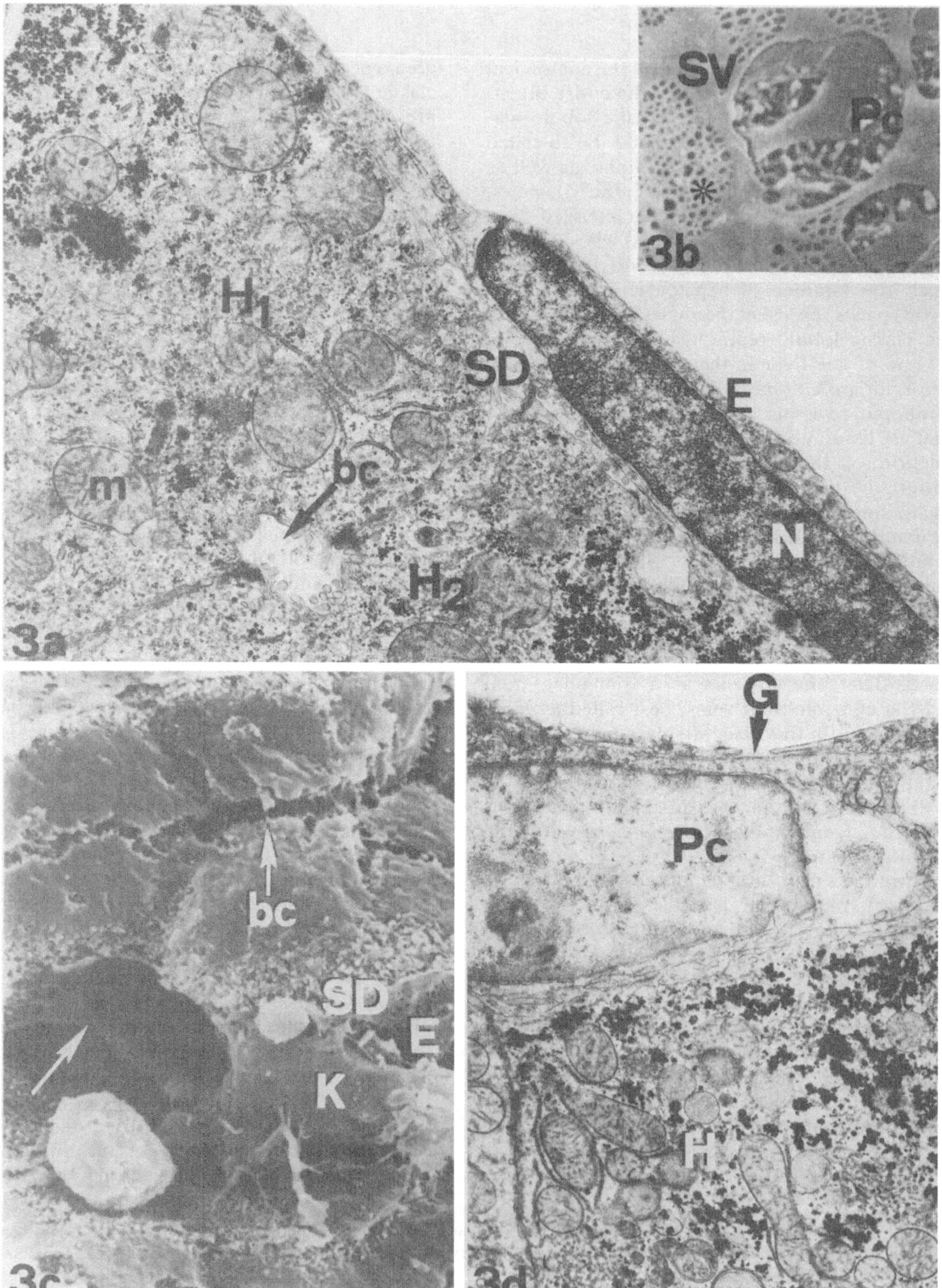
3a
H1
SD
E
m
bc
N
H2
Sv
Pc
3b
bc
SD
E
K
3c
G
Pc
H
3d

2.3. *The liver sinusoidal environment as the fulcrum of hepatic function*

Since the hepatic architecture is strictly related to its blood supply, full comprehension of liver function will depend greatly upon precise knowledge of its microvascular architecture. In fact, the hepatic capillary and its environment (e.g., the sinusoidal cells, the space of Disse, and the "vascular" facets of the hepatocyte) is the area in which, whether directly or indirectly, the liver's main roles are regulated.

The hepatic microcirculation is characterized by a capillary net, known as the *sinusoids*, (Fig. 10-2a) that represents an especially suitable structure for the metabolic and secretory activities carried out by the hepatocytes. The cells populating the hepatic capillaries have morphological features that permit them to cooperate in the control of blood-hepatocyte exchanges [1, 16]. This is facilitated by a special arrangement of liver capillaries that can be well visualized by SEM (Fig. 10-2b). In fact, the extraordinary labyrinthine distribution of the sinusoids among the hepatic laminae is evident in fractured samples [10, 11, 16, 17] as well as in vascular casts [18]. In these preparations, the sinusoids appear to form a three-dimensional network of vascular and perivascular spaces, creating an extraordinarily wide surface area for blood-hepatocyte exchanges [19, 20]. The spatial relationship of the sinusoids with the hepatic laminae reflects the functionally polyhedral nature of the liver. The sinusoids enter the parenchyma in such a way that each hepatocyte is reached by a capillary (Fig. 10-2c). This allows the parenchymal cells to "communicate" with the vascular compartment and, at the same time, with the neighboring hepatocytes. Thus, the hepatocyte may secrete its products both in the biliary tree and in the blood capillaries (amphicrine capability of the liver). The sinusoidal cells play a key role in the regulation of these activities. In fact, all these different cells, thanks to their close spatial relationships, together form a sort of functional system operating in the control of the various activities carried out at the level of the sinusoidal spaces [21–23].

2.4. *The sinusoidal wall, the sinusoidal cells, and the control of blood-hepatocyte exchanges*

The hepatic capillary environment is therefore a complex structure characterized by the variety of its cellular population and its roles. This area is delimited by the sinusoidal wall, which separates the interstitial spaces from the bloodstream. The sinusoidal wall can be regarded as a three-layered formation in which the endothelial lining represents the inner part, the perisinusoidal cells embracing the sinusoids constitute the medial part, and the argyrophilic connective fiber net is the outer part. Attached to this structure are the Kupffer cells, which are often situated within the sinusoid, strategically contacting the endothelium with their long and thin cytoplasmic prolongations or even with their cell bodies [24, 25].

2.4.1. *The sinusoidal endothelium*

The sinusoidal enthodelium (Figs. 10-3a and 10-3b) is the most studied, and probably more important, section of this structure. In fact, it not only constitutes the physical delimitation of the capillary, but also works as a regulatory barrier that selectively filters any substance directed from the blood to the hepatocyte and vice versa [21]. The endothelial *sieve plates* [2], a system of clustered small perforations (Fig. 10-3b) situated on the extensive, flat endothelial-cell prolongations, carry out this function. The size of the endothelial fenestrae has been an unresolved problem for a long time. Today, it is believed that the majority of endothelial fenestrae are homogeneous in size under physiological conditions, although they may

←

Fig. 10-3. Liver sinusoidal wall. **a**: TEM of endothelial cell (E) with a flat nucleus (N). Note the space of Disse (SD) and the relation with two hepatocytes (H1, H2) rich in mitochondria (m); bc = bile canaliculus (rodent, x 10,000). **b**: SEM of sinusoidal sieve plates (SV) with small fenestrations (*). Note, through the artifactual break in the endothelium, flat processes of perisinusoidal cells (Pc) located within the space of Disse (rodent x 32,000). **c**: SEM of a Kupffer cell (K) within a liver sinusoid. Note the relation between the Kupffer cell prolongation (arrow) and the endothelium (E). SD = space of Disse; bc = bile canaliculus (rodent, x 12,000). **d**: TEM of a perisinuoidal cell (Pc) within the space of Disse. Note a large endothelial gap (G) similar to those of Fig. 10-3b that exposed a perisinusoidal cell; H = hepatocyte (rodent, x 15,000).

vary in response to various pathological stimuli, or even in certain paraphysiological conditions involving cell turnover or in relation to other sinusiodal cells [11, 17]. However, much evidence obtained in normal, experimental, and pathological situations seems to confirm the dynamic nature of all endothelial cell types [7, 20] and, in this regard, the liver sinusoidal endothelium must also be considered [7]. Sinusoidal endothelial cells also show endocytotic activity, characterized by the high number of pinocytotic vesicles usually found at the level of the plasmalemma. In normal conditions, they do not show phagocytosis, but they frequently possess lysosomes and may take up molecules such as contrast materials, immune complexes, ferritin, and many others [26]. Such characteristics suggest that the sinusoidal endothelium may play an active role in liver clearance activity [27].

2.4.2. The Kupffer cell

The Kupffer cell (Fig. 10-3c), with its large and irregular body, lies on the sinusoidal endothelium. This cell seems to contribute towards the formation of the sinusoidal wall, mostly by contacting and/or penetrating the endothelium with lamellipodia and filopodia [10, 19]. The Kupffer cell shows the morphological features of the macrophage and is quite different from the endothelial cell. Kupffer and endothelial cells are two ontogenetically distinct cell populations. In fact, Kupffer cells likely originate from monocytes or from immature fetal macrophages [1, 28–30]. They show peroxidase activity at the level of the nuclear envelope, the endoplasmic reticulum, and the annulate lamellae, [31]. These characteristics differentiate the Kupffer cells from endothelial cells, which are negative for peroxidase [1, 32]. Furthermore, Kupffer cells show surface specializations such as blebs, microvilli, ruffles, and other cytoplasmic projections of various size and shape that are morphological signs of cell activation. In particular, they often possess *wormlike structures* [33], which likely represent a membrane reservoir and may contribute toward enlarging the cell surface [34]. The cytoplasm of the Kupffer cell is mainly characterized by a large number of lysosomes and phagosomes varying in size, shape, and density, sometimes related to the Golgi apparatus (GERL) [35], and by a high endocytotic activity, which can be tested by injection of India ink or colloidal gold. These aspects, reported in a large number of studies under both normal and experimental conditions, demonstrate that Kupffer cells — being the largest macrophage population in the body [36] — surely play a fundamental role in protection of the organism from exogenous agents, including bacteria, viruses, and parasites, and from host cells. They also function in the clearance from the blood of most endogenous substances, in the degradation of hemoglobin, and in the storage and release of iron, as shown by the frequent presence in the cytoplasm of phagocytosed erythrocytes and iron-containing pigments [1]. Furthermore, the Kupffer cell's cytoplasmic prolongations seem to form a net within the sinusoidal lumen that creates a kind of microanatomical barrier that allows this cell to contact blood cells or other materials that may elicit immunological responses [7]. The Kupffer cell is often seen to contact other sinusoidal cells, such as the pit cells within the sinusoid [36], or even the perisinusoidal cells within the space of Disse, with its cell projections, which are sometimes seen to pass through the endothelial fenestrae [37]. The frequent contacts occurring between the Kupffer cells and other cells in these areas strongly suggest that functional interactions exist among sinusoidal cells. Finally, it has also been suggested that the intravascular arrangement of the Kupffer-cell prolongations is dynamic, capable of acting as an intrasinusoidal "scaffold" deputed to regulate the sinusoidal caliber and, thus, the intracapillary blood flow [11].

2.4.3. Liver sinusoidal blood flow

In terms of regulation of liver sinusoidal blood flow, the problem is complex and unresolved. In fact, to date it has not yet been established which is (are) the structure(s) controlling hepatic microcirculation, and the site of resistance to portal blood flow in the liver is still a matter of debate. Recent evidence seems to confirm that the major resistance to portal flow is postsinusoidal in cats and dogs, but not in rats, where the sinusoidal pressure seems to be inferior to the portal pressure, which would seem to indicate that the area of resistance is presinusoidal [38]. The smooth-muscle cell components of the vascular wall of both presinusoidal and postsinusoidal venules are surely involved in these mechanisms; however, no morphological evidence of a precise structure controlling portal blood flow within the liver has been reported. At the sinusoidal level and from a morphological point of view, only the perisinusoidal cells (Figs. 10-3b and 10-3d) may be actively involved in the regulation of the sinusoidal caliber, thanks to their topographic relationship with the sinusoidal wall. The recent introduction of chemical digestion techniques, which facilitate exposure of the outer surface of the vascular wall, permits a better view of the three-dimensional architecture of the perisinusoidal cells [37]. These cells, recognized in light microscopy as stellate elements by the Golgi method [24], are easily identified in SEM after NaOH digestion. They clearly appear to embrace the endothelial wall with primary and secondary flat cell projections, as described for the perivascular cells (or pericytes) of other vascular districts. Further, the perisinusoidal cells contain cytoplasmic filaments that are similar to those observed in smooth-muscle cells and are sometimes seen to form junctional complexes with neighboring cells [2, 3]. These features suggest that liver perisinusoidal cells may form the *contractile* portion of the sinusoids. [24].

2.4.4. *The perisinusoidal cell*

The role of the perisinusoidal cell is not only related to the control of capillary hemodynamics. Indeed, it is primarily dependent upon the cell's lipid cytoplasmic inclusion and vitamin A content. By TEM, the perisinusoidal cell usually shows a large number of lipid droplets of varying size. For this reason, it has been called the *fat-storing cell* [3]. The amount and histochemical characteristics of these inclusions may vary greatly among different mammals. In humans, two kinds of perisinusoidal cells have been described in relation to their lipid-droplet content. One type contains lipid droplets that show an intense fluorescence for vitamin A. The second type, called the *empty Ito cell*, identified by its perisinusoidal position, is devoid of lipid droplets [39]. From a functional point of view, these "empty" cells can probably be considered resting cells. All liver cells show cytoplasmic lipids. In mammals, in fact, the metabolism of dietary cholesterol and of many other lipids occurs mainly in the liver. Furthermore, the liver stores most of the lipid-soluble vitamin A. Hepatocytes are the major site in the liver of dietary lipid metabolism, as shown by the high content of triglycerides and cholesterol (esters), whereas the perisinusoidal cells are involved mainly in the storage of retinyl esters [40]. It has also been demonstrated that hepatocytes provide for the uptake of vitamin A, while the perisinusoidal cells seem to be exclusively involved in the storage of retinoids [41]. Perisinusoidal cells seem to be involved in the metabolism of collagen fibers. This is shown by many experimental and pathological conditions in which perisinusoidal cells show the capability to synthesize and degrade collagens. In vitro, this cell is able to produce collagen of type I, III, and IV [25]. Further, in electron micrographs, perisinusoidal cells often appear to be surrounded by collagen fibers [10].

As clearly shown, perisinusoidal cells also have a central role in the regulation of the sinusoidal environment, since they cooperate in liver metabolism, in the mechanical support of the sinusoidal wall, and in the control of sinusoidal blood flow. The anatomical contact of perisinusoidal cells by desmosomes and/or other junctions with the hepatocyte, the Kupffer cell, and the endothelium demonstrate that this cell is a structural and functional complement of the entire sinusoidal wall [7, 11, 25, 37].

2.4.5. *The pit cell*

As far as the pit cell is concerned, recent data show that this element may be classified as the tissue equivalent of large granular lymphocytes (LGL) [40]. LGL are a class of lymphocytes capable of cytolytic activity and include K cells, lymphokinin-activated killer cells, and natural killer cells. Morphological aspects, such as the polarization of cytoplasmic organelles and the presence of azurophilic (LM) and electron-dense granules, as well as the lytic activity against YAC-1 lymphoma cells, suggest that pit cells belong to the class of natural killer cells [6, 40, 42]. However, little is known about the significance of these elements in the immune response. It has been suggested that they are involved in the control of spontaneous neoplasms and thus they may play a role in the process of so-called *immune surveillance* [43]. Pit cells may also be involved in some immunological reactions during hepatitis [44]. The frequent contact of these cells with endothelium and Kupffer cells suggests some sort of functional interaction among them. In addition, the presence of the pit cells in the space of Disse and in the interhepatocytic recesses suggests that these cells are able to migrate from the blood towards the interstitial spaces [44].

2.4.6. The sinusoidal wall and the Disse space

The sinusoidal wall and the pool of sinusoidal cells is the anatomical structure that regulates blood-hepatocyte exchanges. This control is actually a mulifactorial mechanism in which structural relationships between liver cells and the interstitial space play an important part. Liver parenchymal cells are characterized by ultrastructurally different sides that are commonly named *biliary* and *vascular facets*. Vascular facets are always in contact with the sinusoidal wall and appear to be completely covered by a dense microvillous carpet that also extends up to the peripheral areas of biliary facets. Thus, microvilli almost totally line the parenchymal pole of the interstitial spaces. Such spaces are represented by the space of Disse and its extensions in the *interhepatocytic recesses* [19, 45]. Blood-hepatocyte exchanges are mediated by the perisinusoidal and intercellular spaces or the *space of Disse* (Fig. 10-3a). The structures delimiting these spaces are extremely specialized in order to regulate such exchanges [10, 11,19]. Actually, the space of Disse seems to be morphologically polarized. In fact, it is delimited by two walls that are structurally different. On one side, there is the sinusoidal wall; on the other side, there is the parenchymal pole, represented by the vascular facets of the hepatocytes and by the peripheral areas of the biliary facets (the latter delimit the interhepatocytic recesses) [10, 11, 19, 45]. This side is totally covered by microvilli that highly increase the cell surface [46, 47] and thus significantly regulate the secretive/absorptive activities of the hepatocytes [10, 19]. As studied by means of SEM, the space of Disse and the interhepatocytic recesses seem to form a three-dimensional network of communicating lacunae, that is likely to be connected with the space of Mall [10]. Therefore, liver perisinusoidal spaces may represent a lacunar "lymphatic" system [19] (which takes the place of typical lymphatic capillaries, which are not observed in proximity to liver cells and sinusoids) [1]. In such a system, the blood filtrate flows towards the portal space, where it is finally drained by the lymphatic portal vessels. The spaces of Disse, as well as all the interstitial spaces, are not empty spaces, but are occupied by a series of molecules, such as collagen, fibronectin, laminin, elastin, and proteoglycans, which represent the main components of the extracellular matrix. It has been suggested that this structure has a precise significance in several cell processes, including cell exchanges. In fact, it not only serves as a cellular support, but also in determining the phenotypic expression of cells. In addition, it cannot be excluded that the extracellular matrix may cooperate in determining cell shape, allowing cell adhesion and migration, regulating cell differentiation, and, finally, in controlling the quality and cellular traffic within the extracellular spaces [48–52].

3. The metabolic apparatus of the liver

3.1. General aspects

We have seen in the above paragraphs how the liver structure is characterized by the heterogeneity of its cellular components. This is strongly related to the wide spectrum of roles that the liver must perform. Such morphofunctional heterogeneity can also be recognized among hepatocytes. In fact, due to the particular blood supply of this organ, the liver is different from other exocrine glands (e.g., the pancreas) in which the exocrine units are entirely bathed by a qualitatively homogeneous capillary blood (*in parallel* or concurrent blood supply), and hence all acinar cells show identical functional behavior. On the contrary, in the liver, according to Rappaport's concept of the tissue unit, the blood flows from one end (portal space, zone 1) of the hepatic acinus to the other (central vein, zone 3), bathing the hepatocytes one after another (*in serial* or sequential blood supply). This causes the hepatocytes of zone 1 and

2 to be metabolically different from those of zone 3 and, therefore, may present different arrangements and amounts of cytoplasmic organelles.

In a short review like this, it is not possible to adequately report all data concerning hepatocyte zonal variations, and so we refer the reader to more comprehensive literature [15, 53–54]. Thus, in the present paragraph, an "average" description of the morphofunctional features of the cytoplasmic organelles involved in hepatocyte metabolic activity will be given.

The liver plays many specific roles in different fundamental metabolic processes. Its metabolic activity on proteins is essential to life. In fact, the liver is responsible for the deamination of amino acids, the production of urea, and the synthesis of about 90% of plasma proteins (up to 50 grams per day), as well as of many amino acids (the so-called non-essential amino acids). It controls the glucose blood concentration by actively entering in the metabolism of carbohydrates with the following: 1) glycogen storage, 2) glyconeogenesis, 3) formation of glucose from galactose and fructose and 4) formation of other components deriving from the intermediate products of carbohydrate metabolism. It metabolizes almost 50% of the organism's lipids operating in the beta oxidation of fatty acids; in the formation of acetoacetic acid, glycoproteins, and cholesterol; and in controlling the conversion in lipids of a large amount of carbohydrates and proteins. The liver also produces most of the substances necessary for the clotting processes and is one of the most important sites for the storage of vitamins (A, B_{12}, D) and iron (ferritin).

It therefore appears evident that the hepatocyte must have a very rich cell machinery, representative of all these roles.

3.2. *Cell organelles involved in liver metabolism*

3.2.1. *Nucleus*

The hepatocyte nucleus, which coordinates all cell activities, is usually located at the center of the cell. Binuclear cells are frequent (about 25% of cell population) [53, 55], but multinucleated hepatocytes are rare and result from cell fusion [54] (Fig. 10-4a). Nuclei are large and round, containing scattered chromatin clumps characterized by perichromatin granules and one or more nucleoli (Fig. 10-4a) consisting of fine fibrils and granules [1]. They are different from the nuclei of other liver cells because they are larger and usually are less intensely stained. Adult mammalian hepatocytes are often polyploid. This seems to be correlated with aging [56]. In normal conditions, mitotic activity is very low [53], but rapidly increases under appropriate stimuli [54, 56]. Nuclear inclusions are rare and are mainly related to pathological conditions. Monoparticulate glycogen may be observed in hepatocyte nuclei from children, elderly persons, and diabetic patients [54, 57].

3.2.2. *Mitochondria*

Mitochondria (Fig. 10-4b) are very numerous (about 18% of the cell volume) [15]. They take part in the oxidative phosphorylation and oxidation of fatty acids, and are directly involved in all metabolic processes of the hepatocyte. They are ubiquitously dispersed within the cells, but are more concentrated near major sites of ATP utilization [58]. They are often associated with the rough endoplasmic reticulum [59] (Fig. 10-4b). Such a relationship is probably important during formation of cytoplasmic membranes (smooth endoplasmic reticulum) and cytochromes [53, 59]. Hepatocyte mitochondria may be round or elongated. These are bounded by the outer and inner membranes (the former functions as a corset to hold the inner membrane, whose surface is highly increased by typical folding cristae) [58]. Mitochondria show a relatively low-density matrix [1] in which lamellar or tubular cristae, and a variable amount of small dense granules, can be observed.

3.2.3. *Lysosomal apparatus*

The lysosomal apparatus (Figs. 10-4c and 10-4d) is a subcellular compartment biochemically characterized by hydrolase activity that is morphologically represented by a heterogeneous population of spherical organelles, that includes single-membrane-bounded dense bodies, autophagic vacuoles, multivesicular bodies, and coated vesicles. In addition, as part of this apparatus, it can be considered the GERL: a structure

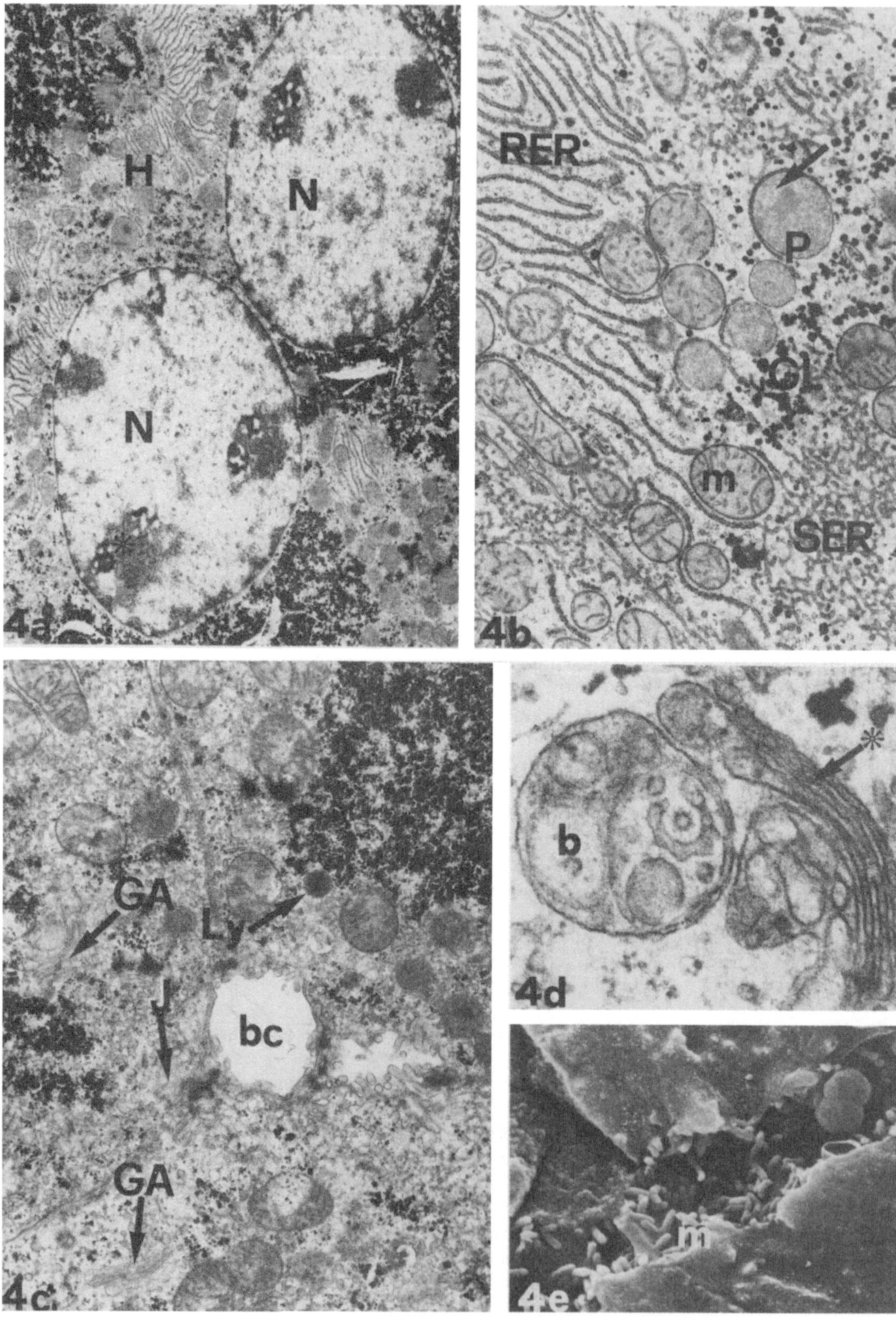

H
N
N
4a
RER
P
CL
m
SER
4b
GA
Ly
J
bc
GA
4c
b
*
4d
m
4e

proximal to the Golgi (G) complex (but not part of it). It consists of smooth-surfaced membranes (likely a specialized area of smooth endoplasmic reticulum) (ER) with the same hydrolase activity as the lysosomes (L) (but without the typical morphology of spherical organelles) that probably play a major role in the formation of lysosomes and, particulary in hepatocytes, in lipoprotein metabolism [60]. Liver lysosomes are involved in numerous metabolic activities and pathological events (mainly digestion of non-self elements). Several classes of lysosomes can be recognized in the liver [57]: 1) primary, the smaller in size and in resting phase; 2) secondary, functionally activated; 3) autophagic vacuoles, containing parts of degradating cyoplasmic organelles; and 4) residual bodies, larger than primary and secondary lysosomes, more frequent in elderly persons, containing residues of nondigested material or pigments such as lipofuscins (which are considered to be undigestible permanent residues). All of these lysosomal structures are often found near the plasma membrane, proximal to the bile canaliculus, forming the so-called peribiliary dense bodies of early histological descriptions [53].

3.2.4. Rough endoplasmic reticulum, smooth endoplasmic reticulum, and Golgi apparatus

The hepatocytic cytoplasmic membranes (Figs. 10-4b and 10-4c) which take part in liver metabolism are represented by the rough endoplasmic reticulum (RER), smooth endoplasmic reticulum (SER), and Golgi apparatus. Their main roles consist of synthesis and conjugation of proteins (albumin, fibrinogen, etc.); metabolism of lipids (formation of cholesterol and bile salts, esterification of FFA to tryglicerides); metabolism of steroids; detoxification and metabolism of drugs; and breakdown of glycogen. Due to the complexity and variety of liver roles, RER and SER are both abundantly represented in mammalian hepatocytes. RER is arranged in aggregates of flat cisternae that are widely distributed in the whole cytoplasm. It is easily recognizable by the high number of attached membrane-bounded ribosomes. Free ribosomes and polyribosomes are also present within the hepatocyte cytoplasm. There is less SER and it has a more complex arrangement. Its tubules, irregular in size, are very tortuous and are often closely connected with the RER and Golgi membranes, as well as with glycogen inclusions [53]. Protein synthesis is primarily performed at the level of the RER (ribosomal component). RER products are addressed towards the SER and then are transferred within the Golgi complex and are finally released at the level of the plasma membrane. Two different pathways of protein release have been suggested. Very-low-density lipoproteins (VLDL), for example, are formed by proteins synthesized at the level of the RER and by SER-derived lipids. After synthesis, these particles are transferred as smooth vesicles to the plasma membrane for secretion (via exocytosis) or are sequestered in Golgi vesicles and then secreted as a packet. SER is mainly involved in the metabolism of steroids and the production of cholesterol, as well as in detoxification processes. The Golgi complex seems to be involved in the transport and traffic control of most cellular products. In fact, it is the final common pathway of most cell activities and plays a role in sorting and recycling most of the substances derived from cellular synthesis or endocytosis. The hepatocyte Golgi complex is characteristically heterogeneous; nevertheless, it can be morphologically identified as a stack of four to six parallel cisternae that often show a convex part facing the nucleus and the endoplasmic reticulum (*cis*-Golgi), and a concave part (*trans*-Golgi) where vesicles and vacuoles (secretory granules) are present. *Cis*- and *trans*-Golgi are connected by means of the medial Golgi , which is thought to represent an intermediate station between the endoplasmic reticulum and the Golgi

←

Fig. 10-4. Ultrastructure of mammalian hepatocyte. **a**: a binucleated liver cell (H). N = nucleus. (TEM, rodent, x 6,000). **b**: TEM of hepatocyte cytoplasm. m = mitochondria; RER = rough endoplasmic reticulum; SER = smooth endoplasmic reticulum; P = peroxisome containing a crystalline inclusion (arrow); GL = glycogen (rodent, x 22,000). **c**: hepatocytes converging to form a bile canaliculus (bc) closed by a junctional complex (J). Note the Golgi apparatus (GA) in proximity to a bile canaliculus. Ly = lysosome (rodent, x 10,000). **d**: high magnification TEM of Golgi apparatus of mammalian liver. Note the membranes (*) and the vesicular body (b). **e**: high magnification SEM (x 30,000) of a fractured bile canaliculus. Note the microvilli (m).

products such as secretory granules or secondary lysosomes (GERL) [61].

3.2.5. *Cytoplasmic inclusions*

The hepatocyte is also rich in cytoplasmic inclusions, which are functionally related to liver metabolism, being the morphological aspect of the liver's storage capability. Glycogen granules (Fig. 10-4b) are the most representative cytoplasmic inclusions in normal mammalian liver. In EM, when stained by lead salts, they occur either in a monoparticulate form (beta glycogen) or, as a rule, arranged in clusters to form the so-called rosettes (alpha glycogen). Glycogen granules are dispersed in the cytoplasm but are often associated with the SER. Glycogen may greatly change in quantity during certain pathological conditions, or, as in storage diseases, it may be found within the organelles (such as lysosomes in Pompe's disease) with the formation of typical glycogenosomes [62]. Lipid inclusions appear as empty vacuoles or osmiophilic droplets, not bounded by membranes and greatly varying in size and quantity. In addition, related to the apoferritin-ferritin system (the so-called hepatic iron buffer), a variable amount of iron-containing granules can also be detected. Although it has been demonstrated that liver iron metabolism occurs in hepatocytes, nevertheless, the pathway of iron transport from the blood to the hepatocytes has not been clarified. Hepatocytes, in fact, do not show receptors for transferrin (the glycoprotein that transports iron). This glycoprotein is only present in the endothelium [63, 64], suggesting that iron transport is essentially transendothelial (transcytosis) [63].

3.2.6. *Cytoskeleton and cytomatrix*

The cytoplasm is a dynamic structure capable of many intrinsic movements. Such movements, as well as the internal cell organization and cell shape, are directly related to the cytoskeleton. This structure depends upon the three-dimensional arrangement of its three, principal, morphofunctionally different components: the microfilaments, the intermediate filaments, and the microtubules. These three filament types are regularly distributed in the cytoplasm and characterize the cytomatrix [65], which, together with other finer filaments (microtrabeculae), is thought to give the gel consistency to cell cytoplasm. Microfilaments are formed by actin; microtubules consist of tubulin and are both related to intracellular motility. Microtubules are thought to be involved in determining cell shape, in mitosis, and in regulating the intracellular transport of vesicles. In the liver they seem to have a role in the secretion of lipoproteins and albumin, and in the release of lipids into the bile. Microfilaments seem to be more directly related to bile secretion. Microfilaments are normally disposed around the bile canaliculi (pericanalicular web) [66]. Many experimental models have suggested that microfilaments play an active role in the dilatation and contraction of bile canaliculi. Thus, they may control the bile canalicular caliber and the bile flow. Intermediate filaments have a more complex composition. They correspond to epithelial cell *tonofilaments* (the old nomenclature) [67–69] and show relationships with the Mallory bodies (a morphological marker of human alcoholic liver disease) [70–72]. They are present around the nucleus, near the cell border, in the cytoplasmic network and around the bile canaliculi. [71].

4. The bile secretory apparatus

4.1. *General aspects*

Hepatocytes continuously produce a small amount of bile that is released into the bile canaliculi. Bile is an aqueous substance that contains bile salts, cholesterol, fatty acids, plasma electrolytes (Na^+, K^+, Ca^{2+}, Cl^-, HCO_3^-), and a pigment called *bilirubin*. Bile salts derive from the metabolism of cholesterol, which is transformed into cholic acid. This is conjugated with glycine and taurine, forming two acids: the glycocholic and taurocholic acids, which compose the biliary salts. Bile salts are excreted with the bile in the gallbladder and then in the gut, where they play two main roles: 1) emulsion of the dietary fats and 2) increase, through their bile salt ions, of the solubility of lipids and liposoluble vitamins. Thus, rapid passage of these substances through the intestinal mucosa is allowed. More than 90% of bile salts pass through the intestinal mucosa along with lipids and reach, via the portal blood, the liver sinusoids, where they are taken up by the hepatocytes for a new cycle. These are the main steps of the mechanisms that regulate the so-called enterohepatic circulation. The hepatocyte is the key cell in these events and performs the roles of the bile formation and secretion, thanks to a pool of specialized structures that can be defined as the *bile secretory apparatus*.

4.2. The bile secretory apparatus

4.2.1. Structures

The structures involved in bile formation and secretion are the following: 1) the sinusoidal wall, 2) the space of Disse, 3) the vascular facet of the hepatocyte, 4) the hepatocyte metabolic apparatus, 5) the hepatocyte cytoskeleton, 6) the hepatocyte biliary facet with the plasma membrane junctional complexes, and 7) the intralobular biliary tree with the peribiliary capillary plexus.

As far as the first five structures are concerned, they have been widely described in the above paragraphs. Here it is worth mentioning the following: 1) the sinusoidal wall is a multicellular formation essential to the regulation of the blood-hepatocyte exchanges; 2) the space of Disse is the area of drainage of interstitial fluids that is lined by the sinusoidal wall and by the hepatocyte vascular side; 3) the latter, thanks to its wide microvillar covering, which tremendously increases the cell surface, is structurally specialized in absorptive processes; 4) a few cell organelles have a strategic position in the hepatocyte. The Golgi complex, in fact, is more abundant in periportal than pericentral hepatocytes, and lysosomes are often seen in close association with bile canaliculi (peribiliary dense bodies). In addition, secretory vesicles are often seen in proximity to the vascular and biliary poles of the plasma membrane; 5) the cytoskeletal filaments are distributed in the cytoplasm in proximity to bile canaliculi and seem to actively contribute towards the release of substances in the biliary tree.

4.2.2. Biliary pole

A main component of the bile secretory apparatus is the biliary pole (Figs. 10-2c and 10-3a) of the hepatocyte plasma membrane. As seen by electron microscopy, it is formed by two distinct parts (or hepatocyte plasmalemma morphofunctional domains) [54]: 1) the central part, consisting of the bile canaliculus, where bile is initially secreted (biliary domain), which is completely sealed by junctional complexes between adjacent hepatocytes (lateral hepatocyte plasmalemma domain), and 2) the lateral part (basolateral or juxtasinusoidal plasmalemma domain), smooth or furnished with microvilli, placed at the sides of the bile canalicular groove facing the intercellular recesses of the interstitial spaces and thus bathed by the blood filtrate. In normal conditions, these two areas are not in communication. In fact, the passage of bile into the blood spaces turns into a pathological condition characterized by jaundice and related to cholestasis and hepatocyte necrosis.

Bile canaliculi have been recognized in both light (LM) and transmission electron microscopy (TEM) for a long time, however, SEM allowed the best visualization of these structures [11] (Fig. 10-4e). In fact, in fractured liver samples, dissociation usually occurs between the biliary facets, permitting easy identification of the longitudinal grooves formed by the bile hemicanaliculi (Fig. 10-2c). The hemicanaliculus, which is one of the two halves of the bile canaliculus, runs down the center of the entire facet. It ranges between 0.5 μm and 2.5 μm in width, being largest near the portal area [11, 73–78]. Bile hemicanaliculi are in continuity with those of neighboring cells. They are straight or, more rarely, tortuous. They are usually unbranched, but they may show a few short branches. Bile hemicanaliculi are bordered by two dense rows of microvilli, which are also seen within the canalicular bed. In the bed of bile, canaliculi sacculations and diverticula may be discovered. These structures may correspond to the so-called intracellular bile canaliculi already described in the past [11]. At the sides of the bile grooves, it is present as a narrow, smooth area, about 0.4 μm in width, which may present few studs and holes. These bands represent the junctional area between two adjacent hepatocytes [77].

By TEM, these interhepatocyte contacts consist of tight junctions (Fig. 10-4c) and desmososmes, as seen from the bile canaliculus toward the interstitial space. In addition, gap junctions (nexus, macula communicans) are also present, often in association with tight junctions strands. The tight junction is a continuous area of close contact between adjacent hepatocyte plasma membranes and represents the sealing device of the bile canaliculus. It forms a barrier that avoids regurgitation of bile into the vascular compartment and regulates the passage of water and electrolytes, i.e., the paracellular pathway of bile excretion. Actually, the main quantity of water and solutes (including macromolecules) that passes from the sinusoid to the bile canaliculus is excreted through the transcellular pathway (i.e., vascular hepatocyte plasmalemma, cytoplasm, and biliary hepatocyte plasmalemma), but many evidences suggest that the paracellular pathway may function to transport water and small molecules [79]. Desmosomes form a continuous band which links the hepatocyte membranes. They are separated by a space of 15–25 nm and are associated with the microfilaments containing actin (see above). The gap junctions are especially structured to allow communication between adjacent cells. These cell-to-cell contacts permit the passage of substances [80] and electrical activities, representing an important pathway for hepatocyte relationships.

4.2.3. Intralobular biliary tree

Before reaching the gallbladder, bile probably undergoes further treatment, consisting of reabsorption of some of its components and additional secretion of others. This is suggested by the structural aspects of the cell forming the wall of the intralobular biliary tree.

Bile flows within the bile canaliculi toward the interlobular bile ducts present in the portal space; however, direct connection between bile canaliculi and interloublar bile ducts is only occasional [10]. Usually, bile canaliculi join the bile ducts through the interposition of the bile ductules (*cholangioles* or *canals of Hering*) [81]. The connection existing between bile canaliculi and interloublar ducts [10, 74, 82–84], which was only partially detected in LM and TEM sections [85–88], may be fully appreciated by SEM [88]. Hepatocytes proximal to the portal space show large bile canaliculi [45]. In this area, several bile canaliculi converge toward the ductules of Hering, forming the canalicular-ductular junction [83]. Two kinds of canalicular-ductular junctions have been described [84, 89]: 1) bile canaliculi, prior to reaching the junction, form a characteristic dilatation, corresponding to the *preductular ampulla* recognized in LM and TEM sections; and 2) bile canaliculi may also join the ductules without ampullary dilatation. The hepatocytes bordering the portal space may be found connected with cuboid epithelial cells, similar to those of the interlobular bile ducts, to form the roots of the ductules of Hering. The ductular wall is monolayered, consisting of cuboid epithelial cells that rest on a thin basal lamina. The epithelial-cell luminal surface is almost totally covered by microvilli, which probably take part in secretive/absorptive processes performed on the bile. Long cilia arising from the center of the epithelial cells have also been described [82]. Their functional significance can be related to a kinetic role for the propulsion of bile. In addition, if cilia have some chemoreceptorial capability, the control of bile quality and quantity may be exerted by these structures in association with the bile secretion regulatory activity of the peribiliary vascular system [10, 11]. This consists of a peribiliary capillary plexus surrounding the interlobular biliary ducts that, according to some authors, has been suggested to exert a countercurrent mechanism for the reabsorption of substances from bile [90]. This structure has been optimally visualized by means of vascular corrosion cast methods. In these preparations, SEM visualizes a peribiliary capillary plexus supplied by branches of the hepatic arteries of the portal space [91, 92]. These vessels anastomose sinusoidal channels or portal veins with many variations among different species [91, 93]. The peribiliary plexus, due to its close connection with interlobular biliary ducts, is thought to exert a feedback control on bile production [10, 91, 94], but further evidence is needed.

5. Structural aspects of liver function during development

5.1. General aspects

The human liver anlage is formed at the level of the distal part of the foregut just before it changes into the midgut. During the third week, an anular thickening of the endoderm (hepatopancreatic ring) appears in this zone whose ventral part originates from the liver area. This area, initially flattened, rapidly proliferates and (when the embryo is 3 mm C-R length, approximately 25 days) generates a bud that remains linked to the foregut through a thin pedunculus. This bud gives rise in the septum transversum (at the stage of 4–5 mm C-R length, approximately 32 days) to a cranial anlage (forming the proliferating hepatic cell cords) and a ventral anlage (forming the cystic duct and the gallbladder). Their common pedunculus originates from the choledocus. While epithelial cell cords proliferate within the mesenchyme of the septum transversum, they are penetrated by a growing capillary plexus derived from the vitelline veins, thus forming the primitive hepatic capillaries. Epithelial cords and capillaries proliferate at the same time. The former give rise to the hepatocyte plates and the latter to the liver sinusoids. Scattered mesenchymal cells, derived from the mesodermal plate of the septum transversum proliferate among the epithelial cords and capillary plexus. These elements probably form the connective cells of the liver tissue, including hemopoietic cells, Kupffer cells, and other sinusoidal elements, as well as the connective capsule. The origin of Kuppfer cells is, however, still a matter of debate. In fact, it is not clear whether Kuppfer cells are formed locally or derive from sources other than the liver, such as the bone marrow. The fact that 1) Kupffer cells may be seen within embryonal liver before initiation of bone marrow hemopoiesis [95, 96] and 2) sinusoidal cells may show mitotic figures [97] suggests that these elements may be locally self-generating. On the other hand, it has been demonstrated that Kupffer cells (like all macrophages) may arise from bone-marrow deriving monocytes [98]. The mesoderm resting on the liver surface proliferates into mesothelial cells, forming the visceral peritoneum which covers the liver capsule. The epithelium forming the wall of the intrahepatic biliary ducts is likely to arise from the proximal part of the hepatic cords. However, biliary duct-cell origin is not clear. It has been suggested that the epithelial cell-layer contacting the periportal vein mesenchyme (under the influence of the latter) [99] gives rise to primitive bile-duct cells. These cells may be transformed into bile-duct epithelial cells forming (at the stage of 8 mm C-R, 5–6 weeks of gestation) the primitive bile duct (ductal epithelial plate) [100, 101]. The primitive bile duct is gradually subjected to transformations so that from it (at the stage of 22–30 mm) originate the intrahepatic biliary ducts present in the portal spaces. It is not clear, however, whether the direction of intrahepatic duct formation is from the periphery towards the hepatic hilus or if the proliferation proceeds from the hilar area towards the periphery. It has also been suggested that intrahepatic bile ducts may originate from epithelial cells present around the portal vein branches that subsequently are confluent, but not from the cells of the hepatic ducts [102]. In addition, intermediate forms between ductular cells and hepatocytes have not been well demonstrated [99, 102, 103]. In the adult, the ductules connecting bile canaliculi with interlobular bile ducts (cholangioles) present both hepatocytes and duct biliary cells, but with distinct morphological features. This suggests that, despite the common origin from the primitive epithelial cellular cords, ductular epithelium and hepatocytes are morphofunctionally distinct.

5.2. *Morphofunctional aspects*

During prenatal life, the liver is involved in two main activities: 1) the rapid establishment of the secretory apparatus and 2) the hemopoietic function. Although these two events are functionally distinct, they are influenced by each other. As shown below, the structure of the fetal liver is in fact highly affected by the presence of numerous hemopoietic elements. In addition, the development of hepatic hemopoiesis is related to liver histogenetic development [104]. Hemopoietic cell proliferation and differentiation appears also to depend upon the microenvironment in which these cells are resident [105]. In particular, Kupffer cells play a key role in controlling hemopoietic cell production by means of their phagocytotic activity.

The excretory apparatus of the liver seems to be completely formed by the end of the third month. In fact, by this time, the gallbladder is connected by means of the cystic duct with the hepatic duct to form the common bile duct. Bile canaliculi are present beginning the fourth week and assume the structural characteristics of the

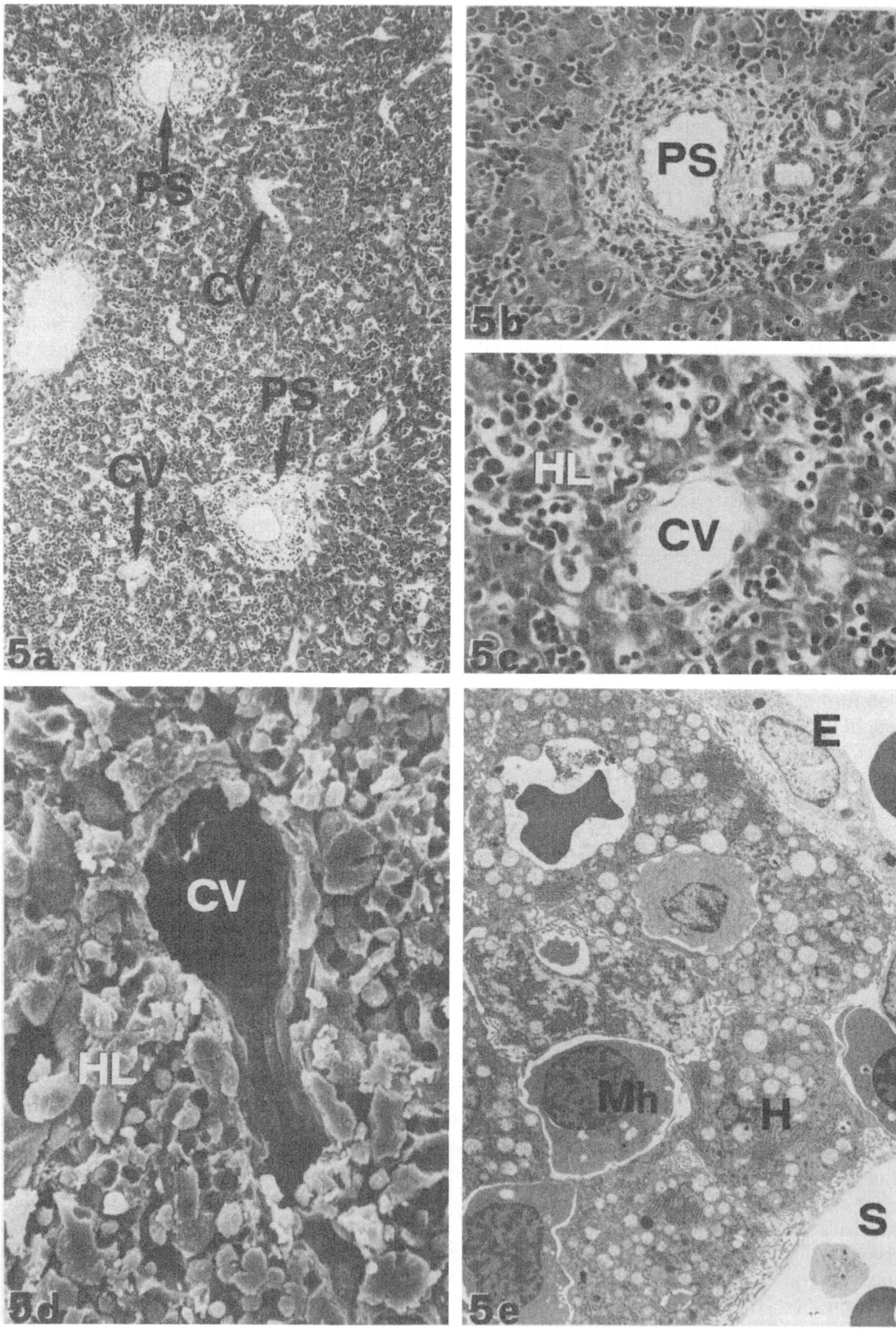
PS
CV
PS
CV
5a
PS
5b
HL
CV
5c
CV
HL
5d
E
Mh
H
S
5e

adult liver after the second month. The hepatocytic Golgi apparatus, which appears near the nucleus around the fourth week, is often noted near the bile canaliculi in the third month. Rough endoplasmic reticulum appears by the seventh week, but it completes its organization in the third month, when smooth endoplasmic reticulum, glycogen inclusions, and iron deposits are well represented within the hepatocytic cytoplasm [100, 106–108].

The liver is the main hemopoietic organ during the second trimester of fetal development. This activity, which normally is exhausted by the end of the seventh month of gestation, may be restored in adult life during certain diseases such as congenital anemia, myeloproliferative disorders, advanced leukemia, and lymphoma. Adult liver hemopoiesis is an expression of illness and often turns into impairment of hepatocyte function due to mechanical displacement of normal liver architecture induced by the hepatic infiltration of hemopoietic cells [109]. Therefore, liver hemopoiesis is normally exclusive of fetal liver. However, the production of hemopoietic cells affects the tissue architecture, even in the normal fetal liver. This causes an alteration of the morphology of both parenchymal and sinusoidal cells, and must also have some effect on cell functions.

5.3. Three-dimensional ultrastructure of fetal liver cells during hemopoiesis

The study of hemopoiesis could be facilitated by three-dimensional observations of fetal liver. SEM, in fact, allows a better interpretation of the liver cell relationship, offering ready views of cell surfaces as related to their morphofunctional features. The ultrastructure of the fetal liver has been widely studied in humans and other mammals by means of TEM [11, 28, 30, 104, 108, 110–113]. Most of these data concern the structure and development of hemopoietic cells. On the other hand, only a few studies have been performed by means of SEM [11, 29, 114–116]. These observations mainly concern the vascular component [117].

As seen by LM and TEM, mammalian fetal hepatocyte (hepatoblast) laminae present a different arrangement than observed in the adult [100, 108, 118] (Figs. 10-5a to 10-5e). SEM studies performed on human embryos (6–20 weeks' gestation) [11, 119] confirmed that fetal hepatocytes (hepatoblasts) are commonly disposed in several thick laminae intermingled with scattered hemopoietic elements and surrounded by an extensive capillary network (Figs. 10-5d and 10–5e). During the stages studied, the general liver architecture is gradually developing towards the adult appearance. In fact, hepatic laminae are seen radiating from the primitive central vein. Portal tracts are well defined from the 12th week, showing spaces resembling the biliary duct and portal vessels (Fig. 10-5b). Developing hepatocytes are differentiated early. In fact, from the sixth week of gestation, they are characterized by a polyhedral shape and show several facets with different surface features. Differentiation of structural surface polarity seems to be an early event in SEM views as well [120] (Figs. 10-6a to 10-6d). Bile canalicular grooves with short and thick microvilli, as well as microvilli-rich sides facing vascular compartments, were recognized in fractured samples at 7–8 weeks of gestation. As seen by TEM, human hepatocyte bile canaliculi are usually formed by two facing hepatocytes (Fig. 10-6e), sealed by junctional complexes. In addition, the confluence of more than two hepatocytes to form the bile canaliculus is frequently observed. Communication between biliary and interstitial spaces has not been demonstrated. Hepatocyte shape and surface are often influenced by intimate contact with hemopoietic elements present in the Disse space. Parallel TEM and SEM observations (Figs. 10-6a, 10-6b, and 10-6e) show that hemopoietic elements are often placed in large impressions of the hepatocyte surface.

←

Fig. 10-5. Liver development. **a**: human fetal liver (12 weeks' gestation) parenchyma in a light microscopic view. Note the large vascular spaces resembling portal spaces (PS) and central veins (CV) (x 200), **b**: higher magification of a portal space (PS) from the same sample as Fig. 10-5a (x 400). **c**: higher magnification of a central vein (CV) from the same sample as Fig. 10-5a. HL = hepatic laminae (x 800). **d**: human embryonal liver (8 weeks' gestation) SEM of a central vein (CV) and hepatic laminae (HL) (x 800). **e**: TEM of 18-week-old human fetal liver. Note the pluricellular laminae formed by the hepatoblasts (H) and mesenchymal-hemopoietic cells (Mh), E = endothelium; S = sinusoid (x 3,800).

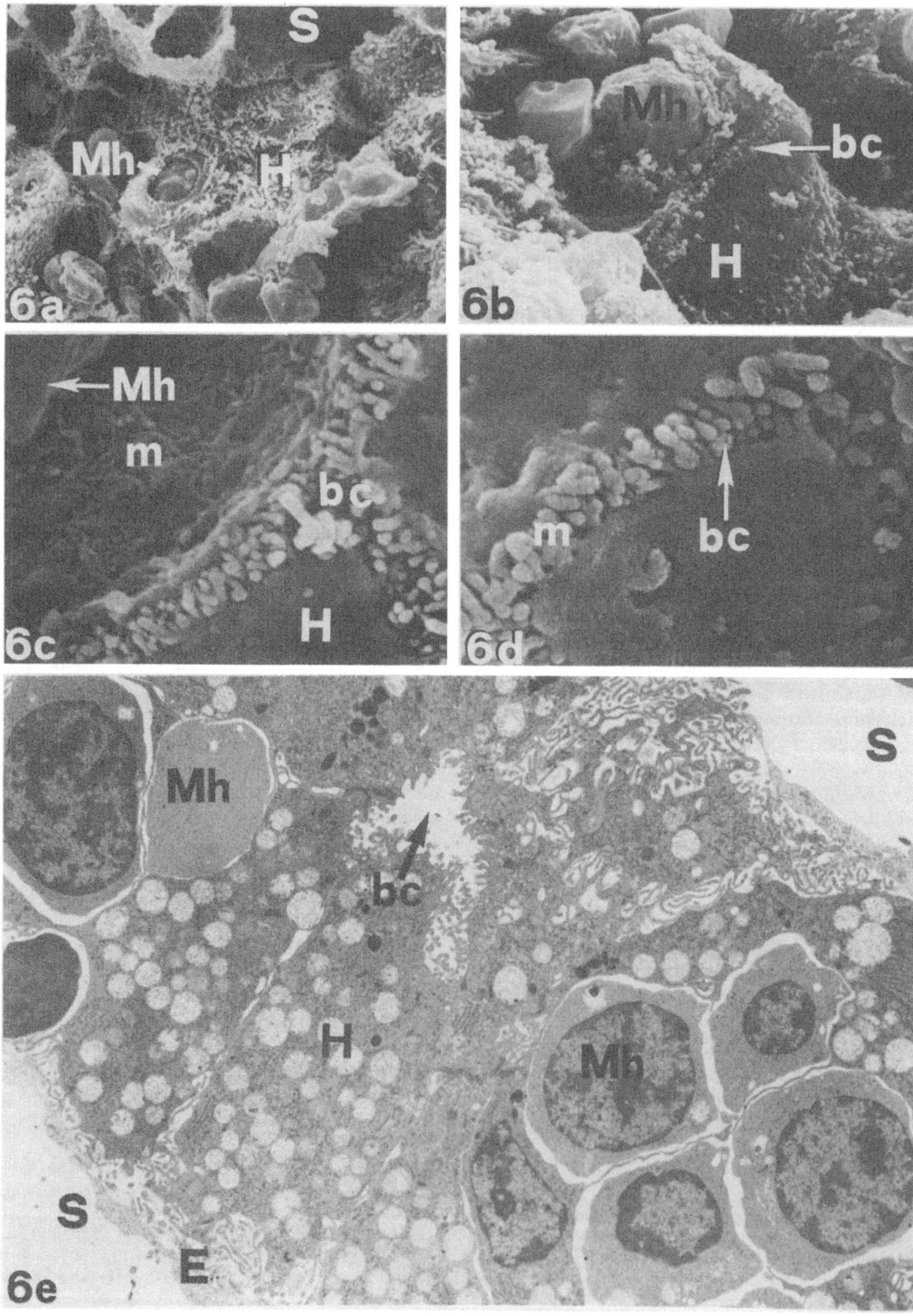

S
Mh
H
6a
Mh
bc
H
6b
Mh
m
bc
H
6c
m
bc
6d
S
Mh
bc
H
Mh
S
E
6e

These surfaces, which can be considered to be a third structural pole of the hepatocyte plasma membrane (hemopoietic pole), are morphologically differentiated towards absorptive/secretive functions, presenting ultrastructural features similar to those characterizing the hepatocytic vascular facets and consisting of a dense microvillous covering. Hemopoietic elements, however, are never seen invading the biliary compartment. Therefore, it can be speculated that the areas of the hepatocyte plasma membrane facing hemopoietic elements, soon after liver hemopoietic activity is terminated, will probably differentiate morphofunctionally into vascular facets.

SEM views show a greater extension of vessels than usually seen in the adult [11, 114–116].

The sinusoidal wall (Figs. 10-7a to 10-7f) of the fetal liver sinusoid is mainly composed of the endothelial lining [108, 110, 111, 114, 115]. Kupffer cells contribute towards the formation of the capillary lining by contacting and/or penetrating the endothelial wall with cytoplasmic prolongations (Figs. 10-7d and 10-7e). SEM seems to confirm (see above) that endothelial cells and Kupffer cells are distinct cell types, even during early stages of fetal development. In fact, these two elements always present different three-dimensional characteristics. [111, 114, 115].

In fetal liver, the space of Disse shows more recesses than in adults, and its distribution within the lobule is extremely variable, being very narrow in some areas and large in others. This is related to the perisinusoidal-space cell population. Under SEM, the wider spaces are occupied by a large amount of hemopoietic elements. These appear as a morphologically homogeneous group of cells having a rounded shape and smooth surface. Their classification may actually be obtained only by means of TEM (Figs. 10-5d, 10-5e, 10-6b, and 10-6e) [104, 108]. The fact that fetal liver interstitial spaces are mainly occupied by hemopoietic elements contributes to render highly difficult the SEM identification and classification of the perisinusoidal cell without using appropriate techniques (e.g., NaOH dissociation) [37]. Fetal perisinusoidal cells (Fig. 10-7a), when seen by TEM, appear to occupy a position analogous to that observed in the adult but seem to contain a smaller amount of lipid inclusions [110, 112, 117].

In early stages (6–8 weeks in humans, 12–14 days in the rat), SEM allows identification of Kupffer cells that show aspects comparable to those of the analogous adult elements [114, 115]. These cells have macrophagic features, including cytoplasmic membrane specializations and frequent contact with blood elements, and, when observed by TEM, erythrophagocytotic activity (Figs. 10-7d and 10-7e).

Endothelial cells present varying ultrastructural appearances during the liver's hemopoietic period and are usually different from the adult liver sinusoidal endothelial cells [118]. During early stages, endothelial cells rarely present clustered fenestrations. Endothelium is rather continuous, except for a few isolated small fenestrae and the frequent occurrence of large intercellular interruptions (Figs. 10-7b, 10-7c, and 10-7f) related to diapedesis of the hemopoietic elements and newly formed blood cells across the sinusoidal wall [111]. Passage of cells through the fetal liver endothelium has been frequently demonstrated by means of TEM in both humans and animals [108, 111, 114, 115, 117]. In the adult liver, such particular interruptions of the endothelium, being crossed by cellular elements, have only been observed in pathological or experimental conditions, or in normal adult hemopoietic organs (e.g., bone marrow) [121–124]. During late developmental stages, the sinusoidal endothelium is more frequently fenestrated [110, 111, 117]. As shown by TEM, these small pores are often occluded by diaphragms. These structures, described in the

←

Fig. 10-6. Ultrastructure of liver development. **a**: SEM of 8-week-old human liver. Note the relation of between the mesenchymal-hemopoietic cell (Mh) and hepatoblasts (hepatocytes) (H); S = forming sinusoid (x 1,000). **b**: SEM of 18-week-old human fetal liver. H = hepatoblast; bc = bile canaliculus; Mh = mesenchymal-hemopoietic cell. (x 6,000). **c**: SEM of the microvilli (m)-rich side of a hepatoblast (H) facing the hemopoietic cell (Mh →); bc = bile canaliculus. (x 18,000 human fetal liver, 15 weeks). **d**: SEM of bile canaliculus (bc) from an 8-week-old human fetal liver; m = microvilli. (x 35,000). **e**: TEM of a 18-week-old human fetal liver. Several hepatoblasts are seen to converge to form a bile cancliculus (bc). H = hepatoblast; Mh = mesenchymal-hemopoietic cells; S = sinusoids; E = endothelium. (x 7,000).

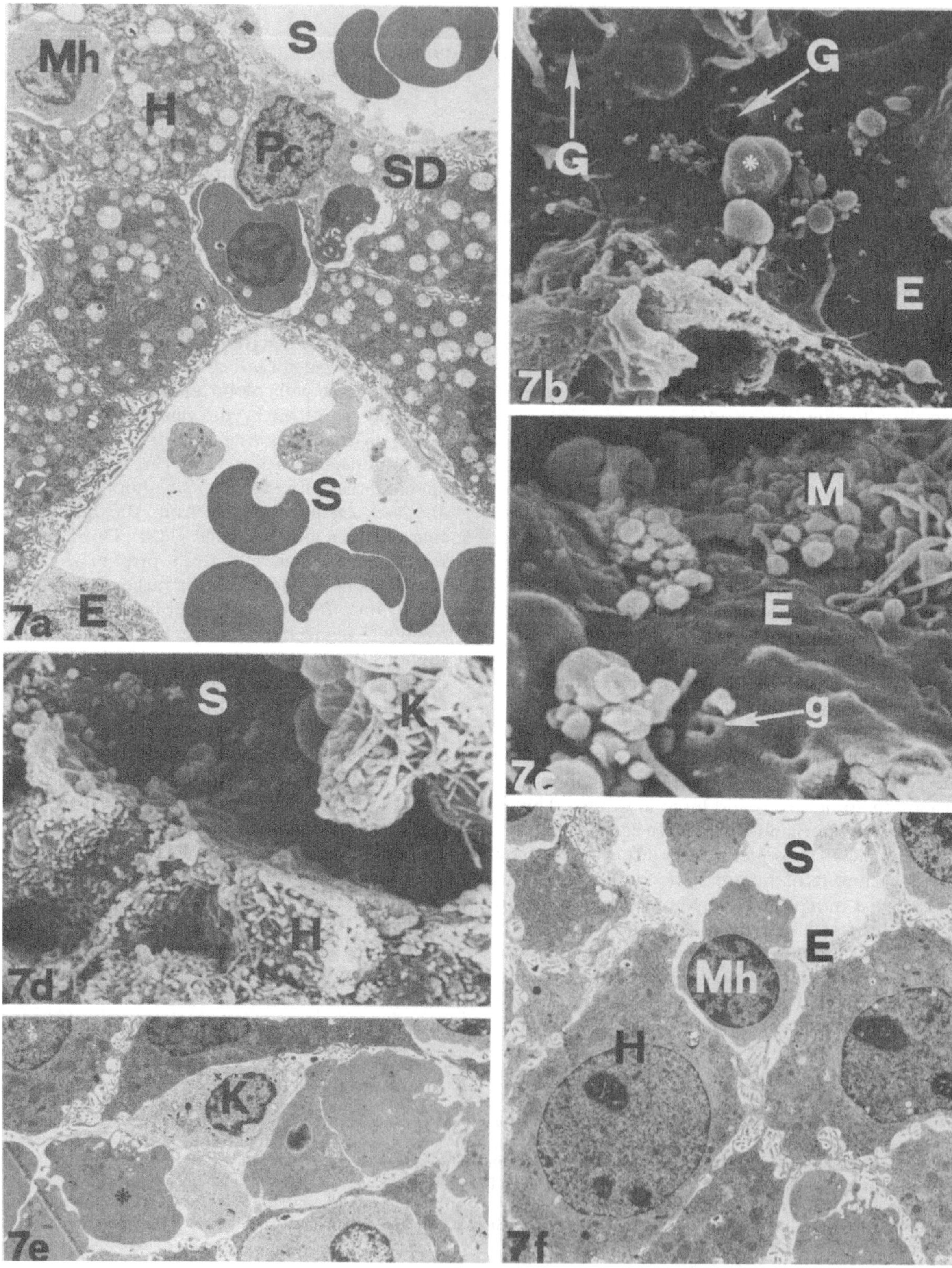

7a
Mh
S
H
Pc
SD
S
E
7b
G
G
G
E
7c
M
E
g
7d
S
K
H
7e
K
7f
S
E
Mh
H

fetal rat liver [111], are similar to those demonstrated in other capillary districts of the adult [11, 125]. The occurrence of diaphragmed fenestrations and the reduced number of large intercellular gaps (in the human fetus of 18–20 weeks of gestation) [117] suggest that the formation of sieving sinusoids is not completed and that the morphofunctional differentiation of liver endothelial cells is ongoing. This can be related to the gradual evolution of mammalian liver function during this developmental stage [11]. These ultrastructural features demonstrate that the liver undergoes a gradual modification of its main roles, which are: a decrease of hemopoietic activity and parallel transformation of its structure for the typical glandular role in the adult [111, 117].

6. Conclusions

Over the past decade, electron microscopic techniques have been increasingly improved and adapted by more and more investigators. These and related methods have helped to provide impressive and detailed views of cells, tissues, and organs that have been routinely employed both in the field of basic research and clinical diagnosis.

Studies on the liver have been greatly facilitated by the continuing refinement of these recent methods, which have led to a more complete and "realistic" three-dimensional concept of hepatocytes and other liver tissues under both normal and pathological conditions [54].

It is our conviction that future progress in the field of liver research, especially in the search for a more accurate picture of the microorganization of this organ, will depend greatly on the integration of fine structural data with more gross physiopathological aspects. Therefore, the modest scope of this review was to present the more salient ultrastructural data obtained by the application of such a correlative approach to investigate the liver's function.

Acknowledgements

The original work of the authors included in this chapter was supported by CNR and MPI grants.

The authors wish to thank Prof. E. Gaudio for preparing some of the light microscopic slides, Dr. D. Finn and Miss A. Martinotti for helping with the manuscript, and Mr. G.F. Franchitto for some of the photographic work.

References

1. Fawcett DW: Bloom and Fawcett — A Textbook of Histology, 11th ed. Philadelphia: WB Saunders Company, 678–715, 1986.
2. Wisse E: An electron microscopic study of the fenestrated endothelial lining of rat liver sinusoids. *J Ultrastruct Res* 31: 125–150, 1970.
3. Wake K: Perisinusoidal stellate cells (fat-storing cells, interstitial cells, lipocytes), their related structure in and around the liver sinusoids, and vitamin A-storing cells in extrahepatic organs. *Int Rev Cytol* 66: 303–353, 1980.
4. Kirn A, Knook DL, Wisse E (eds): Cells of the Hepatic Sinusoid, Vol 1. Rijswijk: The Kupffer Cell Fundation, 1986.
5. Wisse E, van't Noorden JM, van der Meulen J, Daems WTN: The pit cell: description of a new type of cell occurring in rat liver sinusoids and peripheral blood. *Cell Tissue Res* 173: 423–435, 1976.
6. Kaneda K, Wake K: Distribution and morphological characteristics of the pit cells in the liver of the rat. *Cell Tissue Res* 233: 485–505, 1983.
7. Macchiarelli G, Motta PM, Fujita T: Scanning electron microscopy of the liver cell. In: Biopathology of the Liver. An Ultrastructural Approach. PM Motta (ed), Dordrecht: Kluwer Academic Publishers, 37–58, 1988.
8. Morin O, Normand C: Cellular cooperation between sinusoidal liver cells and hepatocytes. In: Cells of the Hepatic Sinusoid, Vol 1, A Kirn, DL Knook, E Wisse (eds), Rijswijk: The Kupffer Cell Foundation, 49–50, 1986.

Fig. 10-7. Liver development. **a**: TEM of a 18-week-old fetal liver. H = hepatoblast; Mh = mesenchymal-hemopoietic cell; E = endothelium; SD = space of Disse; S = sinusoid. Note a perisinusoidal cell (Pc). (x 3,800). **b**: SEM of a 15-week-old fetal liver sinusoid. E = endothelium. Note the large interruptions (G →) and cellular elements (*) seeming to pass through the sinusoidal wall (see also Fig. 10-7F) (x 2,000), **c**: SEM of an 8-week-old human fetal liver sinusoid. E = endothelium. Note small endothelial gaps (g →), probably related to the formation of fenestrae (x 4,000). A macrophage (M) is located in the sinusoidal lumen. **d**: 8-week-old human fetal liver sinusoid (S). H = hepatoblast; K = Kupffer cell. (x 3,500). **e**: 10-week-old human fetal liver sinusoid. Note a Kupffer cell (K) in contact with a red blood cell (*) (x 3,000). **f**: 7-week-old human fetal liver sinusoid (S). Note a mesenchymal-hemopoietic cell (Mh) passing through the sinusoidal wall. E = endothelium; H = hepatoblast. (TEM, x 3,000).

9. Berg T, Blomhoff R, Eskild W, Norum KR: Cooperation between cells in hepatic endocytosis. In: Cells of the Hepatic Sinusoid, Vol 1. A Kirn, DL Knook, E Wisse (eds), Rijswijk: The Kupffer Cell Foundation, 137–138, 1986.
10. Motta P, Muto M, Fujita T: The Liver. An Atlas of Scanning Electron Microscopy, Tokyo/New York: Igaku Shoin, 1978.
11. Motta PM: The three-dimensional microanatomy of the liver. *Arch Histol Jpn* 47: 1–30, 1984.
12. Kiernan F: The anatomy and physiology of the liver. *Phil Trans Roy Soc Lond* 123: 711–770, 1833.
13. Mall FP: A study of the structural unit of the liver. *Am J Anat* 5: 227–308, 1906.
14. Rappaport AM, Borowy ZJ, Lougheed WM, Lotto WN: Subdivision of hexagonal liver lobules into a structural and functional unit. Role in hepatic physiology and pathology. *Anat Rec* 119: 11–34, 1954.
15. Jones AL, Schmucker DL: Current concepts of liver structure as related to function. *Gastroenterology* 73: 833–851, 1977.
16. Macchiarelli G, Motta PM: The three-dimensional microstructure of the liver. A review by scanning electron microscopy. *Scanning Electron Microscopy* III: 1019–1038, 1986.
17. Motta PM, Fujita T, Nishi M: Scanning electron microscopy of the mammalian liver. In: Basic and Clinical Hepatology. PM Motta, LJA DiDio (eds). The Hague: Martinus Nijhoff, 31–50, 1982.
18. Ohtani O, Murakami T, Jones AL: Microcirculation of the liver, with special reference to the peribiliary portal system. In: Basic and Clinical Hepatology. PM Motta, LJA DiDio (eds), The Hague: Martinus Nijhoff, 85–96, 1982.
19. Motta P, Porter KR: Structure of rat liver sinusoids and associated tissue spaces as revealed by scanning electron microscopy. *Cell Tiss Res* 148: 111–125, 1974.
20. Motta PM: A scanning electron microscopic study of the rat liver sinusoid: endothelial and Kupffer cells. *Cell Tiss Res* 164: 371–385, 1975.
21. Wisse E, De Zanger RB, Charles K, Van Der Smissen P, McCuskey RS: The liver sieve: considerations concerning the structure and function of endothelial fenestrae, the sinusoidal wall and the space of Disse. *Hepatology* 5: 683–692, 1985.
22. Fraser R, Bosanquet AG, Day WA: Filtration of chylomicrons by the liver may influence cholesterol metabolism and atherosclerosis. *Atherosclerosis* 29: 113–123, 1978.
23. Bradfield JWB: Liver sinusoidal cells. *J Pathol* 142: 5–6, 1984.
24. Wake K, Senoo H: Morphological aspects of the differentiation of stellate cell line in the vertebrates. In: Cells of the Hepatic Sinusoid, vol I. A Kirn, DL Knook, E Wisse (eds), Rijswijk: The Kupffer Cell Fundation, 215–220, 1986.
25. Wake K: Liver perivascular cells revealed by gold- and silver-impregnation methods and electron microscopy. In: Biopathology of the Liver. An Ultrastructural Approach. PM Motta (ed), Dordrecht: Kulwer Academic Publishers, 23–36, 1988.
26. Frenzel H. Kremer B, Hucker H: The liver sinusoids under various pathological conditions. A TEM and SEM study of rat liver after respiratory hypoxy, tele-cobalt-irradiation and endotoxin application. In: Kupffer Cells and Other Sinusoidal Liver Cells. E. Wisse, DL Knook (eds), Amsterdam: Elsevier North Holland Biomedical Press, 213–222, 1977.
27. Praaning-Van Dalen D, Brouwer A, Knook D: Clearance capacity of rat liver Kupffer, endothelial and parenchymal cells. *Gastroenterology* 81: 1036–1044, 1981.
28. Naito M, Yamamura F, Takeya M, Takahashi K: Ultrastructural analysis of Kupffer cell progenitors. In: Cells of the Hepatic Sinusoid. A Kirn, DL Knook, E Wisse (eds), Rijswijk: The Kupffer Cell Foundation, 13–20, 1986.
29. Deimann W, Fahimi HD: Peroxidase cytochemistry and ultrastructure of resident macrophages in fetal rat liver. A development study. *Develop Biol* 66: 43–56, 1978.
30. Naito M, Takahashi K, Takahashi H, Kojima M: Ontogenetic development of Kupffer cells. In: Sinusoidal Liver Cells. K Knook, E Wisse (eds), Amsterdam: Elsevier Biomedical Press, 155–164, 1982.
31. Wisse E: An ultrastructural characterization of the endothelial cell in the rat liver sinusoid under normal and various experimental conditions, as a contribution to the distinction between endothelial and Kupffer cells. *J Ultrastruct Res* 38: 528–562, 1972.
32. Jones EA, Summerfield JA: Kupffer cells. In: The Liver: Biology and Pathobiology. IM Arias, WB Jakoby, H Popper, D Schachter, DA Shafritz (eds), New York: Raven Press, 683–704, 1988.
33. Wisse E: Ultrastructure and function of Kupffer cells and other sinusoidal cells in the liver. In: Kupffer Cells and Other Liver Sinusoidal cells. E Wisse, DL Knook (eds), Amsterdam: Elsevier North Holland Biomedical Press, 33–66, 1977.
34. Tanuma Y, Ohata M: Transmission electron microscope observation of epithelial cells with single cilia in intrahepatic biliary ductules of bats. *Arch Histol Jpn* 41(4): 367–376, 1978.
35. Novikoff AB: GERL, its form and function in neurons of rat spinal ganglia. *Biol Bull* 127: 358, 1964.
36. Bouwens L, Geerts A, Wisse E: Structure and function of the hepatic sinusoidal wall. In: Biopathology of the Liver. An Ultrastructural Approach. PM Motta (ed), Dordrecht: Kluwer Academic Publishers, 13–22, 1988.
37. Takahashi-Iwanaga H, Fujita T: Application of an NaOH maceration method to a scanning electron microscopic observation of Ito cells in the rat liver. *Arch Histol Jpn* 49: 349–357, 1986.
38. Lautt WW, Greenway CV, Legare DJ, Weisman H: Localization of intrahepatic portal vascular resistance. *Am J Physiol* 251: 375–381, 1986.
39. Tanuma Y, Ito T, Shibasaki S: Further electron microscope studies on the human hepatic sinusoidal wall with special reference to the fat-storing cells. *Arch Histol Jpn* 45: 263–274, 1982.
40. Brouwer A, Hendriks HFJ, Brekelmans P, Buijtenhek M, Knook DL: Lipid analysis of different types of isolated liver cells. In: Cells of the Hepatic Sinusoid. A

Kirn, DL Knook, E Wisse (eds), Rijswijk: The Kupffer Cell Foundation, 175–180, 1986.
41. Elhanany E, De Leeuw AM, Hendriks HFJ, Brouwer A, Knook DL: Uptake and storage of vitamin A in rat liver studied by electron microscopic autoradiography. In: Cells of the Hepatic Sinusoid. A Kirn, DL Knook, E Wisse (eds), Rijswijk: Kupffer Cell Foundation, 187–192, 1986.
42. Kaneda K, Wake K: Pit cells in extrahepatic organs of the rat. *Anat Rec* 21: 192–197, 1985.
43. Ashwell JD: Lymphocytes: properties and strategies. In: The Liver: Biology and Pathobiology. IM Arias, WB Jakoby, H Popper, D Schachter, DA Shafritz (eds), New York: Raven Press, 867–884, 1988.
44. Bioulac-Sage P, Latry P, Dubroca J, Quinton A, Balabaud C: Pit cells in human liver. In: Cells of the Hepatic Sinusoid. A Kirn, DL Knook, E Wisse (eds), Rijswijk: The Kupffer Cell Foundation, 415–420, 1986.
45. Motta PM: The three-dimensional fine structure of the liver as revealed by scanning electron microscopy. *Int Rev Cytol* 6: (Suppl) 347–397, 1977.
46. Grisham JW, Nopanitaya W, Compagno J, Nagel AEH: Scanning electron microscopy of normal rat liver: the surface structure of its cells and tissue components. *Am J Anat* 144: 295–322, 1975.
47. Grisham JW, Nopanitaya W, Compagno J: Scanning electron microscopy of the liver: a review of methods and results. In: Progress in Liver Diseases, H Popper, F Schaffner, (eds), New York: Grune & Stratton, 1–23, 1976.
48. Wicha MS, Lowrie G, Kohn E, Bagavandoss P, Mahn T: Extracellular matrix promotes mammary epithelial growth and differentiation in vitro. *Proc Natl Acad Sci USA* 79: 3213–3217, 1982.
49. Tung PS, Fritz IB: Extracellular matrix promotes rat Sertoli cells histotypic expression in vitro. *Biol Reprod* 30: 213–229, 1984.
50. Bade EG, Nitzgen B: Extracellular matrix (ECM) modulates the EGF-induced migration of liver epithelial cells in serum-free, hormone-supplemented medium. *In Vitro Cell Dev Biol* 21: 245–248, 1985.
51. Madri JA, Pratt BM: Endothelial cell-matrix interactions: in vitro models of angiogenesis. *J Histochem Cytochem* 34: 85–91, 1986.
52. Rojkind M: Extracellular matrix. In: The Liver: Biology and Pathobiology. IM Arias, WB Jakoby, H Popper, D Schachter, DA Shafritz (eds), New York: Raven Press, 707–716, 1988.
53. Jones AL, Spring-Mills E: The liver and gallbladder. In: Histology. Cell and Tissue Biology. L Weiss (ed), New York: Elsevier Biomedical Press, 708–748, 1983.
54. Motta PM (ed): Biopathology of the Liver, an Ultrastructural Approach. Dordrecht: Kluwer Academic Publishers, 1988.
55. Wheatley DN: Binucleation in mammalian liver. *Exp Cell Res* 74: 455, 1972.
56. DiDio LJA, Gaudio E, Correr S: Microanatomy of the liver: physiopathological, clinical and surgical aspects. In: Biopathology of the Liver. An ultrastructural approach. PM Motta (ed), Dordrecht: Kluwer Academic Publishers, 99–112, 1988.
57. Modesti A, Torrisi MR, Caramia FG: The human hepatocyte: ultrastructural features of adult and fetal liver. In: Biopathology of the Liver. An Ultrastructural Approach. PM Motta (ed), Dordrecht: Kluwer Academic Publishers, 1–12, 1988.
58. Johnson LV, Walsh ML, Chen LB: Localization of mitochondria in living cell with rhodamine 123. *Proc Natl Acad Sci USA* 77: 990–994, 1980.
59. Cascarano J, Montesano DF, Pickett CB, James TW: Rough endoplasmic reticulum-mitochondrial complexes from rat liver. *Exp Cell Res* 139: 39–50, 1982.
60. Novikoff AB, Novikoff PM: Lysosomes. In: The Liver: Biology and Pathology. IM Arias, WB Jakoby, H Popper, D Schachter, DA Shafritz (eds), New York: Raven Press, 227–239, 1988.
61. Novikoff AB, Novikoff PM: Cytochemical contributions to differentiating GERL from Golgi apparatus. *Histochem J* 9: 525–551, 1977.
62. Ghadially FN: Lysosomes. In: Ultrastructural Pathology of the Cell and Matrix. FN Ghadially (ed), London: Butterworths, 435–580, 1982.
63. Kishimoto T, Tavassoli M: Transendothelial transport (transcytosis) of iron-transferrin complex in the rat liver. *Am J Anat* 178: 241–249, 1987.
64. Soda R, Tavassoli M: Liver endothelium and not hepatocytes or Kupffer cells have transferrin receptors. *Blood* 63: 270–276, 1984.
65. Porter KR, Becherle M, McNiven MA: The cytoplasmic matrix. In: Spatial Organization of Eukaryotic Cells. Modern Cell Biology, Vol 2. GR McIntosh (ed), New York: Alan R Liss, 259–302, 1983.
66. Arias IM, Jakoby WB, Popper H, Schachter D, Shafritz DA (eds): The Liver. New York: Raven Press, 1988.
67. Oshio C, Phillips MJ: Contractility of bile canaliculi, implications for liver function. *Science* 212: 1041–1042, 1981.
68. Lazarides E, Revel JR: The molecular basis of cell movement. *Sci Am* 240: 100–113, 1979.
69. Adelstein RS, Scordilis SP, Trotter JA: The cytoskeleton and cell movement: general considerations. *Meth Archive Exp Pathol* 8: 1–41, 1979.
70. Mallory FB: Cirrhosis of the liver: five different types of lesions from which it may arise. *Bull Johns Hopkins Hosp* 22: 69–74, 1911.
71. Phillips MJ, Satir P: The cytoskeleton of the hepatocyte: organization, relationships, and pathology. In: The Liver: Biology and Pathobiology. IM Arias, WB Jakoby, H Popper, D Schachter, DA Shafritz (eds), New York: Raven Press, 11–28, 1988.
72. Phillips MJ: Mallory bodies and the liver. *Lab Invest* 47: 311–313, 1982.
73. Phillips MJ, Oshio C, Miyairi M, Smith CR: Intrahepatic cholestasis as a canalicular motility disorder: evidence using cytochalasin. *Lab Invest* 48: 205–211, 1983.
74. Brooks SEH, Haggis GH: Scanning electron microscopy of rat's liver. Application of freeze-fracture and freeze-drying techniques. *Lab Invest* 29: 60–64, 1973.
75. Myai K, Wagner RM, Richardson AL: Preparation of liver for combined SEM and TEM study. *Scanning*

Electron Microscopy I: 283–290, 1974.

76. Vonnahme FJ: A scanning electron microscopic study of the liver of the monkey *Macaca speciosa*. II. Intra and extrahepatic biliary system. *Cell Tiss Res* 215: 207–214, 1981.
77. Motta PM, Fumagalli G: Structure of rat bile canaliculi as revealed by scanning electron microscopy. *Anat Rec* 182: 499–514, 1975.
78. Nopanitaya W, Grisham JW: Scanning electron microscopy of mouse intrahepatic structures. *Exp Mol Pathol* 23: 441–458, 1975.
79. Metz J, Robenek H: Normal and pathological liver cells as studied in freeze-fracture replicas. In: Biopathology of the Liver. An Ultrastructural Approach. PM Motta (ed), Dordrecht: Kluwer Academic Publishers, 149–166, 1988.
80. Loewenstein W: Permeability of the junctional membrane channel. In: International Cell Biology. BR Brinkley, K Porter (eds), New York: Rockefeller University Press, 70–82, 1977.
81. Hering E: Über den Bau der Wirbelthierleber, *Arch F Mikr Anat* 3: 88–114, 1867.
82. Motta P, Fumagalli G: Scanning electron microscopy demonstration of cilia in rat intrahepatic bile ducts. *Z Anat Entwickl-Gesch* 145: 223–226, 1974.
83. Marinozzi G, Muto M, Correr S, Motta P: Scanning electron microscopy observations of intrahepatic biliary tree. I. Canaliculo-ductular junction. Bile ductules and ducts. *J Submicrosc Cytol* 9: 127–143, 1977.
84. Itoshima T, Kiyotoshi S, Kawaguchik K, Yoshino K, Munetomo F, Ohta W, Shimada Y, Nagashima H: Scanning electron microscopy of rat bile canalicular-ductular junction. *Scanning Electron Microscopy* III: 373–378, 1980.
85. Lettule M: Les canalicules biliaries du foie humain leur repartition a l'interieur du lobule hepatique et leur modes d'abouchement dans les canaux perii-lobularies. *J Physiol Pathol Gen* 1: 789–801, 1915.
86. Daems WT: The microanatomy of the smallest biliary pathway in mouse liver tissue. *Acta Anat* 46: 1–24, 1961.
87. Steiner JW, Carruthers JS: Studies on the fine structure of the terminal branches of the biliary tree. I. The morphology of normal bile canaliculi, bile pre-ductules (ducts of Hering) and bile ductules. *Am J Path* 38: 639–661, 1961.
88. Motta PM: Scanning electron microscopy of the liver. In: Progress in Liver Disease. H Popper, F Schaffner (eds), Vol VII, New York: Grune & Stratton, 1–16, 1982.
89. Itoshima T, Yoshino K, Yamamoto K, Ohata W, Kubota M, Ukida M, Ito T, Hirakawa H, Munetomo F, Shimada Y: Scanning electron microscopy of the bile ductule. *Gastroenterol Jpn* 12: 476–482, 177.
90. Henderson JR, Daniel PM: Portal circulations and their relations to countercurrent systems. *Q J Exp Physiol* 63: 355–369, 1978.
91. Murakami T, Itoshima T, Shimada Y: Peribiliary portal system in the monkey liver as evidenced by the injection replica scanning electron microscope method. *Arch Histol Jpn* 37: 245–260, 1974.
92. Ohtani O, Murakami T: Peribiliary portal system in the rat liver studied by the injection-replica scanning electron microscope method. *Scanning Electron Microscopy* II: 241–245, 1978.
93. Nopanitaya W, Lamb JC, Grisham JW, Carson JL: Effect of hepatic venous outflow obstruction of pores and fenestrations in sinusoidal endothelium. *Br J Exp Path* 57: 604–609, 1976.
94. Ohtani O: The peribiliary portal system in the rabbit liver. *Arch Histol Jpn* 42: 153–167, 1979.
95. Fukuda T: Ultrastructure of fetal hemopoiesis. Morphological studies on the hepatic hematopoiesis of human fetuses. *Acta Haem Jpn* 41(6): 1204–1230, 1978.
96. Naito M, Yamamura F, Takahashi K: Ontogenesis of macrophages in the developing murine liver anlage. *Proc XIth Int Congr on Electron Microscopy*, Kyoto, 2835–2836, 1986.
97. Naito M, Wisse E: Observation on the fine structure and cytochemistry of sinusoidal cells in fetal and neonatal rat liver. In: Kupffer Cells and Other Liver Sinusoidal Cells. E Wisse, DL Knook (eds), Amsterdam: Elsevier North Holland Biomedical Press, 497–505, 1977.
98. Van Furth R, Blusse' Van Oud Alblas D: New aspects on the origin of Kupffer cells. In: Sinusoidal Liver Cells. DL Knook, E Wisse (eds), Amsterdam: Elsevier North Holland Biomedical Press, 173–183, 1982.
99. Enzan H, Okhita T, Fujita H, Iijima S: Light and electron micrsocopic studies on the development of periportal bile ducts of the human embryo. *Acta Path Jpn* 24(4): 427–444, 1974.
100. Elias H, Sherrick JC: Morphology of the Liver. New York: Academic Press, 1969.
101. Gerber MA, Thung SN: Histology of the Liver. *Surg Pathol* 11: 709–722, 1987.
102. Shiojiri N, Katayama H: Secondary joining of the bile ducts during the hepatogenesis of the mouse embryo. *Anat Embryol* 177: 153–163, 1987.
103. Koga A: Morphogenesis of intrahepatic bile ducts of the human fetus. Light and electron microscopic study. *Z Anat Entwickl-Gesch* 135: 156–184, 1971.
104. Fukuda T: Fetal hemopoiesis. II. Electron microscopic studies on human hepatic hemopoiesis. *Virchows Arch B* (Cell Pathol) 16: 249–270, 1974.
105. Medlock ES, Haar J: The liver hemopoietic environment: II. Peroxidase reactive mouse fetal liver hemopoietic cells. *Anat Rec* 207: 43–53, 1983.
106. Sadler TW: Langman's Medical Embryology, 5th ed. Baltimore: Williams & Wilkins, 1985.
107. Tanikawa K: Ultrastructural Aspects of the Liver and its Disorders, 2nd ed. Tokyo: Igaku-shoin, 338, 1979.
108. Zamboni L: Electron microscopic studies of blood embryogenesis in humans. I. The ultrastructure of the fetal liver. *J Ultrastruct Res* 12: 509–524, 1965.
109. Erslev AJ: Hemopoietic functions of the liver. In: The Liver: Biology and Pathology. IM Arias WB Jakoby, H Popper, D Schachter, DA Shafritz (eds), New York: Raven Press, 1073–1081, 1988.
110. Bartok I, Toth J, Remenar E, Viragh S: Fine structure of perisinusoidal cells in developing human and mouse liver. *Acta Morphol Hung* 31: 337–352, 1983.
111. Bankston PW, Pino RM: The development of the sinusoids of fetal rat liver: morphology of endothelial cells,

Kupffer cells, and the transmural migration of blood cells into sinusoids. *Am J Anat* 159: 1–15, 1980.

112. Kobayashi K, Iwasaki S, Takahashi K, Ueno M, Igarashi A, Nagai H, Suzuki K, Katsuya T: On the fine structure of Ito's fat-storing cells in the liver of the mouse fetus after maternal treatment with vitamin A. *Congr Anat* 25: 165–174, 1985.
113. Zamboni L: Electron microscopic studies of blood embryogenesis in humans. II. The hemopoietic activity in the fetal liver. *J Ultrastruct Res* 12: 525–541, 1965.
114. Makabe S, Motta PM: Scanning electron microscopy of foetal liver sinusoids in humans. *Folia Morphol Praha* 28: 85–87, 1980.
115. Motta PM, Makabe S: Foetal and adult liver sinusoids and Kupffer cells as revealed by scanning electron microscopy. In: The Reticuloendothelial System and the Pathogenesis of Liver Disease. H Liehr, M Grun (eds), Amsterdam: Elsevier/North Holland Biomedical Press, 11–16, 1980.
116. Overton J, Meyer R: Aspects of liver and gut development in the chick. *Scanning Electron Microsc* II: 737–746, 1984.
117. Macchiarelli G, Makabe S, Motta PM: Scanning electron microscopy of adult and fetal liver sinusoids. In: Sinusoids in Human Liver, Health and Disease. P Bioulac-Sage, C Balabaud (eds), Rijswijk: Kupffer Cell Foundation, 63–85, 1988.
118. Fukuda T: Ultrastructure of fetal hemopoiesis. Morphological studies on the hepatic-hematopoiesis of human fetuses. *Acta Haem Jpn* 41(6): 1204–1230, 1978.
119. Makabe S, Macchiarelli G, Motta PM: The three-dimensional organization of the human fetal liver as seen by the scanning electron microscopy. *Acta Anat* 130(1): 57, 1987.
120. Macchiarelli G, Makabe S, Motta PM: The three-dimensional arrangement of the fetal liver cells. *8th Int Symposium on Morph Sc.* In: *Quaderni di Anatomia Pratica, Serie XLIV*: 239, 1988.
121. Walker RM, Racz WJ, McElliot TF: Scanning electron microscopic examination of acetaminophen-induced hepatotoxicity and congestion in mice. *Am J Pathol* 113: 321–330, 1983.
122. Ohata M, Ito T: Liver parenchyma with special references to extrasinusoidal macrophages and sinusoidal blood cells. Part 1. Sinusoidal cells and macrophages in normal and India-ink perfused livers. *Arch Histol Jpn* 49: 83–103, 1986.
123. Becker RP, DeBruyn PPH: The transmural passage of blood cells into the myeloid sinusoids and the entry of platelets into the sinusoidal circulation: a scanning electron microscope investigation. *Am J Anat* 145: 183–206, 1976.
124. Muto M: A scanning electron microscopic study on rat bone marrow sinuses and transmural migration of blood cells. *Arch Histol Jpn* 39: 51–66, 1976.
125. Rhodin IAG: The diaphragm of capillary endothelial fenestrations. *J Ultrastruct Res* 6: 171–185, 1962.

Author's address:
Dr. G. Macchiarelli,
Department of Anatomy
Faculty of Medicine
University "La Sapienza" of Rome
Via A. Borelli, 50
00161 Rome
Italy

CHAPTER 11

Ultrastructural approach to liver pathology

FIORENZA BONVICINI & RENZO LASCHI

1. Introduction

The microambient of the hepatocyte, as it appears at the level of electron microscopy (EM), reflects the complexity of liver functions. Studying the ultrastructure of the hepatic cell means capturing and fixing a moment of its intense metabolic life. To evaluate the normal or pathological conditions of the hepatocyte, it is therefore necessary to take into account its morphological variability in relation to its function. It is also important to know the topographical context of the hepatocytes being studied, particularly their localization in the Rappaport acinous structure. The contribution of an ultrastructural study of the liver in pathological conditions provides more detailed knowledge of the structural changes forming the substratum of the functional changes, thus allowing us to better understand the pathogenetic mechanisms of the diseases and to increase the diagnostic sensitivity of liver biopsy.

2. Cholestasis

Stasis of the bile is correlated with a number of cellular and subcellular changes, which may be primary or secondary to cholestasis itself. The interest of hepatologists in the ultrastructural study of the cholestatic liver has been twofold: to gain and to develop pathogenetic informations of possible clinical-diagnostic use, especially with regard to the diagnostic differences between intrahepatic and extrahepatic cholestasis. However, no ultrastructural marker for either type of cholestasis has been found in humans. Conversely, Miyai et al. described "specific" cholestasis lesions induced by lithocolic acid in the rat, which were subsequently confirmed by other authors [1, 2].

2.1. Essential alterations

The main alterations common to all types of cholestasis, which are defined as *essential*, are those localized in the biliary pole of the hepatocytes [3] (Figs. 11-1 to 11-3):

- a reduction in the number of microvilli up to their complete disappearance, with or without formation of cytoplasmatic blebs, which at times seem to occlude the bile canaliculus lumen: these are the earliest changes, and they appear even before the icterus and the presence of bile pigment is visible with light microscopy;
- dilatation of the bile canaliculus that can not be correlated with the entity or type of cholestasis;
- diverticulation and invagination of the canaliculi that have a winding course (so-called canalicular varicosis) and formation of new ones: these aspects are similar to what is found in newborn liver and suggest the formation of new compensatory-type canaliculi;
- alterations in the pericanalicular ectoplasm, which appears widened and at times is homogenous and displays clumping of the microfilaments. This is indicative of an alteration of the cytoskeleton of the biliary pole. The functional

Riva, A. and Motta, P.M., (eds.), Ultrastructure of the Extraparietal Glands of the Digestive Tract.
ISBN 978-1-4612-8211-2.

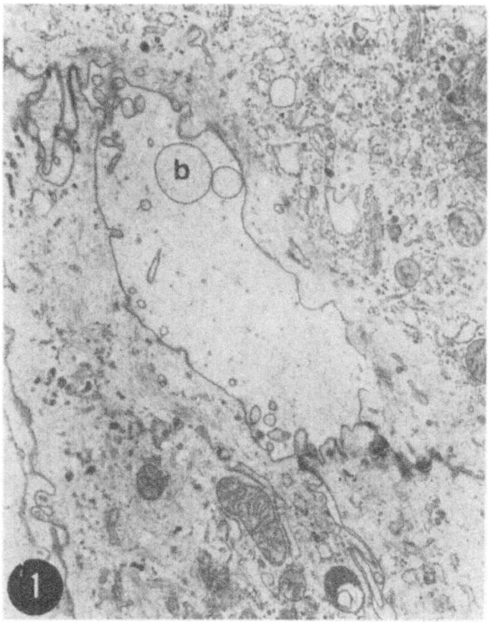

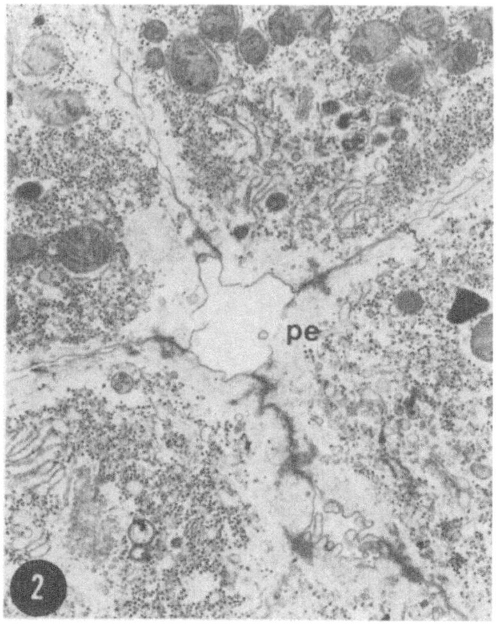

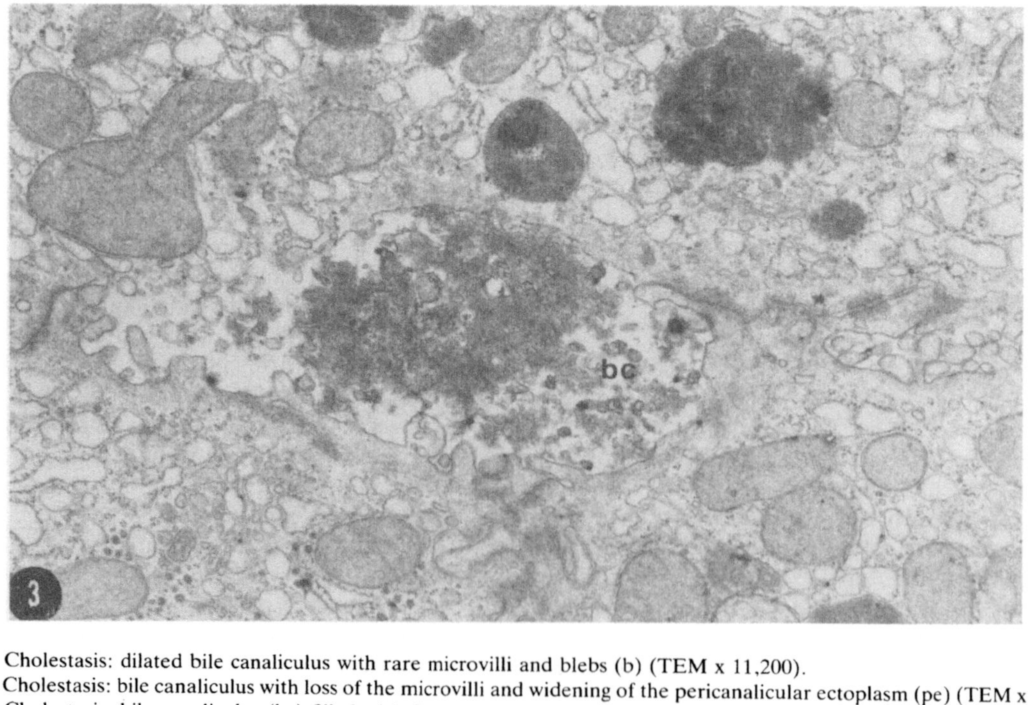

Fig. 11-1. Cholestasis: dilated bile canaliculus with rare microvilli and blebs (b) (TEM x 11,200).
Fig. 11-2. Cholestasis: bile canaliculus with loss of the microvilli and widening of the pericanalicular ectoplasm (pe) (TEM x 7,300).
Fig. 11-3. Cholestasis: bile canaliculus (bc) filled with fine granular and lamellar material (biliary plug) (TEM x 28,200).

correspondent is the loss of the tone and peristaltic capacity of the biliary canaliculus, with a consequent negative effect on the biliary flow itself. The changes of the microvilli may also be induced by alterations of the membrane microambient, primary or secondary to the cholestasis. For example, primitive alterations of the membrane are responsible for estrogen-induced cholestasis [4, 5]. The role of the pericanalicular cytoskeleton in maintaining the biliary flow is demonstrated by both experimental studies and by morphological human-pathology studies. Phalloidin and cytochalasin B, which act on the contractile myosin-associated actin filaments, cause experimental intrahepatic cholestasis [6–10]. Alterations in the microfilaments of the hepatocyte biliary pole have also been described in norethandrolone cholestasis [11]. Cholestasis with accumulation of actin microfilaments has been described in a form of familial liver disease in North American Indians [12]. The alterations of the cytoskeleton, visible at the ultrastructural level, and of the membrane, identified more easily biochemically, may also be secondary to a cholestatic process through the detergent action of the biliary salts.
the presence of electron-dense material of mainly biliary origin in the lumen of canaliculi with variable morphology: granular, homogeneous, fibrillar, and lamellar. Sometimes it appears to occlude the canalicular lumen, i.e., forming the so-called biliary plugs or thrombi.

Under normal conditions, the tight junctions between the hepatocytes are relatively impermeable. In freeze-fracture images, they appear to be constituted by parallel strands (up to five). A reduction in the number of strands can, however, be observed in cholestasis [13, 14]. There is also a widening of the intercellular spaces. These changes could explain the increased permeability of the biliary canaliculus. However, even in severe prolonged cholestasis, the cohesion of tight junctions is maintained.

2.2. *Secondary alterations*

Secondary alterations [3] are those changes that arise late and that partly represent the adaptation of the cell to the new pathological situation:

- an alteration of the smooth endoplasmic reticulum that appears to be hypertrophic. Smooth endoplasmic reticulum (SER) is the site of formation of biliary acids. The importance of the biliary salts in determining the bile flow (the so-called bile-salt-dependent flow) [15] led in the past to the hypothesis that cholestasis could be the result of a primary alteration in the metabolism of the biliary acids themselves, in particular a reduction in primary trihydroxylate in favor of the dihydroxylate and monhydroxylate forms such as the above-mentioned lithocholic acid. It is known that the biliary acids are formed from cholesterol by the microsomal system in the smooth endoplasmic reticulum. Ultrastructural studies have shown an excess of biochemically inactive membrane formation (hypoactive, hypertrophic endoplasmic reticulum) in cholestasis [16]. However, these SER alterations appear late, and it is therefore probable that they are secondary to cholestasis. They could, however, favor the maintenance of a cholestatic condition.

Alterations of other organelles include:

- dilatation, degranulation, and fragmentation of the cisternae of the rough endoplasmic reticulum;
- hypertrophy of the Golgi complex;
- an increase in the number of secondary lysosomes, namely, autophagolysosomes;
- an increase in the number of peroxisomes;
- curling of the mitochondrial cristae;
- pigment inclusions. There are two basic types of pigment inclusions: 1) bile pigment inclusions, which initially have a fine granular and fibrillar aspect and are not surrounded by a membrane. They represent deposits of bilirubin in a degenerative area of the cytoplasm. In longstanding cholestasis, these inclusions are captured in autophagic vacuoles, which merge with the lysosomes. The lysosomal enzymes deconjugate the bilirubin, which therefore reacts to stains for unconjugated bilirubin. 2) Pseudobiliary inclusions. These are the so-called myelin-like figures, probably consisting of phospholipids. They appear as vacuoles surrounded by a membrane and contain osmiophilic lamellar material arranged in spirals and combined with variable amounts of granular material. These inclusions

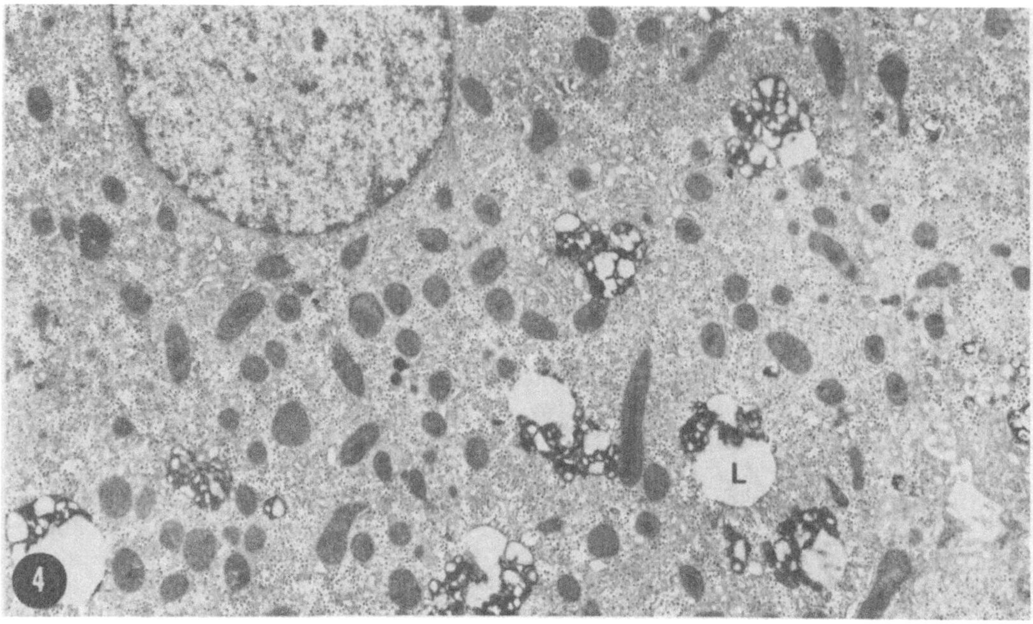

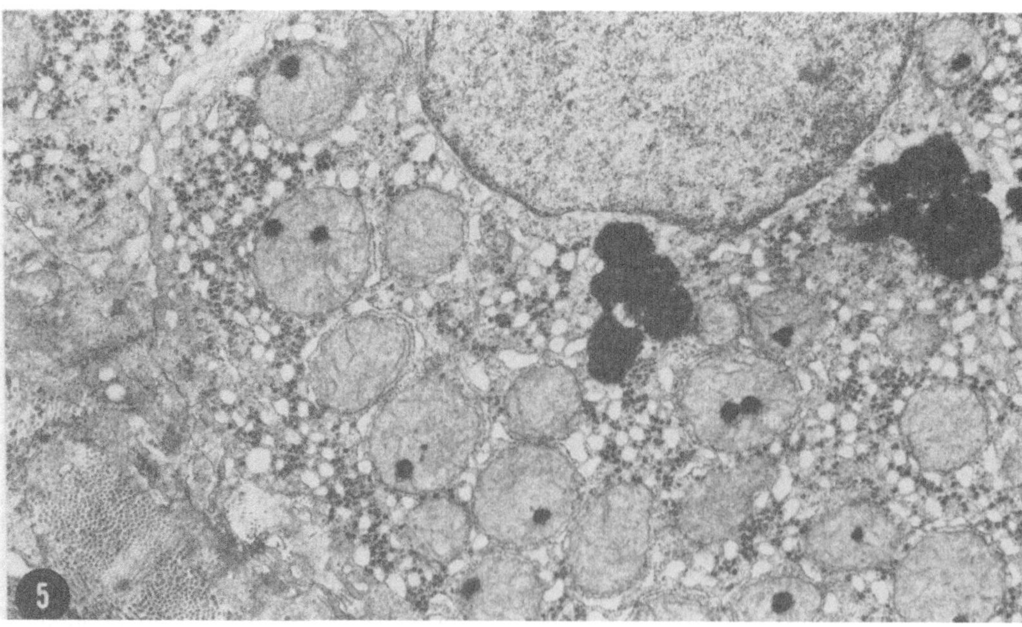

Fig. 11-4. Gilbert's syndrome: hepatocyte with pigment inclusions. The globular-lipidic (L) aspect prevails on the electron-dense granulo-filamentous component (TEM x 9,100).
Fig. 11-5. Dubin–Johnson syndrome: hepatocyte with electron-dense pigment inclusions (TEM x 30,000).

are particularly numerous in the cytoplasm of the cells presenting a "feathery" degeneration at light microscopy.

3. Constitutional hyperbilirubinemia

The term *nonhemolytic constitutional hyperbilirubinemia* (NHCH) includes some syndromes with more or less ascertained genetic transmission, at the basis of which is a reduced hepatic clearance of bilirubin in the absence of cholestasis [17]. This is due to alterations that involve various stages in bilirubin metabolism. Two basic groups of NHCH are recognized:

- nonconjugated bilirubin hyperbilirubinemia: 1) from reduced hepatic uptake of bilirubin (defective synusoidal membrane carrier system) and reduced glucuronyltransferase activity (Gilbert's syndrome); 2) from absent or reduced glucuronyltransferase activity (type I and type II Crigler–Najjar syndrome, respectively);
- conjugated bilirubin hyperbilirubinemia: 1) due to altered secretion, probably caused by a defect in the canalicular membrane carrier system (Dubin–Johnson syndrome); 2) from possible alterations in the uptake, storage, and secretion of bilirubin (Rotor's syndrome).

At light microscopy, the liver presents only minimal changes. As there are no important hepatocyte lesions, the ultrastructural picture is dominated by the presence of accumulations of pigment, the nature of which is not yet clear. The morphological aspect can not therefore be correlated with the functional aspect, even if the characteristics of the pigment vary in the different syndromes [18].

3.1. Gilbert's syndrome

The histopathological aspect of the liver is quite normal; minimal changes, such as a mild chronic infiltrate in the portal tract and a lipofuscin-like pigment at the biliary pole of perivenular hepatocytes, can be observed. At the ultrastructural level pigment inclusions appear to consist of two components: one is small and granulo-filamentous with medium to low electron density (proteins), and the other is prominent and has a globularlipidic aspect (Fig. 11-4). The same components can be observed in the Dubin–Johnson syndrome, but in inverse proportion. These inclusions seem to be associated with lysosomal structures. They are not limited to these syndromes, as Gilbert's inclusions are quite similar to lipofuscinic granules.

The presence or absence of hypertrophy of the smooth endoplasmic reticulum distinguish two types of patients, the so-called Gilbert EM positive and the Gilbert EM negative, the former having a significant increase in unconjugated bilirubin following calorie restriction [19, 20].

3.2. Dubin–Johnson syndrome

The liver has a characteristic macroscopic dark grey, or even black, appearance. The histopathological picture is not changed any further, except for the presence of coarse-granular intracellular brown pigments. At the submicroscopic level, pigment granules may be found throughout the hepatocytes but are generally localized mainly in the pericanalicular ectoplasm. Pigment may also be found in Kupffer cells. The granules have a strong electron density, being almost entirely made up of a granulo-filamentous component (Fig. 11-5). They are mainly found within membrane-bound vacuoles, 0.5–2.5 μm in diameter, of lysosomal origin. Their chemical composition is still unknown. The presence of melanin and urocoproporphyrin has been excluded. Studies with electron spin-resonance spectroscopy strongly suggest the presence of polymers of adrenaline metabolites [21]. The entity of the pigment inclusions is variable and is not correlated with the grade of hyperbilirubinemia. Minor ultrastructural changes include fragmentation and vesiculation of the rough endoplasmic reticulum, paracrystalline inclusions in the mitochondrial matrix, and slight hypertrophy of smooth endoplasmic reticulum.

In the Crigler–Najjar syndrome, the ultrastructural changes are quite aspecific. In Rotor's syndrome, the liver has a macroscopically and histologically normal appearance, with no evident pigmentation. At the ultrastructural level, the entity of pigment inclusions is variable. When present, they share common features with those of the Gilbert's and Dubin–Johnson syndromes.

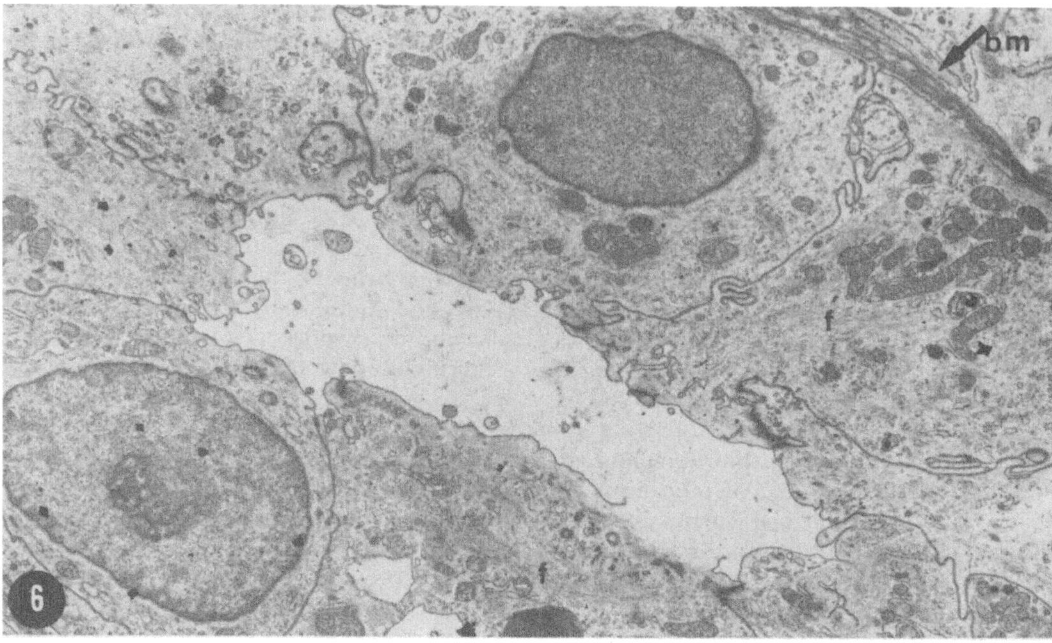

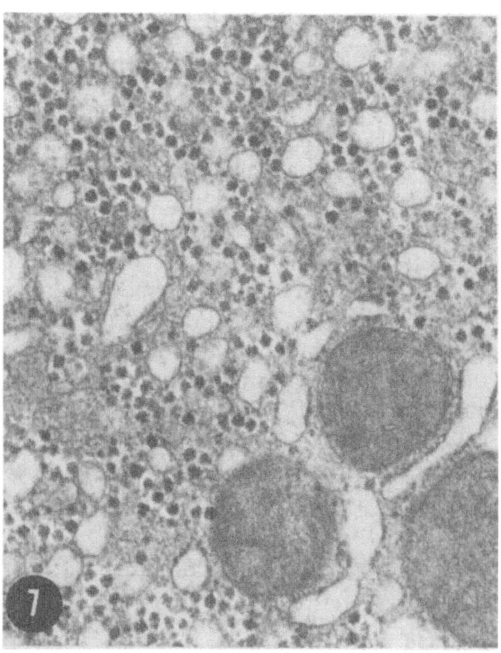

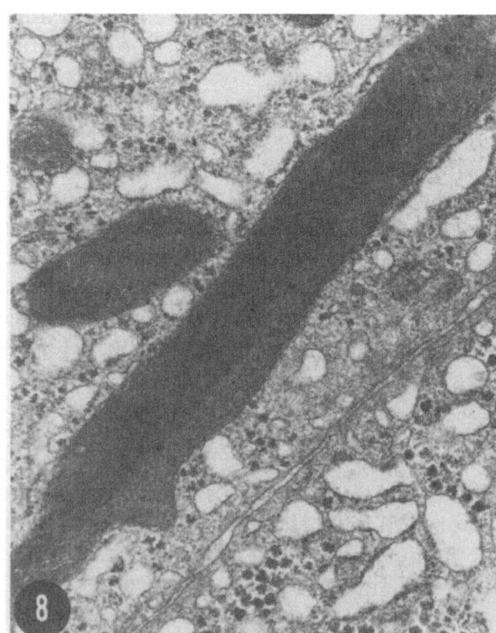

Fig. 11-6. Primary biliary cirrhosis: biliary ductule with bundles of microfilaments (f), thickening, and splitting of the basement membrane (bm) (TEM x 6,000).

Fig. 11-7. Alcoholic liver disease: hypertrophy of smooth endoplasmic reticulum (SER) (TEM x 32,000).

Fig. 11-8. Alcoholic liver disease: megamitochondrion with a crystalline inclusion (TEM x 28,000).

4. Primary biliary cirrhosis

Primary biliary cirrhosis is a chronic liver disease in which the clinical picture is dominated by cholestasis, which starts slowly and evolves gradually [22]. The etiology is still unknown, even if some aspects strongly suggest an immunological pathogenesis, i.e., the presence of mitochondrial antibodies (the antigen has been localized by immunocytochemical ultrastructural studies on the internal face of the mitochondrial membrane and on the cristae) [23], high levels of serum immunoglobulin M, the histological aspect of lymphocytic aggression towards bile ducts, etc. The main immunological event should actually be mediated by cytotoxic lymphocytes damaging the bile-duct epithelium. The disease may remain undiagnosed for a long time when the only clinical symptom is pruritus, which may be present for many years before the appearance of the icterus. For this reason, sufficient knowledge is not available on the early morphological changes of the liver. The histological staging of the disease includes: 1) florid duct lesion, 2) ductular proliferation, 3) scarring, 4) cirrhosis.

It is in the first stage that the pathognomonic alterations occur at the level of the septal and larger bile ducts. These are infiltrated by lymphocytes, which are found in abundant number in the portal tracts, together with an infiltrate of plasma cells, large histiocytes or epithelioid cells, and a few eosinophils. Granulomas can also be found in portal tract. Changes in bile-duct epithelium, such as swelling and stratification of epithelial cells and finally rupture, are seen. These findings must be differentiated from those of chronic aggressive hepatitis, in which the involved ducts are those of medium size and in which the rupture of the epithelium is never observed. Unfortunately, due to their localization, the so-called pathognomonic lesions described are more easily found in large bioptical-surgical fragments than in small-needle biopsy fragments.

Moreover, the hepatic alterations are patchy and may not evolve in parallel in different liver areas. Staging of the disease on a histological level must not be considered absolute. Ultrastructural investigation shows ducts with dilated lumen and rarefaction up to the loss of the microvilli. The epithelial cells may show cytophagosomes and lysosomes with myelinic figures. A few filamentous structures are also present in the hyaloplasm. The hepatocytes often show the presence of numerous cytophagosomes, alterations of the mitochondria, such as curling of the cristae, and crystalline inclusions in their matrix. Glycogen bodies and peroxisomes that are increased in volume have also been reported. Bile canaliculi appear normal. The destruction of the bile ducts is followed by replacement with fibrous tissue, while there is an attempt at compensation by ductular proliferation (stage 2). In this stage, a decrease in cytophagosomes at both the ductular and hepatocytary level is observed, on one hand, with an increase in microfilaments that appear grouped in bundles, on the other hand. Degranulation of rough endoplasmic reticulum can also be seen. The basement membrane may be normal or reduplicated [24].

In stages 3 and 4, the histological picture may be merely indicative and not pathognomonic. On the other hand, electron-microscopic study shows typical alterations of these stages, i.e., thickening, doubling, and subsequent breakage of the basement membrane of residual bile ductules (Fig. 11-6). Other alterations observed are a reduction in the ductule lumen, an absence of microvilli, the presence of ductular epithelial cells that are increased in volume with electron-lucent cytoplasm alternated with cells with dense cytoplasm that are packed with organelles, a further increase in filamentous structures, hypertrophy of the Golgi complex, rarefaction and vacuolization of rough endoplasmic reticulum with degranulation, and an increased number of large cytophagosomes and lysosomes [25].

5. Reye's syndrome

This syndrome prevalently affects children and often occurs after a viral infection. Aspirin has also been incriminated. Encephalopathy with cerebral edema and fatty infiltration of the viscera are hallmarks of this disease, causing the death of approximately 40% of patients. Liver histopathology reveals diffuse microglobular steatosis, while no cell necrosis is found. Early in the course of the disease, electron microscopy shows a diffuse alteration in mitochondria, which are swollen and pleomorphic. The same applies to brain mitochondria, suggesting that the metabolic alterations occurring in this syndrome are dependent of impaired mitochondrial function, which correlates with the ultrastructural damage [26].

6. Alcoholic liver disease

Hepatic damage caused by alcohol can be translated into the following anatomoclinical pictures: steatosis (or fatty liver), alcoholic hepatitis, and cirrhosis. All three can coexist. The strict correlation between the extent and duration of alcoholic intake and the severity of hepatic damage has now been extensively demonstrated, both in humans and experimentally. Genetic factors predisposing individuals toward a certain type of hepatic damage can not, however, be excluded, above all with regard to alcoholic hepatitis and cirrhosis [27].

6.1. Fatty liver

The histopathological picture is macroglobular steatosis localized prevalently in the perivenular hepatocytes. The nucleus of the hepatocytes is in an eccentric position with respect to the large fat droplets. When the membrane of two contiguous hepatocytes breaks, so-called fatty cysts are formed. Small areas of hepatocytic necrosis surrounded by mononuclear infiltrate make up the *lipogranulomas*. At the ultrastructural level, alcohol-dependent alterations are observed that reflect the involvement of certain metabolic pathways. Hypertrophy of the smooth endoplasmic reticulum (Fig. 11-7) correlates well with the activation of the microsomal ethanol oxidizing system (MEOS). The mitochondrial alterations are clearly visible and consist of disorientation and shortening of the cristae, an increase in the dense granules of the matrix, the presence of vacuoles with fibrillar and lamellar inclusions, and breaks in the membrane. Spherical megamitochondria (up to 12 μm in diameter), which are lengthened (up to 15 μm in diameter), containing crystalline inclusions (Fig. 11-8), or having an irregular morphology with many cristae and crystalline inclusions, are clearly visible. The mitochondrial alterations (like the SER alterations) are not specific, but are probably correlated with the metabolic commitment of the mitochondria in the metabolism of the acetaldehyde formed by the oxidation of ethanol. The acetaldehyde probably depresses important mitochondrial functions. Direct damage of the mitochondrial membrane by alcohol has also been suggested; for the moment, however, the alcohol/hepatocyte membrane (plasmatic and others) relationships have still to be defined. A peculiar type of lysosome was described in the liver in subjects with alcohol abuse and associated vitamin A deficiency. The lysosomes, called *multivesicular*, appear limited by a single membrane with a homogeneous electron-dense crescent and contain lipid-like particles of different sizes [28].

6.2. Alcoholic hepatitis

Alcoholic hepatitis has a histopathological picture characterized by ballooning of the hepatocytes with different degrees of steatosis and necrosis. The most altered hepatocytes are the perivenular ones. In the portal tract and in the parenchyma, an infiltrate of polymorphonuclear leucocytes is present. Fibrosis is initially localized in the perivenular area with so-called sclerosing hyaline necrosis. Capillarization of sinusoids also occurs, i.e., formation of a basement membrane in the space of Disse. In the cytoplasm of some hepatocytes, Mallory bodies can be observed, which appear in hematoxylin-eosin (H/E) as reniform intracytoplasmatic inclusions with an intense eosinophil color. Ultrastructural study has contributed to the identification of three types of Mallory bodies: type 1, made up of fibrils in a parallel arrangement with a diameter of 14.1 nm; type 2, consisting of filaments orientated at random, with an average diameter of 15.2 nm. These are the most frequently found; type 3, with almost complete disappearance of fibrillar structures and consisting of granular or homogeneous electron-dense material.

The nature of the Mallory-body microfilaments is still being studied: it is known that their chemical composition and their structure are different from that of actin filaments, of vimentin, and of microtubules; nevertheless, the experimental administration of antitubulin agents, such as colchicine and griseofulvine, induces the appearance of Mallory bodies. One hypothesis is that they consist of prekeratin [29–31].

Mallory bodies have always been considered the hallmark of alcoholic hepatitis. We should, however, remember that they are also found in other diseases, though to a lesser extent, such as primary biliary cirrhosis, Wilson's disease, and cholestasis. Their number is correlated with the severity of the necrosis, steatosis, and cholestasis.

Various data suggest that alcohol or its metabolites are capable of evoking a cell-mediated immunological response. For example, lymphocytes from patients with alcoholic hepatitis are cytotoxic towards autologous hepatocytes. Mallory bodies are capable of evoking an immunological response, which has been thought to play a role in the pathogenesis of the chronic development of hepatic damage. A chemotactic effect of Mallory bodies with respect to the polymorphonuclear leucocytes has also been demonstrated. This effect, however, is not always present in vivo, since immunochemical techniques have also shown the presence of Mallory bodies in the fatty liver without neutrophil infiltrate.

The problem of the mechanisms leading to fibrosis, and finally to cirrhosis, is a basic point that still remains unsolved. In fact, although it is known that necrosis and inflammation induce scarring, these do not always precede cirrhosis. The first deposition of fibrous tissue is around the terminal hepatic veins (perivenular), and this is observed above all, to be associated with necrotizing processes known as *sclerosing hyaline necrosis*. However, perivenular fibrosis may also be present in simple fatty liver, as can be seen in both human and experimental studies. In this case, it is an unfavorable prognostic sign of evolution towards cirrhosis. Ultrastructural and immunocytochemical studies have made, and will probably continue to make, a great contribution to the knowledge of the mechanisms of fibrogenesis. Some authors have shown that the first event of the fibrosis process is an increase in the number of perivenular myofibroblasts [32]. It is suggested that alcohol and its metabolites induce the proliferation of these cells, which produce collagen. The fat-storing cells (Ito cells, lipocytes) have been considered possible precursors of myofibroblasts. Alcohol transforms them into transitional cells (activated cells) containing fewer lipids and more rough endoplasmic reticulum, which also have fibrogenetic capacity [33].

7. Drug-induced liver injury

Drug-induced liver injury represents a phenomenon of growing proportions. The liver is often the target organ, since it plays a primary role in the metabolism and detoxification of the majority of drugs. This process takes place in the microsomes of the smooth endoplasmic reticulum. The hepatic damage caused by drugs is polymorphous and can therefore mimic diseases with different etiologies (acute and chronic hepatitis, biliary obstruction, granulomatosis, fibrosis, vascular changes, tumors). The pathogenesis of the alterations found in the liver has not yet been fully explained. Drugs (and toxins) can be divided into two groups: those that cause true and predictable damage (this, in turn, may be the result of a direct effect or of interference with a metabolic pathway) and those that cause unpredictable damage (or hypersensitivity-related idiosyncratic reactions or metabolite-dependent toxic reactions) [34]. There are more than 400 drugs that can induce acute or chronic hepatic damage. The histopathological picture may be similar to that found in acute viral hepatitis. The lack of necrotic aspects and the presence of an eosinophil infiltrate may suggest drug-induced damage; drugs such as alothan and isoniazid can cause fulminant hepatitis, which is morphologically similar to the viral form. Other drugs give rise to a cholestatic form of hepatitis, to chronic hepatopathies, acute or chronic intrahepatic cholestasis, etc. Macroglobular steatosis is a frequent result of prolonged treatment with corticosteroids, while the microglobular form can be caused by tetracyclines. Amiodarone therapy causes hepatic injuries resembling those of alcoholic hepatitis. This shows how difficult it is to make a correct diagnosis, even when the histopathological picture of the liver is considered together with the clinical and laboratory data. The contribution of electron microscopy to the study of the liver in drug-induced injury is not universally acclaimed, although there is a great deal of literature on the subject [35]. The major weakness of ultrastructural study in this field is the aspecificity of the alterations. Its importance has, however, recently been reproposed and stressed. Some authors [36] have proposed a series of major and minor criteria, which, when combined, can be highly indicative of drug-induced hepatic damage and can therefore provide a valid support for the histopathological and clinical picture of a correct diagnosis. The major criteria include marked proliferation of smooth endoplasmic reticulum due to the induction of microsomal enzymes, the presence of concentric membranous lamellar formations derived from the endoplasmic reticulum, and lysosomal phospholipidosis. The latter is the morphological expression of the drug-phospholipid interaction with the formation of a complex in which the phospholipids can not be metabolized. Under light microscopy, the hepatocytes have a foamy appearance (they simulate congenital lysosomal diseases); at the ultrastructural level, they appear to contain membrane-bound crystalloid or lamellated residual bodies. Other alterations include giant mitochondria with bizarre shapes, an increased number of peroxisomes, and nuclear alterations such as *nucleolar segregation*.

Minor changes include vesiculation of endoplasmic reticulum, lipofuscinosis, an increased number of lipolysosomes, glycogen bodies, a decrease in glycogen particles, and an increase in Ito cells.

8. Viral hepatitis

Electron microscopy and immunocytochemistry at the ultrastructural level have played a determinant role in the morphological knowledge of the various antigenic components of the hepatitis viruses, as well as their localization in the hepatic cell.

8.1. Hepatitis type A virus (HAV)

This is a picorna virus (25–28 nm) with cubic symmetry and no external envelope or subunits. Electron microscopy has identified two types of particles, full (unpenetrated by stain) and empty (penetrable by stain). Empty particles probably represent virions devoid of nucleic acid. Immunoelectron microscopy has confirmed the presence of 27 nm particles in cytoplasmic vesicles and lysosomes of hepatocytes and Kupffer cells. No particles were found in the nucleus [37]. It seems likely that the virus has a prevalent cytopathic effect.

8.2. Hepatitis type B virus (HBV)

Negative staining and immunoelectron microscopy first identified the "Dane particle" in the serum of patients. This 42 nm particle consists of a 28 nm inner core that corresponds immunologically to the hepatitis-B core antigen (HBcAg) and an outer shell corresponding to hepatitis-B surface antigen (HBsAg). Dane particles were localized by electron microscopy in the cisternae of endoplasmic reticulum and in intercellular and Disse spaces.

Hepatitis-B core antigen (HBcAg) was identified both in the nucleus and cytoplasm of the infected hepatocytes as fully or empty spherical particles (Fig. 11-9) whose outer coat is formed by 7–10 nm subunits. The infected nucleus is enlarged and has a rarefied chromatin and nucleoplasm (sanded nuclei at light microscopy). Core particles have also been localized around the nuclear pores and the perinuclear cisternae; in the cytoplasm, they may be aggregated or spread. Sometimes they are found in close proximity to the plasma membrane [38–40].

Hepatitis-B surface antigen (HBsAg) appears as spherical or tubular particles, both in serum and hepatocytes (Fig. 11-10). In the hepatocytes these particles are seen only within the endoplasmic reticulum membranes, where they are synthesized. They may be spherical in shape (35 nm in diameter with a dense core) or tubular and of variable length, depending on the type of section. Excess antigen gives the light microscopy aspect of "ground glass" cells.

Immunoelectron microscopy made it possible to characterize both core and surface antigens and to determine their exact localization within the hepatocyte. On this basis, hypotheses have been formulated on the replication, assembly, and release of the viral particles. The virus attaches to a hepatocyte plasma membrane receptor and reaches the nucleus. Core particles are formed and pass through the nuclear pores towards the endoplasmic reticulum cisternae. Here they are coated with the surface antigen, resulting in the

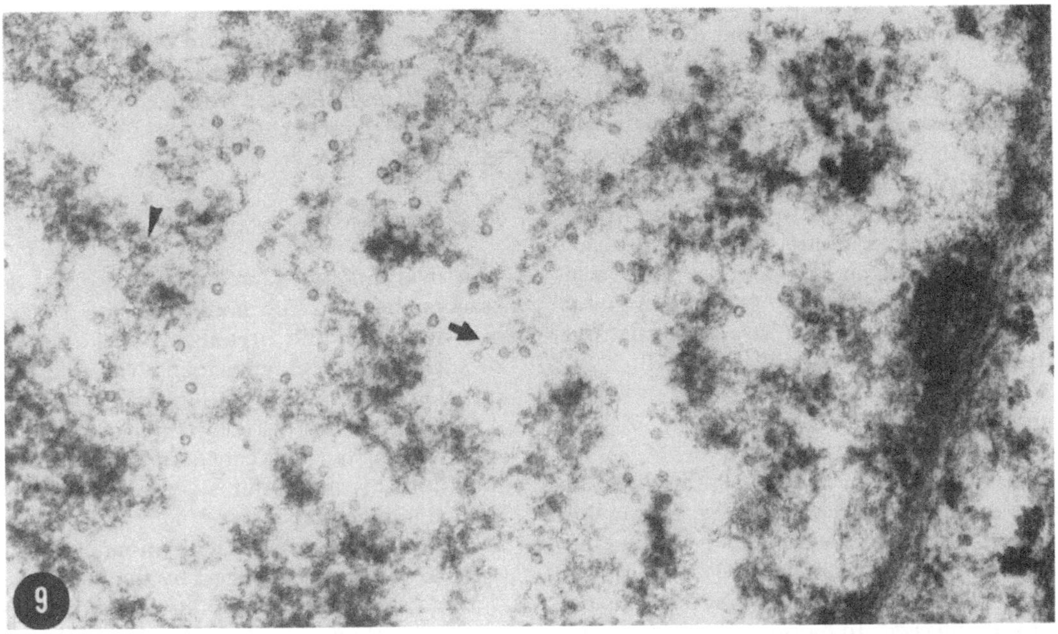

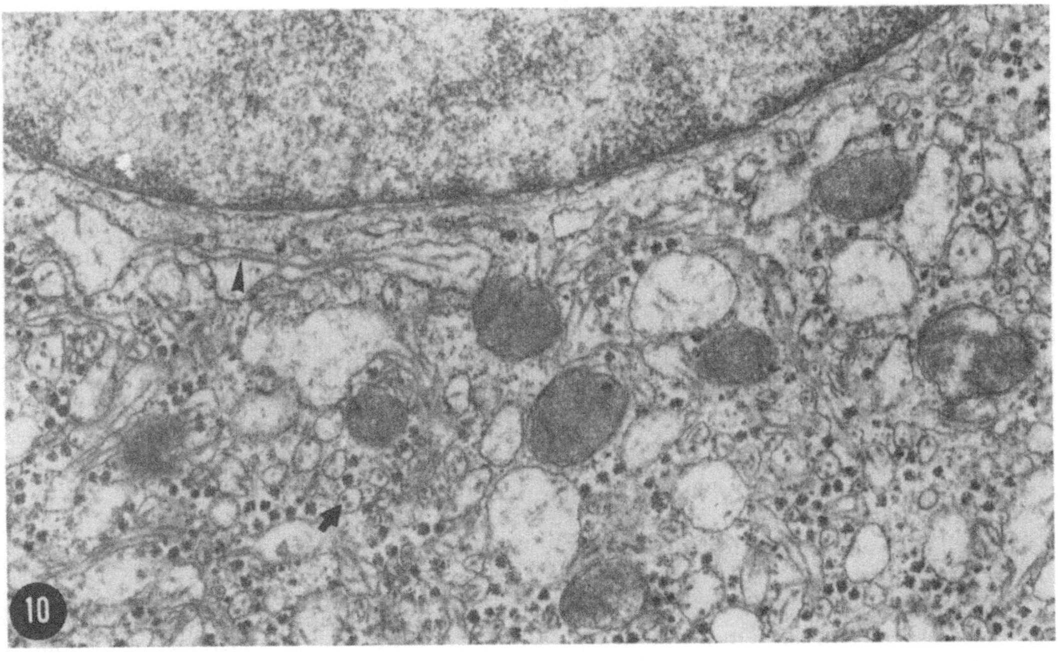

Fig. 11-9. HBV-hepatitis: empty (arrow) and full (arrowhead) HBcAg particles in the nucleus of an hepatocyte (TEM x 80,000).
Fig. 11-10. HBV-hepatitis: spherical (arrow) and tubular (arrowhead) HBsAg particles within the cisternae of smooth endoplasmic reticulum (TEM x 30,000).

complete virion. Virion and incomplete viral particles are then released into the extracellular space.

The presence and localization of the various antigens of hepatitis-B virus in the hepatocyte correlate with the clinical picture and the evolution of the disease. A cell-mediated immunoreaction to HBsAg is thought to play a role in the pathogenesis of hepatic injury [41–43].

Hepatitis-B e antigen (HBeAg) is a protein subunit of the core responsible for ongoing viral synthesis, i.e., for infectivity. Its ultrastructural localization has still to be established [44].

8.3. Hepatitis delta virus (HDV)

The Delta virus was initially interpreted as an antigen of the hepatitis-B virus, since it was detected by immunofluorescence techniques in the nucleus of HBsAg-positive patients [45]. It is now known that it is a defective virus that depends on HBsAg for its survival. In blood the viral particles are 35–37 nm in diameter and show an inner and outer component, the latter being provided by HBsAg. Recently 20–30 nm particles were found by immunoelectron microscopy in patients with chronic B hepatitis, mainly within the nucleus [46]. However, the exact ultrastructural features of the delta virus in the hepatocytes are unknown.

8.4. Non-A, non-B hepatitis virus (NANBV)

The diagnosis of NANB hepatitis is still made clinically by exclusion, since the virus has not yet been identified. Two types of hepatocellular changes were described in chimpanzees experimentally infected [47, 48]: intranuclear clusters of particles of 15–27 nm in diameter and peculiar tubular structures made up of two parallel walls (in longitudinal section) or appearing circular (in cross section). This was thought to indicate the existence of different agents for non-A, non-B hepatitis, as the two types of alterations were never found in the same animal. On the contrary, similar changes were found together in the human liver [49]. Cytoplasmic changes were interpreted as being due to a derangement of the endoplasmic reticulum. Intranuclear particles were also considered to be induced by the NANB virus: they do not correspond to the virus itself [50–53]. Electron-dense cytoplasmic fibrillar inclusions are also frequently found in the liver of NANB hepatitis and resemble the viroplasmic aggregates of togavirus, paramyxovirus, etc. [54].

9. Storage diseases

The liver functions include the storage of substances destined for nonimmediate use or that are surplus in the organism. The accumulation of such substances may lead to severe morphofunctional alterations of the liver.

There are various congenital disorders of the metabolism that can cause an intracellular accumulation of metabolites, mainly through the following mechanisms:

- synthesis of molecules with anomalous structure that are not metabolized by the cell, since they are not recognized by its catabolic systems: congenital or acquired reduction of the capacity that some substrates have for binding certain metabolites; congenital anomaly in the regulation of the synthesis of macromolecules, with resulting hyperproduction of the same;
- reduction in catabolic processes.

The storage diseases can be subdivided on the basis of the localization of the substances that accumulate, extralysosomal or lysosomal. Electron microscopy has been of fundamental importance for the characterization of the so-called lysosomal diseases; in fact, ultrastructural investigation makes it possible not only to know the subcellular topography of the deposits, but also to recognize their composition on the basis of their ultrastructure. This accounts for an important diagnostic application of the electron microscope in this field [55].

9.1. Congenital lysosomal diseases

These diseases are commonly classified on the basis of the type of substance accumulated (i.e., mucopolysaccharidosis, glycolipidosis, etc.). The major limitation of this classification lies in the fact that it does not take into account the forms of mixed accumulation resulting from multiple and/or nonidentified enzymatic defects.

9.1.1. Mucopolysaccharidosis.

Light microscopy shows vacuolated hepatocytes. Electron microscopy shows the characteristic presence of hypertrophic secondary lysosomes (5–10 μm in diameter) in which a flocculent or granulofilamentous material with low electron density is present. These lysosomes are present in Kupffer cells and in the hepatocytes.

9.1.2. Niemann–Pick disease.

Three forms of this disease can be distinguished on the basis of the degree of enzymatic defect and clinical signs: acute infantile, subacute, and adult, all three of which have hepatic (and splenic) involvement. Light-microscopic diagnosis is based on the presence of the so-called Pick cells, i.e., large cells with a diameter at times exceeding 40 μm, containing one or two often pycnotic nuclei. These cells present a honeycomb aspect when treated with lipid solvents. Histochemical staining has shown the presence of sphingomyelin. The ultrastructural correspondent is the presence of lysosomal inclusions containing electron-dense material sometimes alternating with transparent lamellae. The hepatocytes also show lysosomal inclusions, with a matrix containing membranous structures. The storage product is often arranged in dark and light lamellae with the characteristic periodicity of complex lipids arranged in parallel or concentric form. Kupffer cells are larger and contain lysosomes with a prevalently granulo-filamentous material.

9.1.3. Gaucher's disease.

Three types of clinical presentation are also recognized in this form of glycolipidosis: acute or infantile, subacute or juvenile, and adult. All these show splenomegaly. This disease is in fact characterized by an accumulation of glucosylcerebrosides, which fill the histiocytic cells. Under light microscopy, Gaucher's cells are large (20–80 μm in diameter) and mononucleate, and contain parallel streaks. At the ultrastructural level, these streaks are found within large lysosomes (0.6–4 μm in diameter). They are made up of tubular structures (13 and 75 nm thick), are tightly packed, and are parallel to the main axis of the lysosomal vacuoles (Fig. 11-11). These structures identify glucocerebrosides, which are the only complex lipids that do not form pseudomyelinic figures. Gaucher's cells may also contain erythrophagosomes, and this accounts for the content of ferritin detectable with microanalytical methods. The hepatocytes are not affected.

9.1.4. Gangliosidosis GM 1 (Norman–Landing disease).

The hepatocytes appear vacuolated due to the presence of lysosomes (0.3–11 μm in diameter) that contain flocculent or medium granular material along with disarranged filamentous material (10–20 nm thick and 1–1.5 μm long) and rare amorphous homogeneous spherules. The same lysosomal structures are found in endothelial cells and Kupffer cells.

9.1.5. Glycogenosis type II or Pompe's disease.

In this disease the glycogen is not broken down in the lysosomes, due to the lack of the lysosomal enzyme, acid alpha-1.4 glucosidase. This leads to the accumulation of glycogen in various parenchymal cells. The most evident clinical alterations are weakness of the skeletal muscles, hepatomegaly, cardiomegaly, and macroglossia. Light microscopy can not differentiate this disease from other glycogenosis such as Von Gierke's disease. Electron microscopy shows an increased glycogen content in the hyaloplasm of the hepatocytes and glycogen-filling,lysosomal structures (glycogen residual bodies) in the form of both rosettes and single particles (lysosomal glycogen) (Fig. 11-12). This differentiates Pompe's disease from Von Gierke's, where glycogen accumulates outside the lysosomes in the hyaloplasm and also in the nucleus. Lysosomes loaded with glycogen are also found in endothelial cells and Kupffer cells, and in epithelial cells of bile ductules.

9.2. Wilson's disease

This disease may also be considered as a lysosomal disease, i.e., a defective excretion of copper from the lysosomes to the bile [56].

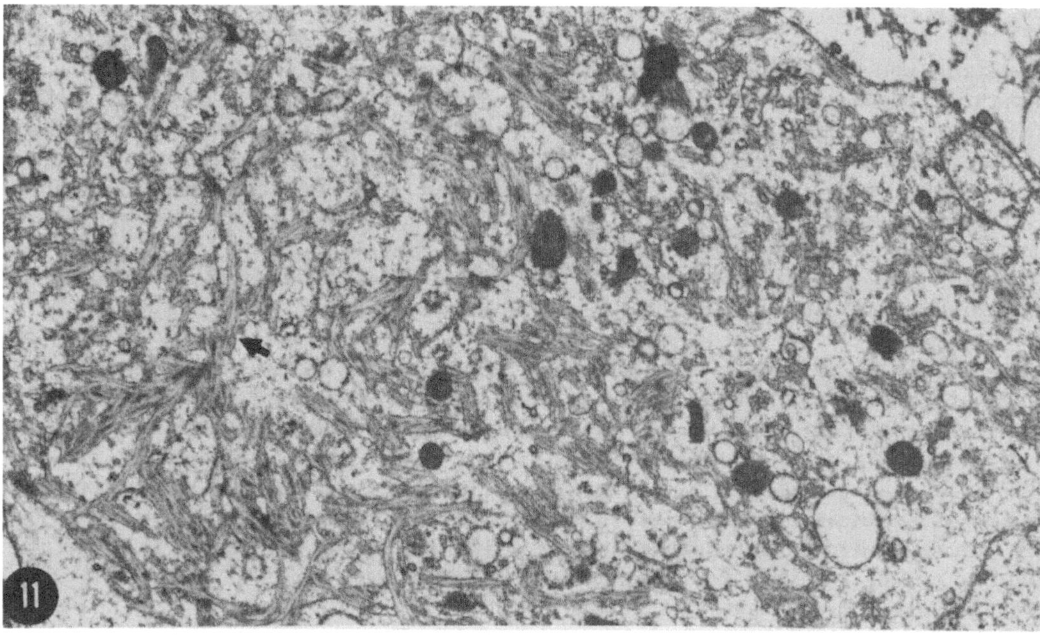

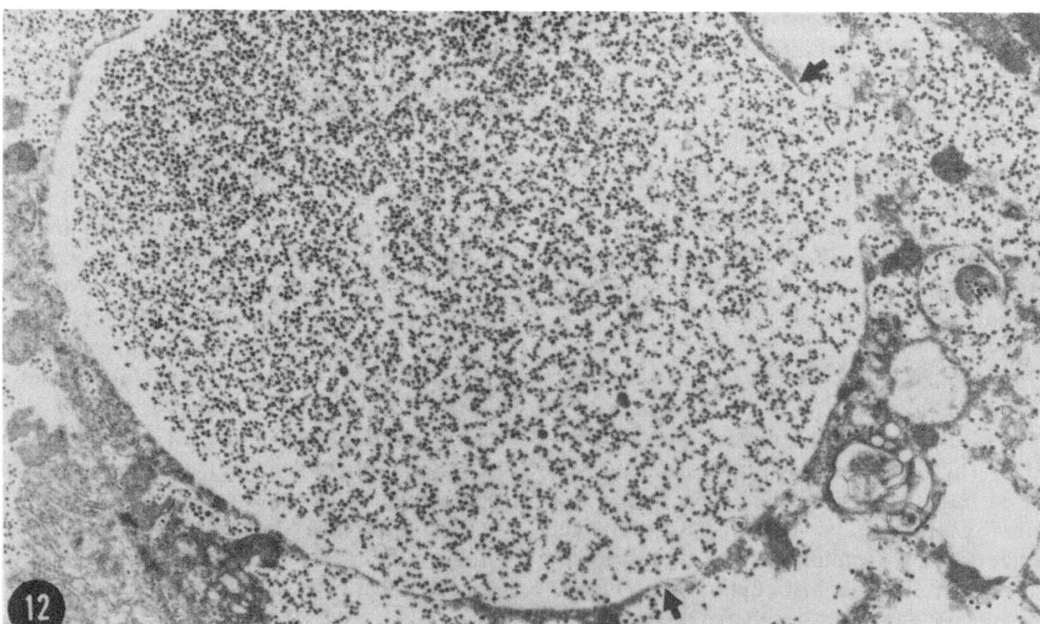

Fig. 11-11. Gaucher's disease: Kupffer cell filled with tubular structures (arrow) (TEM x 11,800).
Fig. 11-12. Pompe's disease. Hepatocyte with a large lysosomal vacuole (arrows) filled with glycogen particles (TEM x 9,000).

This is an autosomal recessive disease in which degeneration of basal ganglia and cirrhosis of the liver occur as a consequence of abnormal copper deposition. The Kayser–Fleischer ring in the cornea is also due to copper deposits. In addition to excess copper in the liver, the other primary alteration is a deficiency of caeruloplasmin, a copper-binding glycoprotein. Lysosomes of patients affected by Wilson's disease were found to contain 43 times the concentration of copper found in normal subjects. In the early stages of the disease, the liver shows only fatty changes under light microscopy. Transmission electron microscopy shows aspecific alterations such as mitochondria alterations (pleomorphism, separation of outer from inner membranes, enlarged dense granules); these changes are probably related to steatosis, as well as to changes in peroxisomes. These organelles, in fact, increase in number and are enlarged, with matrices of different electron density. In the later stages, hepatocytes show normal intracytoplasmic organelles, except for the presence of numerous lysosomes with a dense osmiophilic matrix and clear round vacules (lipolysosomes). In the early stages of the disease, electron-probe x-ray microanalysis shows the presence of an abnormal amount of copper in both the lysosomes and the cytoplasm [57]. With the progression of the disease, copper accumulates only in the lysosomes. An excess of copper in the liver is also found in primary biliary cirrhosis. A combination of clinical, laboratory, and histopathological data are therefore necessary for the diagnosis. Electron microscopy and related techniques such as microanalysis may be useful, providing additional information in the early stages of the disease.

10. Concluding remarks

This chapter reviews the main diseases and syndromes in which electron microscopy has made an important contribution to a better understanding of the data supplied by light microscopy, knowledge of pathogenetic mechanisms, and diagnosis. As far as the latter point is concerned, hepatocyte ultrastructural pathology is often expressed by aspecific changes that can otherwise become relevant when evaluated in a clinical context. This implies continuous collaboration between pathologists practicing electron microscopy and clinicians.

To date, the majority of studies on liver ultrastructural pathology have been performed with traditional transmission electron microscopy (TEM) techniques [58–62]; application of immunoelectronmicroscopy is now growing. We can expect future important contribution from the systematic use of new high resolution-power instruments and new technologies applied to both transmission and scanning electron microscopy.

References

1. Miyai K, Price VM, Fisher MM: Bile acid metabolism in mammals: ultrastructural study on the intrahepatic cholestasis induced by lithocholic and chenodeoxycholic acids in the rat. *Lab Invest* 24: 292–302, 1971.
2. Bonvicini F, Gautier A, Gardiol D, Borel G-A: Cholesterol in acute cholestasis induced by taurolithocholic acid. A cytochemical study in transmission and scanning electron microscopy. *Lab Invest* 38: 487–495, 1978.
3. Desmet VJ: Cholestasis; extrahepatic obstruction and secondary biliary cirrhosis. In: Pathology of the Liver. RNM MacSween, PP Anthony, PJ Scheuer (eds), London: Churchill Livingstone 272–305, 1979.
4. Davis RA, Kern F, Showalter R, Sutherland E, Sinensky M, Simon FR: Alterations of hepatic (Na^+K^+)-ATPase and bile flow by estrogen: effects on liver surface membrane lipid structure and function. *Proc Natl Acad Sci USA* 75: 4130–4134, 1978.
5. Boelsterli UA, Rakhit G, Balazs T: Modulation by S-adenosyl-methionine of hepatic Na^+, K^+-ATPase, membrane fluidity, and bile flow in rats with ethinyl estradiol-induced cholestasis. *Hepatology* 3: 12–17, 1983.
6. Oda M, Price VJ, Fisher MM, Phillips MJ: Ultrastructure of bile canaliculi, with special reference to the surface coat and pericanalicular web. *Lab Invest* 31: 314–323, 1974.
7. Phillips MJ, Oda M, Mak E, Fisher MM, Jeejeebhoy KN: Microfilament dysfunction as a possible cause of intrahepatic cholestasis. *Gastroenterology* 69: 48–58, 1975.
8. Adler M, Chung KW, Schaffner F: Pericanalicular hepatocytic and ductular microfilaments in cholestasis in man. *Am J Pathol* 98: 603–616, 1980.
9. Elias E, Hruban Z, Wader JB, Boyer JL: Phalloidin induced cholestasis: a microfilament-mediated change in junctional complex permeability. *Proc Natl Acad Sci USA*: 2229–2233, 1980.
10. Mesland DAM, Los G, Spiele H: Cytochalasin B disrupts the association of filamentous web and plasma membrane in hepatocytes. *Exp Cell Res* 135: 431–435, 1981.
11. Phillips MJ, Oda M, Funatsu K: Evidence for microfilament involvement in norethandrolone-induced intrahepatic cholestasis. *Am J Pathol* 93: 729–744, 1978.

12. Weber AM, Tuchweber B, Yousef I, Brochu P, Turgeon C, Gabbiani G, Morin CL, Roy CC: Severe familial cholestasis in North American Indian children: a clinical model of microfilament dysfunction? *Gastroenterology* 81: 653–662, 1981.
13. De Vos R, Desmet VJ: Morphology of liver cell tight junctions in ethinyl estradiol induced cholestasis. *Pathol Res Pract* 171: 381–383, 1981.
14. Desmet VJ, De Vos R: Tight junctions in the liver. In: Progress in Liver Diseases, Vol 7. H Popper, F Schaffner (eds), New York: Grune and Stratton, 31–50, 1982.
15. Schaffner F, Popper H: Classification and mechanisms of cholestasis. In: Liver and Biliary Disease. R Wright, GH Millward–Sadler, KGMM Alberti, S Karran (eds) London: Baillière Tindall, WB Saunders, 359–386, 1985.
16. Schaffner F, Popper H: Cholestasis is the result of hypoactive hypertrophic smooth endoplasmic reticulum in the hepatocyte. *Lancet* 2: 355–359, 1969.
17. Gollan G, Schmid R: Bilirubin metabolism and hyperbilirubinaemic disorders. In: Liver and Biliary Disease: R Wright, GH Millward-Sadler, KGMM Alberti, S Karran (eds), London: Baillière Tindall, WB Saunders, 301–357, 1985.
18. Minio-Paluello F, Gautier A, Magnenat P: L'ultrastructure du foie humaine lors d'ictères idiopathiques chroniques. III. Inlusions pigmentoires dans les syndromes de Gilbert, de Rotor et de Dubin-Johnson. *Z Zellforsch Mikrosk Anat* 72: 168–183, 1966.
19. Dawson J, Carr-Locke DL, Talbot JC, Rosenthal FD: Gilbert's syndrome: evidence of morphological heterogeneity. *Gut* 20: 848–853, 1979.
20. Gasbarrini G, Bonvicini F, Corazza GR, Miglio F, Corinaldesi R, Bernardi M, Pesa O, Marcacci R: Ultrastructural studies of the liver in Gilbert's disease before and after barbiturate therapy. *Il fegato* 19: 743–762, 1973.
21. Varma RR, Sarna T: Hepatic pigments in Dubin–Johnson syndrome and mutant Corriedale sheep are not melanin. *Gastroenterology* 84: 1401, 1983.
22. Sherlock S: Primary biliary cirrhosis. In: Liver and Biliary Disease. R Wright, GH Millward-Sadler, KGMM Alberti, S Karran (eds), London: Baillière Tindall, WB Saunders, 861–879 1985.
23. Bianchi FB, Penforld PL, Roitt IM: Mitochondrial antibodies in primary biliary cirrhosis. V. Ultrastructural localization of the antigen to the inner mitochondrial membrane using a direct peroxidase conjugate. *Br J Exp Pathol* 54: 652–657, 1973.
24. Chedid A, Spellberg MA, De Beer RA: Ultrastructural aspects of primary biliary cirrhosis and other types of cholestatic liver disease. *Gastroenterology* 67: 858–869, 1974.
25. Busachi CA, Badiali De Giorgi L, Bianchi FB: Increased ductular microfilaments in primary biliary cirrhosis. *J Submicrosc Cytol* 11: 409–412, 1979.
26. Partin JC, Schubert WK, Partin JS: Mitochondrial ultrastructure in Reye's syndrome (encephalopathy and fatty degeneration of the viscera). *N Engl J Med* 285: 1339–1343, 1971.
27. Lieber CS, Salaspuro MP. Alcoholic liver disease. In: Liver and Biliary Disease. R Wright, GH Millward–Sadler, KGMM Alberti, S Karran (eds), London: Baillière Tindall, WB Saunders, 881–947, 1985.
28. Leo MA, Sato M, Lieber CS: Effects of hepatic vitamin A depletion on the liver in humans and rats. *Gastroenterology* 84: 562–572, 1983.
29. Borenfreund E, Higgins PG, Peterson E: Intermediate-sized filaments in cultured rat liver tumor cells with Mallory body-like cytoplasm abnormalities. *J Natl Cancer Inst* 64: 323–333, 1980.
30. Franke WW, Denk H, Schmid E, Osborne M, Weber K: Ultrastructural, biochemical and immunologic characterization of Mallory bodies in livers of griseofulvin-treated mice. Fimbriated rods of filaments containing prekeratin-like polypeptides. *Lab Invest* 40: 207–226, 1979.
31. Kimoff RJ, Huang S: Immunocytochemical and immunoelectron microscopic studies on Mallory bodies. *Lab Invest* 45: 491–503, 1981.
32. Lieber CS, Nakano M, Worner TM. Ultrastructure of the initial stages of hepatic perivenular fibrosis after alcohol. *Transact Assoc Am Physicians* 94: 292–330, 1981.
33. Mak KM, Leo MA, Lieber CS: Alcoholic liver injury in baboons: transformation of lipocytes to transitional cells. *Gastroenterology* 87: 188–200, 1984.
34. Ishak KG: Pathology of the drug-induced toxic diseases. In: The Liver. RH Ridell (ed), New York: Churchill-Livingstone 457–514, 1982.
35. Jezequel AM: Ultrastructural changes induced by drugs in the liver. In: The Hepatobiliary System. W Taylor (ed), London: Plenum Press, 179–203, 1976.
36. Phillips MJ: Electron microscopy of drug effects in human liver disease. In: The Liver. An Atlas and Text of Ultrastructural Pathology. MJ Phillips, S Poucell, J Patterson, P Valencia, (eds). New York: Raven Press, 159–238, 1987.
37. Shimizu YK, Shikata T, Beninger PR, Sata M, Setoyama H, Abe H, Tanikawa K: Detection of hepatitis A antigen in human liver. *Infect Immun* 36: 320–324, 1982.
38. Gudat F, Bianchi L, Sonnabend W, Thiel G, Aenishaenslin W, Stalder GA: Patterns of core and surface expression in liver tissue reflect state of specific immune response in hepatitis B. *Lab Invest* 31: 1–9, 1975.
39. Hirschan SZ, Gerber M, Garfinkel E: Observations on the core particles of hepatitis B virus and the DNA polymerase associated with hepatitis B antigen. *Am J Med Sci* 270: 141–149, 1975.
40. De Vos R, Ray MB, Desmet VJ: Electron microscopy of hepatitis B virus components in chronic active liver disease. *J Clin Pathol* 32: 590–600, 1979.
41. Ray MB, Desmet VJ, Bradburne AF, Desmyter J, Fevery J, De Groote J: Differential distribution of hepatitis B surface antigen and hepatitis B core antigen in the liver of hepatitis B patients. *Gastroenterology* 71: 462–469, 1976.
42. Busachi CA, Ray MB, Desmet VJ: An immunoperoxidase technique for demonstrating membrane localized HBsAg in paraffin sections of liver biopsies. *J Immunol Methods* 19: 95–99, 1978.
43. Yamada G, Nakane PK: Hepatitis B core and surface antigens in liver tissue. *Lab Invest* 36: 649–659, 1979.
44. Arnold W, Niesen JO, Hardt F, Meyer Zum Buschenfeld KH: Localization of HBe antigen in nuclei of hepatocytes

in HBsAg-positive liver disease. *Gut* 18: 994–996, 1977.

45. Rizzetto M, Canese MG, Arico' S, Crivelli O, Trepo C, Bonino F, Verme G: Immunofluorescence detection of new antigen-antibody system (δ/anti δ) associated to hepatitis B virus in liver and serum of HBsAg carriers. *Gut* 18: 997–1003, 1977.
46. Kojima T, Callea F, Desmyter J, Desmet VJ: Immuno-electron microscopy of hepatitis δ antigen in hepatocytes. *Hepatology* 55: 217–225, 1986.
47. Shimizu YK, Feinstone SM, Purcell RH, Alter HJ, London WT: Non-A, non-B hepatitis: ultrastructural evidence for two agents in experimentally infected chimpanzees. *Science* 205: 197–200, 1979.
48. Tsiquaye KN, Bird RG, Tovery G, Wyke RJ, Williams R, Zuckerman AJ: Further evidence of cellular changes associated with non-A, non-B hepatitis. *J Med Virol* 5: 63–71, 1980.
49. Busachi CA, Realdi G, Badiali De Giorgi L, Alberti A: Hepatocellular ultrastructural changes in patients with acute and chronic non-A, non-B hepatitis. *J Submicrosc Cytol* 12: 681–686, 1980.
50. Busachi CA, Realdi G, Alberti A, Badiali De Giorgi L: Ultrastructural changes in the liver of patients with chronic non-A, non-B hepatitis. *Med Virol* 7: 205–212, 1981.
51. De Vos R, De Wolf-Peeters C, Van Stapel MJ, Callea F, De Groote G, Desmyter J, Mortelmans J, Fevery J, De Groote J, Desmet VJ: New ultrastructural marker in hepatocytes in non-A, non-B viral hepatitis. *Liver* 2: 35–44, 1982.
52. Kunzer KD, Kemmer C, Porst H, Renger F. Non-A, non-B hepatitis: ultrastructural findings in human liver biopsies. *Exp Pathol* 21: 36–45, 1982.
53. Busachi CA, Badiali De Giorgi L, Alberti A, Tremolada F, Laschi R, Realdi G, Pisi E: Intranuclear particles in non-A, non-B hepatitis. *Hepatology* 4: 571–574, 1984.
54. Phillips MJ, Poucell S, Patterson J, Valencia P: Viral hepatitis. In: The Liver. An Atlas and Text of Ultrastructural Pathology. MJ Phillips, S Poucell, J Patterson, P Valencia, New York: Raven Press, 37–100, 1987.
55. Dustin P, Tondeur M, Libert J: Metabolic and storage diseases. In: Electron Microscopy in Human Medicine, Vol 2. JV Johannessen (ed), London: McGraw–Hill International, 151–221, 1979.
56. Sternlieb I, Scheinberg IH: Wilson's disease. In: Liver and Biliary Disease. R Wright, GH Millward–Sadler, KGMM Alberti, S Karran (eds), London: Baillière Tindall, WB Saunders, 949–961, 1985.
57. Goldfischer S, Moskal J: Electron probe microanalysis of liver in Wilson's disease. Simultaneous assay for copper and for lead deposition by acid phosphatase activity in lysosomes. *Am J Pathol* 48: 305–315, 1966.
58. Laschi R: Diagnostic ultrastructural du foie. *Arch Anat Cytol Path* 26: 267–270, 1978.
59. Laschi R, Busachi CA: Electron microscopy in human liver pathology. In: Basic and Clinical Hepatology. PM Motta, LJA DiDio (eds), The Hague: Martinus Nijhoff 137–161, 1982.
60. Johannessen JV: Electron microscopy in human medicine. In: The Liver, Vol 8. JV Johannessen (ed), London: McGraw–Hill International, 1979.
61. Jezequel AM: Some aspects of ultrastructural pathology. In: Liver and Biliary Disease. R Wright, GH Millward–Sadler, KGMM Alberti, S Karran (eds), London: Baillière Tindall, WB Saunders, 517–536, 1985.
62. Phillips MJ, Poucell S, Patterson J, Valencia P: The Liver. An Atlas and Text of Ultrastructural Pathology. New York: Raven Press, 1987.

Author's address:
Fiorenza Bonvicini
Institute of Clinical Electron Microscopy
University of Bologna
Policlinico S. Orsola
40138 Bologna
Italy

CHAPTER 12

The human gallbladder

LILIANA LUCIANO & ENRICO REALE

1. Introduction

The human gallbladder (*vesica fellea*) is a pear-shaped sac inserted between the under surface of the right lobe of the liver and the peritoneal serous membrane. It consists of three parts or regions: the neck, body (representing the main part of the organ), and fundus (a blind end). The neck is in continuity with the cystic duct, which connects the gallbladder with the common bile duct (*ductus choledochus*). The lack of morphological information concerning the normal human gallbladder, especially at the level of electron microscopy, was mentioned by Wallraff [1] and remains virtually unchanged to the present day. The reason for this lies in the difficulty or even impossibility of 1) obtaining healthy organs and 2) processing them quickly enough in order to achieve a satisfactory morphology. Most of our present knowledge on the structure and function of the gallbladder stems from studies on animals. Nevertheless, the question arises as to how far these findings on animals can be transferred to humans, especially in consideration of the wide anatomical and physiological differences encountered down the vertebrate scale. Schmidt and Ivy [2] demonstrated that a very close relationship exists between hepatic bile and the function of the gallbladder. In addition, they observed that the relative anatomical capacity (i.e., the volume of the organ), the physiological capacity (i.e., the number of hours in which the entire hepatic secretion can be stored in the gallbladder), and the sphincteric resistance (i.e., the resistance at the sphincter of Oddi) vary consistently from one species to another.

The capacity of the human gallbladder is subject to individual variability. The full volume values, according to x-ray cholecystography, range from 14 ml to 60 ml, with an average of 33 ml [3]. In turn, this should be related to individual variability by the emptying process of the organ, as recently demonstrated by cholescintigraphy in normal subjects [4]. The human gallbladder concentrates between 10% and 20% of the original liver bile volume (800–1000 ml of hepatic bile during 24-hour production) by selective reabsorption of water and solutes [5]. As a consequence, there is an increase in the dry content from 1%–3% (hepatic bile) to 14%–20% (gallbladder bile) without alterations of the gallbladder volume [reviewed in 1]. Its filling depends on the high sphincteric resistance [6], which is required for the gallbladder to fill up. Thus, the gallbladder can regulate the intraluminal pressure of the biliary tract in both animals [2] and humans [7].

2. The wall of the gallbladder in light and electron microscopy

The wall of the human gallbladder (normal adult subjects) measures from less than 1 mm to 3 mm. This is considered to be the upper normal value [8]; estimates higher than 3.5 mm are regarded as predictive for gallbladder disease [9]. The wall of the gallbladder is composed of three layers: *mucosa*, *muscularis*, and *adventitia* (also called

Riva, A. and Motta, P.M., (eds.), Ultrastructure of the Extraparietal Glands of the Digestive Tract.
ISBN 978-1-4612-8211-2.

perimuscular connective tissue). The inferior surface of the organ is covered by another layer, the *serosa* (peritoneum).

2.1. *Mucosa (tunica mucosa)*

Numerous folds of the mucosa subdivide the internal surface into irregularly shaped and extended areas. The folds depend on the degree of distension (from full to empty) of the gallbladder. Rokitansky–Aschoff sinuses, often erroneously interpreted as glands, are deep invaginations of the mucosa, which may even penetrate through the meshes of the muscle layer. These are believed to represent initial signs of pathological changes in the gallbladder wall [10]. Glands, probably secreting mucus, should only be found in the neck region.

The gallbladder mucosa consists of a surface epithelium and a lamina propria.

2.1.1. *Epithelium.*

The luminal surface of the gallbladder is covered by a single layer of cells of different type (Fig. 12-1). These are (1) principal cells, (2) pencil-shaped (or rod-shaped) cells, and (3) wandering (or basal) cells. Brush cells (4) and granular cells (5) have been described in other mammals or in pathological human gallbladder, respectively.

2.1.1.1. Principal cells. These are the most numerous gallbladder cells and have been found in all species examined. They are columnar (about 25 μm high, and 8 μm wide in the middle of the cell body and 4 μm wide in the basal region) and are conspicuously polarized. The nucleus is located in the lower portion of the cell, and is large and ovoidal in shape (about 8 μm long and 5–6 μm wide). Its major axis is parallel to that of the cell, and its nucleolus is prominent.

The cytoplasm in the supranuclear region is characterized by the occurrence of zones in which different organelles predominate in regular succession (Fig. 12-2a). Secretory granules form the first zone immediately below the terminal web. The granules are ordered into one or more rows parallel to the cell surface. A second zone follows, about 3 μm high, where numerous mitochondria are assembled. The third and largest zone contains the Golgi complex, mitochondria frequently flanked by small, flattened cisternae of the rough endoplasmic reticulum, dense bodies with a heterogeneous content and form, and secretory granules. The size of these granules is very different, but their content is morphologically comparable to that of the first apical zone. The number of secretory granules and mitochondria, as well as the extension of the Golgi complex, show variations from one region to another of the same gallbladder. Therefore, the subdivision into zones of the apical cytoplasm is not always clearly recognizable.

The Golgi complex — the site of cellular secretion — is usually prominent (Fig. 12-3). It consists of several stacks of sacculi that, in a longitudinal section, delineate a basally open V-shaped region of the cell. In cross section this region is roundshaped. Each stack is composed of two to five flattened sacculi. *Cis*- and *trans*-located sacculi are easily differentiated, since the *trans* (innermost or exoplasmic) sacculus displays thicker membranes than the *cis* (outermost or endoplasmic) one. Numerous small ovoidal vesicles with an electron-dense content (presumably cross sections of GERL), precursors of secretory granules, and secretory granules lie inside of the region surrounded by the Golgi stacks (Fig. 12-3). Both vesicles and granules are frequently in continuity with the innermost sacculi.

Previous investigations on the human gallbladder have demonstrated positive PAS and Alcian blue (pH 1.2) reactions in the Golgi area of the principal cells, in their apical granules, and in the brush border [11]. The PA-TCH-silver proteinate method [12] for localization of neutral mucosubstances and the colloidal iron method [13] for localization of acid mucosubstances at the electron microscopical level have confirmed the light microscope findings. More exactly, the reactions were positive 1) in the innermost sacculi of the Golgi complex, 2) in the secretory granules of the Golgi area and of the apical zone, and 3) in the surface coat and on and between the microvilli [11]. These results demonstrate that the principal cells of the human gallbladder epithelium can synthesize, store, and secrete neutral and acid mucosubstances in the same granule. The acid mucosubstances are, at least in part, sulfated. Secretion of the granule content into the gallbladder lumen occurs through a very active exocytosis, a process frequently seen in the principal cells [11]. Mucosubstances are also produced in cultures of epithelial cells of the human gallbladder [14].

Dense bodies are the most conspicuous organelles of the human principal cells (Figs. 12-2, 12-3, and 12-9a). They are largely composed of assemblies of electron-dense granules (about 30 nm in diameter), and lucent vesicles and vacuoles (ranging from 40 nm to 1 μm or even more). In addition to these structures, the dense bodies contain myelin figures (concentric bilayer, Fig. 12-3) and bundles of filaments (Fig. 12-9a). The dense bodies are surrounded by a discontinuous membrane. It is conceivable that they originate from the confluence of cytoplasmic granules and vesicles into assemblies. A second type of dense body, clearly bounded by a membrane, is round-shaped and shows a content that is remarkably electron dense, especially in areas displaying stacks of bilayers. Abundant dense bodies with a heterogeneous aspect characterize not only the principal cells of the human gallbladder, but also that of other mammals. In the mouse, for instance, their number increases with age; their heterogeneity is also visible in freeze-fracture replicas [15]. Some of the dense bodies show a positive acid-phosphatase reaction and therefore they are lysosomes [15].

It may be interesting to note that in cases of cholesterosis, the dense bodies decrease, in the epithelial cells covering extensive areas of the lamina propria containing the foam cells and lie in more apical position. In these epithelial cells also, the secretory granules of the first zone are decreased or almost absent (Fig. 12-8).

Mitochondria are again numerous in the basal region of the principal cells (Fig. 12-2a). Thus, these organelles are assembled in the second apical zone and in the basal region; they are scattered elsewhere in the cytoplasm. This typical distribution can also be seen by light microscopy (Fig. 12-1), as already shown by the schematic drawing of Nagahiro [16].

From the apical region down to the base, especially along the side of the principal cells, the cytoplasm contains an extensive smooth endoplasmic reticulum, ribosomes, glycogen granules (in the β and α forms, which is unusual outside of the liver parenchymal cells), and filaments. All these components impart an electron-dense aspect to the principal cells. The occurrence of "light cells" and "dark cells," both with landmarks of the principal cells, is a fixation artifact, since the dark cells present the morphological characteristics of shrinkage [17]. Light and dark cells, however, may also show a different structure, i.e., they may belong to different cell types (see below).

The luminal surface of the principal cells bears a tight brush border, with microvilli about 1.5 μm long and 0.12 μm thick. The surface is covered by a very electron-dense coat, forming a continuous layer in which the microvilli are embedded (Fig. 12-2a). Yamada [18] named *antennulae microvillares* the *lace-like* filaments that he observed on the microvilli of the mouse gallbladder epithelium.

After double fixation with aldehyde(s) and OsO_4, a bundle of filaments (mainly actin filaments) can be seen extending from the axial part of each microvillus to the terminal web. In this region, the filaments mingle with the tightly interwoven filaments of the terminal web. After primary fixation in OsO_4, both the axial part of the microvilli and the terminal web have a homogeneous aspect, due to the incomplete preservation of the actin by OsO_4 solutions [19]. Microvilli with a large base, bifurcations, and other morphological irregularities (malformations?) are seen in some areas and suggest an altered function of the epithelium. The lateral plasma membrane, immediately below the junctional complex, forms numerous folds. These run close and parallel to the lateral cell surface, filling the intercellular space up to the nuclear region. Below this the intercellular space may be large (Fig. 12-2a).

2.1.1.2. Pencil-shaped (or rod-shaped) cells. Since their first description by Pfuhl [20], the occurrence of these cells among those of the human gallbladder epithelium has been repeatedly confirmed. Indeed, independent of the fixative solution used, light microscopy reveals single, deeply stained cells extending from the luminal surface of the epithelium to the basement membrane. The lower third of these cells is usually thin and branched (Fig. 12-1).

According to previous studies [21], in thin sections the pencil cells differ from the principal cells in their slender form and their more electron-dense cytoplasm (Fig. 12-2b). In addition, they are characterized by 1) a large number of densely packed mitochondria with a uniform distribution in the whole cell body and 2) numerous folds and winding invaginations of the basolateral plasma membrane, especially in the basal region (Fig. 12-2b). We do not agree that pencil cells are immature forms or transitional forms between basal cells and the principal cells [22], yet we would suggest that pencil cells belong to a distinct cell type. So-called *mitochondria-rich cells* have been described in various epithelia and have been related to a special ion-transporting function [reviewed in 23]. It is conceivable that the pencil cells of the human gallbladder epithelium are also involved in a special ion transport, which differs from that of the principal cells.

2.1.1.3. Wandering (or basal) cells. Different types of cells that occur in the gallbladder surface epithelium are included in this category. Their usually basal location, form, and size, as well as the aspect of their nucleus, can be easily seen by light microscopy (Fig. 12-1). Electron microscopy displays at least three types of basal cells, all originating from the connective tissue.

The *first type* consists of small round-shaped cells, with a central, ovoid nucleus, cytoplasm of low electron density, and few organelles (Fig. 12-2a). Most frequently, and especially in light microscopic investigations, these cells are considered to be less differentiated and devised to replace shed elements. It is our opinion that these cells, which in the mouse originate from the lamina propria, do *not* change their appearance into that of principal cells: transitional forms have never been seen, and division stages of the principal cells, although infrequent, have been described [17]. The cells of this first type could be either lymphocytes or macrophages. This uncertainty is due to the presence in our specimens of numerous cells inside the surface epithelium with structural characteristics of both lymphocytes and macrophages.

The *second type* of intraepithelial wandering cells is represented by the *macrophages* occurring in the proximity of the epithelial surface (Fig. 12-1). They show numerous, long, cytoplasmic processes, which can also be found away from the cell body. This suggests that the cell may establish a functional relationship with many principal cells. The cytoplasmic organelles of these cells are numerous. The Golgi complex has a moderate extension but encompasses a large number of electron-dense granules, conceivably primary lysosomes (Figs. 12-4a and 12-4b). Cisternae of the rough endoplasmic reticulum (ER) in continuity with the smooth ER, and probably filled with cholesterol (Figs. 12-4a to 12-4c, occur in the cell body and in the cytoplasmic processes.

The *third type* of intraepithelial wandering cell consists of mast cells. They are numerous in the connective tissue of the human gallbladder wall and may migrate in the epithelium. The same staining of the cytoplasmic granules — in the lamina propria and inside of the epithelial layer — suggested to Wallraff and Dietrich [24] that they were mast cells. The granules of human mast cells may contain several typical scrolls that are missing from the mast cells of other mammals. Granules with scrolls can frequently be seen in the mast cells of the lamina propria, yet are infrequently seen in those of the intraepithelial elements (Figs. 12-5a and 12-5b).

Intraepithelial mast cells have been further described in the gallbladder of cattle and sheep [25], and thus they are found in normal, healthy animals. This finding is of interest because intraepithelial mast cells can be observed in the nasal mucosa in cases of allergy [26], and they are often detectable in the epithelium of the fundus region of the human stomach [27]. It is conceivable that mast cells influence the activity of neighboring cells through the release of their granule content inside the epithelium.

2.1.1.4. Brush cells and granular cells. In the mouse gallbladder, two additional types of cells, the *brush cells* [reviewed in 28] and the *granular cells* [17] are present. Until now, the first type has never been described in human gallbladder epithelium. These cells, however, are also rare in laboratory animals, apart from some particular organs such as, for example, the *ductus choledochus* [29].

The second type of cells (the granular cells) has already been observed in the epithelium of the human biliary ducts and are named *entero-*

chromaffin [30]. The granular cells are extremely rare in human gallbladder epithelium; they can be seen in the cystic duct and are abundant in the distal region of the *ductus choledochus* [31]. These results, which indicate a particular topographical distribution of the granular cells, might explain their absence from our specimens (see note on p. 246). Laitio and Nevalainen [32] reported the occurrence of two types of endocrine cells in the metaplastic epithelium of the neck region of a human gallbladder removed for gallstones. They did not observe these cells in the neck region of the healthy gallbladder.

The paucity of endocrine cells or even their absence, in the epithelium of the human gallbladder raises doubts about the origin of the hormone cholecystokinine (which induces contractions of the human gallbladder, see below) from cells located in this epithelium.

2.1.1.5. Junctions between epithelial cells. In thin sections, a junctional complex comparable to that of all single-layered columnar epithelia is seen: *a zonula occludens,* a *zonula adhaerens,* and a *macula adhaerens* (desmosome). These components are located in the apical region of the epithelial cells bordering the lumen of the human gallbladder. Additional desmosomes in variable number can be found between adjacent principal cells (especially in the apical two thirds of the cell body), as well as between these cells and the pencil-shaped cells. The three types of wandering cells described above do not show junctions.

Freeze-fracture replicas of the human gallbladder epithelium (Fig. 12-6) show that the *zonula occludens* is more extended than is suggested by thin sections. Most frequently, it is composed of five to eight strands forming a tight network towards the luminal border. In the apical basal direction, the network becomes larger with single-strand ramifications. Small *maculae communicantes* (gap junctions) can be seen between the strands of the zonula occludens and also below the junctional complex (Fig. 12-6).

Variations in the number of the strands of the *zonulae occludentes* and of their arrangement in gallbladder epithelium have been described in different animal species [33]. Usually, extensions of the zonulae occludentes and the degree of interconnections among their strands represent key parameters for the evaluation of paracellular transport [34]. Therefore, only a quantitative assessment of these parameters may be useful for transport studies on human gallbladder epithelium.

2.1.2. Lamina propria.

A lucid layer (*lamina rara externa*), usually optically empty but known to contain glycoproteins and proteoglycans — at least in laboratory animals [35], separates the *lamina densa* of the basement membrane from the epithelial plasma membrane. The lamina densa is followed by an indistinct layer (*zona diffusa, lamina fibro-reticularis*) connecting the lamina densa to the underlying connective tissue.

As in other regions of the intestinal tract, the lamina propria of the human gallbladder shows an extracellular matrix in which collagen fibrils, microfibrils, and proteoglycans predominate. The collagen fibrils are usually thin (up to 40 nm in diameter). They probably belong to genetic type III collagen, as in the lamina propria of other parts of the gastrointestinal tract [36]. Close to the epithelial layer, there are microfibrils but not elastic fibers. These occur in the proximity of the muscle layer. Fibroblasts, granulocytes (eosinophils and neutrophils), plasma cells, macrophages, and mast cells are seen among blood and lymphatic vessels. Numerous Weibel–Palade bodies (up to ten in one cell section) characterize the endothelial cells of the venules. Close to the epithelial layer, there are numerous components of the peripheral nervous system. In general, these are unmyelinated nerve fibers. Small bundles of these nerve fibers run parallely to the epithelial sheet. Fibroblasts may lie in close apposition to the Schwann cells.

2.2. Muscular layer

The muscular layer is composed of bundles of smooth muscle cells with a spiral orientation, from the right hand to the left and *vice versa,* according to a "Scherengitter" arrangement. Light microscopy shows that these bundles cross each other obliquely. The thickness of the muscular layer undergoes individual variations; it relates, however, to the total thickness of the gallbladder wall [8]. Comparative light microscopic investigations

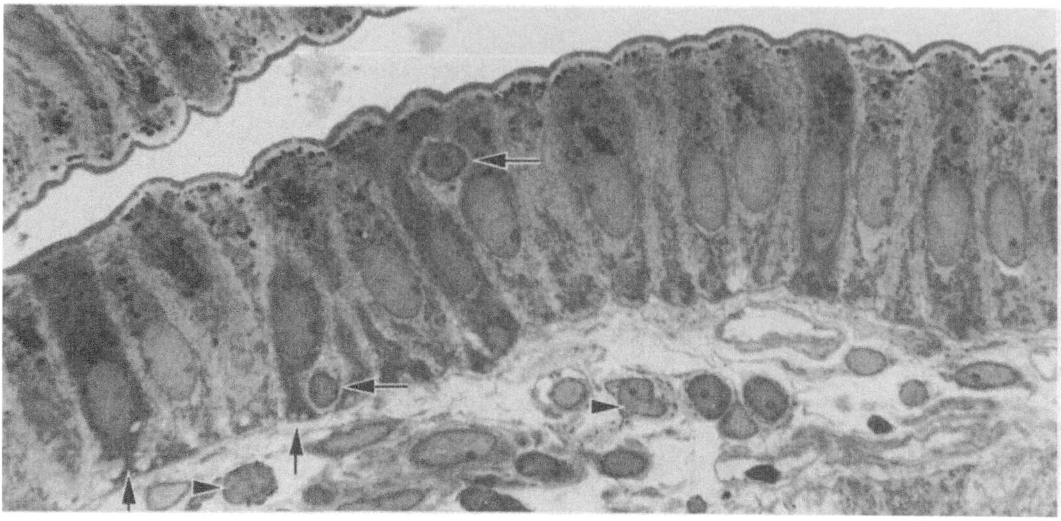

Fig. 12-1. Principal cells, some pencil-shaped cells (the vertical arrows point to their basal, branched region) and two wandering cells (horizontal arrows). In the lamina propria blood capillaries and connective tissue cells; the arrowheads indicate mast cells. (x 1,000).

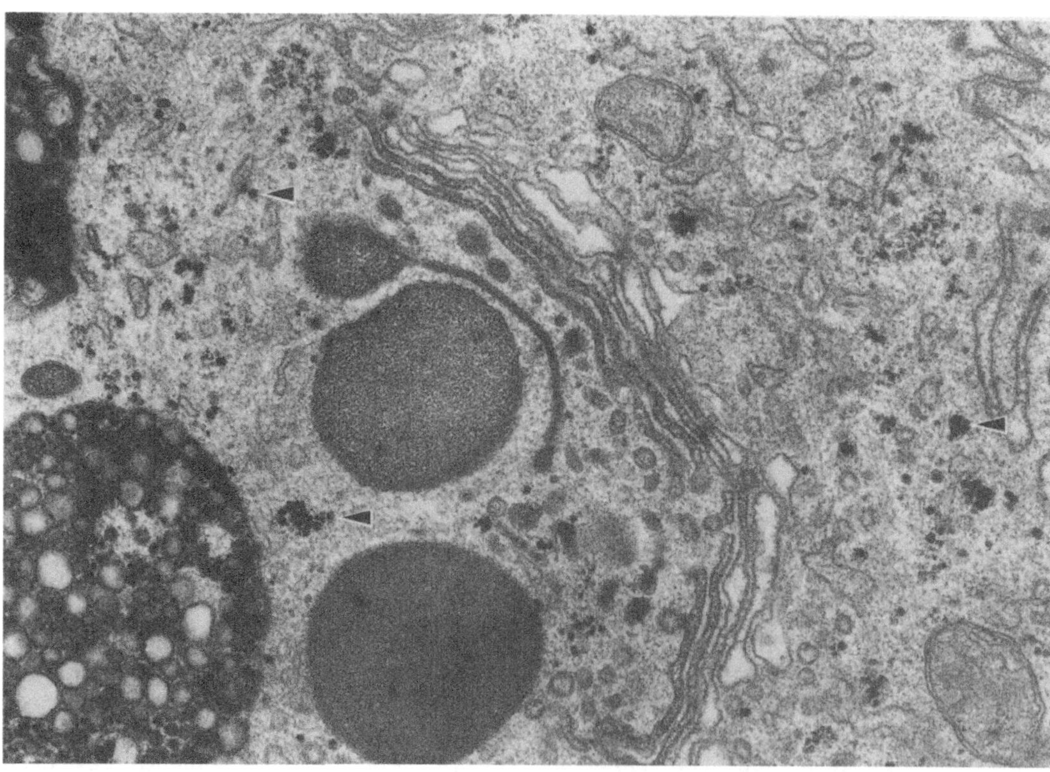

Fig. 12-3. Region of the Golgi complex in a principal cell. A stack of sacculi is flanked on the right by components of the smooth-surfaced endoplasmic reticulum and on the left by components of the GERL, precursors of secretory granules, and a secretory granule. The dense bodies consist mainly of lucent granules (low body) and concentric lamellae (upper body). Arrowheads indicate some glycogen granules. (x 40,000).

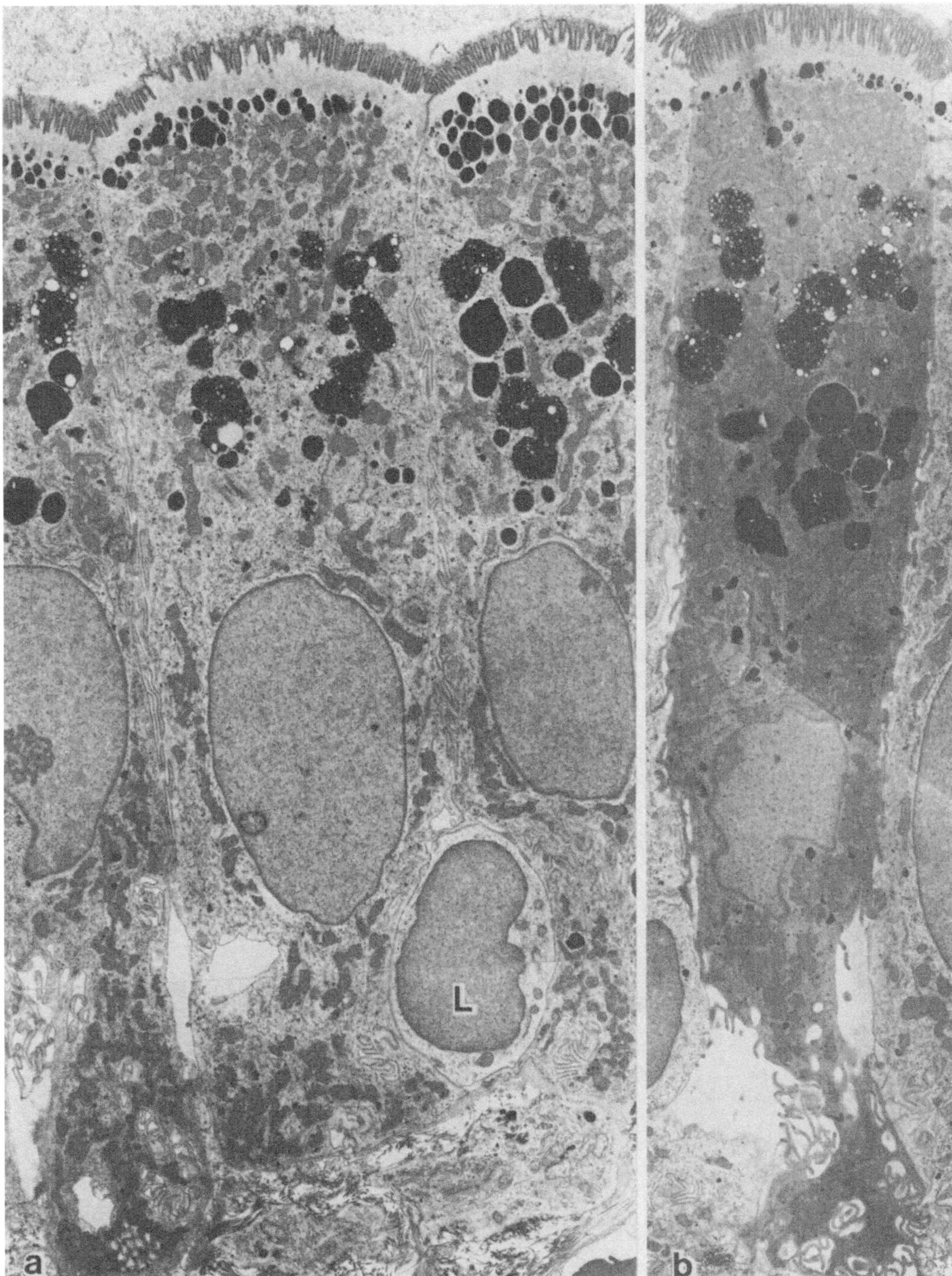

Fig. 12-2. Principal cells (**a**) and pencil-shaped cell (**b**). In the principal cell, the microvilli of the brush border are irregularly shaped. Below the terminal web, the zones (1) with rows of secretory granules, (2) with mitochondria, and (3) with dense bodies, secretory granules, and the Golgi complex. In the pencil-shaped cell (**b**), a great number of mitochondria uniformly fills the cell. Especially below the nucleus, the plasma membrane gives rise to numerous deep and winding invaginations. A wandering cell (presumably a lymphocyte, L) is visible on **a**. **a**, **b**: x 6,500).

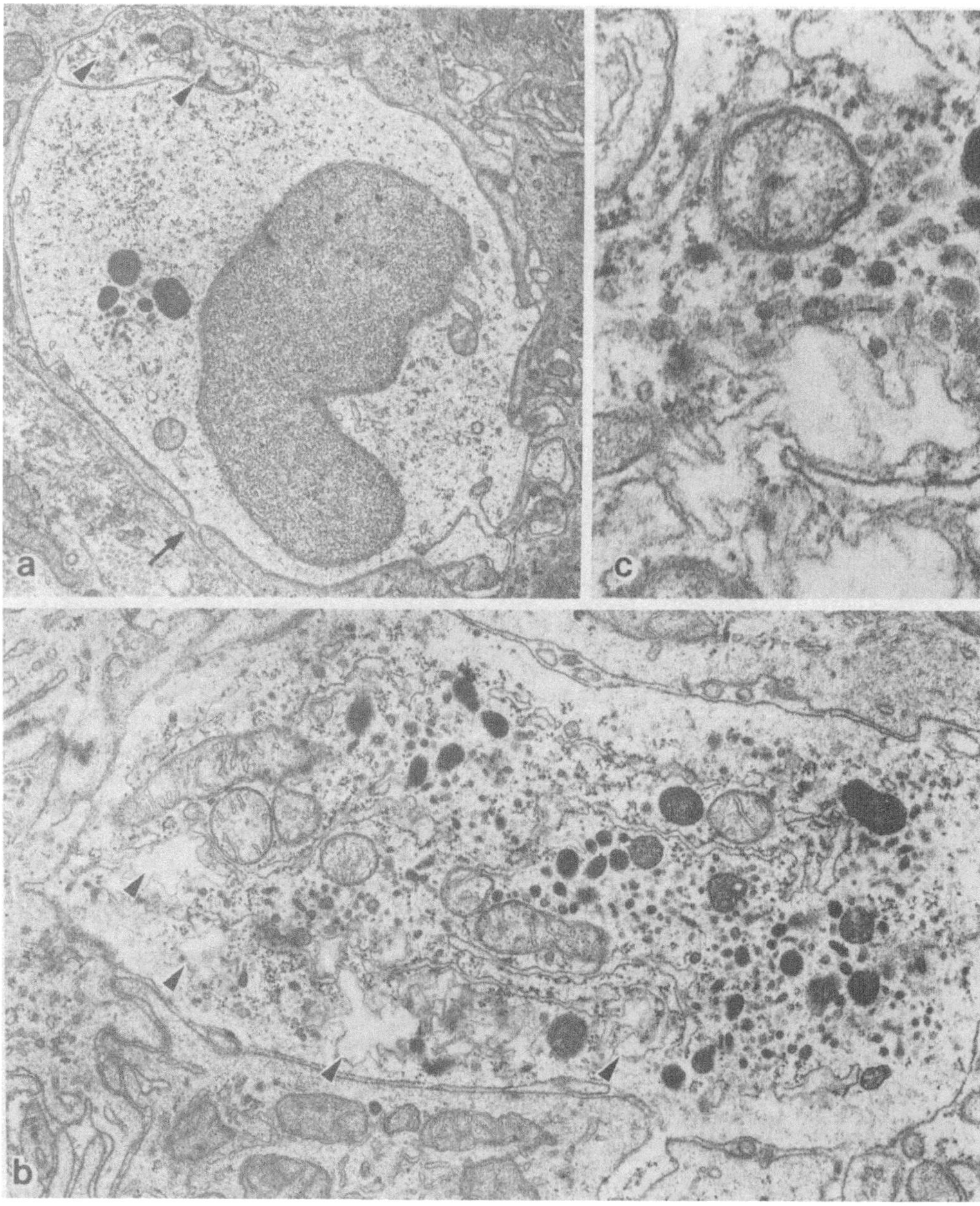

Fig. 12-4. Wandering cells. **a**: An intraepithelial macrophage with just a few dense bodies. In one of its process large polymorphic cisternae are present (arrowheads). A narrow space (arrow) separates the cell from the basement membrane. **b**: Part of an intraepithelial macrophage with numerous dense bodies and some polymorphic cisternae (arrowheads). **c**: Higher magnification of polymorphic cisternae, presumably containing cholesterol. **a**: x 12,000; **b**: x 20,000; **c**: x 60,000.

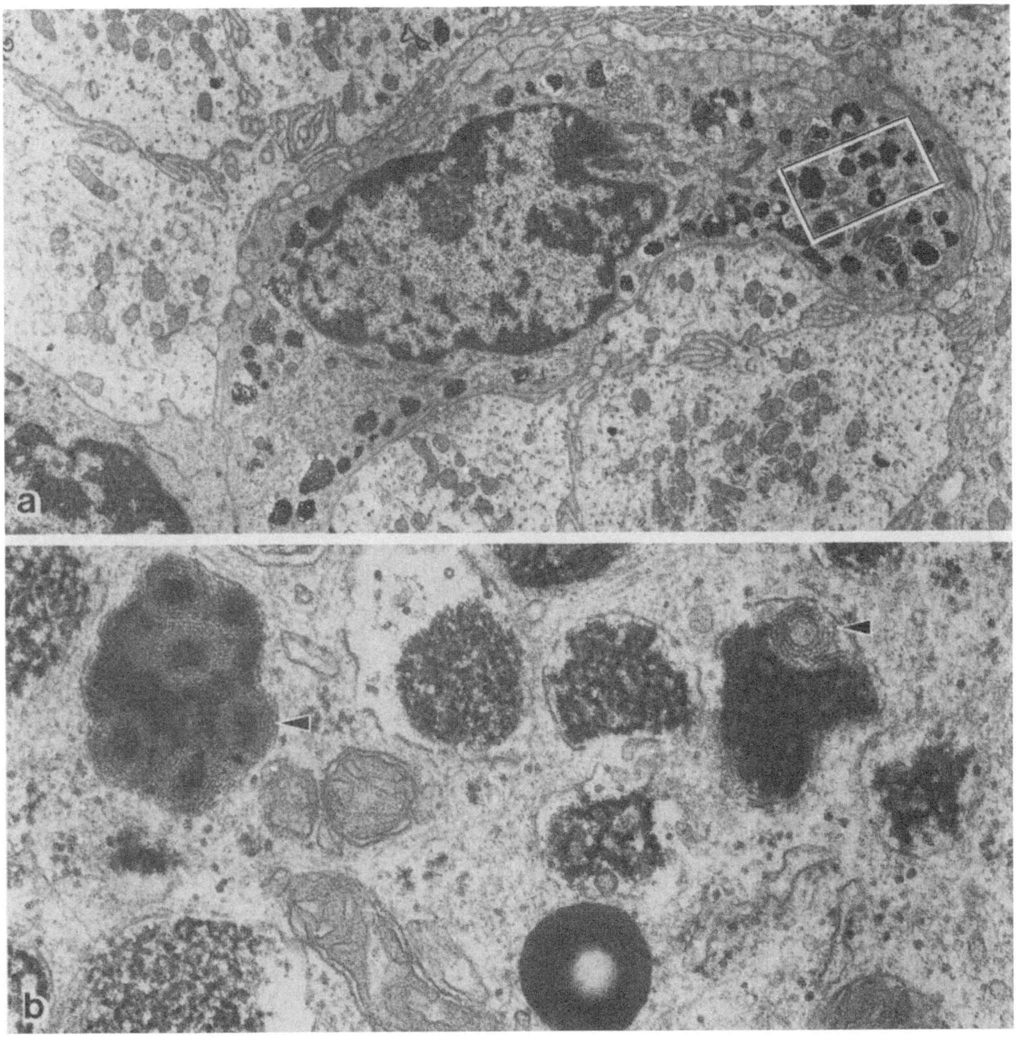

Fig. 12-5. Wandering cell, intraepithelial mast cell. The framed area in **a** is shown under higher magnification in **b**. Granules with scrolls are indicated by arrowheads. **a**: x 8,000; **b**: x 54,000.

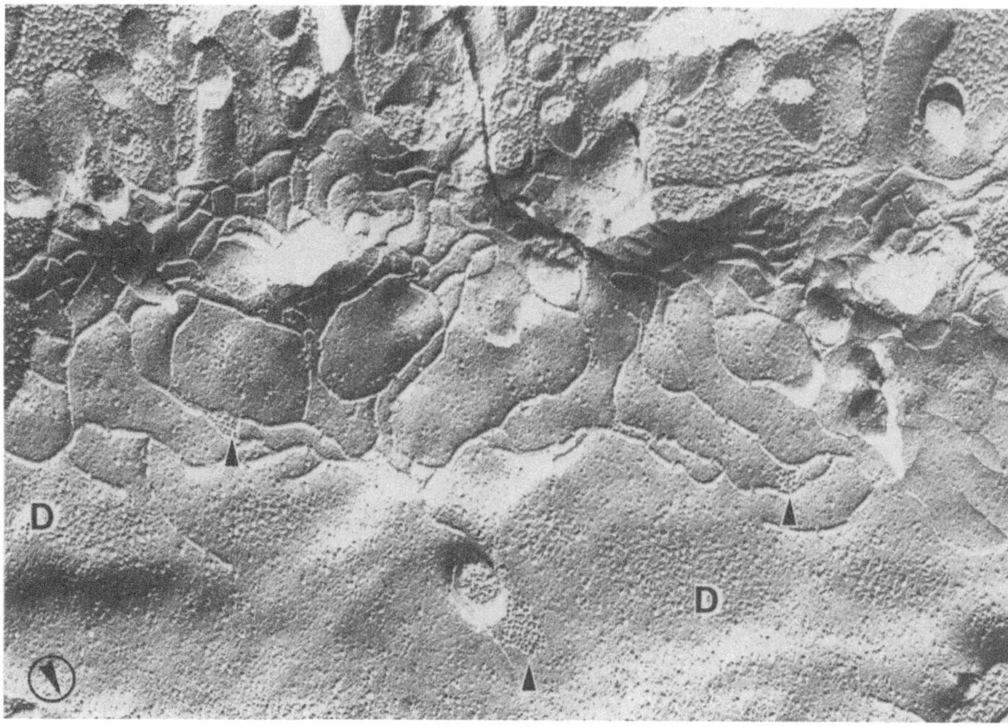

Fig. 12-6. Junctional complex between principal cells. The apical strands of the zonula occludens are closely interconnected, forming a network, the meshes of which become larger in the apical-basal direction. Within the meshes and outside of them are small gap junctions (arrowheads). D = desmosomes. Encircled arrowhead indicates direction of shadowing. (x 53,000).

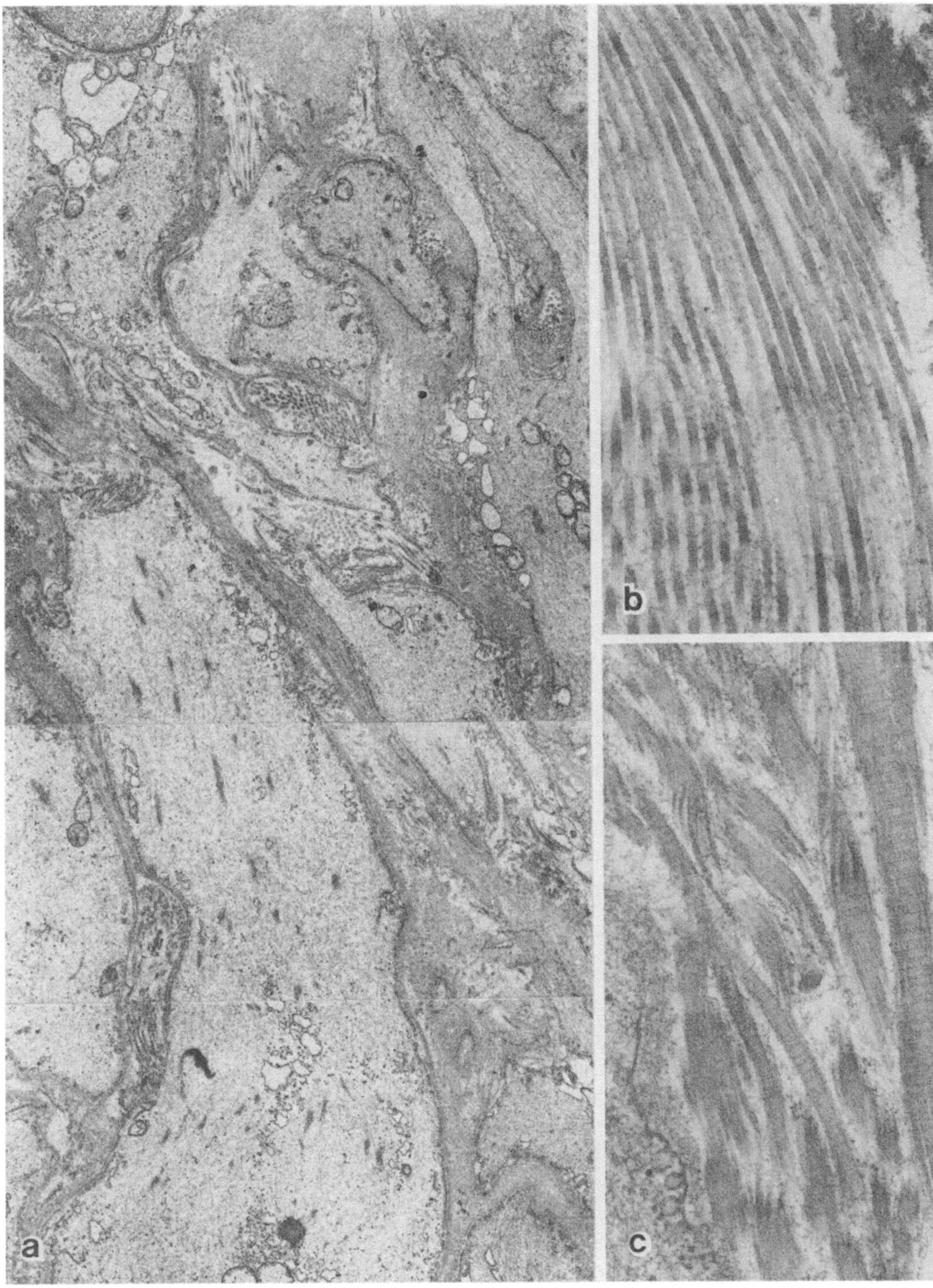

Fig. 12-7. Some smooth-muscle cells of the muscular layer (**a**) extensively surrounded by a basement membrane. The collagen fibrils encompassing this layer (**b**) or located in between the muscle cells (**c**) have a different aspect; their repeating period is the same (about 64 nm). **a**: x 8,000; **b**, **c**: x 40,000).

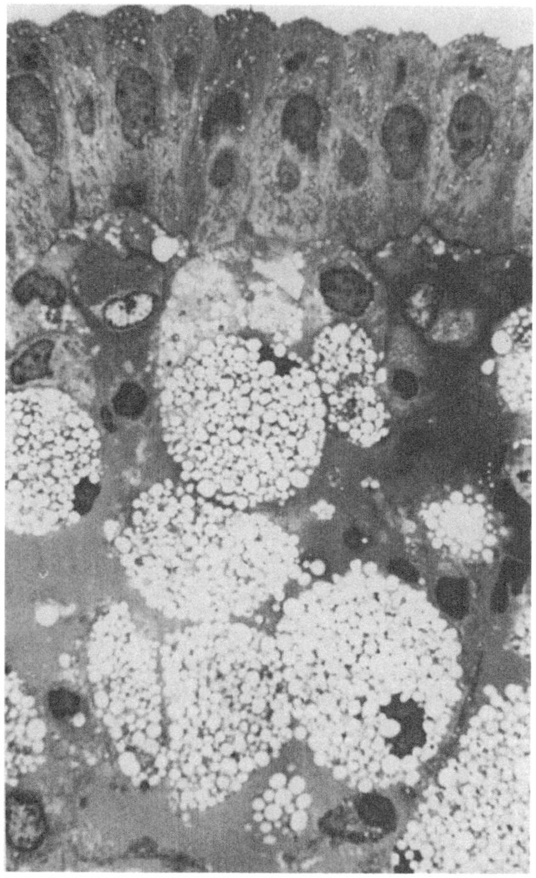

Fig. 12-8. Several, large foam cells are seen in the lamina propria. The epithelial cells overlying this region show numerous cytoplasmic droplets, probably containing cholesterol; their secretory granules are consistently reduced in number (compare with Fig. 12-1). x 1,000.

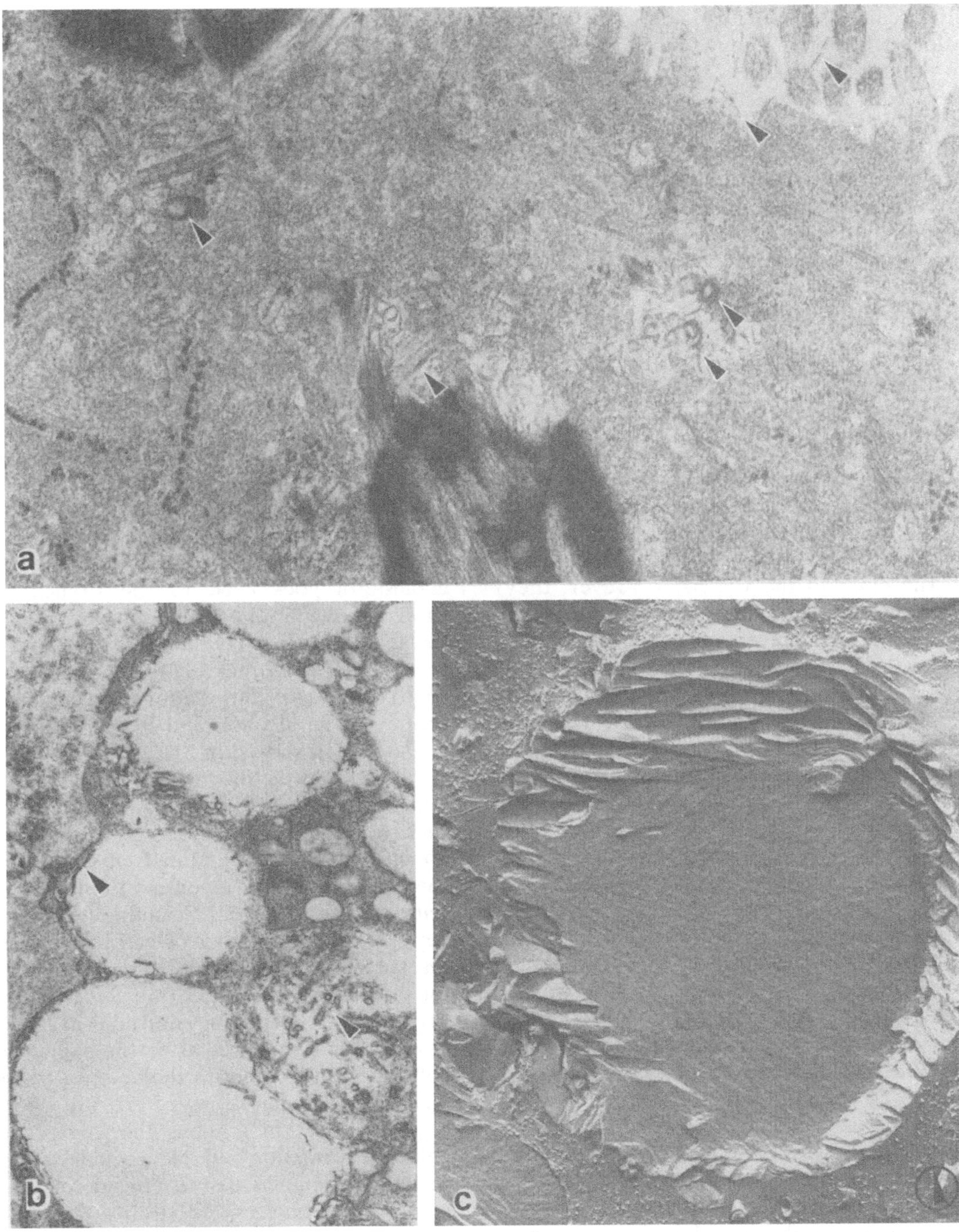

Fig. 12-9. Cholesterol localization by addition of 2% digitonin to the fixative solution. Same specimen as in Fig. 12-8. In thin sections, spicular and cylindrical structures (arrowheads), i.e., cholesterol-digitonin complexes, are visible in the cytoplasm of principal cells (**a**) and foam cells (**b**). In freeze-fracture replicas (**c**), the membrane of a granule of a foam cell is marked by the complexes, not the granule content. **a**: x 40,000; **b**: x 16,000; **c**: x 40,000.

demonstrate that the muscular layer of the human gallbladder is thinner (about one third thinner) than in other animals [1].

The smooth muscle cells are extensively surrounded by a basement membrane (Fig. 12-7a), which is only interrupted where two cells are in close apposition. Otherwise, the basement membrane of one cell continues without interruption with that of the adjacent cell. The space between facing basement membranes is occupied by connective tissue components such as collagen fibrils, microfibrils, and proteoglycans. The collagen fibrils may be thicker than those outside of the muscle bundles (up to 150 nm or more) and may display their microfibrillar substructure (compare Figs. 12-7b with 12-7c). In the connective tissue outside of the muscle bundles, and of the large septa between these bundles, connective tissue cells (such as fibroblasts and mast cells), vessels of different sizes, and small unmyelinated nerve fibers can be seen. These fibers usually run close to the vessels. However, they can also lie close to the smooth muscle cells of the gallbladder muscular layer. Here some terminal endings with assemblies of small vesicles and moderately electron-dense granules are bordered on the side facing the smooth muscle cells by only the Schwann-cell basement membrane. Conceivably, this arrangement facilitates transmitter diffusion.

3. Contraction

Ivy and Oldberg [37] gave the name *cholecystokinine* (CCK) to the substance they discovered that is capable of inducing a contraction of the gallbladder. The polypeptide hormone CCK was later isolated from the porcine gastrointestinal tract [38]. Its widespread actions have recently been reviewed [39, 40]. Numerous investigations demonstrate that gallbladder motility undergoes neural control; however, CCK plays a major role in gallbladder contraction and acts as a relaxant on Oddi's sphincter [40]. Until now, the sites of CCK production inside the gallbladder have not been identified, nor have eventual interactions between CCK and other hormones influencing gallbladder motility been determined [41].

A close relationship has been found between contraction of the gallbladder and circulating concentrations of CCK in humans, both after fat ingestion and after exogenous administration of CCK [42–44], although variations due to sex and age are known [4, 45, 46]. Recent investigations show that CCK induces contraction of the human gallbladder through a direct interaction on receptors of the smooth muscle cell [47]. Comparative studies in healthy volunteers and patients with gallstones reveal a decrease in CCK receptors in the patients with gallstones [47]. These observations might be important in explaining the origin of gallstones, since altered contraction may represent a prerequisite for their formation [48].

4. Some pathological aspects

The human gallbladder is one of the organs in which cholesterol appears in an impressive pathological aspect: the gallstones. The discovery of cholesterol goes back to the French chemist Chevreul, who in the last century [49] first identified this substance in human gallstones and named it *cholesterine* (Greek: *chole* = bile, *steros* = solid). Berthelot [50] demonstrated that cholesterine was an alcohol, and it was later renamed *cholesterol* [reviewed in 51].

Sonography is the most suitable method to detect the presence of gallstones [52]. Although these have been found in Egyptian mumies dating about 1500 B.C. [53], statistical studies indicate that cholelithiasis is a common affliction of contemporary societies [54]. Scanning electron microscopy on a large number of gallstones reveals remarkable morphological differences depending on their composition [55, 56]. Several investigations on the epidemiology and clinical associations of cholelithiasis, as well as on the pathogenesis of gallstones and potential therapeutic treatments, have recently been reviewed [57, 58]. Cholesterol crystal nucleation remains the key problem in gallstone formation [59]. New evidence exists that gallbladder mucins play a crucial role in cholesterol crystal nucleation [60].

An increase of cholesterol in gallbladder bile is not always associated with gallstone formation. An excess of cholesterol can be stored in the gallbladder wall, as in cases of cholesterosis. Synonyms of this disorder are *cholesteatose, lip-*

oidosis, cholesterol polyposis, and strawberry gallbladder (or vesicule fraise, Stippchengallenblase). The mucosa of the gallbladder is studded with minute, yellow spots, which are similar to the pale yellow-gray seeds that punctate the surface of a strawberry. Histologically, the characteristic features consist of enlargement and distention of the mucosal folds into club shapes, with clustering of round to polyhedral foamy cells. Few electron microscopical investigations have been performed on gallbladders with cholesterosis [61–63]. A mechanism for the uptake and storage of cholesterol in these gallbladders has been proposed [62]. According to this suggestion, the smooth endoplasmic reticulum of the principal cells accumulates the substance. Intraepithelial macrophages take up cholesterol and transport it into the lamina propria, where they remain as foam cells. These aspects can be seen in the region of the mucosa where the process is evolving, i.e., where single foam cells or small assemblies of these can be recognized. In the regions of the wall showing advanced stages of the disease, foam cell deposits invade the lamina propria and the muscular layer, which are no longer visible. The epithelial cells covering these regions lie on a thick basement membrane and show just a few, if any, apical secretory granules and a remarkably decreased number of dense bodies; several pale vesicles appear, especially in the apical cytoplasm (Fig. 12-8). Secretory granules and dense bodies are abundant in apparently unaltered regions of the mucosa, away from the invaded areas (Fig. 12-1).

Digitonin can be used as a marker for extracellular and intracellular free cholesterol [64]. Electron microscopy of human gallbladders with cholesterosis reveals spicular and circular structures in the principal cells that are the result of the reaction (precipitation) between cholesterol and digitonin [65]. We have examined human cholesterosic gallbladders after fixation in the presence of digitonin, in thin sections and in freeze-fracture replicas. The first method shows cytoplasmic complexes, which agrees with the work of Koga et al. [65]. They are scattered across the whole cell, are associated with dense bodies (Fig. 12-9a), and are interposed between the microvilli of the free cell surface (Fig. 12-9a). The electron density of the plasma membrane of the microvilli is irregular. In replicas, the membrane of both the microvilli and the lateral cell surface displays characteristic alterations due to cholesterol-digitonin-complexes [66], indicating a high cholesterol content in these membrans. Comparable complexes can be found assembled in the cytoplasm. After treatment with digitonin, the plasma membrane and the membrane of cytoplasmic granules of the foam cells are marked in thin section (Fig. 12-9b)and in replicas (Fig. 12-9c) by the cholesterol-digitonin complexes; however, the granule content has the same aspect as in controls (i.e., fixed in the absence of digitonin). This observation suggests that cholesterol inside of these granules is not free; it might occur here in esterified form and therefore can not be revealed by the procedure used.

5. Concluding remarks

The gallbladder is an organ fulfilling different physiological functions, the precise number of which is unknown [67]. Among these functions, the most extensively studied is that of absorption. Our present information about the transport mechanisms in the gallbladder comes largely from in vitro studies on gallbladders known for peculiar absorbing properties, belonging to different animal species [5, reviewed in 67].

However, it has been known for a long time that some animal species have gallbladders that have greater efficiency in fluid absorption and storage of bile than do others. Moreover, both hepatic bile output and sphinteric resistance, which are closely related to gallbladder function, vary consistently. For example, the gallbladder of the rabbit, although characterized by a high absorptive capability, is unable to concentrate bile because of the great output of diluted hepatic bile; this animal has a very low sphinteric resistance.

However, the human gallbladder is characterized by its ability to concentrate bile; hepatic bile output is comparably low, and sphinteric resistance is high. Therefore, the question arises as to what extent the above-mentioned different functional parameters can be correlated to morphology.

If we use TEM to compare the fine structure of the principal cells of the rabbit gallbladder [68] and of the human gallbladder (this chapter), some structural diversities can been observed: in humans the cells, in addition to morphological aspects indicating absorptive activity, also display aspects indicating a pronounced secretory function. The last function is not as evident in the rabbit. Therefore, the number of secretory granules and the extension of the Golgi apparatus differ in the two species. When SEM of the internal surface of the organ is considered, the rabbit gallbladder reveals a rather regular network of folds [69]. In humans, similar folds are present but their arrangement is irregular [70]. The two species also show structural differences, which might reflect the above mentioned functional diversities. These differences mainly concern the epithelium. Conceivably, the other layers of the gallbladder wall (lamina propria, muscular layer), as well as the vessels and their distributions, should also be considered. Therefore, on the basis of few morphological studies, a definitive answer to the above mentioned question is not possible at the present time. Further morphological comparative studies on the gallbladders of appropriate species are necessary in order to correlate functional parameters with morphological ones or vice versa.

Note

The specimens used for the present review were collected from gallbladders of six women, 28 to 67 years of age, undergoing surgery because of polyposis. The body of the organ was used, both in the region of the pathological alteration and in other areas where tissue looked normal by macroscopy. The tissue blocks were removed 5 minutes after blood supply ligature, were fixed in either OsO_4 or aldehyde and OsO_4 solutions, and were embedded in Epon. Sections were stained with toluidine blue for light microscopy and uranyl acetate-lead citrate for electron microscopy. Digitonin 2% was added to the aldehyde solution in order to localize cholesterol. For replicas, aldehyde fixed specimens were used with and without digitonin.

Acknowledgment

Supported in part by the Deutsche Forschungsgemeinschaft (En 65/15/1)

We are indebted to Dr. C. Wolpers (Lübeck) for providing the surgical specimens and the clinical data on the patients. We would also like to thank Prof. Dr. J. Thiele (University of Cologne) for his critical review of the manuscript.

References

1. Wallraff J: Gallengangsystem, Gallenblase und Galle. In: Handbuch der Mikroskopischen Anatomie, Vol V/4. W Bargmann, (ed), Berlin: Springer, 277–363, 1969.
2. Schmidt CR, Ivy AC: The general function of the gall bladder. *J Cell Comp Physiol* 10: 365–383, 1937.
3. Edholm P: Gallbladder evacuation in the normal male induced by cholecystokinin. *Acta Radiol* 53: 257–265, 1960.
4. Mackie CR, Baxter JN, Grime JS, Hulks G, Cuschieri A: Gall bladder emptying in normal subjects — a data base for clinical cholescintigraphy. *Gut* 28: 137–141, 1987.
5. Diamond JM: Transport mechanisms in the gallbladder. In: Handbook of Physiology: Alimentary Canal. CF Code (ed), Washington DC: American Physiological Society, 2451–2482 1969.
6. Geenen JE, Hogan WJ, Dodds WJ, Stewart ET, Arndorfer RC: Intraluminal pressure recording from the human sphincter of Oddi. *Gastroenterology* 78: 317–324, 1980.
7. Torsoli A, Ramorino ML, Alessandrini A: Motility of the biliary tract. *Rendic R Gastroenterol* 2: 67–80, 1970.
8. Whitehouse RW, Martin DF: Contrast-enhanced computed tomography of the normal and abnormal gallbladder. *Br J Radiol* 59: 1083–1085, 1986.
9. Engel JM, Deitch EA, Sikkema W: Gallbladder wall thickness: sonographic accuracy and relation to disease. *Am J Roentgenol* 134: 907–909, 1980.
10. Fawcett DW: A Textbook of Histology, 11th ed. Philadelphia: WB Saunders, 1986.
11. Luciano L, Reale E, Wolpers C: Die Feinstruktur der Gallenblase und der Gallengänge. V. Histochemische Lokalisierung von Mukosustanzen im menschlichen Gallenblasenepithel. *Histochemistry* 38: 57–70, 1974.
12. Thiéry JP: Mise en évidence des polysaccharides sur coupes fines en microscopie électronique. *J Microscopie* 6: 987–1018, 1967.
13. Mowry RW: Improved procedure for the staining of acidic polysaccharides by Müller's colloidal (hydrous) ferric oxide and its combination with the Feulgen and the periodic acid-Schiff reactions. *Lab Invest* 7: 566–576, 1958.
14. Yoshitomi S, Miyazaki K, Nakayama F: Demonstration and maintenance of mucus secretion in cultured human gallbladder epithelial cells. *In Vitro Cell Develop Biol* 23: 559–566, 1987.
15. Luciano L, Wermbter G, Reale E: Die Feinstruktur der Gallenblase und der Gallengänge. III. Beobachtungen an gefriergeätzten Präparaten der Gallenblase der Maus unter besonderer Berücksichtigung der dichten Körper

und des Verbindungskomplexes. *Cytobiologie* 7: 76–88, 1973.
16. Nagahiro K: Zytologische Untersuchungen über die Epithelzelle der Gallenblase des Menschen. *Cytologia* (Tokyo) 9: 132–163, 1938.
17. Luciano L: Die Feinstruktur der Gallenblase und der Gallengänge. I. Das Epithel der Gallenblase der Maus. *Z Zellforsch* 135: 87–102, 1972.
18. Yamada E: The fine structure of the gall bladder epithelium of the mouse. *J Biophys Biochem Cytol* 1: 445–458, 1955.
19. Hayat MA: Fixation for Electron Microscopy. New York: Academic Press, 1981.
20. Pfuhl W: Die Gallenblase und die extrahepatischen Gallengänge. In: Handbuch der Mikroskopishen Anatomie des Menschen, vol 5. von W Möllendorff (ed), Berlin: Springer Verlag, 426–462, 1932.
21. Evett RD, Higgins JA, Brown AL Jr: The fine structure of normal mucosa in human gall bladder. *Gastroenterology* 47: 49–60, 1964.
22. Bader G: Die submikroskopische Struktur des Gallenblasenepithels. III. Mitteilung. Das Epithel der Steingallenblase des Menschen. *Frank Z Path* 74: 501–511, 1965.
23. Bartels H, Welsch U: Freeze fracture study of the turtle lung. 2. Rod-shaped particles in the plasma membrane of a mitochondria-rich pneumocyte in *Pseudemys (Chrysemys) scripta*. *Cell Tissue Res* 236:453–457, 1984.
24. Wallraff J, Dietrich KF: Zur Morphologie and Histochemie der Steingallenblase des Menschen. *Z Zellforsch* 46: 155–231, 1957.
25. Toledo OMS, Morales CR, Pereyra LA, Jordao T, Montes GS: Migrating mast cells in the gallbladder epithelium of cattle and sheep. *Histochemistry* 72: 433–442, 1981.
26. Enerback L: Mast cell heterogeneity: the evolution of the concept of a specific mucosal mast cell. In: Mast Cell Differentiation and Heterogeneity. AD Befus, J Bienenstock, JA Denburg (eds). New York: Raven Press, 1–26, 1986.
27. Solcia E, Capella C, Vassallo G, Buffa R: Endocrine cells of the gastric mucosa. *Int Rev Cytol* 42: 223–286, 1975.
28. Luciano L, Reale E: A new morphological aspect of the brush cells of the mouse gallbladder epithelium. *Cell Tissue Res* 201: 37–44, 1979.
29. Luciano L, Castellucci M, Reale E: The brush cells of the common bile duct of the rat. Thin section, freeze-fracture and scanning electron microscopy. *Cell Tissue Res* 218: 403–420, 1981.
30. Erspamer V: Die enterochromaffinen Zellen der Gallenwege in normalen und pathologischen Zuständen. (Nach Untersuchungen beim Menschen und bei Säugetieren.) *Virchows Arch* 296: 70–92, 1936.
31. Erspamer V: Sulla presenza di cellule argentofile (preenterocromaffini) nelle vie biliari dell'uomo e di alcuni mammiferi. *Anat Anz* 85: 272–284, 1938.
32. Laitio M, Nevalainen T: Ultrastructure of endocrine cells in metaplastic epithelium of human gall bladder. *J Anat* 120: 219–225, 1975.
33. Claude P, Goodenough DA: Fracture faces of zonulae occludentes from "tight" and "leaky" epithelia. *J Cell Biol* 58: 390–400, 1973.
34. Bullivant S: Possible relationships between tight junction structure and function. In: Epithelial Ion and Water Transport. ADC Micknight, JP Leader (eds), New York: Raven Press, 265–275, 1981.
35. Castellucci M, Familiari G, Caggiati A: Heterogeneous distribution of ruthenium red-proteoglycan aggregates in the basement membrane (BM) of the gallbladder epithelium. In: The Extracellular Matrix: Chemistry, Biology, Pathology. J Jeffry, B Mecham (eds), St Louis: Washington University Press, 1982.
36. Montes GS, Bezerra MSF, Junqueira LCU: Collagen distribution in tissues. In: Ultrastructure of the Connective Tissue Matrix. A Ruggeri, PM Motta (eds), Boston: Martinus Nijhoff Publishers, 65–88, 1984.
37. Ivy AC, Oldberg E: A hormone mechanism for gallbladder contraction and evacuation. *Am J Physiol* 86: 599–613, 1928.
38. Mutt V, Jorpes JE: Structure of porcine cholecystokinin-pancreozyimin. 1. Cleavage with thrombin and with trypsin. *Eur J Biochem* 6: 156–162, 1968.
39. Morley JE: The ascent of cholecystokinin (CCK) — from gut to brain. *Life Sci* 30: 479–493, 1982.
40. Walsh JH: Gastrointestinal Hormones. In: Physiology of the Gastrointestinal Tract, 2nd ed. LR Johnson (ed), New York: Raven Press, 181–253, 1987.
41. Ryan JP: Motility of the gallbladder and biliary tree. In: Physiology of the Gastrointestinal Tract, 2nd ed. LR Johnson (ed), New York: Raven Press, 695–721, 1987.
42. Byrnes DJ, Borody T, Daskalopoulos G, Boyle M, Benn I: Cholecystokinin and gallbladder contraction: effect of CCK infusion. *Peptides* 2: 259–262, 1981.
43. Wiener I, Inoue K, Fagan CJ, Lilja P, Watson LC, Thompson JC: Release of cholecystokinin in man. Correlation of blood levels with gallbladder contraction. *Ann Surg* 194: 321–325, 1981.
44. Lilja P, Fagan CJ, Wiener I, Inoue K, Watson LC, Rayford PL: Infusion of pure cholecystokinin in humans. Correlation between plasma concentrations of cholecystokinin and gallbladder size. *Gastroenterology* 83: 256–261, 1982.
45. Sacchetti G, Mandelli V, Roncoroni L, Montanari C: Influence of age and sex on gallbladder emptying induced by a fatty meal in normal subjects. *Am J Roentgenol* 119: 40–50, 1973.
46. Kaene P, Colwell D, Baer HP, Clanachan AS, Scott GW: Effects of age, gender and female sex hormones upon contractility of the human gallbladder *in vitro*. *Surg Gynecol Obstet* 163: 555–560, 1986.
47. Upp JR, Nealon WH, Singh P, Fagan CJ, Jonas AS, Greeley GH, Thompson JC: Correlation of cholecystokinin receptors with gallbladder contractility in patients with gallstones. *Ann Surg* 205: 641–648, 1987.
48. LaMorte WW, Schoetz DJ, Birkett DH, Williams LF: The role of the gallbladder in the pathogenesis of cholesterol gallstones. *Gastroenterology* 77: 580–592, 1979.
49. Chevreul M: Recherches chimiques sur les corps gras, et particulièrement sur leurs combinaisons avec les alcalis. Sixième Mémoire. Examen des graisses d'homme, de mouton, de boeuf, de jaguar et d'oie. *Ann Chim Phys* 2:

339–372, 1816.
50. Berthelot M: Sur plusieurs alcools nouveaux. Combinaisons des acides avec la cholestérine, l'éthal, le camphre de Bornéo et la méconine. *Ann Chim Phys* 56: 51–98, 1859.
51. Mead JF, Alfin-Slater RB, Howton DR, Popjak G: Biosynthesis of cholesterol and related substances. In: Lipids. Chemistry, Biochemistry, and Nutrition. New York: Plenum Press, 295–367, 1986.
52. Cooperberg PL, Gibney RG: Imaging of the gallbladder, 1987. *Radiology* 163: 605–613, 1987.
53. Glenn F: Biliary tract disease since antiquity. *Bull NY Acad Med* 47: 329–350, 1971.
54. Gracie WA, Ransohoff DF: The natural history of silent gallstones. The innocent gallstone is not a myth. *N Engl J Med* 307: 798–800, 1982.
55. Wolpers C, Blaschke R: Electron microscopy of human gallstones. In: 29th Ann Proc Electron Microscopy Soc Amer. CJ Arceneaux (ed), Baton Rouge, LA: Claitor's Publishing Division, 296–297, 1971.
56. Wolpers C: Gallenblasensteine. Ihre Morphogenese und Auswahl zur Litholyse. Basel: Karger Verlag, 1987.
57. Myant NB: Cholesterol gallstones; plasma cholesterol in liver disease. In: The Biology of Cholesterol and Related Steroids. London: Heinemann Medical Books, 853–887, 1981.
58. L Capocaccia, G Ricci, F Angelico, M Angelico, AF Attili (eds): Epidemiology and Prevention of Gallstone Disease. Lancaster: MTP Press, 1984.
59. Holzbach RT: Recent progress in understanding cholesterol crystal nucleation as a precursor to human gallstone formation. *Hepatology* 6: 1403–1406, 1986.
60. Smith BF: Human gallbladder mucin binds biliary lipids and promotes cholesterol crystal nucleation in model bile. *J Lipid Res* 28: 1088–1097, 1987.
61. Nevalainen T, Laitio M: Ultrastructure of gallbladder with cholesterolosis. *Virchows Arch Abt B Zellpath* 10: 237–242, 1972.
62. Luciano L, Wolpers C: Die Feinstruktur der Gallenblase und der Gallengänge. IV. Die Cholesterose der menschlichen Gallenblasenschleimhaut. *Virchows Arch Abt B Zellpath* 14: 147–158, 1973.
63. Koga A: Fine structure of the human gallbladder with cholesterosis with special reference to the mechanism of lipid accumulation. *Br J Exp Path* 66: 605–611, 1985.
64. Williamson JR: Ultrastructural localization and distribution of free cholesterol (3β hydroxysterols) in tissues. *J Ultrastruct Res* 27: 118–133, 1969.
65. Koga A, Todo S, Nishimura M: Electron microscopic observations on the cholesterol distributed in the epithelial cells of the gallbladder. *Histochemistry* 44: 303–306, 1975.
66. Elias PM; Goerke J, Friend DS: Freeze-fracture identification of sterol-digitonin complexes in cell and liposome membranes. *J Cell Biol* 78: 577–596, 1978.
67. Rose RC: Absorptive functions of the gallbladder. In: Physiology of the Gastrointestinal Tract, 2nd ed. LR Johnson (ed), New York: Raven Press, 1455–1468, 1987.
68. Kaye GI, Wheeler HO, Whitlock RT, Lane N: Fluid transport in the rabbit gallbladder. A combined physiological and electron microscopic study. *J Cell Biol* 30: 237–268, 1966.
69. Castellucci M, Caggiati A: Surface aspects of the rabbit gallbladder mucosa and their functional implications. *J Submicrosc Cytol* 12: 375–390, 1980.
70. Nielsen OV, Nielsen ML, Lauritzen KEF: Examination of the mucosa of the normal human gall bladder by scanning electron microscopy. *Micron* 5: 281–291, 1975.

Author's address:
Liliana Luciano
Medizinische Hochschule Hannover
Abteilung Zellbiologie und Elektronenmikroskopie
Kostanty-Gutschow-Str. 8
3000 Hannover 61
Federal Republic of Germany

CHAPTER 13

Microvascularization of the extraparietal glands of the alimentary canal

OSAMU OHTANI, TAKEHITO TAGUCHI & TAKURO MURAKAMI

1. Introduction

The vascular corrosion casting/scanning electron microscopy (SEM) method [1] has been extensively employed to study the microvascular organization of various tissues and organs, including the extraparietal glands, and so our knowledge of microvascular architecture in general has markedly increased over the last decade [2]. SEM observations of tissues whose connective tissue fibers have been removed with HCl or NaOH have not only helped to reveal microvasculatures, but have also contributed toward our understanding of the three-dimensional organization of cellular elements [3, 4]. The recently introduced cell-maceration/SEM method, a combination of cell removal with NaOH (25°C) and SEM observation [5], can demonstrate collagen fibrillar networks ensheathing blood vessels, and thus the microvascular organization of some organs and tissues. This chapter will review the findings on extraparietal glands obtained by these scanning electron microscopic methods.

2. Methods

2.1. *Vascular corrosion casting/SEM method*

Vascular corrosion casts were made using the method described by Murakami [1]. A polyethylene catheter was inserted into the aorta or the main feeding artery of the organ and, following blood washout with saline, the partially polymerized methacrylate (Mercox, Dainippon Inc, Tokyo, sold by Oken Shoji, Tokyo) was injected through the catheter. The perfusion and injection were performed by hand with a syringe. Injection was continued until resin free of saline droplets began to flow from the outflow points. Injected animals or organs remained in situ at room temperature for 30 minutes or longer and then were put into a hot water bath for several hours. The tissues were corroded in a concentrated solution of NaOH (15%–20%, w/v). The vascular corrosion casts obtained were washed under running water, frozen in water, and cut with a razor blade into small blocks suitable for SEM observation. They were rinsed in water, airdried, coated with gold, and observed under SEM at an accelerating voltage of 5 kV. In order to observe particular structures, some casts were dissected with fine tweezers and needles under a stereo-light microscope prior to SEM observation.

2.2. *HCl/collagenase treatment*

In order to observe the three-dimensional architecture of the parenchymal elements and its relationship to the microvasculature, connective tissue fibers were removed using the method introduced by Evan et al. [3].

Tissues were immersion fixed in 2.5% glutaraldehyde in phosphate buffer (pH 7.4) for 3 hours or longer. The fixed tissues were then hydrolyzed in 8N HCl for about 50 minutes at 60°C, followed by immersion in collagenase solution at a concentration of 10 mg/10 ml of 0.1 M

Riva, A. and Motta, P.M., (eds.), Ultrastructure of the Extraparietal Glands of the Digestive Tract.
ISBN 978-1-4612-8211-2.

phosphate buffer (pH 6.8) for 8 hours at 37°C. After rinsing in phosphate buffer, the digested tissues were dehydrated in a series of graded concentrations of ethanol, critical-point dried, mounted on metal stubs, coated with gold, and observed under SEM at an accelerating voltage of 20 kV.

2.3. *Cell-maceration/SEM method*

In order to demonstrate collagen fibrillar networks ensheathing blood vessels and those surrounding cellular elements, cellular elements were macerated away and collagen fibrils were demonstrated in their natural locations with a newly introduced cell-maceration/SEM method [5,6]. Tissue pieces were fixed in 2.5% glutaraldehyde and immersed in a 10% aqueous solution of NaOH for 3–7 days at about 25°C. The tissue pieces were then extensively washed in distilled water for 3–4 days at about 25°C. The extracted collagen fibrillar mass was put in a 1% aqueous solution of tannic acid for 3 hours, rinsed in distilled water overnight, and postfixed in a 1% aqueous solution of OsO_4 for 2–3 hours (conductive staining [7]). The specimens were dehydrated in a series of graded concentrations of ethanol and freeze-cracked with a razor blade in liquid nitrogen. They were then critical-point dried using liquid CO_2, mounted on metal stubs with double sticky tapes, coated with gold, and observed under SEM at an accelerating voltage of 15–20 kV.

3. Salivary glands

The saliva undergoes extensive modification during its passage along the striated intralobular duct [8,9]. An important role for the microcirculation of the ducts has also been implied in this modification process. Whether there is any connection between acinar and ductal circulation in the salivary gland is the subject of continuing controversy. Kowalewsky [10] first suggested (and was later supported by Fraser and Smaje [11]) that acinar and ductal capillary networks are arranged in parallel. In contrast, Burgen and Seeman [12] described a countercurrent portal system, with blood flowing from the capillary bed around the striated duct to that of the acinar portion. Blair-West et al. [13], on the other hand, reported a concurrent portal system in which the periacinar capillaries drain into a sinusoidal capillary network that is closely associated with the striated intralobular ducts.

We will describe here the general structure of the rat submandibular gland and its microvascular architecture. Particular attention will be given to the vascular connections between the periacinar and periductal capillary network.

3.1. *General structure of the salivary gland*

The salivary glands consist of a great number of a glandular structures that secret saliva. The smaller glands lie in the oral mucous membrane. The larger glands (the major salivary glands) are paired structures situated away from the oral cavity, to which their secretions are transported through the excretory ducts. The major salivary glands are the parotid, submandibular (or submaxillary), and sublingual glands.

The major salivary glands are made up of a great number of lobules surrounded by connective tissues. The three-dimensional structure of rat salivary gland is clearly demonstrated by scanning electron microscopy after removal of stromal connective tissues by acid hydrolysis plus enzymatic digestion [14–16]. The following description is restricted to the submandibular glands of the rat.

The surface of the submandibular gland is made up of numerous acini clustered into groups of various sizes and shapes (Fig. 13-1). The stromal surface of the acinar cells is embraced by myoepithelial cells. Between the acini run blood vessels, mainly of capillary size, around which nerve fibers are closely entwined (Fig. 13-1). On the superficial aspect of the lobule, a plexus of nerve fibers is seen embracing each acinus. Occasionally, nerve fibers with enlarged round terminals are observed closely associated with the acini [16]. In between the apical part of acinar cells, there are fine intercellular canaliculi, into which the primary saliva is secreted.

Each cluster of four to five polymorphous acini is connected with the intercalated duct, which is lined with low cuboidal epithelium. The intercalated ducts are connected with the convoluted ducts, which frequently take sigmoid courses

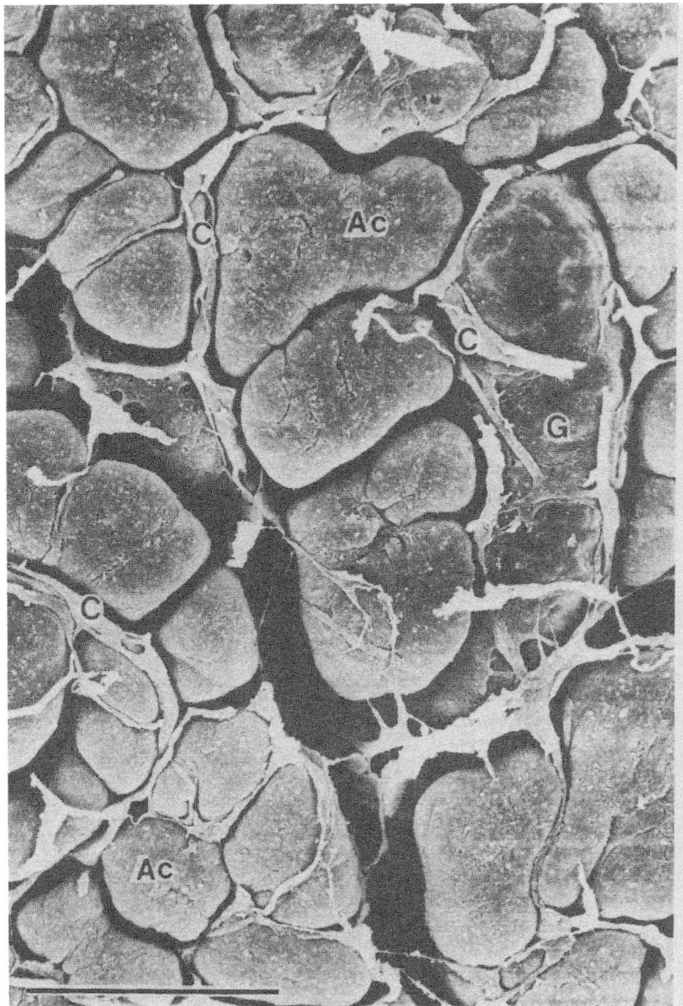

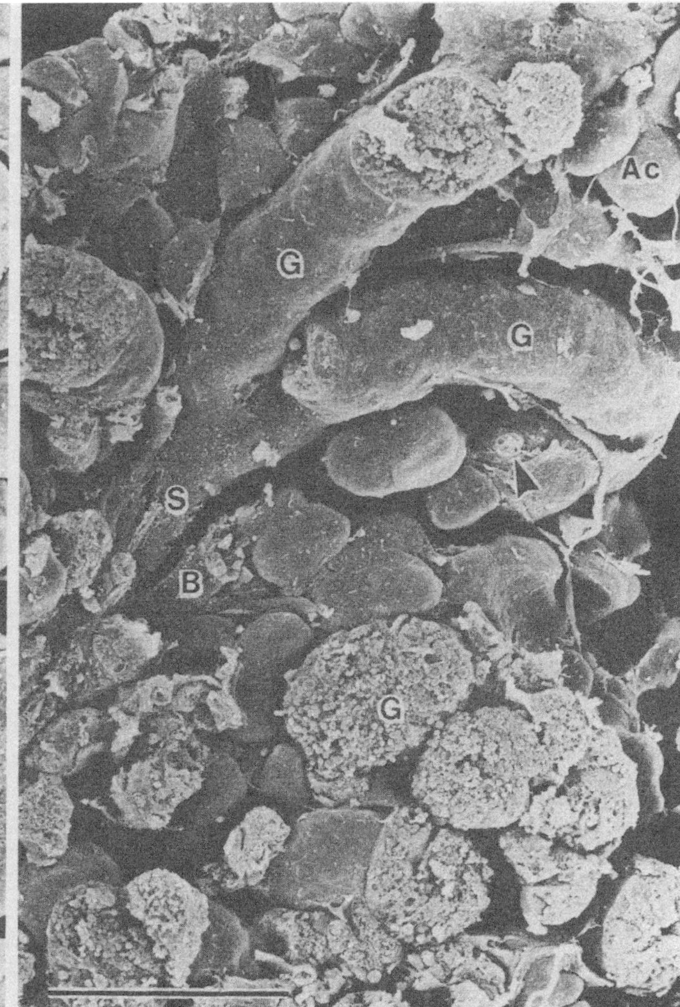

Fig. 13-1. SEM view of the rat submandibular gland treated with HC1/collagenase. Capillaries entwined with the nerve fibers (C) are seen around the clusters of polymorphous acini (Ac) and around a convoluted granular duct (G). Bar = 50 μm.

Fig. 13-2. SEM view of the rat submandibular gland treated with HC1/collagenase. Convoluted granular ducts (G) lead into the striated intalobular duct (S). An arrowhead indicates a fractured intercalated duct. Ac = acinus; B = blood vessel. Bar = 50 μm.

(Fig. 13-2). It has been reported during the past few decades that there exist in the convoluted duct cells a number of biologically active polypeptides such as kallikreins, renin, nerve growth factor (NGF), epidermal growth factor, etc. (reviewed by Barka [17]). These substances are secreted into the saliva, but have also been shown to exist in the systemic circulation.

The convoluted duct is connected with the intralobular striated duct (also called the salivary duct) (Fig. 13-2). The intralobular striated duct is lined with cuboidal cells, which have rodlike mitochondria oriented perpendicular to the base of the cell and deep infoldings of the basal plasma membrane into the cytoplasm. This duct is known to contribute to the process of salivary secretion. The intralobular striated duct takes a rather straight course and leads into the interlobular excretory duct. The excretory duct is lined with pseudostratified columnar epithelium, with occasional goblet cells. The cytological aspect of the salivary glands was extensively reviewed by Pinkstaff [18]. The general structure of the rat submandibular gland is schematically illustrated in Fig. 13-3.

3.2. Intralobular microcirculation of the salivary gland

In general, the interlobular arteries and their venous counterparts of the salivary glands run along the interlobular duct system. The intralobular arterioles and venules travel along the duct system, which runs from the central portion of the lobule to its periphery (Fig. 13-4). The intralobular arterioles break up into capillaries and loop around the acini, intercalated ducts, and convoluted ducts. The distance between the adjacent branching points of these capillaries is about 25–45 μm, as measured by vascular corrosion casts. The striated duct is surrounded by a sinusoidal capillary network. Although the vascular plexus surrounding the striated duct is occasionally supplied directly by the arterioles, it receives blood from the capillary network surrounding the acini, the intercalated duct, and the convoluted ducts through two types of vessels (Fig. 13-4). First, the capillaries of the secretory portion drain into the sinusoidal capillary network surrounding the striated ducts through capillaries or venules, which may constitute a portal system. Second, the venules from the secretory portion join those running closely along the striated ducts, forming part of the vascular plexus around the ducts. The circulation of the secretory portion of the salivary glands is thus linked with the intralobular striated duct circulation through capillaries and venules, and thereby blood in the secretory portion drains into the sinusoidal capillary plexus around the striated duct [16]. This pattern of circulation is also reported in sheep parotid glands [13] and in canine parotid glands [12].

It is generally accepted that the saliva, which is isotonic with respect to plasma when secreted at the acini, is extensively modified during its passage along the striated intralobular duct to a hypotonic fluid by reabsorption of sodium and chloride, together with secretion of potassium [8, 19, 20]. The fact that the striated duct is surrounded by a well-developed sinusoidal capillary plexus seem to be appropriate for such physiological activities of the duct.

3.3. Excretory duct microcirculation of the salivary gland

The interlobular ducts and the main excretory ducts are surrounded by a close-meshed network of subepithelial capillaries that is more dense than that of the secretory portion. The capillary plexus around the excretory duct is directly supplied by branches of the interlobular arteries and drains into venules leading into the interlobular veins (Fig. 13-5). However, near the commencement of the interlobular duct the subepithelial capillary plexus also receives capillaries and venules from the secretory portion. The circulation of the remainder of the interlobular ducts and the main excretory duct is independent of that of the secretory portion. The excretory duct is reported to play an active role in the transformation of the secreted fluid into final saliva by the process of ion transport [21].

It has been suggested that the main excretory duct of the human submandibular gland has both secretory and absorptive activities [22]. Thus, the close-meshed subepithelial capillary plexus of the excretory duct seems to be appropriate for its secretory/absorptive functions.

Fig. 13-3. Schematic presentation showing the three-dimensional organization of the rat submadibular glands with its blood vessels. Note the portal vessels (arrowheads) that collect capillaries around the acini (Ac), intercalated duct (I), and convoluted granular duct (G) and join the sinusoidal capillary network around the striated intralobular duct (S). E = interlobular excretory duct with its capillary network; A = artery; V = vein.

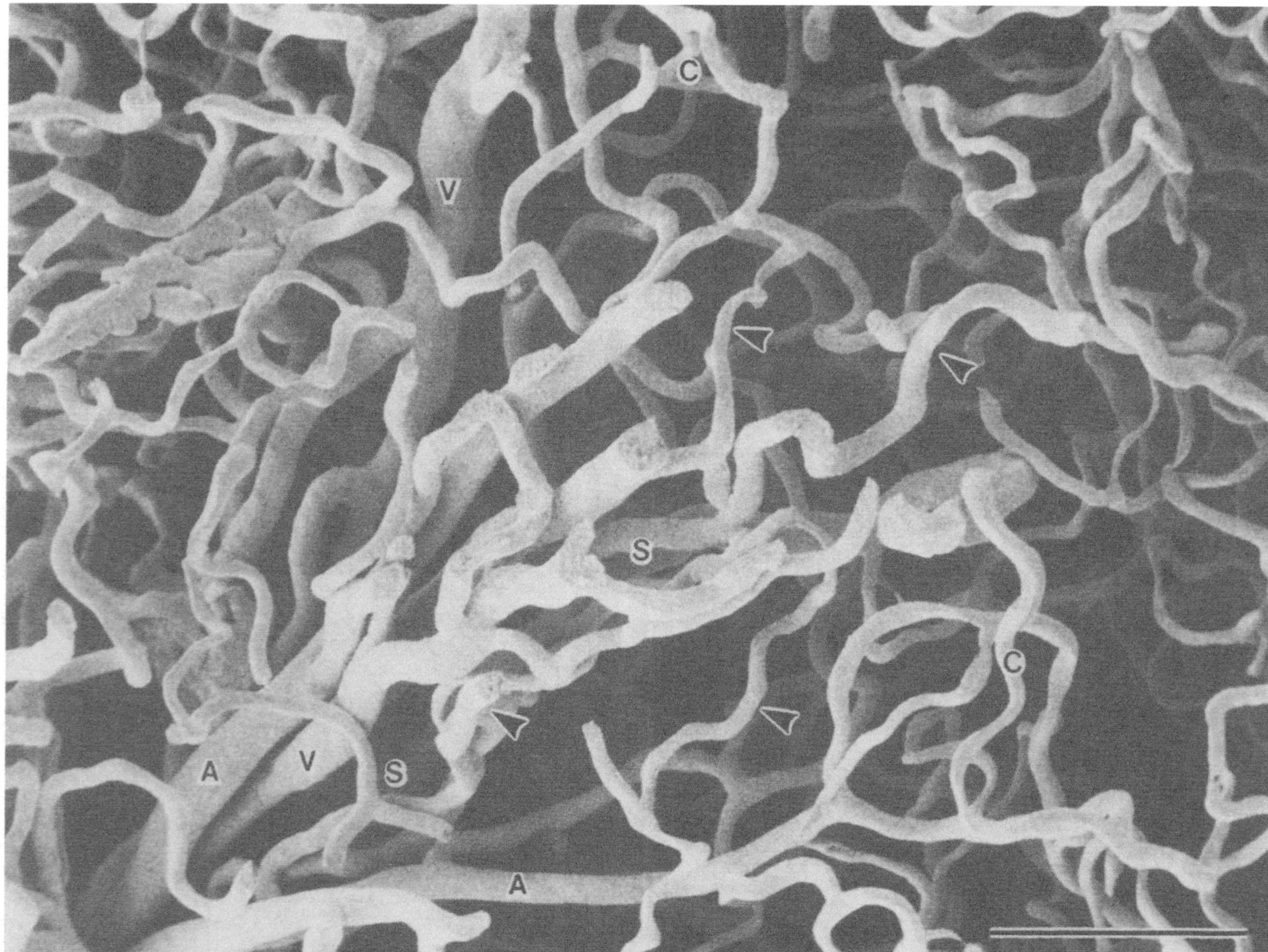

Fig. 13-4. SEM view of the vascular corrosion cast of the rat submandibular gland. Intralobular arterioles (A) and venules (V) run along the vascular plexus of the striated duct (S). Arrowheads indicate capillaries that collect those of the secretory portion (C) and drain into the sinusoidal capillary plexus around the striated duct. Bar = 100 μm.

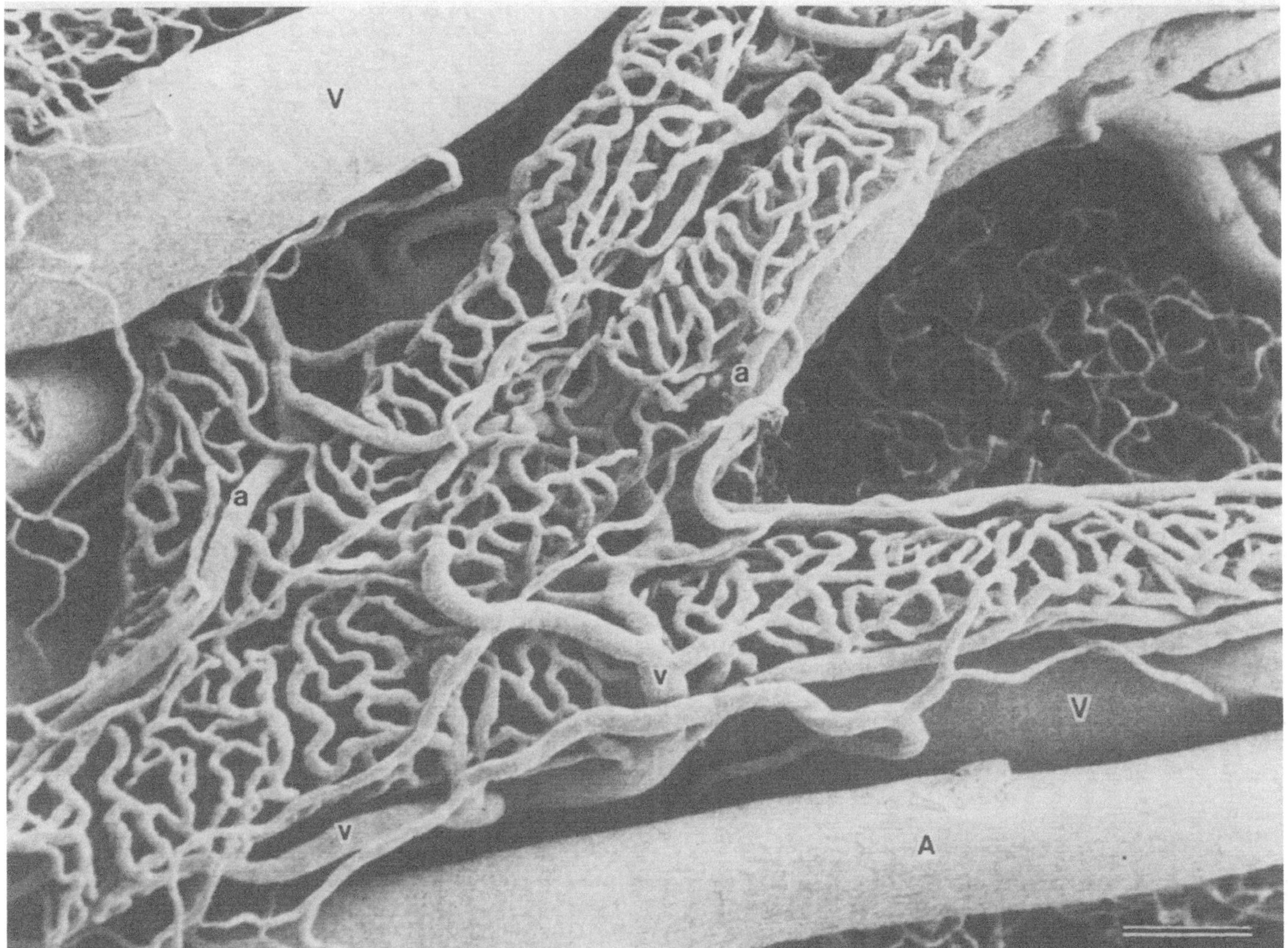

Fig. 13-5. SEM view of the vascular corrosion cast of the rat submandibular gland showing the capillary plexus surrounding the interlobular excretory duct. A,V = interlobular artery and vein, a,v = arterioles and venules of the interlobular excretory duct. Bar = 100 μm.

Arteriovenous anastomoses (AVAs) were described by early investigators [23]. Suddick and Dowd [24], using intravascular injection of either India ink or silicon rubber, reported that AVAs are only present in the intralobular duct microcirculation, but not in the periacinar circulation. Later studies, however, failed to confirm the presence of any AVAs in the salivary glands [11, 25, 26]. Instead, our SEM studies of vascular corrosion casts have revealed that arterioarterial (AAAs) and venovenous anastomoses (VVAs) do exist along the excretory ducts [16]. The larger the duct becomes, the more frequently the AAAs and VVAs appear.

Marked constrictions are frequently observed on vascular corrosion casts of the salivary glands at the confluent sites of the efferent venules of the periductal plexus with the interlobular veins and at the commencement of the afferent arterioles of the plexus [16]. These constrictions suggest the existence at these sites of sphincters or some fibrous structures that prevent the vessels from dilating. At this time these structures are not well known, however, together with AAAs and VVAs, they seem to participate in regulating ductal circulation.

4. Pancreas

Why are the islets of Langerhans found in the pancreas? This is a question that has attracted the interest of many researchers. Some authors [27, 28] have thought that a clue to the answer lies in the vascular connections between the islets and the exocrine pancreas.

Since the works of Wharton [29] and Thiel [30], the portal connections, now called the *insuloacinar portal system* [28], between the capillary network in the islets and that in the acinar tissue have been confirmed by light microscopy of living tissues [31–34] and by scanning electron microscopy of vascular corrosion casts in a variety of animals [33–38]. The portal connections between the islets and the pancreatic ductules, and between the exocrine acini and the pancreatic ductules, have also been described in the rat and rabbit [37].

4.1. Terminal distribution of arterioles

Lobules of various size and shape are obvious upon gross inspection of vascular corrosion casts of the pancreas. The interlobular artery runs with its venous counterparts, sending off intralobular arterioles (Figs. 13-6 and 13-7). The intralobular arterioles fall into three categories, according to their destinations: 1) afferent vessels of the islet (Figs. 13-6 to 13-8) 2) acinar arterioles supplying the acini (Fig. 13-6) and 3) arterioles supplying the duct system by forming the periductular plexus (Fig. 13-7).

The afferent vessels of the islets account for about 10%–20% of the total terminal arterioles in the pancreas [34]. The acinar arterioles preferentially supply the exocrine pancreas far removed from the islet. The venules tend to commence farther away from the islet.

4.2. Microcirculation of the islet

In the rat and rabbit, the afferent arterioles, after reaching an islet, divide into swollen sinusoidal capillaries running along the islet cortex where A and D cells exist. The cortical sinusoidal capillaries frequently take U-shaped turns and give off secondary capillaries, that enter the islet core, which is largely occupied by B cells [34]. In the monkey and horse, the afferent arterioles penetrate deep into the islet core occupied by A and D cells, and there they radiate sinusoidal capillaries towards the islet cortex, where B cells are found [28, 35]. The islets of the guinea pig and the dog (Fig. 13-8), in which A and D cells are rather evenly dispersed among B cells, tend to have a vascular pattern intermediate between that of the monkey and the rat. Our recent study, using the cell-maceration/SEM method, has demonstrated the three-dimensional architecture of collagen fibrillar sheaths enveloping the islet capillaries in the human pancreas [5] (Fig. 13-9). In the smaller islets, a relatively larger sheath for the blood vessel often enters the islet core, from which smaller sheaths radiate to the islet cortex. In such smaller islets, non-B cells are located mainly adjacent to the centrally located larger vessels [39]. In larger islets, however, vessels

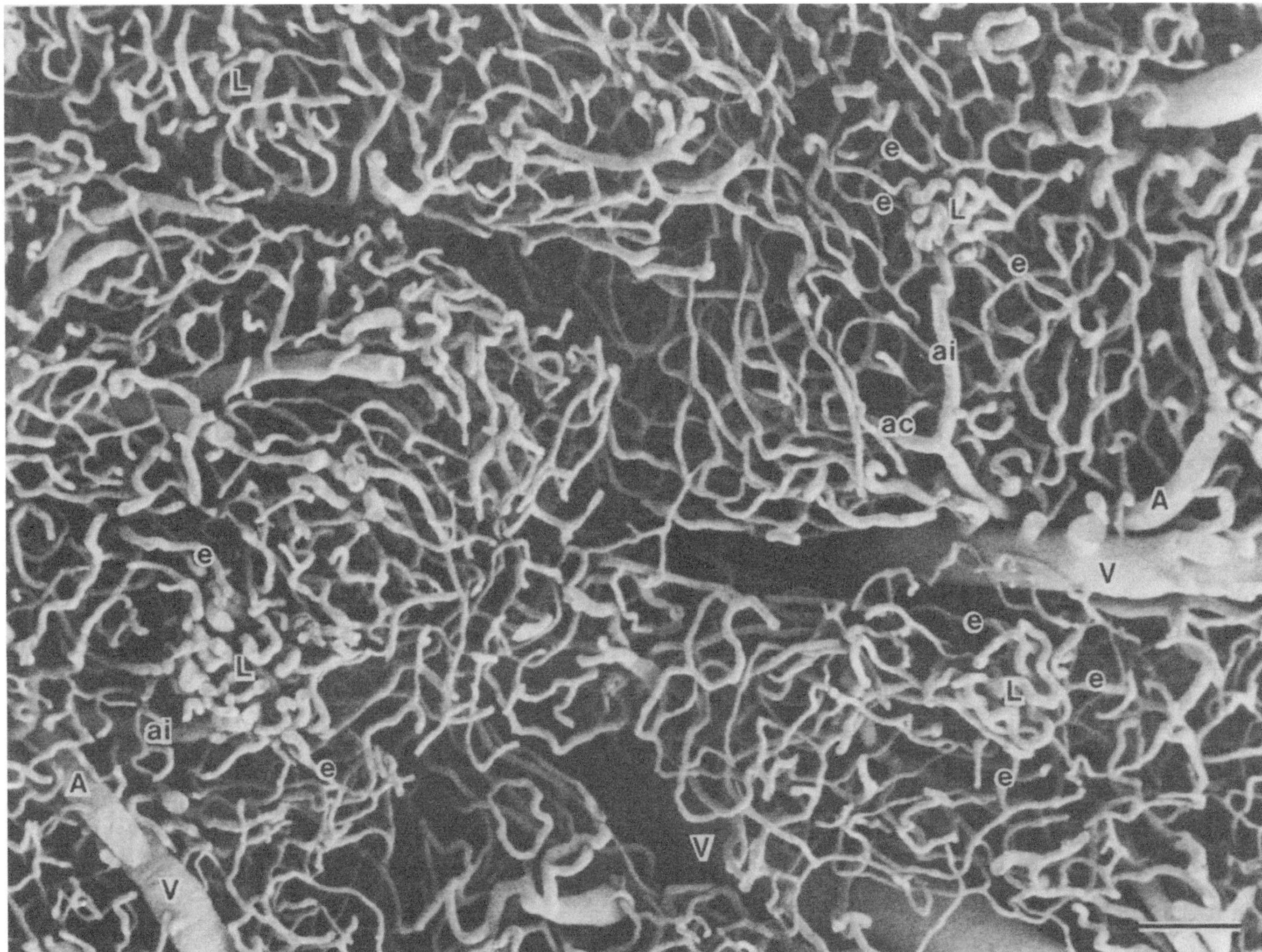

Fig. 13-6. SEM view of the vascular corrosion cast of the dog pancreas. The interlobular artery (A) runs with its venous counterpart, the interlobular vein (V), and gives off many intralobular arterioles. Two types of arterioles are visible: the afferent arterioles (ai) of the sinusoidal capillary networks of the islets (L) and the acinar arterioles (ac) that directly break up into the periacinar capillaries. The acini in the vicinity of the islets are supplied preferentially by the efferent vessels (e) of the islets, i.e., by the insulo-acinar portal vessels, whereas the acini located farther away from the islets are supplied directly by the acinar arterioles. Bar = 100 μm.

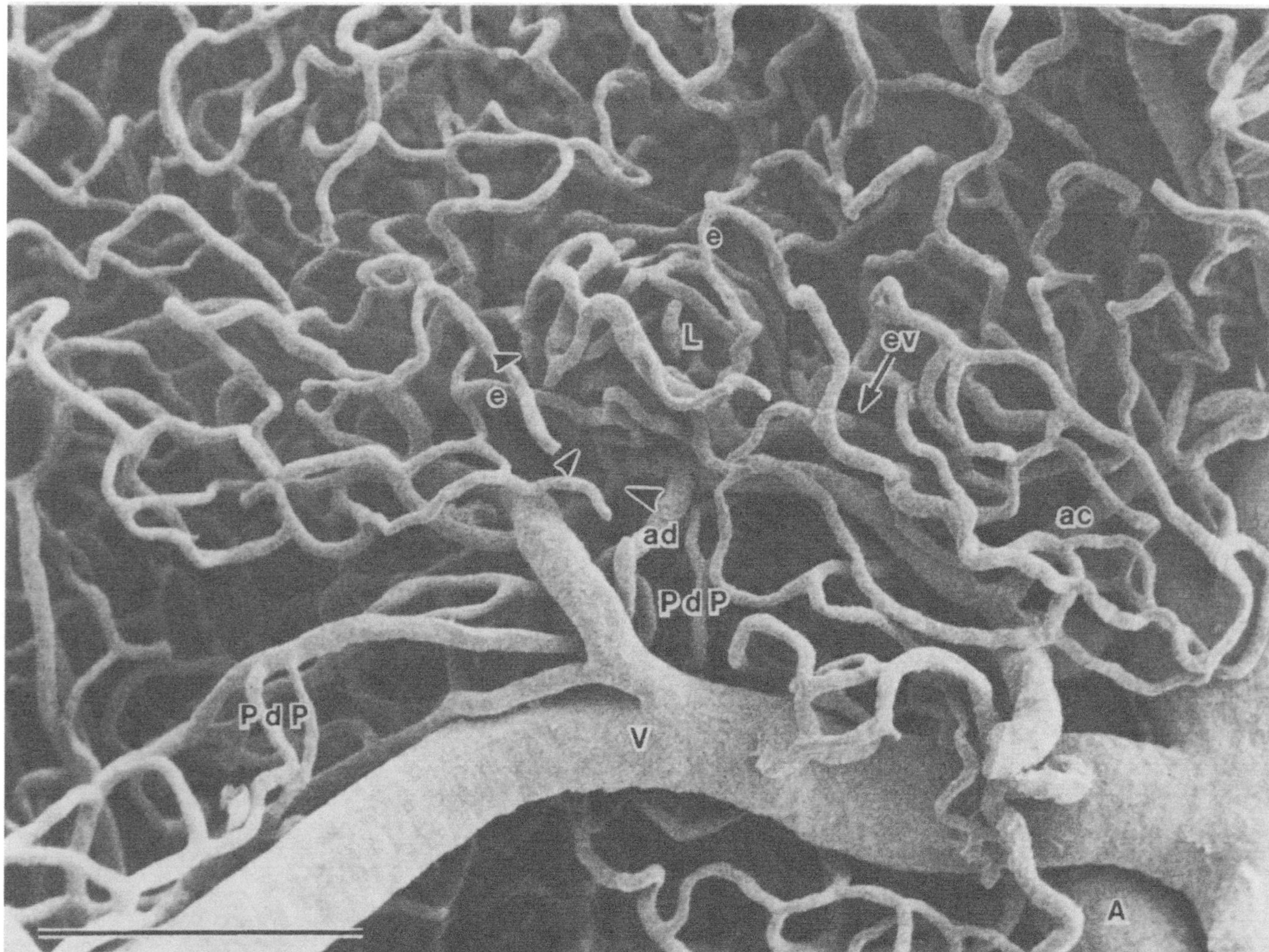

Fig. 13-7. SEM view of the vascular corrosion cast of the rat pancreas. Three types of intralobular arterioles can be seen: an arteriole (small arrowhead) supplying the islet (L), an arteriole (ac) breaking up into periacinar capillaries, and an arteriole (ad) supplying the periductular plexus (PdP). The islet possesses efferent vessels that branch out into the periacinar capillaries (insuloacinar portal vessels) (e), an efferent vessel that joins the periductular plexus (insulo-ductular portal vessel) (large arrowhead), and an emissary vein (ev with an arrow) leading into the interlobular vein (V). A = interlobular artery. Bar = 100 μm.

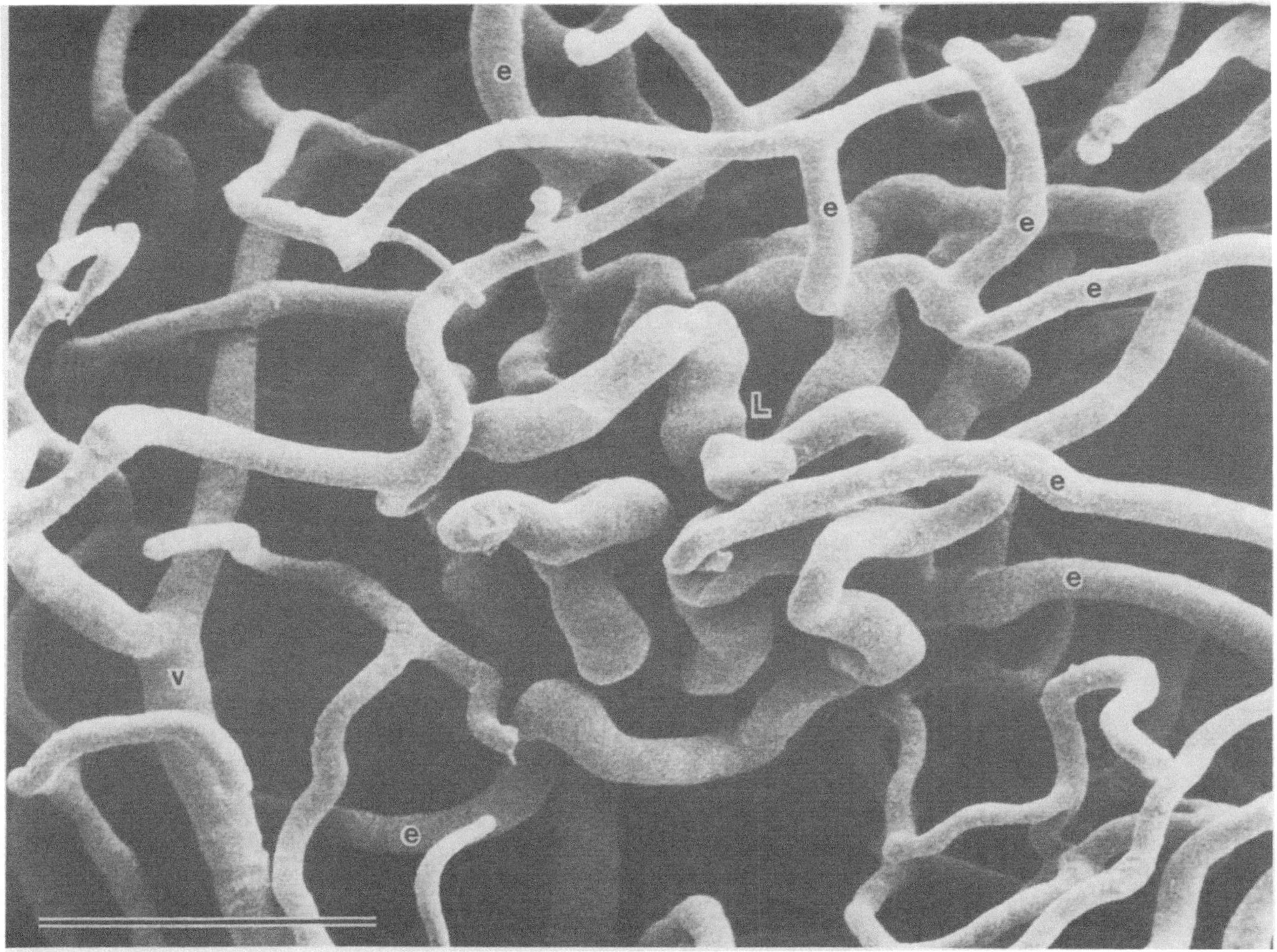

Fig. 13-8. SEM view of the vascular corrosion cast of the dog pancreas showing the glomerulus of the islet capillaries (L) and its efferent vessels (e), the insulo-acinar portal vessels, breaking up into the periacinar capillaries. v = intralobular venules collecting the periacinar capillaries. Bar = 5 μm.

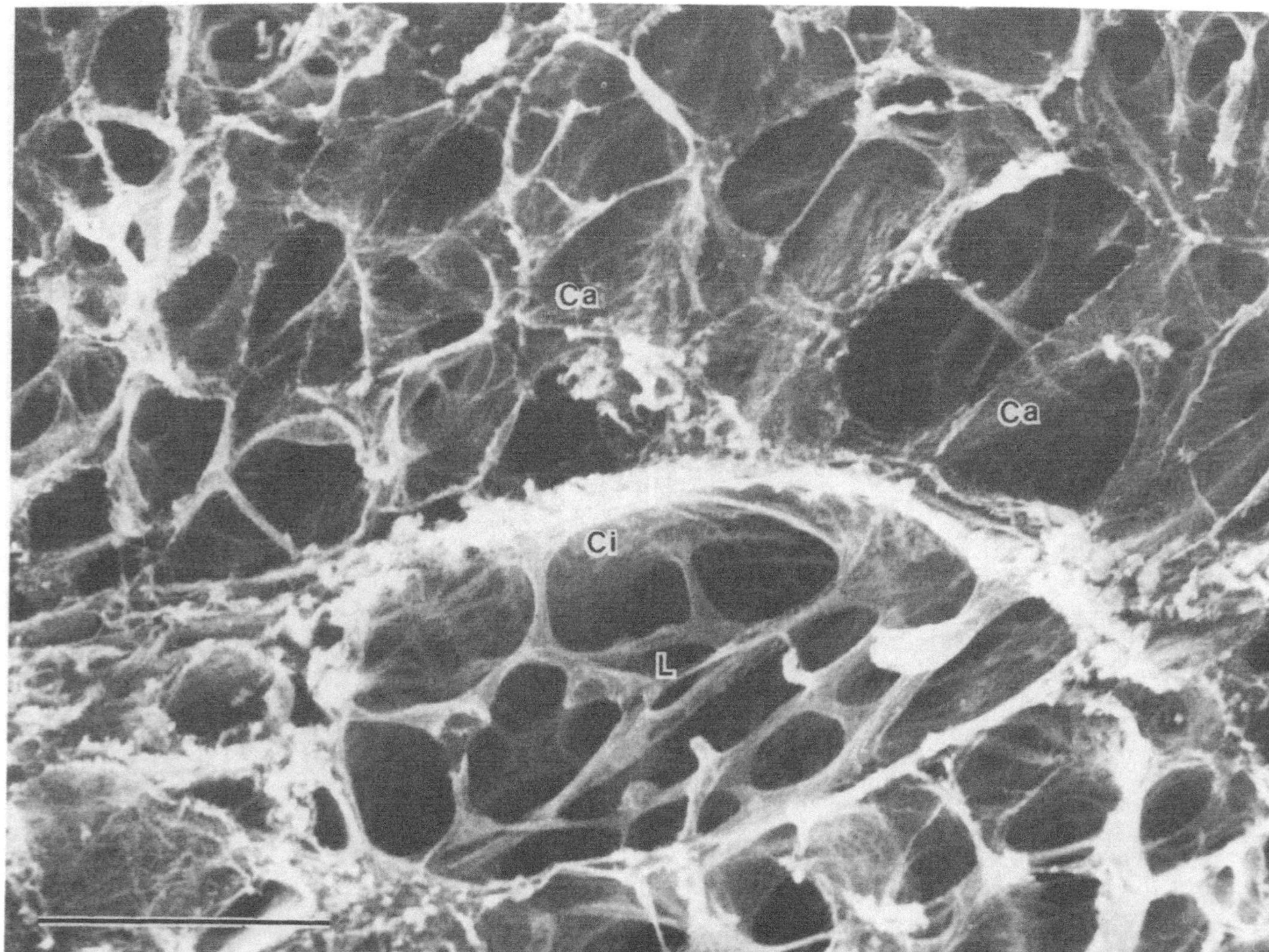

Fig. 13-9. SEM view of the human pancreas treated with NaOH. The collagen fibrillar network delimits the spaces for housing the acini (Ca). An islet (L) is also surrounded by a collagen fibrillar capsule (Ci) in which collagen fibrillar sheaths around the islet capillaries form a three-dimensional network. Bar = 50 μm.

preferentially entering the islet core have not been described.

Such interspecies differences in the vascular pattern within the islet suggest that blood flows from the area of A and D cells to that of B cells. This seems to be a device by which glucagon released from A cells and somatostatin released from D cells can be transferred to their primary target, the B cells [28]. This hypothesis is consistent with the fact that glucagon stimulates the secretion of insulin, while somatostatin inhibits its secretion.

4.3. Insulo-acinar portal system and emissary vein of the islet

The sinusoidal vascular plexus of the islet radiates numerous efferent vessels, which form the capillary plexus of the exocrine acini in the vicinity of the islet. Thus, the vascular route leading from the islet through its efferent vessels to the periacinar capillary plexus properly constitutes a *portal system* and has been referred to as the *insulo-acinar portal system* [28]. Most of the efferent vessels of the islet in the monkey, dog, and guinea pig are of capillary size, whereas the islet of the rabbit and rat frequently has wider efferent vessels, which run a rather long distance before breaking up into the second, periacinar capillary plexus. These vessels may better deserve the name of *portal* vessels. Our recent study using the cell-maceration/SEM method has revealed the presence of the insulo-acinar portal system in the human pancreas [5]. The insulo-acinar portal system is schematically illustrated in Fig. 13-10.

Intravital microscopy of the living pancreas in situ has revealed that blood leaving the islet flows through the portal vessels to the exocrine acini [32–34, 40].

It is thought that insular hormones and neurohormones [41] are transported in high concentrations through the insulo-acinar portal vessels to the exocrine pancreas, where they exert control over its secretory activities. Some evidence favoring this hypothesis has now accumulated (reviewed by Henderson et al. [42]).

The rat islets frequently possess direct venous routes draining the insular blood into the interlobular vein (Fig. 13-7). Such efferent vessels of the islet have been designated as the emissary veins of the islet [34]. In the rat, islets with emissary veins account for 60% of the islets, while in the rabbit, dog, and cat, such islets are rare. The well-developed emissary veins of the islet are therefore characteristic of the rat, although they are few in number in other animals examined so far. The significance of the emissary vein of the islet is not known at this time. The emissary veins of the islet may provide a short passageway for blood in the islet to reach the general circulation. It seems to be a device by which insular hormones are quickly conveyed to the whole body, possibly playing an important role in emergencies such as hyperglycemia and hypoglycemia.

4.4. Microcirculatory unit of the pancreas

The microvascular organization of the dog pancreas strongly suggests that the pancreatic lobule consists of subdivisions or microcirculatory units, the islet being located in the center of each unit (Fig. 13-6). The exocrine acini in the vicinity of the islet, i.e., in the peri-insular zone, are preferentially supplied by the insulo-acinar portal vessels, while those in the periphery of the unit, i.e., in the tele-insular zone, are mainly supplied directly by arterioles. The venules commence at the periphery of the unit. Such a microcirculatory pattern is well consistent with the "halo" phenomenon.

4.5. Periductular circulation

The duct system of the pancreas is surrounded by the sinusoidal capillary plexus (Fig. 13-11). Arterioles form polygonal meshes of capillaries around the duct system, and these are gradually collected into thicker venules that anastomose with each other and accompany the ducts to form intralobular or interlobular veins.

The vascular connections between the periacinar and the periductular capillary plexi are of particular interest. Two types of vascular connections have been noted [37] (Fig. 13-11). First, blood in the venules of the lobular portion flows into the larger caliber periductular vessels, which are probably of a venular nature. Second, blood in the periacinar capillary plexus flows through capillaries into the periductular plexus (Fig. 13-11).

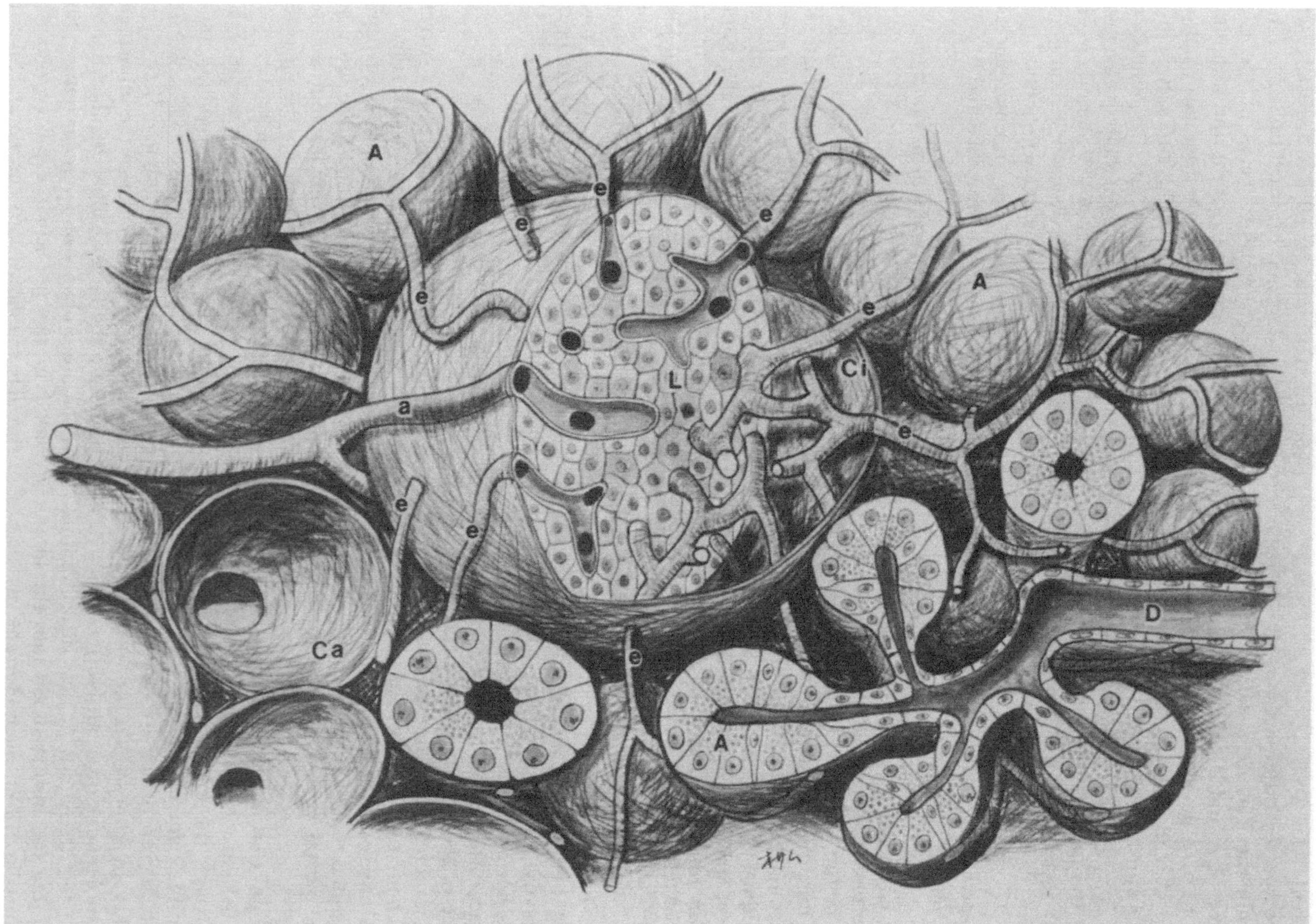

Fig. 13-10. Schematic presentation showing the microvascularization in and around the islet (L) of the pancreas. a,e = afferent and efferent vessels of the islet; A = acinus; D = ductule; Ca, Ci = collagen fibrillar network surrounding the acinus and the islet.

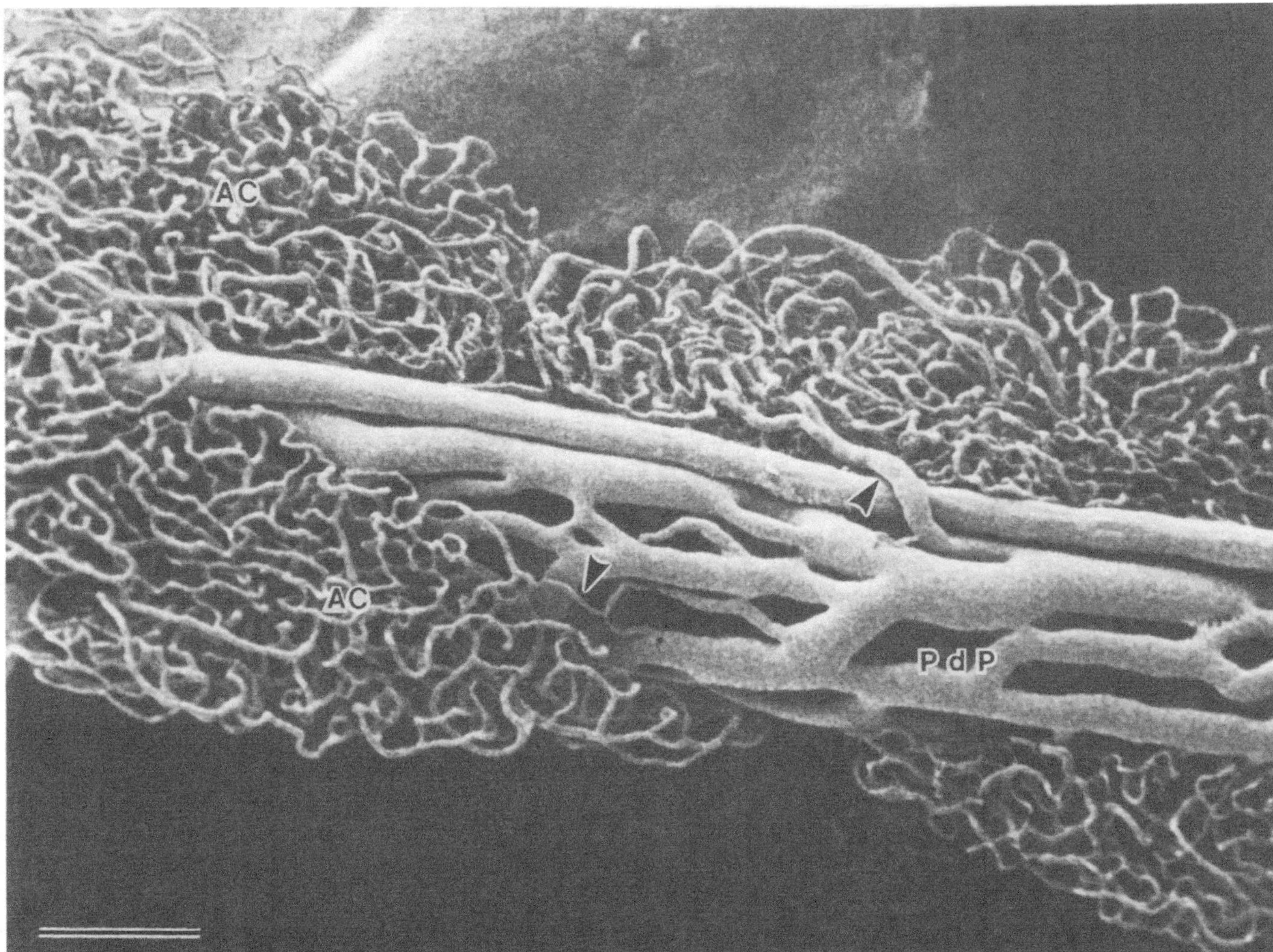

Fig. 13-11. SEM view of the vascular corrosion cast of the rabbit pancreas. The periductular plexus (PdP) receives venules and capillaries (arrowheads) from the periacinar capillary network (AC). Bar = 100 μm.

Particularly noteworthy is that the islets, especially those in the rat, often occur close to the ductules. Such islets usually receive short arterioles, and their efferent vessels partly enter the periductular plexus [37] (Fig. 13-7). This vascular route, from the islet through its efferent vessels to the ductular plexus, may also deserve the name of the *insulo-ductular portal system* [37]. Intravital microscopy of living pancreas in situ has confirmed that blood flows from the islet through its efferent vessels to the ductular plexus [33]. These microcirculatory patterns indicate that the duct system of the pancreas receives blood rich in islet secretions.

The effects of islet hormones on the pancreatic duct system are not well documented at this time. Administration of substances inhibiting insulin release (e.g., alloxan, streptozotocin) inhibits bicarbonate secretion, suggesting that islet hormones influence the ductules of the gland.

5. Liver

5.1. General aspects of liver vasculature

The liver is made up of a great number of small lobules. The lobule represents the anatomical unit of structure of the liver. This concept of the "classical hepatic lobule" was initially reported by Kiernan [43].

The liver has a dual blood supply: Approximately 80% of the blood entering the liver is poorly oxygenated, but nutrient-rich, venous blood supplied by the portal vein, while the remainder is highly oxygenated and is supplied by the hepatic artery.

The interlobular portal vein gives off numerous short side branches or inlet venules, which rather abruptly branch out into the hepatic sinusoids at the margin of the hepatic lobule.

The terminal arterioles that come off the interlobular hepatic artery fall into four categories, depending on their terminations: 1) the terminal arterioles directly draining into the hepatic sinusoids at the periphery of the hepatic lobule, 2) those connecting with the portal venules (i.e., the arterio-portal anastomoses), 3) those contributing to the formation of the peribiliary plexus surrounding the intrahepatic bile duct, and 4) those forming the periportal capillary plexus supplying the tissues in the portal tract [44] (Figs. 13-12 and 13-13).

5.2. Microvascularization of the hepatic lobule

The hepatic sinusoids, extensively interconnecting en route, run centripetally to the central venule, located roughly in the center of the hepatic lobule. The central venule leads into the thicker sublobular vein, which in turn drains into the hepatic vein.

The hepatic sinusoids are ensheathed by a collagen fibrillar network (Fig. 13-14). The sheath lies in the space of Disse. Collagen fibrillar bundles forming a coarse network frequently stretch between adjacent collagen fibrillar sheaths of the hepatic sinusoids (Fig. 13-14). In the portal tract, there are condensations of bundles of collagen fibrils running along the longitudinal axis of the portal tract. Collagen fibrils surrounding the hepatic sinusoids are continuous with those in the portal tract and with those making up the liver's thick capsule. Thus, the collagen fibrillar network constitutes a skeletal framework for the liver.

There are some regional differences in the organization of the sinusoids. Near their origins from the portal venules, the sinusoids are slightly narrower, more tortuous, and more extensively interconnected than their terminations in the central venule. However, in most mammals the peripheral boundaries of the lobules are not well defined, except in pigs, where a definitive layer of connective tissues circumscribes the lobule. Because of such interconnections of the sinusoids between adjacent lobules and regional differences in intralobular oxygenation, metabolic functions, and responses to disease, the concept of the acinus was proposed to define the hepatic functional unit [45, 46]. The axis of the acinus is a portal tract containing terminal portal venules and hepatic arterioles. Its peripheral boundaries are circumscribed by an imaginary line containing the neighboring central venules.

5.3. Peribiliary portal system

As noted by Mall [47] and confirmed by later injection studies [48, 49], most branches of the

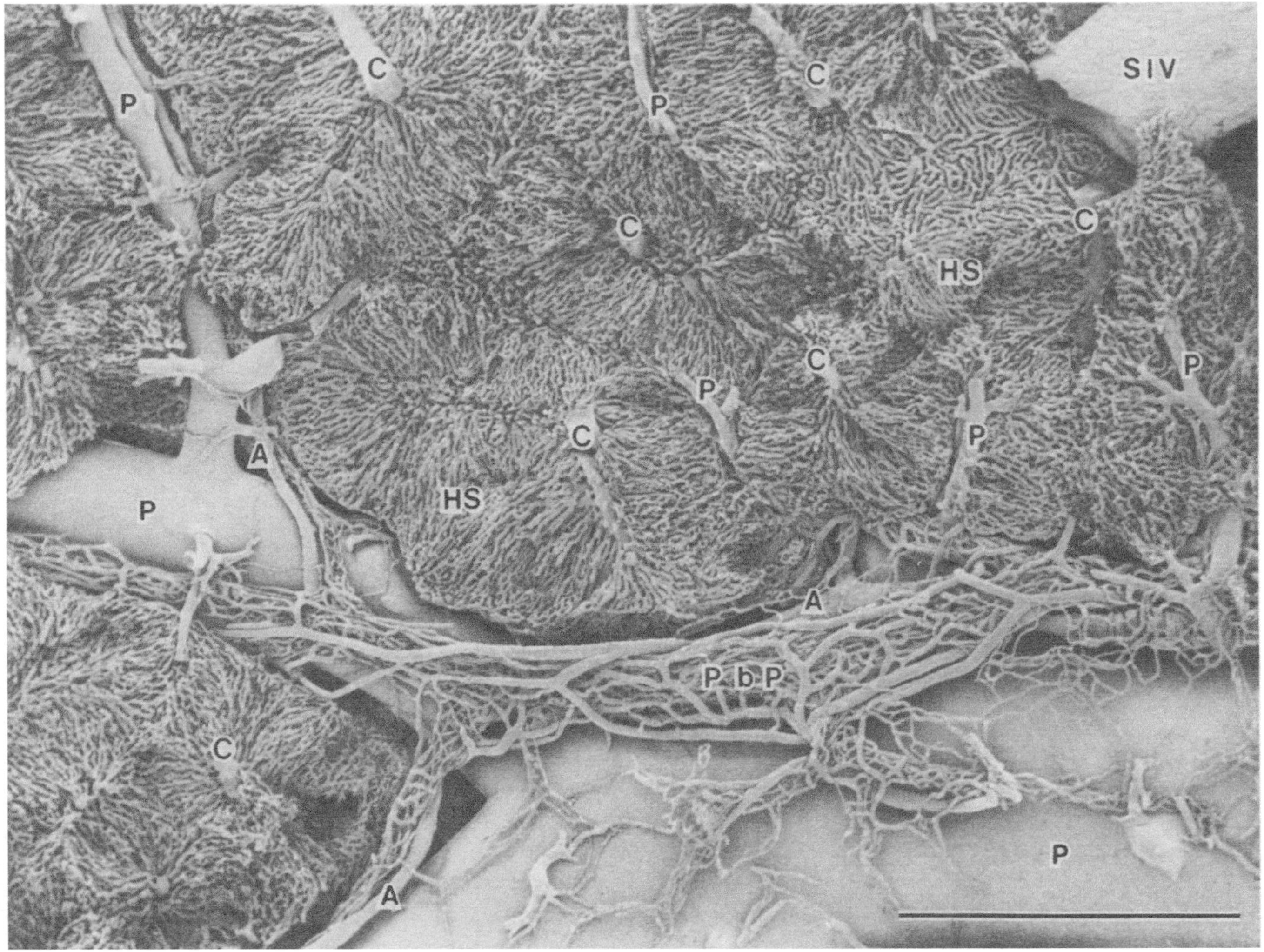

Fig. 13-12. SEM view of the vascular corrosion cast of the rat liver. The interlobular portal vein (P), hepatic artery (A), and peribiliary plexus (PbP) run parallel in the portal tract. There are many portal venules that break up into the hepatic sinusoids (HS). The extensively interconnected sinusoids drain into the central venules (C). SIV = sublobular vein. Bar = 1 mm.

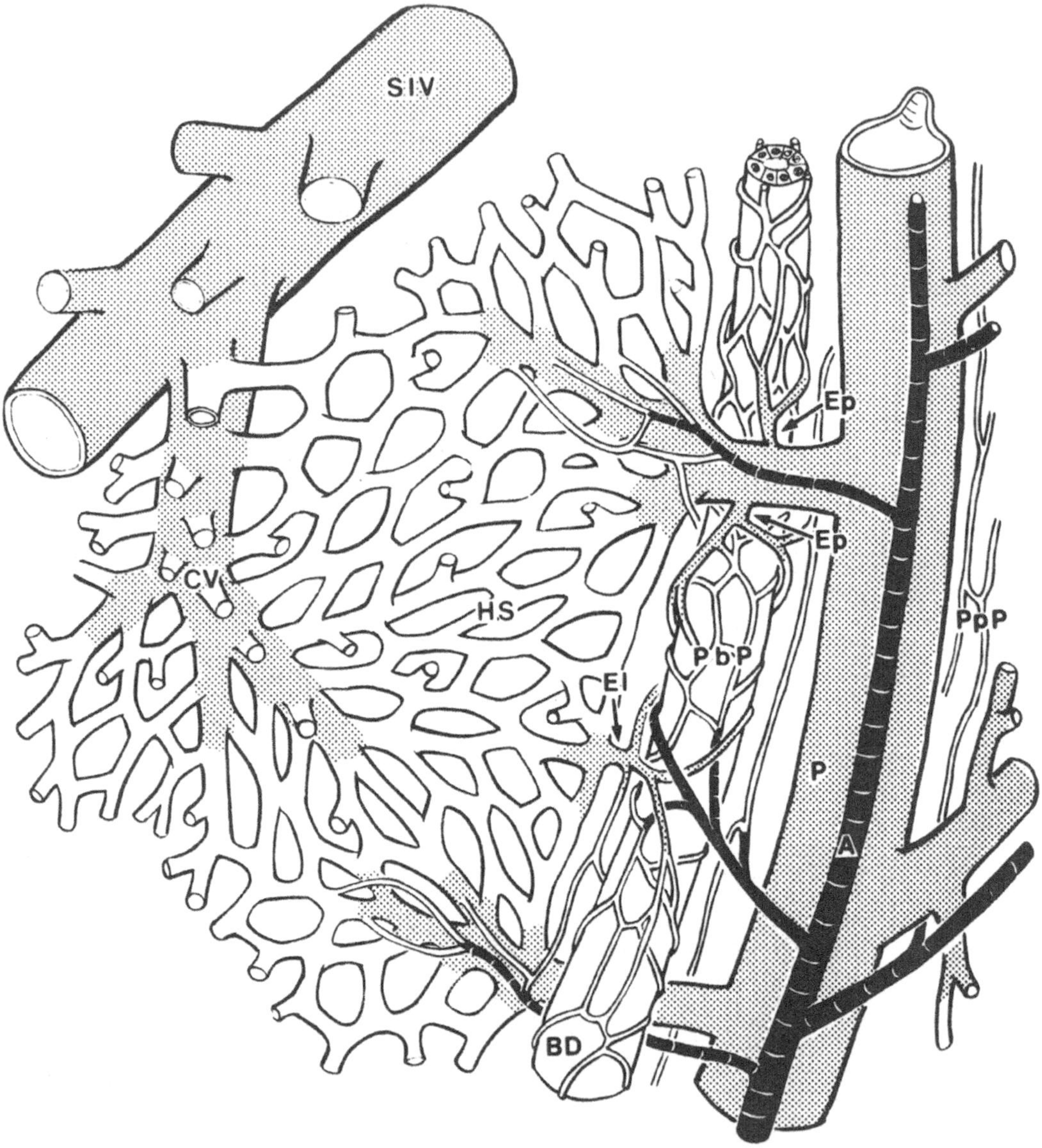

Fig. 13-13. Schematic presentation showing the general idea of the microvascularization of the liver. A,P = interlobular hepatic artery and portal vein; HS = hepatic sinusoids; CV = central venule; SIV = sublobular vein; BD = bile duct; PbP = peribiliary plexus; Ep,El = "prelobular" and "lobular" efferent vessels of the peribiliary plexus; PpP = periportal plexus.

interlobular arteries contribute to the formation of an elaborate capillary network surrounding the bile duct system, termed the *peribiliary plexus*. The bile ductule or small bile duct accompanies one or two capillaries with interconnections (Figs. 13-12 and 13-15). As the duct becomes larger, it is surrounded by a network of capillaries, and finally by the plexus, consisting of two layers: an inner network of subepithelial capillaries and an outer layer of venules and arterioles.

The peribiliary plexus is collected into fairly independent efferent vessels leading either into the hepatic sinusoids or into the interlobular portal vein (Fig. 13-15). The former route, di-

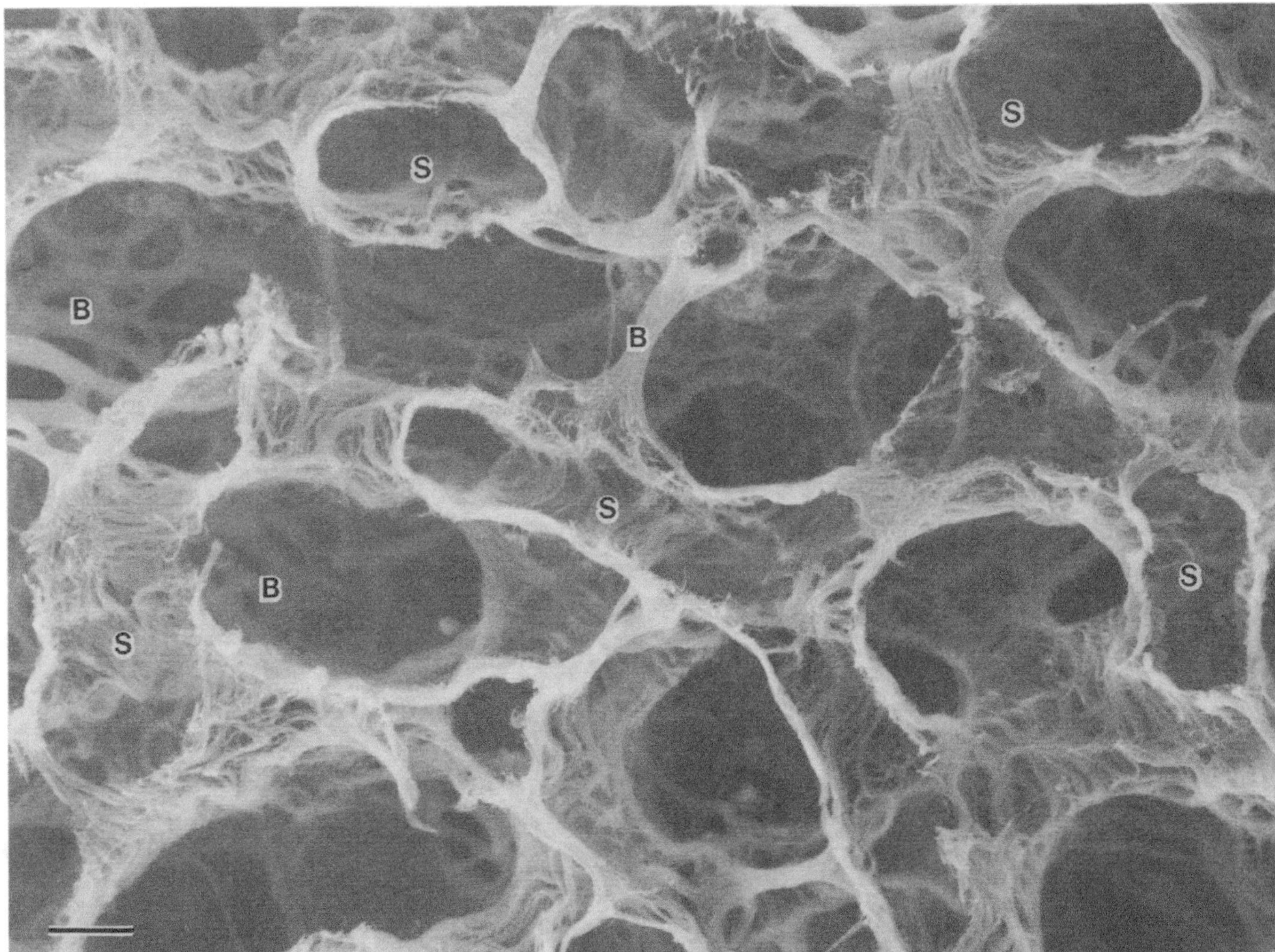

Fig. 13-14. SEM view of the human liver treated with NaOH. The collagen fibrillar network forms a three-dimensional anastomosing sheath (S) for housing the hepatic sinusoids. Bundles of collagen fibrils also form networks (B) that stretch between adjacent sheaths of the hepatic sinusoids. Bar = 10 μm.

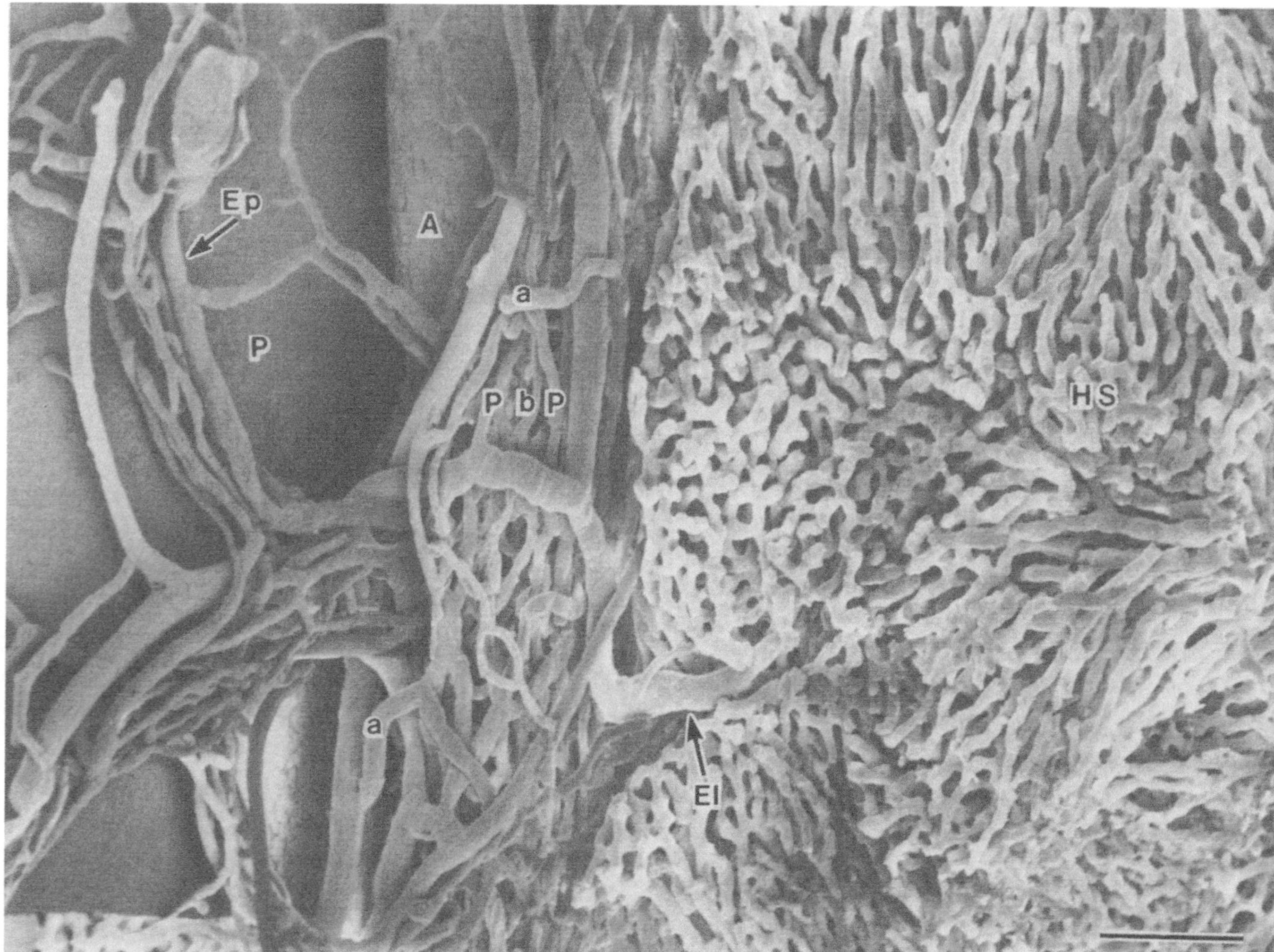

Fig. 13-15. SEM view of the vascular corrosion cast of the rat liver showing the "prelobular" (Ep) and "lobular" (El) efferent vessels of the peribiliary plexus (PbP). A,P = interlobular hepatic artery and portal vein; a = afferent arteriole of the peribiliary plexus; HS = hepatic sinusoids. Bar = 100 μm.

rectly draining into the hepatic sinusoids, is designated as the *lobular branch*, while the latter, ending in the interlobular portal venules or veins, is designated as the *prelobular branch* [44]. In the rabbit and human, both branches occur at almost the same frequency [44,50]. In the rat, the prelobular branches are seen more frequently than the lobular branches [51], whereas in the monkey the lobular branches occur almost exclusively, with few prelobular branches [52].

Since the peribiliary plexus has an elaborate network of capillaries with efferent vessels reentering the hepatic sinusoids, the vascular route from the plexus through its efferent vessels to the hepatic sinusoids qualifies to be described as a *portal* system and should be referred to as the *peribiliary portal system* [44,51,52].

A possible function of the peribiliary portal system has been considered in terms of a countercurrent mechanism by which reabsorbed substances from the bile could be transported back to the hepatocyte.

5.4. Microvascularization of the extrahepatic bile duct and the gallbladder

Arterial branches, on reaching the extrahepatic bile duct, ramify into many arterioles that, giving off the capillaries in the adventitia en route, enter the lamina propria of the duct. There these arterioles break up into capillaries and form a closely meshed network immediately below the epithelium (Fig. 13-16). This capillary network resembles that of the intrahepatic bile duct system. The capillaries in the lamina propria and in the adventitia are collected into venules that repeatedly anastomose and branch to form a venular plexus in the adventitia (Fig. 13-16). The venular plexus drains into thicker veins that ultimately lead into the hepatic portal vein.

The extenstive dense network of capillaries in the lamina propria suggests its involvement in reabsorbing some substances from the bile and in taking up any hormonal substances released by endocrine cells.

The gallbladder is a modified bile duct for concentrating and storing the bile. The microvascular organization of the gallbladder is quite similar to that of the extrahepatic bile duct. In the adventitia, arterial branches give off many smaller branches and arterioles. Some of them break up into capillaries in the adventitia and in the muscular layer, but most of them enter the lamina propria, where arterioles break up into capillaries to form a closely meshed reticular network immediately below the epithelium (Fig. 13-17). The extremely dense network of subepithelial capillaries would be appropriate for absorptive functions of the gallbladder, as well as for supplying vital nutrients and oxygen to the overlying epithelium.

The efferent veins of the gallbladder lead either into the hepatic portal vein or directly into the sinusoids in the hepatic lobules surrounding the gallbladder [53] (Fig. 13-18). The vascular route from the subepithelial capillaries in the gallbladder to the liver also qualifies to be called a *portal* system, analogous to the peribiliary portal system in the liver. Thus, we propose the term *peribiliary portal system* for the portal system associated with the extrahepatic bile ducts and the gallbladder, as well as the intrahepatic bile ducts.

The portal system would play an important role in concentrating the bile and in transporting absorbed substances and any hormonal substances produced by the gallbladder wall back to the hepatic sinusoids in order to control the microcirculation of the liver, and to the hepatocytes, presumably to act on them. Recent studies have shown that motilin-, substance P-, and somatostatin-containing cells exist in the extrahepatic biliary system [54–56]. A vasoactive intestinal peptide (VIP) nerve supply of the gallbladder, papilla of Vater, and sphincter of Oddi has also been reported [57,58]. VIP may be secreted into the microcirculation of these organs, as proposed in the islet of Langerhans of the pancreas by Fujita and Kobayashi [41].

The portal veins directly linking the gallbladder and the hepatic sinusoids may be particularly important clinically, as they may provide a quick route of spread to the liver for some gallbladder diseases.

6. Concluding remarks

SEM of vascular corrosion casts has clearly demonstrated that the extraparietal glands of the

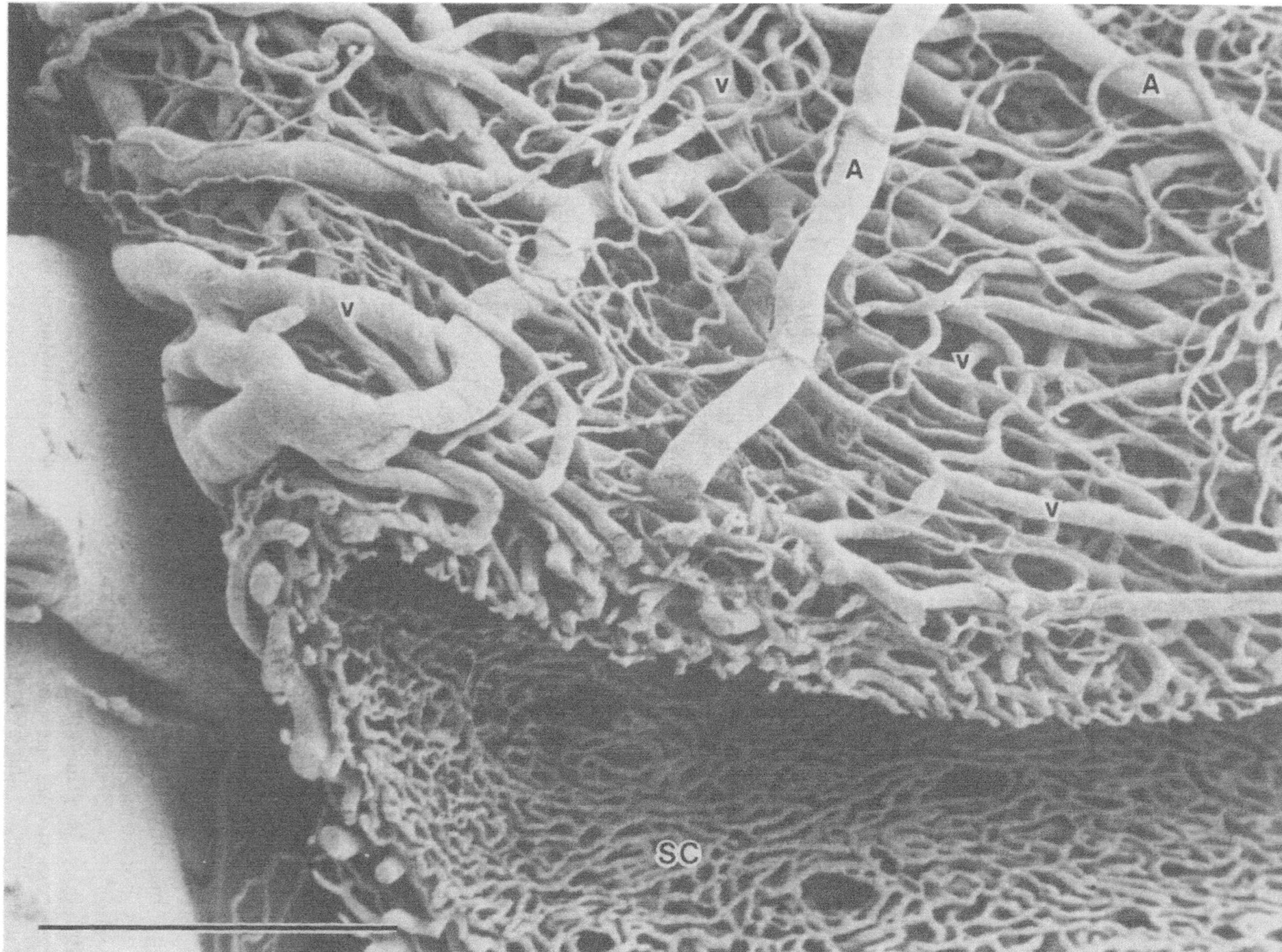

Fig. 13-16. SEM view of the vascular corrosion cast of the extrahepatic bile duct in the rat. There is a close-meshed network of subepithelial capillaries (SC), which are gathered into the venular plexus (v) located in the adventitia. A = artery. Bar = 500 μm.

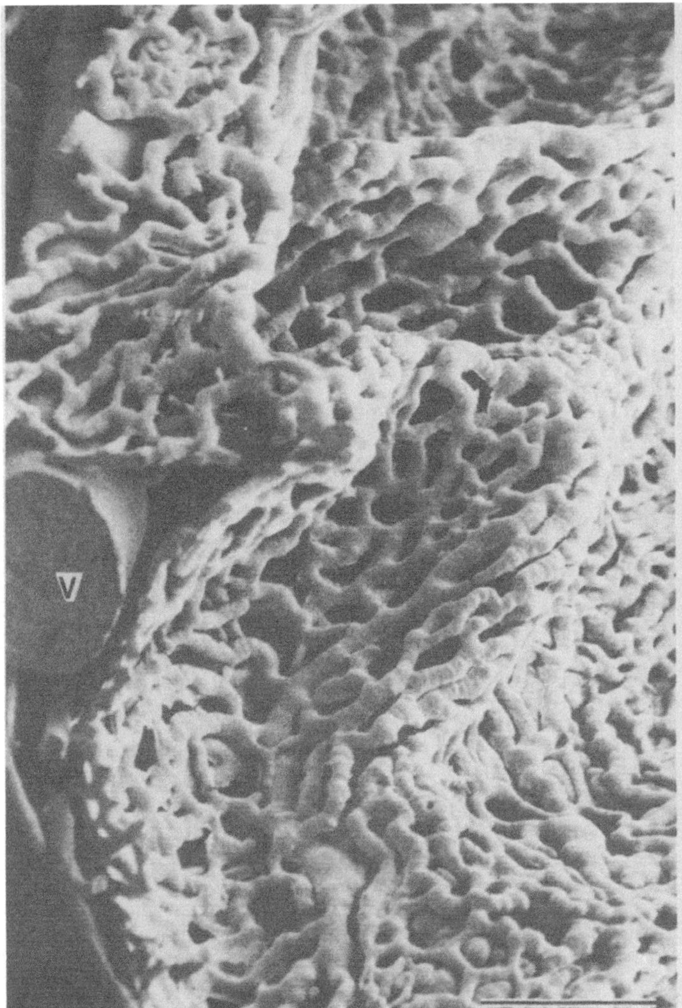

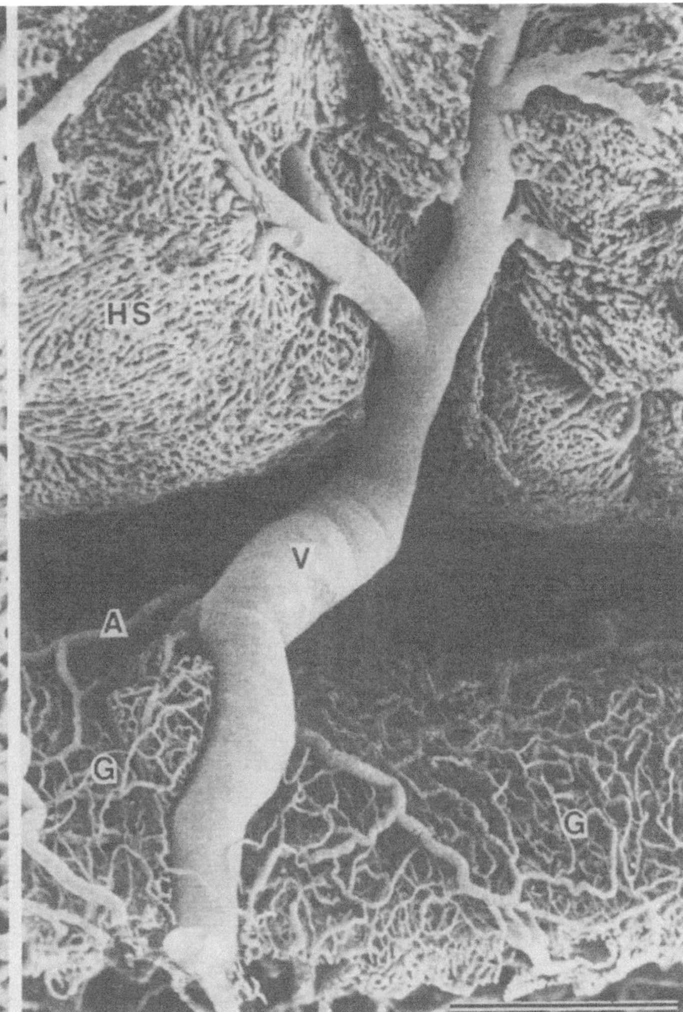

Fig. 13-17. SEM view of the vascular corrosion cast of the rabbit gallbladder, showing an extensive dense network of subepithelial capillaries. V = vein. Bar = 100 μm.
Fig. 13-18. SEM view of the vascular corrosion cast of the rabbit gallbladder (G), showing its efferent vein (V), which breaks up into the hepatic sinusoids (HS). A = artery supplying the gallbladder. Bar = 500 μm.

alimentary canal are richly vascularized. It has also confirmed that these glands possess their own specially differentiated microvascular patterns. This difference suggests that the functional significance of the microcirculation in secretory/absorptive activities differs in each gland.

In the salivary gland, the microcirculation of the intralobular striated duct is downstream from that of the secretory portion, while the interlobular duct circulation is independent from the lobular one. Such microcirculatory patterns would be appropriate for modifying the saliva during its passage along the intralobular and interlobular duct.

The microcirculation of the pancreatic lobule consists of subdivisions or units, the islet being located in the center of each unit. The exocrine acini in the peri-insular zone are preferentially supplied by the insulo-acinar portal vessels, while those in the tele-insular zone are mainly supplied by arterioles. Although there are some venules directly draining the islet, most of the venules commence in the tele-insular zone. This microcirculation pattern strongly suggests that the peri-insular zone is under the influence of the insular hormones and neurohormones.

In the liver, the efferent vessels of the reticular network of capillaries surrounding the biliary duct system, including the extrahepatic bile duct and the gallbladder, branch out into the hepatic sinusoids and thus form the *peribiliary portal system*. This portal system would play an important role in transporting absorbed substances and all hormonal substances produced by the biliary system back to the hepatic sinusoids, in order to control the microcirculation of the liver, and to the hepatocytes, presumably to act on them.

SEM of HCl/collagenase-treated tissues has helped in revealing the microvasculature and has also contributed toward our understanding of the organization of the cellular elements, such as the complicated structure of the salivary glands.

The recently introduced, cell-maceration/SEM method has revealed the continuum of the collagen fibrillar network, which not only constitutes the skeletal framework of the organ, but also provides those cells that are housed there with a specific microenvironment suitable for their activities. This new method is not only useful for the observation of the arrangements manifested by collagen fibrils, but is also expected to facilitate studying alterations in fibrillar arrangements in the pathological state and in age-related changes.

In the subendothelial spaces of Disse of the liver, many collagen fibrils form a dense network surrounding the sinusoids. In the pancreas, thin networks spun by collagen fibrils surround the acini and ducts, thus delimiting an interstitial compartment. The islet of Langerhans is surrounded by a collagen fibrillar capsule within which collagen fibrillar sheaths for islet capillaries form a three-dimensional anastomosing network.

References

1. Murakami T: Application of the scanning electron microscope to the study of the fine distribution of the blood vessels. *Arch Histol Jpn* 32: 445–454, 1971.
2. Ohtani O, Kikuta A, Ohtsuka A, Taguchi T, Murakami T: Microvasculature as studied by the microvascular corrosion casting/scanning electron microscope method. I. Endocrine and digestive system. *Arch Histol Jpn* 46: 1–42, 1983.
3. Evan AP, Dail WG, Dammrose D, Palmer C: Scanning electron microscopy of cell surface following removal of extracellular material. *Anat Rec* 185: 433–447, 1976.
4. Miller BG, Woods RI, Bohlen HG, Evan AP: A new morphological procedure for viewing microvessels: A scanning electron microscopic study of the vasculture of small intestine. *Anat Rec* 203: 493–503, 1982.
5. Ohtani O: Three-dimensional organization of the connective tissue fibers of the human pancreas. A scanning electron microscopic study of NaOH-treated tissues. *Arch Histol Jpn* 50: 557–566, 1987.
6. Ohtani O, Ushiki T, Taguchi T, Kikuta A: Collagen fibrillar networks as skeletal frameworks: A demonstration by cell-maceration/scanning electron microscope method. *Arch Histol Cytol* 51:249–261, 1988.
7. Murakami T: A revised tannin-osmium method for non-coated scanning electron microscope specimens. *Arch Histol Jpn* 36: 189–193, 1974.
8. Mason DK, Chisholm DM: Salivary Glands in Health and Disease. Philadelphia: WB Saunders, 1975.
9. Henderson JR, Daniel PM: Portal circulation and its relation to countercurrent system. *Q J Exp Physiol* 63: 355–369, 1978.
10. Kowalewsky N: Über das Blutgefässystem der Speicheldrüsen. *Arch Anat Physiol Anat Abt* 8: 385–395, 1885.
11. Fraser PA, Smaje LH: The organization of the salivary gland microcirculation. *J Physiol* 272: 121–136, 1977.

12. Burgen ASV, Seeman P: The role of the salivary duct system in the formation of the saliva. *Can J Biochem Physiol* 36: 119–143, 1958.
13. Blair-West JR, Coghlam JP, Denton DA, Nelson J, Wright RD, Yamauchi A: Ionic, histological and vascular factors in the reaction of the sheep's parotid to high and low mineralocorticoid status. *Physiology* 205: 563–579, 1969.
14. Brocco SL, Tamarin A: The topography of rat submandibular gland parenchyma as observed with the scanning electron microscope. *Anat Rec* 194: 445–460, 1979.
15. Nagato T: Scanning electron microscopical image of myoepithelial cells. *J Electron Microsc* 27: 235–236, 1978.
16. Ohtani O, Ohtsuka A, Lipsett J, Gannon B: The microvasculature of rat salivary glands. A scanning electron microscopic study. *Acta Anat* 115: 345–356, 1983.
17. Barka T: Biologically active polypeptides in submandibular glands. *J Histochem Cytochem* 28: 836–859, 1980.
18. Pinkstaff C: The cytology of salivary glands. *Int Rev Cytol* 63: 141–261, 1980.
19. Magnos JA, Braum G, Hamann KF: Micropuncture study of sodium and potassium excretion in the rat parotid saliva. *Pflügers Arch Ges Physiol* 291: 99–106, 1966.
20. Dawes C: The chemistry and physiology of saliva. In: Textbook of Oral Biology. Shaw, Sweeney, Cappucino, Meller (eds), Philadelphia WB Saunders, 593–616, 1978.
21. Schneyer LH, Young JA, Schneyer CA: Salivary secretion of electrolytes. *Physiol Rev* 52: 720–777, 1972.
22. Testa-Riva F, Puxeddu P, Riva A, Diaz G: The epithelium of the excretory duct of the human submandibular gland. A transmission and scanning electron microscopic study. *Am J Anat* 160: 381–393, 1981.
23. Holzlohner E, Neissing C: Über Kapillardrosselung bei vermehrter Organdruckblutung. *Z Biol* 97: 108–112, 1936.
24. Suddick RP, Dowd FJ: The microvascular architecture of the rat submaxillary salivary gland. Possible relationship to secretory mechanism. *Archs Oral Biol* 14: 567–576, 1969.
25. Haggendal E, Sivertsson R: About arterio-venous shunts in salivary glands. A study with krypton elimination technique in dogs. *Acta Physiol Scand* 71: 85–88, 1967.
26. Nikolow S: Vergleichend-anatomische Untersuchungen über die Mikrovascularisation der Speicheldrüsen. *Anat Anz* 130: 545–551, 1972.
27. Henderson JR: Why are the islets of Langerhans? *Lancet* ii: 469–470, 1969.
28. Fujita T: Insulo-acinar portal system in the horse pancreas. *Arch Histol Jpn* 35: 161–171, 1973.
29. Wharton GK: The blood supply of the pancreas, with special reference to that of islands of Langerhans. *Anat Rec* 53: 55–81, 1932.
30. Thiel A: Untersuchungen über das Gefass-system des Pancreaslappchens bei vershiedenen Saugern mit besonderer Berucksichtigung der Kapillarknauel der Langerhansschen Inseln. *Z Zellforsch* 39: 339–372, 1954.
31. McCuskey RS, Chapman TM: Microscopy of the living pancreas in situ. *Am J Anat* 126: 395–407, 1969.
32. Fraser PA, Henderson JR: The arrangement of the endocrine and exocrine pancreatic microcirculation observed in the living rabbit. *J Exp Physiol* 65: 151–171, 1980.
33. Ohtani O: Microcirculation of the pancreas: A correlative study of intravital microscopy with scanning electron microscopy of vascular corrosion casts. *Arch Histol Jpn* 46: 315–325, 1983.
34. Ohtani O, Ushiki T, Kanazawa H, Fujita T: Microcirculation of the pancreas in the rat and rabbit with special reference to the insulo-acinar portal system and emissary vein of the islet. *Arch Histol Jpn* 49: 45–60, 1986.
35. Fujita T, Murakami T: Microcirculation of monkey pancreas with special reference to the insulo-acinar portal system. A scanning electron microscope study of vascular casts. *Arch Histol Jpn* 35: 255–263, 1973.
36. Fujita T, Yanatori Y, Murakami T: Insulo-acinar axis, its vascular basis and its functional and morphological changes caused by CCK-PZ and caerulein. In: Endocrine Gut and Pancreas. T Fujita (ed), Amsterdam: Elsevier, 347–357, 1976.
37. Ohtani O, Fujita T: Microcirculation of the pancreas with special reference to the periductular circulation. A scanning electron microscope study of vascular casts. *Biomed Res* 1: 130–140, 1980.
38. Syed Ali S: Angioarchitecture of the pancreas of the cat. Light-, scanning- and transmission electron microscopy. *Cell Tissue Res* 235: 675–682, 1984.
39. Grube D, Bohn R: The microanatomy of human islets of Langerhans with special reference to somatostatin (D-) cells. *Arch Histol Jpn* 46: 327–353, 1983.
40. Nishino H, Ozawa K, Takeishi M, Nagata K, Watanabe Y: Pancreatic microcirculation in rats (particularly the islet of Langerhans). In: Microcirculation Annual 1985. M Tsuchiya, M Asano, M Oda, I Okazaki (eds), Amsterdam: Excerpta Medica, 325–337, 1985.
41. Fujita T, Kobayashi S: Proposal of a neurosecretory system in the pancreas. An electron microscope study in the dog. *Arch Histol Jpn* 42: 277–295, 1979.
42. Henderson JR, Daniel PM, Fraser PA: The pancreas as a single organ: The influence of the endocrine upon the exocrine part of the gland. *Gut* 22: 158–166, 1981.
43. Kiernan F: The anatomy and physiology of the liver. *Phil Trans Roy Soc London* 123: 711–770, 1833.
44. Ohtani O: The peribiliary portal system in the rabbit liver. *Arch Histol Jpn* 42: 153–167, 1979.
45. Rappaport AM: The microcirculatory hepatic unit. *Microvasc Res* 6: 212–228, 1973.
46. Rappaport AM, Borowy ZJ, Lougheed WM, Lotto WN: Subdivision of hexagonal liver lobules into a structural and functional unit: Role in hepatic physiology and pathology. *Anat Rec* 119: 11–34, 1954.
47. Mall FP: A study of the structural unit of the liver. *Am J Anat* 5: 227–308, 1906.
48. Andrews WHH, Maegraith BG, Wenyon CEM: Studies on the liver circulation. II. The microanatomy of the hepatic circulation. *Ann Trop Med* 43: 229–237, 1949.
49. Hase T, Brim J: Observation on the microcirculatory architecture of the liver. *Anat Rec* 156: 156–174, 1966.
50. Ohtani O, Murakami T, Jones AL: Microcirculation of the liver, with special reference to the peribiliary portal system. In: Developments in Gastroenterology, Vol 2.

Basic and Clinical Hepatology. PM Motta, LJA DiDio (eds), The Hague: Martinus Nijhoff, 85–96, 1982.
51. Ohtani O, Murakami T: Peribiliary portal system in the rat liver as studied by the injection replica SEM method. *Scanning Electron Microscopy/1978* II: 241–244, 1978.
52. Murakami T, Itoshima T, Shimada Y: Peribiliary portal system in the monkey liver as evidenced by the injection replica scanning electron microscope method. *Arch Histol Jpn* 37: 245–260, 1974.
53. Ohtani O: Microcirculation studies by the injection-replica method with special reference to the portal circulation. In: Three Dimensional Microanatomy of Cells and Tissue Surfaces. DJ Allen, PM Motta, LJA DiDio (eds), New York: Elsevier, North-Holland, 51–70, 1981.
54. Judge OM, Dickman PS, Trapukdi S: Nonfunctioning argyrophilic tumor (APUDoma) of the hepatic duct. Simplified methods for detecting biogenic amines in tissue. *Am J Clin Pathol* 66: 40–45, 1976.
55. Heitz P, Polak JM, Kasper M, Timson CM, Pearse AGE: Immunoelectron cytochemical localization of motilin and substance P in rabbit bile duct enterochromaffine (EC) cells. *Histochemistry* 50: 319–325, 1977.
56. Dancygier H, Klein U, Leuschner U, Hubner K, Classen M: Somatostatin-containing cells in the extrahepatic biliary tract of humans. *Gastroenterology* 86: 892–896, 1984.
57. Sundler F, Alumets J, Hakanson R, Ingemansson S, Fahrenkrug J, Schaffalitzky de Muckadell O: VIP innervation of the gallbladder. *Gastroenterology* 72: 1375–1377, 1979.
58. Alumets J, Schaffalitzky de Muckadell O, Fahrenkrug J, Sundler F, Hakanson R, Uddman R: A rich VIP nerve supply is characteristic of sphincters. *Nature* 280: 155–156, 1979.

Author's address:
Osamu Ohtani
Department of Anatomy
Okayama University Medical School
Okayama, 700
Japan

Index

GPSR Compliance

The European Union's (EU) General Product Safety Regulation (GPSR) is a set of rules that requires consumer products to be safe and our obligations to ensure this.

If you have any concerns about our products, you can contact us on ProductSafety@springernature.com

In case Publisher is established outside the EU, the EU authorized representative is:

Springer Nature Customer Service Center GmbH
Europaplatz 3
69115 Heidelberg, Germany

Batch number: 10684222

Printed by Printforce, the Netherlands